# Pesticides and Insecticides: Development and Use

# Pesticides and Insecticides: Development and Use

Edited by Bobby Jones

SYRAWOOD
PUBLISHING HOUSE
New York

Published by Syrawood Publishing House,
750 Third Avenue, 9th Floor,
New York, NY 10017, USA
www.syrawoodpublishinghouse.com

**Pesticides and Insecticides: Development and Use**
Edited by Bobby Jones



International Standard Book Number: 978-1-68286-503-3 (Hardback)



**Cataloging-in-Publication Data**

Pesticides and insecticides : development and use / edited by Bobby Jones.
p. cm.
Includes bibliographical references and index.
ISBN 978-1-68286-503-3
1. Pesticides. 2. Insecticides. 3. Plant diseases. I. Jones, Bobby.
SB951 .P47 2018
632.95--dc23

Printed in China.

# TABLE OF CONTENTS

# PREFACE

This book has been a concerted effort by a group of academicians, researchers and scientists, who have contributed their research works for the realization of the book. This book has materialized in the wake of emerging advancements and innovations in this field. Therefore, the need of the hour was to compile all the required researches and disseminate the knowledge to a broad spectrum of people comprising of students, researchers and specialists of the field.

Pesticides and insecticides are both used for eradicating pests. The purpose of using these is to protect plants from insects that can damage them and also to control organisms that are harmful and can cause diseases. One of the major reasons behind the increase in yield in the past decade is the use of insecticides and pesticides. This book studies their effects and usage in today's time. As the applications of pesticides and insecticides are emerging at a rapid pace, the contents of this book will help the readers understand the modern concepts and applications of the subject.

At the end of the preface, I would like to thank the authors for their brilliant chapters and the publisher for guiding us all-through the making of the book till its final stage. Also, I would like to thank my family for providing the support and encouragement throughout my academic career and research projects.

**Editor**

# 1

# Does Mortality Risk of Cigarette Smoking Depend on Serum Concentrations of Persistent Organic Pollutants? Prospective Investigation of the Vasculature in Uppsala Seniors (PIVUS) Study

**Duk-Hee Lee[1,2], Lars Lind[3], David R. Jacobs Jr.[4,5], Samira Salihovic[6], Bert van Bavel[6], P. Monica Lind[7]***

**1** Department of Preventive Medicine, School of Medicine, Kyungpook National University, Daegu, Korea, **2** BK21 Plus KNU Biomedical Convergence Program, Department of Biomedical Science, Kyungpook National University, Daegu, Korea, **3** Department of Medical Sciences, Cardiovascular Epidemiology, Uppsala University Hospital, Uppsala, Sweden, **4** Division of Epidemiology and Community Health, School of Public Health, University of Minnesota, Minneapolis, Minnesota, United States of America, **5** Department of Nutrition, University of Oslo, Oslo, Norway, **6** MTM Research Center, School of Science and Technology, Örebro University, Örebro, Sweden, **7** Department of Medical Sciences, Occupational and Environmental Medicine, Uppsala University, Uppsala, Sweden

## Abstract

Cigarette smoking is an important cause of preventable death globally, but associations between smoking and mortality vary substantially across country and calendar time. Although methodological biases have been discussed, it is biologically plausible that persistent organic pollutants (POPs) like polychlorinated biphenyls (PCBs) and organochlorine (OC) pesticides can affect this association. This study was performed to evaluate if associations of cigarette smoking with mortality were modified by serum concentrations of PCBs and OC pesticides. We evaluated cigarette smoking in 111 total deaths among 986 men and women aged 70 years in the Prospective Investigation of the Vasculature in Uppsala Seniors (PIVUS) with mean follow-up for 7.7 years. The association between cigarette smoking and total mortality depended on serum concentration of PCBs and OC pesticides (P value for interaction = 0.02). Among participants in the highest tertile of the serum POPs summary score, former and current smokers had 3.7 (95% CI, 1.5–9.3) and 6.4 (95% CI, 2.3–17.7) times higher mortality hazard, respectively, than never smokers. In contrast, the association between cigarette smoking and total mortality among participants in the lowest tertile of the serum POPs summary score was much weaker and statistically non-significant. The strong smoking-mortality association observed among elderly people with high POPs was mainly driven by low risk of mortality among never smokers with high POPs. As smoking is increasing in many low-income and middle-income countries and POPs contamination is a continuing problem in these areas, the interactions between these two important health-related issues should be considered in future research.

**Editor:** Keitaro Matsuo, Kyushu University Faculty of Medical Science, Japan

**Funding:** This study was supported by the Swedish Research Council (VR) and the Swedish Research Council for Environment, Agricultural Sciences and Spatial Planning (FORMAS), Korea Health technology R&D Project, Ministry of Health & Welfare, Republic of Korea (A111716), and BK21 PLUS KNU Biomedical Convergence Program for Creative Talent. The funders had no role in study design, data collection and analysis, decision to publish, or preparation of the manuscript.

**Competing Interests:** The authors have declared that no competing interests exist.

* E-mail: monica.lind@medsci.uu.se

## Introduction

Cigarette smoking, as a major risk factor for cancer, cardiovascular diseases, and pulmonary diseases, is an important cause of preventable death globally [1]. However, the magnitude of associations between cigarette smoking and mortality varies across countries and calendar time [2,3,4,5]. To date, this variation has been attributed mainly to methodological biases such as misclassification of smoking status, cohort effects, or random variation [2,3,4,5].

In our previous study [6], we formulated a novel hypothesis that persistent organic pollutants (POPs) modify the risk of cigarette smoking on death. POPs are lipophilic chemicals that accumulate in adipose tissue and are associated with the risk of various chronic diseases [7,8,9]. The biological plausibility for our hypothesis was that experimental studies in mice reported that pretreatment with some POPs increased toxicity of important chemicals contained in cigarette smoke, like benzopyrene, dimethylnitrosamine, and N-nitrosodiethlyamine [10,11].

Supporting our prior hypothesis, we observed different associations between cigarette smoking and total mortality depending on serum concentrations of POPs among the elderly in the U.S. [6]. In that study, one surprising finding was that cigarette smoking did not increase the risk of mortality in the lowest category of POPs. Despite the possibility that smoking-related diseases risks are lower among the elderly than in younger persons due to selective survival [12], finding any subgroup with no association between cigarette smoking and mortality was unexpected.

Our earlier study is the only one published on this topic and the number of current smokers was too small [6]. Thus, we studied here whether the finding was replicable in another dataset. We evaluated if there are similar interactions of serum concentrations of POPs with cigarette smoking on the risk of mortality among men and women aged 70 years, living in the community of Uppsala, Sweden (Prospective Investigation of the Vasculature in Uppsala Seniors (PIVUS) study). Because our previous study found that, among various POPs, polychlorinated biphenyls (PCBs) or organochlorine (OC) pesticides showed clear interactions with cigarette smoking and mortality [6], we used POPs burden calculated based on serum concentrations of both OC pesticides and PCBs in this study.

## Materials and Methods

Study subjects at baseline were 1,016 men and women, residents of Uppsala, Sweden and age 70 years at time of examination between April 2001 and June 2004. Among participants at baseline, 986 subjects had valid measurement of PCBs and OC pesticides at baseline. The study was approved by the Ethics Committee of the University of Uppsala. The participants gave written informed consent.

All subjects were investigated in the morning after an overnight fast, with no medication or smoking allowed after midnight. The participants were asked about their health behaviors, medical history, and regular medication. Body mass index (BMI) was derived from measured height and weight (kg/m$^2$). Serum cholesterol and triglyceride concentrations were determined in an enzymatic assay (Abbott, Abbott Park, IL, USA).

POPs were measured in stored plasma samples collected at baseline. Analyses of POPs were performed using a Micromass Autospec Ultima (Waters, Mildford, MA, USA) high resolution chromatography coupled to high resolution mass spectrometry (HRGC/HRMS) system based on the method by Sandau et al [13] with some modifications. All details of POPs analyses were provided elsewhere [14]. Among 16 PCB congeners and 5 OC pesticides, 2 OC pesticides (trans-chlordane and cis-chlordane) with detection rate $<10\%$ were not included in the final analyses. An established summation formula based on serum cholesterol and serum triglyceride concentrations was used to calculate the total amount of lipid in each plasma sample [15]. Thereafter the wet-weight concentrations of the POPs were divided by this lipid estimate to obtain lipid-normalized concentrations. As models based on wet-weight concentrations adjusted for serum cholesterol and triglyceride as covariates showed similar results, we presented the results based on lipid-normalized concentrations for the consistency of analytic strategy with the previous study [6].

The fact of death was ascertained through linkage to the Swedish Register of Death Causes at the National Board of Health and Welfare. Causes of death data were not available. Follow-up time for each person was calculated as the difference between the first examination date and the last known date alive or censored. Persons who survived the entire follow-up period were censored on Jan 1, 2012. Median follow-up time was 7.7 years (range 0.3–9.8 years) and we documented 111deaths.

For the analysis, smoking status was expressed as never, former and current. First, we calculated the summary measure of 16 PCB congeners and 5 OC pesticides by summing the rank orders of the individual POPs for subjects with detectable values of each POP, assigning rank 0 to not detectable values. Three summary measures were made based on both PCBs and OC pesticides, PCBs only, and OC pesticides only.

Next, we checked if the associations between cigarette smoking and mortality differed by tertiles of the summary measures of POPs in predicting total mortality using Cox proportional hazard models. P values for interaction were calculated based on three categories of each of POPs and cigarette smoking. Adjusting covariates were gender, physical activity (none, moderate, and vigorous), BMI (kg/m$^2$), and alcohol consumption (g/day). We further considered medication for diabetes or hypertension and history of myocardial infarction or stroke as possible confounders.

In addition to stratified analyses, adjusted HRs with the common reference group of never smokers within the 1$^{st}$ tertile of POPs were presented. Also, the same analyses were applied to the U.S. elderly within our previous study [6] because only results stratified analyses by POPs levels were previously reported in that study. Methodologic details have been presented [6]. All statistical analyses were performed with PC-SAS version 9.1.

## Results

Baseline characteristics and history of chronic diseases according to cigarette smoking status were shown in Table 1. Compared to never or current smokers, former smokers tended to be men, more obese, alcohol drinkers, and more under diabetes medication. Current smokers were less obese and more physically inactive than never or former smokers. There were no statistical significant differences across smoking categories for prevalence of hypertension medication or history of myocardial infarction or stroke.

Table 2 indicates mean concentrations of individual POPs depending on cigarette smoking status. Among 16 PCBs, PCB074, PCB105, and PCB118 were statistically significant or marginally significantly lower in current vs. never smokers. These three PCBs have weak dioxin activity, however, other PCBs with dioxin activity like PCB126, PCB156, and PCB169 did not show any trend. On the contrary, among 3 OC pesticides, *p,p'*-DDE had higher levels among current than never smokers.

When POPs were not considered in analyses, adjusted hazard ratios (HRs) for all-cause mortality were 1.3 for former smokers and 2.1 (95% CI: 1.2–3.7) for current smokers, compared with never smokers. However, the associations were substantially different depending on summary measures of POPs (Table 3). Compared with 70 year old people within the 1$^{st}$ or 2$^{nd}$ tertiles of summary measures of PCBs and OC pesticides, those within the 3$^{rd}$ tertile showed strong associations between cigarette smoking and total mortality. Adjusted HRs were 3.7 (1.5–9.3) for former smokers and 6.4 (2.3–17.7) for current smokers. Further adjustment for medication for diabetes or hypertension and history of myocardial infarction or stroke did not change the result. Summary measures calculated from PCBs or OC pesticides separately showed similar associations (Table S1).

Figure 1 shows adjusted HRs when never smokers within the 1$^{st}$ tertile of PCBs and OC pesticides were used as the common reference group. Never smokers within the 3$^{rd}$ tertile of PCBs and OC pesticides showed a statistically significantly lower risk of mortality with adjusted HR of 0.3 (0.1–0.9) while current smokers with the same levels of POPs concentrations had a statistically significantly higher risk of mortality with adjusted HR of 2.3 (1.0–5.0). When we applied the same analyses to the U.S. elderly who were included in our previous study, patterns were similar to those in the current study, although no individual HR reached statistical significance (Figure S1). In that study, the summary measures for PCBs and for OC pesticides were separate because PCB vs OC pesticide measurements were performed in different participants.

To better understand possible confounding, Table 4 compared baseline characteristics of 4 subgroups (never smokers & low POPs, current smokers & high POPs, never smokers & low POPs, and current smokers & high POPs), in particular focusing on never smokers with high POPs who had the lowest mortality. Compared

**Table 1.** Baseline characteristics according to the status of cigarette smoking among 986 elderly aged 70, Prospective Investigation of the Vasculature in Uppsala Seniors (PIVUS) study.

| Characteristics | Status of cigarette smoking | | | p value |
|---|---|---|---|---|
| | Never smokers (N = 471) | Former smokers (N = 410) | Current smokers (N = 105) | |
| Men (%) | 44.4 | 57.6 | 47.6 | 0.01 |
| BMI (kg/m$^2$, %) | | | | <0.01 |
| <25 | 34.4 | 26.6 | 51.4 | |
| 25-<30 | 44.8 | 48.1 | 34.3 | |
| ≥30 | 20.8 | 25.4 | 14.3 | |
| Exercise (%) | | | | 0.01 |
| No | 8.5 | 11.7 | 18.1 | |
| Mild | 64.1 | 63.4 | 66.7 | |
| Moderate or vigorous | 27.4 | 24.9 | 15.2 | |
| Alcohol consumption (g/day, %) | | | | <0.01 |
| 0 | 17.2 | 11.5 | 14.3 | |
| 1–14 | 75.2 | 74.8 | 80.0 | |
| ≥15 | 7.6 | 13.7 | 5.7 | |
| Diabetes medication (%) | 4.7 | 8.8 | 4.8 | 0.03 |
| Hypertension medication (%) | 29.5 | 33.4 | 27.6 | 0.34 |
| History of myocardial infarction (%) | 5.7 | 8.3 | 10.7 | 0.13 |
| History of stroke (%) | 3.2 | 3.9 | 4.8 | 0.69 |

to other subgroups, 70 year old participants within the never smokers & high POPs group were the most physically active and had the lowest prevalence of history of myocardial infarction.

## Discussion

Generally, the present study replicated our report in a U.S. elderly population [6]. The association between cigarette smoking and total mortality depended on serum concentration of summary scores reflecting the background mixture of PCBs and/or OC pesticides. When POPs were not considered in the analyses, the risk of mortality among current-smokers was about two times higher than never smokers. However, among elderly with relatively high POPs, former or current smokers had about 4 to 7 times higher mortality than never smokers while the association between cigarette smoking and total mortality much weaker and statistically non-significant among elderly with relatively low serum concentrations of POPs. In addition to the main results, a subsidiary finding of lower concentrations of dioxin-like PCBs among smokers than never smokers was also similar to finding in the previous study [6].

These findings suggested that different concentrations of POPs among populations may partly explain variability in smoking-related total mortality association across previous epidemiological studies [2,3,4,5]. More importantly, the similar weak association between cigarette smoking and mortality among the elderly people in the lowest category of POPs from these two studies could indicate that the presence of certain levels POPs is a necessary factor for cigarette smoking to increase the risk of death in the elderly.

At first, we had expected the interaction between these two factors to be driven by the high risk of death among current smokers with high POPs. However, these two studies showed that the strong association between cigarette smoking and mortality among the elderly with high POPs had a strong component of low mortality in never smokers as well as of high mortality in former or current smokers.

The very low mortality among the 150 never smokers in PIVUS with high POPs is a provocative finding. A similar pattern of reduced mortality was apparent in the National Health and Nutrition Examination Survey (NHANES) (Figure S1), suggesting that this finding is not simply bias or chance. An ecologic finding in the cohort study of male British doctors may be viewed as concordant. In that study, the excess mortality associated with smoking was greater during the second half of follow-up (1971~1991) than the first half (1951~1971) [2], largely because the mortality rate among non-smokers had decreased substantially over time while the mortality rate among smokers had remained about constant [2]. Historically, PCBs and OC pesticides were widely used after World War II and periods with the highest body burden of POPs were 1960s and 1970s [16]. Therefore, any effect due to high POPs may be more strongly reflected in the latter part of cohort study.

It is difficult to explain very low mortality among never smokers with high POPs. One possible explanation is survival bias. However, for survival bias to explain this finding, never smokers with high POPs must have had a higher death rate before reaching age 70 than other subgroups like active smokers with high POP, leaving healthier survivors to participate in the PIVUS and NHANES studies. This would be an odd pattern. Also, compared to other subgroups, the elderly never smokers with high POPs were the most physically active and had the lowest prevalence of history of myocardial infarction, both of which might predispose to reduced mortality rate. Therefore, this type of survival bias would seem to be unlikely even though we cannot totally exclude it.

**Table 2.** Adjusted* serum concentrations (geometric means, ng/g lipid) of individual polychlorinated biphenyls (PCBs) or organochlorine (OC) pesticides according to the status of cigarette smoking, Prospective Investigation of the Vasculature in Uppsala Seniors (PIVUS) study.

| | Status of cigarette smoking | | | |
|---|---|---|---|---|
| **Analytes** | **Never smokers (N = 471)** | **Former smokers (N = 410)** | **Current smokers (N = 105)** | **p for trend** |
| **Polychlorinated biphenyls (PCBs)** | | | | |
| PCB074† | 14.1 | 13.6 | 12.9 | 0.09 |
| PCB099 | 13.4 | 13.7 | 14.3 | 0.38 |
| PCB105† | 5.0 | 5.0 | 4.1 | 0.02 |
| PCB118† | 31.0 | 30.7 | 25.0 | <0.01 |
| PCB126† | 5.9 | 5.9 | 6.5 | 0.44 |
| PCB138 | 123.7 | 124.5 | 135.4 | 0.15 |
| PCB153 | 213.7 | 216.5 | 228.3 | 0.19 |
| PCB156† | 23.7 | 23.1 | 24.7 | 0.78 |
| PCB157 | 4.4 | 4.3 | 4.5 | 0.70 |
| PCB169† | 25.8 | 25.9 | 26.7 | 0.52 |
| PCB170 | 75.5 | 75.3 | 80.1 | 0.30 |
| PCB180 | 177.6 | 177.3 | 187.1 | 0.37 |
| PCB189 | 3.2 | 3.2 | 2.9 | 0.34 |
| PCB194 | 16.0 | 15.2 | 17.2 | 0.88 |
| PCB206 | 4.2 | 3.9 | 4.1 | 0.23 |
| PCB209 | 4.0 | 3.8 | 4.0 | 0.36 |
| **Organochlorines pesticides (OCPs)** | | | | |
| p,p′-DDE | 270.1 | 296.4 | 322.4 | 0.04 |
| Trans-nonachlor | 20.6 | 21.9 | 21.8 | 0.14 |
| Hexachlorobenzene | 40.1 | 39.1 | 39.5 | 0.46 |

*Adjusted for gender, BMI, exercise, and alcohol consumption.
†POPs with dioxin activity.

**Table 3.** Adjusted hazard ratios (HRs)* and 95% confidence intervals (CIs) for all-cause mortality rate by summary measures† of polychlorinated biphenyls (PCBs) or organochlorine (OC) pesticides, Prospective Investigation of the Vasculature in Uppsala Seniors (PIVUS) study.

| | | Status of cigarette smoking | | | p for trend | p for interaction |
|---|---|---|---|---|---|---|
| | | **Never smokers** | **Former smokers** | **Current smokers** | | |
| **All subjects** | | | | | | |
| | Cases/No | 40/471 | 50/410 | 21/105 | | |
| | Adjusted HR(95%CI) | Referent | 1.3(0.9–2.0) | 2.1(1.2–3.7) | <0.01 | |
| **Summary measure of 16 PCBs and 3 OC pesticides** | | | | | | |
| 1st tertile | Cases/No | 16/160 | 19/138 | 5/30 | | |
| | Adjusted HR(95%CI) | Referent | 1.2 (0.6–2.4) | 1.4 (0.5–4.0) | 0.46 | 0.02 |
| 2nd tertile | Cases/No | 18/161 | 10/134 | 5/34 | | |
| | Adjusted HR(95%CI) | Referent | 0.7 (0.3–1.6) | 1.2 (0.5–3.4) | 0.97 | |
| 3rd tertile | Cases/No | 6/150 | 21/138 | 11/41 | | |
| | Adjusted HR(95%CI) | Referent | 3.7 (1.5–9.3) | 6.4 (2.3–17.7) | <0.01 | |

*Hazard Ratios (HRs) adjusted for gender, BMI, exercise, and alcohol consumption.
†Values of compounds belonging to in each summary measure were individually ranked; the rank orders of the individual POPs were summed to calculate summary measures and the summaries were divided into tertiles.

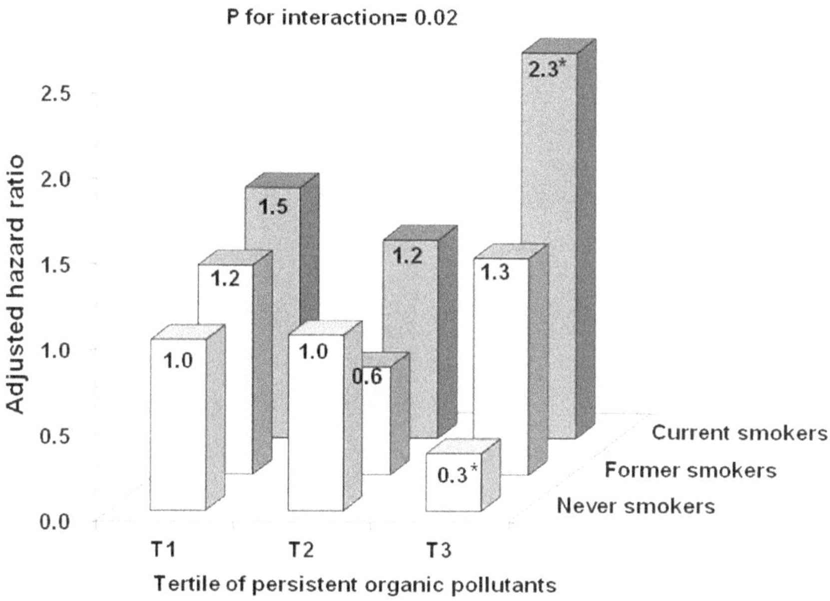


**Figure 1. Interactions of cigarette smoking with summary measure of polychlorinated biphenyls (PCBs) and organochlorines (OC) pesticides on mortality.** Hazard ratios (HRs) were estimated using a common reference group (never smokers & T1), adjusted for gender, BMI, exercise, and alcohol consumption. T1, first tertile; T2, second tertile; T3, third tertile, Prospective Investigation of the Vasculature in Uppsala Seniors (PIVUS) study.

In addition, low mortality in relation to high POPs is biologically plausible under some conditions, a case in point relating to associations between POPs and telomeres. Telomere length has been proposed as a marker of mitotic cell age and as a general index of human aging [17]. In our previous cross-sectional study among apparently healthy Koreans [18], telomere length was increasing across serum concentrations of POPs within the lower range of POPs. In that study, the interpretation was that low dose POPs may act as a tumor promoter in carcinogenesis based on experimental findings on arsenic. For example, low concentrations of arsenic relevant to current human exposure elongated telomeres in vitro and increased *myc* and *ras* oncogenes while high concentrations of arsenic decreased telomere length [19]. As *myc* oncogene can activate telomerase [20], the authors interpreted these experimental findings on arsenic as a role of tumor promoter of low dose arsenic in human.

However, an opposite interpretation of how POPs relate to telomere length is also possible. It is well-known that shorter telomere length is associated with higher risk of early death [21]. There is a higher mortality rate, especially from heart disease and infectious disease, among elderly people who have shorter telomeres in blood DNA [22]. Therefore, in PIVUS and NHANES persons with relatively high serum POPs concentrations within background exposure levels may have had a longer survival than persons with lower serum POPs concentrations. As cigarette smoking is reported to decrease telomere length [23], longer survival with higher POPs could be observed only in never smokers, as we observed in these two studies.

Although molecular mechanisms for low dose POPs to increase telomere length have not been studied, one speculation is that certain levels of POPs might excite production of cytoprotective and restorative proteins including growth factors, phase II and antioxidant enzymes, and protein chaperones, known as hormetic effects [24]. Increased telomerase activity or slow-down of age-dependent telomere shortening has been regarded as one marker of activation of cytoprotective and restorative proteins [25]. Some previous human and experimental studies support possible beneficial effects of certain levels of POPs. For example, decreased risk of soft tissue sarcoma was reported with increased concentrations of dioxins or PCBs with dioxin activity in the general population[26]. Also, in some animal studies of low doses of TCDD or DDT, there was a tendency towards fewer tumors or altered hepatic foci than in controls, indicating an anti-carcinogenic process [27,28]. Furthermore, decreased lipid peroxidation level in rats treated with low dose DDT was reported, compared to control rats [29].

Studies of the levels of POPs in the global environment show that emission sources of a number of POPs in the last 20 years have shifted from industrialized countries of the Northern Hemisphere to less developing countries in tropical and subtropical regions [30]. This is due to a late production ban on OC pesticides: some OC pesticides are still being used in agriculture and for the control of diseases, such as malaria [31]. Also, there is another active exposure source of PCBs through e-waste recycling in these countries [32]. Although the burden of cigarette smoking use is currently greatest in high-income countries, rates of smoking are increasing in many low-income and middle-income countries. Therefore, the interaction between these two important health-related issues, POPs and cigarette smoking, should be further studied from a variety of viewpoints including molecular mechanisms.

The strengths of our study include the homogenous age and community-based sampling of study subjects. There were some limitations to our study. Due to a limited sample size, we could not consider more detailed information on cigarette smoking like total pack-years or duration of smoking cessation in analyses. In addition, analyses focusing on cause-specific mortality were not possible as cause of death information was not available. Finally, even though the consistency of findings across two studies lends credence to our findings and seems to reduce the likelihood that our finding is explainable by chance or bias, our findings still require replication by other cohort studies.

**Table 4.** Comparison of baseline characteristics among 4 subgroups, Prospective Investigation of the Vasculature in Uppsala Seniors (PIVUS) study.

| | Low POPs | | High POPs | | |
|---|---|---|---|---|---|
| | Never smokers (N = 160) | Current smokers (N = 30) | Never smokers (N = 150) | Current smokers (N = 41) | p value |
| Dead (%) | 10.0 | 16.7 | 4.0 | 26.8 | <0.01 |
| Men (%) | 31.9 | 40.0 | 56.0 | 53.7 | <0.01 |
| BMI (kg/m$^2$, %) | | | | | <0.01 |
| <25 | 26.3 | 46.7 | 36.0 | 48.8 | |
| 25-<30 | 40.6 | 36.7 | 51.3 | 46.3 | |
| ≥30 | 33.1 | 16.7 | 12.7 | 4.9 | |
| Moderate or vigorous exercise (%) | | | | | <0.01 |
| No | 10.0 | 10.0 | 10.7 | 26.8 | |
| Mild | 63.8 | 80.0 | 58.7 | 61.0 | |
| Moderate or vigorous | 26.3 | 10.0 | 30.7 | 12.0 | |
| Alcohol consumption (g/day, %) | | | | | 0.03 |
| 0 | 20.6 | 20.0 | 14.0 | 7.3 | |
| 1–14 | 76.3 | 76.7 | 74.0 | 85.4 | |
| ≥15 | 3.1 | 3.3 | 12.0 | 7.3 | |
| Diabetes medication (%) | 3.8 | 0% | 7.3 | 9.8 | 0.17 |
| Hypertension medication (%) | 31.3 | 30.0 | 31.3 | 22.0 | 0.68 |
| History of myocardial infarction (%) | 6.9 | 13.8 | 4.0 | 15.0 | 0.05 |
| History of stroke (%) | 1.9 | 6.7 | 5.3 | 4.9 | 0.35 |

This study has confirmed a strong association between cigarette smoking and mortality among elderly people with relatively high POPs, but a much weaker and not statistically significant smoking-mortality association among those with relatively low POPs. However, the observation that the lowest mortality was seen among never smokers with high POPs is provocative and requires further studies on the role of POPs in longevity. In addition more prospective studies in human, in-vitro and in-vivo experimental studies would help to elucidate potential molecular mechanisms.

## Supporting Information

**Figure S1** Interactions of cigarette smoking with summary measure of polychlorinated biphenyls (PCB) or organochlorine (OC) pesticides predicting mortality in the U.S. general population (reanalyses of published results using the same National Health and Nutrition Examination Survey (NHANES) datasets (Lee, 2013). Number of study subjects and deaths were 610 and 142 for PCB analyses and 702 and 157 for OC pesticide analyses. Hazard ratios (HRs) were estimated using a common reference group (never smokers & T1), adjusted for gender, BMI, exercise, and alcohol consumption. T1, first tertile; T2, second tertile; T3, third tertile. Although none of the HRs was statistically significantly different from the reference group, as previously reported, interaction p-values were 0.008 for PCBs and 0.024 for OC pesticides.

## Author Contributions

Conceived and designed the experiments: DHL LL PML. Performed the experiments: DHL LL PML. Analyzed the data: DHL. Contributed reagents/materials/analysis tools: SS BB. Wrote the paper: DHL LL DRJ SS BB PML.

## References


1. Murray CJ, Lopez AD (1997) Mortality by cause for eight regions of the world: Global Burden of Disease Study. Lancet 349: 1269–1276.
2. Doll R, Peto R, Wheatley K, Gray R, Sutherland I (1994) Mortality in relation to smoking: 40 years' observations on male British doctors. Bmj 309: 901–911.
3. Hunt D, Blakely T, Woodward A, Wilson N (2005) The smoking-mortality association varies over time and by ethnicity in New Zealand. Int J Epidemiol 34: 1020–1028.
4. Jacobs DR Jr, Adachi H, Mulder I, Kromhout D, Menotti A, et al. (1999) Cigarette smoking and mortality risk: twenty-five-year follow-up of the Seven Countries Study. Arch Intern Med 159: 733–740.
5. van de Mheen PJ, Gunning-Schepers LJ (1996) Differences between studies in reported relative risks associated with smoking: an overview. Public Health Rep 111: 420–426; discussion 427.
6. Lee YM, Bae SG, Lee SH, Jacobs DR, Lee DH (2013) Associations between cigarette smoking and total mortality differ depending on serum concentrations of persistent organic pollutants among the elderly. J Korean Med Sci 28: 1122–1128.
7. Carpenter DO (2011) Health effects of persistent organic pollutants: the challenge for the Pacific Basin and for the world. Rev Environ Health 26: 61–69.
8. Ha MH, Lee DH, Jacobs DR (2007) Association between serum concentrations of persistent organic pollutants and self-reported cardiovascular disease prevalence: results from the National Health and Nutrition Examination Survey, 1999–2002. Environ Health Perspect 115: 1204–1209.
9. Lee DH, Lee IK, Song K, Steffes M, Toscano W, et al. (2006) A strong dose-response relation between serum concentrations of persistent organic pollutants and diabetes: results from the National Health and Examination Survey 1999–2002. Diabetes Care 29: 1638–1644.
10. Diwan BA, Ward JM, Kurata Y, Rice JM (1994) Dissimilar frequency of hepatoblastomas and hepatic cystadenomas and adenocarcinomas arising in hepatocellular neoplasms of D2B6F1 mice initiated with N-nitrosodiethylamine

and subsequently given Aroclor-1254, dichlorodiphenyltrichloroethane, or phenobarbital. Toxicol Pathol 22: 430–439.

11. Hutton JJ, Meier J, Hackney C (1979) Comparison of the in vitro mutagenicity and metabolism of dimethylnitrosamine and benzo[a]pyrene in tissues from inbred mice treated with phenobarbital, 3-methylcholanthrene or polychlorinated biphenyls. Mutat Res 66: 75–94.
12. Psaty BM, Koepsell TD, Manolio TA, Longstreth WT Jr, Wagner EH, et al. (1990) Risk ratios and risk differences in estimating the effect of risk factors for cardiovascular disease in the elderly. J Clin Epidemiol 43: 961–970.
13. Sandau CD, Sjodin A, Davis MD, Barr JR, Maggio VL, et al. (2003) Comprehensive solid-phase extraction method for persistent organic pollutants. Validation and application to the analysis of persistent chlorinated pesticides. Anal Chem 75: 71–77.
14. Salihovic S, Lampa E, Lindstrom G, Lind L, Lind PM, et al. (2012) Circulating levels of persistent organic pollutants (POPs) among elderly men and women from Sweden: results from the Prospective Investigation of the Vasculature in Uppsala Seniors (PIVUS). Environ Int 44: 59–67.
15. Phillips DL, Pirkle JL, Burse VW, Bernert JT Jr, Henderson LO, et al. (1989) Chlorinated hydrocarbon levels in human serum: effects of fasting and feeding. Arch Environ Contam Toxicol 18: 495–500.
16. Solomon GM, Weiss PM (2002) Chemical contaminants in breast milk: time trends and regional variability. Environ Health Perspect 110: A339–347.
17. Blackburn EH (2005) Telomeres and telomerase: their mechanisms of action and the effects of altering their functions. FEBS Lett 579: 859–862.
18. Shin JY, Choi YY, Jeon HS, Hwang JH, Kim SA, et al. (2010) Low-dose persistent organic pollutants increased telomere length in peripheral leukocytes of healthy Koreans. Mutagenesis 25: 511–516.
19. Zhang TC, Schmitt MT, Mumford JL (2003) Effects of arsenic on telomerase and telomeres in relation to cell proliferation and apoptosis in human keratinocytes and leukemia cells in vitro. Carcinogenesis 24: 1811–1817.
20. Wu KJ, Grandori C, Amacker M, Simon-Vermot N, Polack A, et al. (1999) Direct activation of TERT transcription by c-MYC. Nat Genet 21: 220–224.
21. Bakaysa SL, Mucci LA, Slagboom PE, Boomsma DI, McClearn GE, et al. (2007) Telomere length predicts survival independent of genetic influences. Aging Cell 6: 769–774.
22. Cawthon RM, Smith KR, O'Brien E, Sivatchenko A, Kerber RA (2003) Association between telomere length in blood and mortality in people aged 60 years or older. Lancet 361: 393–395.
23. Valdes AM, Andrew T, Gardner JP, Kimura M, Oelsner E, et al. (2005) Obesity, cigarette smoking, and telomere length in women. Lancet 366: 662–664.
24. Mattson MP (2008) Hormesis defined. Ageing Res Rev 7: 1–7.
25. Yokoo S, Furumoto K, Hiyama E, Miwa N (2004) Slow-down of age-dependent telomere shortening is executed in human skin keratinocytes by hormesis-like-effects of trace hydrogen peroxide or by anti-oxidative effects of pro-vitamin C in common concurrently with reduction of intracellular oxidative stress. J Cell Biochem 93: 588–597.
26. Tuomisto J, Pekkanen J, Kiviranta H, Tukiainen E, Vartiainen T, et al. (2005) Dioxin cancer risk—example of hormesis? Dose Response 3: 332–341.
27. Sukata T, Uwagawa S, Ozaki K, Ogawa M, Nishikawa T, et al. (2002) Detailed low-dose study of 1,1-bis(p-chlorophenyl)-2,2,2- trichloroethane carcinogenesis suggests the possibility of a hormetic effect. Int J Cancer 99: 112–118.
28. Viluksela M, Bager Y, Tuomisto JT, Scheu G, Unkila M, et al. (2000) Liver tumor-promoting activity of 2,3,7,8-tetrachlorodibenzo-p-dioxin (TCDD) in TCDD-sensitive and TCDD-resistant rat strains. Cancer Res 60: 6911–6920.
29. Shutoh Y, Takeda M, Ohtsuka R, Haishima A, Yamaguchi S, et al. (2009) Low dose effects of dichlorodiphenyltrichloroethane (DDT) on gene transcription and DNA methylation in the hypothalamus of young male rats: implication of hormesis-like effects. J Toxicol Sci 34: 469–482.
30. Tanabe S, Minh TB (2010) Dioxins and organohalogen contaminants in the Asia-Pacific region. Ecotoxicology 19: 463–478.
31. Wong MH, Leung AO, Chan JK, Choi MP (2005) A review on the usage of POP pesticides in China, with emphasis on DDT loadings in human milk. Chemosphere 60: 740–752.
32. Someya M, Ohtake M, Kunisue T, Subramanian A, Takahashi S, et al. (2010) Persistent organic pollutants in breast milk of mothers residing around an open dumping site in Kolkata, India: specific dioxin-like PCB levels and fish as a potential source. Environ Int 36: 27–35.

# 2

# Transgenic *Bacillus thuringiensis* (Bt) Rice Is Safer to Aquatic Ecosystems than Its Non-Transgenic Counterpart

**Guangsheng Li[1], Yongmo Wang[1]*, Biao Liu[2], Guoan Zhang[1]**

**1** Hubei Insect Resources Utilization and Sustainable Pest Management Key Laboratory, College of Plant Science and Technology, Huazhong Agricultural University, Wuhan, Hubei, P.R. CHINA, **2** Nanjing Institute of Environmental Science, Ministry of Environmental Protection of the People's Republic of China, Nanjing, P.R. CHINA

## Abstract

Rice lines genetically modified with the crystal toxin genes from *Bacillus thuringiensis* (Bt) have experienced rapid development, with biosafety certificates for two Bt rice lines issued in 2009. There has still been no commercial release of these lines yet due to public concerns about human health and environmental risks. Some studies confirmed that Bt rice was as safe as conventional rice to non-target organisms when pesticides were not applied, however, pesticides are still required in Bt rice to control non-lepidopteran pests. In this study, we assessed the environmental effects of two Bt rice lines expressing either the *cry1Ab/1Ac* or *cry2A* genes, respectively, by using zooplanktons as indicator species under normal field management practices using pesticides when required. In the whole rice growing season, non-Bt rice was sprayed 5 times while Bt rice was sprayed 2 times, which ensured both rice achieved a normal yield. Field investigations showed that rice type (Bt and non-Bt) significantly influenced zooplankton abundance and diversity, which were up to 95% and 80% lower in non-Bt rice fields than Bt rice fields. Laboratory rearing showed that water from non-Bt rice fields was significantly less suitable for the survival and reproduction of *Daphnia magna* and *Paramecium caudatum* in comparison with water from Bt rice fields. Higher pesticide residues were detected in the water from non-Bt than Bt rice fields, accounting for the bad performance of zooplankton in non-Bt field water. Our results demonstrate that Bt rice is safer to aquatic ecosystems than non-Bt rice, and its commercialization will be beneficial for biodiversity restoration in rice-based ecosystems.

**Editor:** Juan Luis Jurat-Fuentes, University of Tennessee, United States of America

**Funding:** This study was supported by the Major Projects of National Genetically Modified Organisms Breeding (NO. 2014ZX08012005) and the National Natural Science Funds (31370544). The funders had no role in study design, data collection and analysis, decision to publish, or preparation of the manuscript.

**Competing Interests:** The authors have declared that no competing interests exist.

* Email: ymwang@mail.hzau.edu.cn

## Introduction

Rice (*Oryza sativa* L.) is a staple food for more than three billion people [1], and more than 90% of total rice is produced and consumed in Asia [2]. Over 100 species of insects attack and damage rice, and lepidopteran pests such as rice stem borers are the most serious with yield losses ranging from 5% to 25% [3]. Rice has been successfully modified to express the cry genes derived from the *Bacillus thuringiensis* (Bt) bacterium, and several Bt rice lines with high resistance to lepidopteran pests have been developed over the past 15 years in China [4]. The China Ministry of Agriculture (CMOA) has confirmed that safety certificates were issued in November 2009 for two Bt rice lines, Huahui no. 1 and Bt Shanyou 63 [5], both of which express a fusion *cry1Ab/1Ac* gene. However, the safety certificates did not lead to final commercial release because there were fierce public debates on the health and environmental risks of Bt rice [6]. There is therefore an urgent need to provide the public with scientifically based knowledge of the actual environmental effects of genetically modified rice.

Many field and laboratory experiments have assessed the impact of Bt rice on paddy arthropods, including parasitoids, predators and non-target herbivores. In general, Bt rice has no negative impact (in comparison to non-Bt rice) to the individual fitness, population abundance and community diversity of non-target arthropods [7–13]. However, almost all of the field studies have excluded pesticides from both the Bt and non-Bt rice fields. Although Bt rice is effective in reducing pesticide sprays, some pesticide application is still required to control fungal diseases and non-lepidopteran pests if full yield is to be achieved [14,15]. To accurately assess the environmental impact of Bt rice it is therefore necessary to conduct assessments under normal agricultural practices, including pesticide sprays when required.

Previous environmental assessments of Bt rice have mainly focused on terrestrial arthropods, and have not considered the aquatic ecosystem [16]. In most growth stages of rice, a layer of water is sustained in the rice field. In tropic or subtropical areas, this paddy water layer teems with zooplankton such as rotifers, cladocerans and copepods, most of which are shared by adjacent ponds and lakes, and those organisms constitute an important food source for fish and predatory insects [17]. These organisms also carry out active functions in nutrient cycling in agroecosystems [18]. Further, pesticides readily dissolve in water, and the environmental effects of pesticide use are likely to be most acute in these ecosystems [19].

In the present study, we evaluated the environmental effects of two Bt rice lines, one of which has been granted a safety certificate by the CMOA, using paddy zooplanktons as indicator species. Bt rice and non-transgenic rice were planted in plots in a complete randomized block design. Experimental fields were inspected weekly for rice pest occurrence and sprayed with pesticides when required to guarantee that non-Bt and Bt rice achieved equal yields (Fig. 1). The diversity and abundance of three types of zooplankton were compared between Bt and non-Bt treatments. Field water was collected to conduct laboratory rearing experiments of water quality using *Daphnia magna* and *Paramecium caudatum*, two species commonly used to evaluate water pollutants [20,21]. Pesticide residues in field-collected water were also compared between Bt and non-Bt rice fields.

## Materials and Methods

### Ethics Statement

All field studies were carried out at an experimental field which is maintained by Huazhong Agricultural University (HZAU) and located in the campus (latitude 30°34′ N, longitude 114°47′ E). Special permits for the field studies were obtained from Science and Technology Department of HZAU which executes the function of GMO management under the supervision of China National Agricultural GMO Safety Committee (CNAGSC). All materials used during rice planting (pesticides and fertilizer etc.) were registered for legal use, and the products of Bt rice (seed and straw) were disposed according to the regulations formulated by CNAGSC. The field studies did not involve endangered or protected species.

### Bt rice materials and field design

Two Bt rice lines, Bt rice 1 and Bt rice 2, and their common non-transgenic isogenic counterpart, were used as plant materials. Both the Bt rice lines are variety of Minghui 63, an elite indica rice restorer line. Bt Rice 1 was transformed with a synthetic *cry1Ab/1Ac* gene and Bt rice 2 was transformed with the *cry2A* gene [22,23]. Bt rice 1 is also called Huahui no. 1, which was issued with a safety certificate by the CMOA in 2009 [5]. Field experiments were carried out at an experimental field at the HZAU campus in Wuhan (China) in 2013. The experimental field was constructed from a small hill in 2008, and planted with rice for 3 years without pesticide spray. According to a randomized complete block design (RCBD), the field was divided into 3 blocks, and each block divided into 3 plots which were planted with Bt rice 1, Bt rice 2 and non-Bt rice randomly. Each plot covered an area of 220 $m^2$. Solid ridges and drainage channels were built around plots to avoid water flow between the plots. Both Bt and non-Bt rice were sowed in early May and transplanted by hand one month later. Spacing between rows was 25 cm, and hills within rows were 20 cm apart. The field was irrigated with water from a lake nearby. The soil in the experimental field was typically a yellow-brown soil consisting of 25.6% clay, 68.8% silt, 5.6% sand, and 11.2% organic matter. Field management was performed following conventional agronomic practices. Pesticides including avermectins, triazophos, difenoconazole, propiconazole and buprofezin were used when any of the rice pests exceeded action threshold (see Table S1). Action threshold refers to a point at which pest population or environmental conditions indicate that pest control action must be taken to guarantee no economic loss. A windshield was held downwind between plots when spraying.

### Field investigation of zooplanktons

Five litres of water were collected from each plot in the morning before 10 a.m., and were concentrated to 50 ml using plankton net (64 μm mesh size). The concentrated samples were fixed with Lugol's solution, and stored in amber bottles. Three kinds of

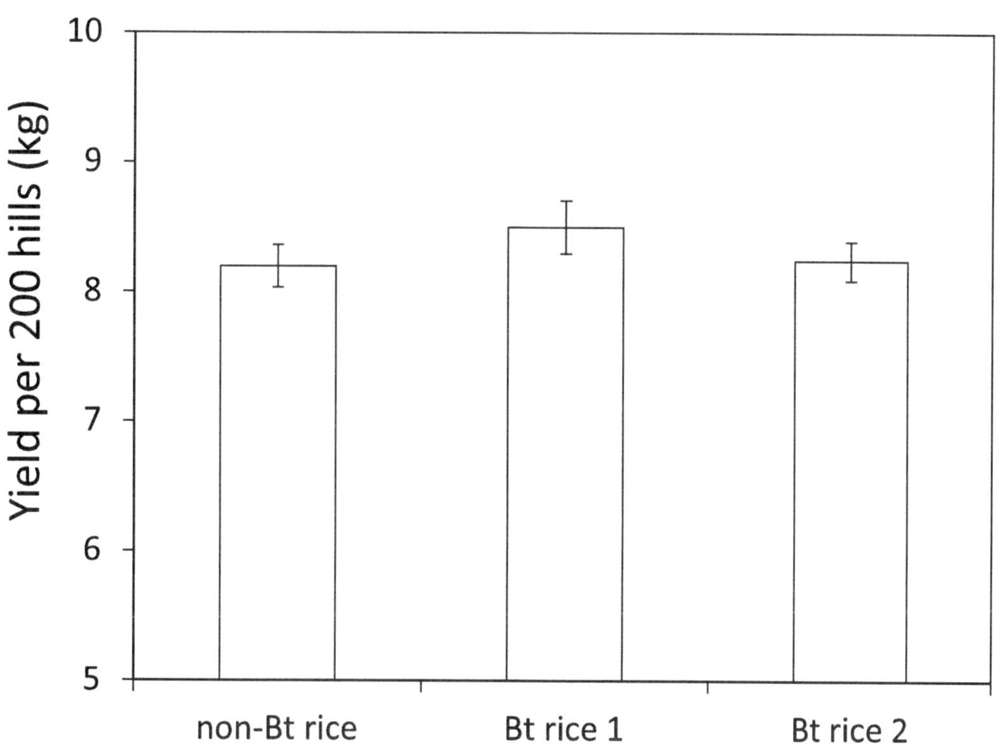


**Figure 1. Yield of Bt and non-Bt rice. Non-Bt rice = Minghui 63, Bt rice 1 = Minghui 63 *cry1Ab/1Ac*, Bt rice 2 = Minghui 63 *cry2A*.** Non-Bt rice as well as Bt rice were sprayed with pesticides when insect or fungal pests reached the action threshold. Error bars indicate the standard error. Planned comparison of means showed that the yield of non-Bt rice and Bt rice did not differ significantly (d.f. = 6; $t = 1.303$; $p = 0.240$).

zooplanktons, Rotifera, Cladocera and Copepoda, were identified down to the lowest possible taxonomic level, and were counted with the drop count method [24]. Each sample was counted twice and the mean was recorded. The zooplankton investigation was conducted twice, 5 and 12 days after the final pesticide spray on August 1st (see table S1).

### *Daphnia magna* culturing in field-collected water

*Daphnia magna* is highly sensitive to toxic substances in aquatic ecosystem, and has a short generation time [25]. *Daphnia magna* used in the experiments were from a single clonal population cultured in laboratory. Field water was collected from each plot and was filtered by 64 μm plankton net to remove suspended particles and plankton. The filtered water was used as culture medium in the water quality tests. A newly hatched *D. magna* (< 24 h old) was transferred to a 100-ml beaker filled with 80 ml filtered field-collected water using a broad-tipped pipette. This was replicated 30 times per plot (9 plots in total). Three millilitres of *Scenedesmus obliquus* solution ($4{\sim}5\times10^5$ cells/ml), cultivated according to Stein 1973 [26], was added to each beaker as food source for the *D. magna* [27]. Every 3 days, each *D. magna* was transferred to a new beaker with new filtered field-collected water and food source of *S. obliquus*. Each beaker was inspected daily to record survival and presence of newly born *D. magna*, and removal of offspring. All cultures were maintained under static conditions at 23±1°C and a controlled photoperiod at 12 h light and 12 h darkness.

### Field-collected water quality test using *Paramecium caudatum*

*Paramecium* are unicellular organisms belonging to Ciliophora that are commonly found in ponds and lakes. They are often used as bioindicators of aquatic environmental pollutants [28]. Stock cultures of *P. caudatum* were cultured in laboratory according to Ishikawa and Hota's procedure [29]. Twenty grams of dry rice straw (from a field without pesticide application) was boiled in 2 L tap water for 5 min and the supernatant was used as liquid food of *P. caudatum*. The experimental unit consisted of a 50 ml centrifuge tube with 10 ml culture medium, 3 ml *P. caudatum* stock culture and 37 ml filtered field-collected water. The initial density in the tubes was 54.3 cells ml. Each plot had 5 experimental replicates. The cultures were maintained in a climate chamber at 23±1°C, with a photoperiod of 12 h light and 12 h darkness. The density in each tube was determined every day in the morning using a stereo microscope and counting chamber. This experiment lasted for 8 days.

### Pesticide residue test in field-collected water

Field water samples were collected from each plot on August 11, 2013 (see Table S1 for pesticide application dates). All samples were filtered through a plankton net (64 μm mesh size) and stored into 100 ml amber glass bottles at 4°C. The samples were sent to Technology Service Centre Co., Ltd (http://www.labfaster.com/) the next day, and subjected to liquid chromatography-mass spectrometry (LC-MS/MS) analysis for residues of 5 pesticides. Those pesticides had been used in the experimental field before the sampling date. Each sample was analyzed twice and the average residue was recorded.

### Rice yield test

After rice grain fully ripened in September, 20 rows of rice each containing 10 hills were selected randomly in each plot. Panicles were threshed and rice grain was dried in sunlight and weighed.

### Data analysis

Yield data were analyzed by one-way RCBD ANOVA, and means were compared between non-Bt rice and Bt rice by a planned comparison. Data of zooplankton in field investigation were recorded as each group abundance and total abundance, and those abundances were transformed by ln (x+1) to meet the assumption of homogeneity of variances and then were subjected to one-way RCBD ANOVA. The means of abundance were separated by Fisher's LSD test. In order to explore the effect of overall rice type on abundance of zooplankton, multivariate analysis of variances (MANOVA) combining each group of zooplankton were also conducted. The survival rates of *D. magna* were arcsine-transformed and subjected to repeated measures ANOVA to estimate rice type and culturing time effect and their interaction; data of total reproduction were square root transformed and analysed by one-way RCBD ANOVA followed by a LSD test. The population growth of *P. caudatum* were also ln (x+1) transformed and subjected to repeated measures ANOVA. All analyses were carried out using general linear model (GLM) in SPSS package (version 16.0, SPSS Inc).

## Results

### Zooplankton abundance and diversity in Bt and non-Bt rice field

Thirty-three species were identified, 17 species in Rotifera, 14 in Cladocera and 2 in Copepoda (see Table S2). In the first investigation, 4, 21 and 18 species were found in non-Bt rice, Bt rice 1 and Bt rice 2 treatment, respectively, and 16, 24 and 19 species in the second investigation (see Table S2). One-way randomized complete block design (RCBD) ANOVAs detected significant rice type effect (d.f. = 2,4; $F>10.659$; $p<0.035$) and non-significant block effect (d.f. = 2,4; $F<4.515$; $p>0.094$) in both investigations on abundance of each zooplankton group (Table 1). The lowest abundance was always found in non-Bt rice plots. In the first investigation, total abundance in Bt rice treatments was around 1,500 individuals $L^{-1}$, which was significantly higher than that in non-Bt rice treatment (below 50 individuals $L^{-1}$) (LSD test, $p<0.01$); the differences between Bt rice 1 and Bt rice 2 were not significant (LSD test, $p = 0.974$). The second investigation found rebound in zooplankton abundance to around 1,600 and 640 individuals $L^{-1}$ in Bt rice and non-Bt rice respectively, but the trend of differences between Bt and non-Bt did not change (Table 1). Multivariate analyses of variance (MANOVA) were used to assess overall rice type effect on abundance of 3 groups of zooplankton, and significant effects were detected in the first investigation (d.f. = 6,4; $F = 9.049$; $p = 0.026$), as well as in the second (d.f. = 6,4; $F = 7.678$; $p = 0.034$). Therefore the water layer in Bt rice fields harbored more abundant and diverse zooplankton than that of non-Bt rice.

### Laboratory rearing of *D. magna* and *P. caudatum*

We tested the survival of *D. magna* and *P. caudatum* in water collected from Bt and non-Bt rice plots in the laboratory. Daphnids (<24 hrs old) were reared in 100 ml field-collected water with *Scenedesmus obliquus* as feed, and every 3 days the culture medium was refreshed with field-collected water. Differences on survival rate were found between Bt and non-Bt field-collected water one day after the onset of this experiment (Fig. 2A). In following days, survival rate in the non-Bt treatment decreased gradually to zero by the end of this 13-day experiment, while survival rates in the two Bt treatments remained relatively stable at 52% and 60% by day 13. Repeated-measures ANOVA within RCBD found significant rice type effect (d.f. = 2,4;

**Table 1.** Abundance (individuals $L^{-1}$) and diversity (number of species) of Rotifera, Cladocera and Copepoda in Bt and non-Bt rice plots.

| Treatment | Rotifera | Cladocera | Copepoda | Total |
|---|---|---|---|---|
| 1st investigation (abundance) | | | | |
| non-Bt rice | 10.33±5.17b | 0±0b | 35.33±25.56b | 45.66±29.55b |
| Bt rice 1 | 172.66±74.99a | 29.00±7.57a | 1501.00±651.00a | 1702.66±688.26a |
| Bt rice 2 | 254.67±119.32a | 56.26±12.33a | 1031.67±335.56a | 1342.50±348.50a |
| 1st investigation (number of species) | | | | |
| non-Bt rice | 1.33±0.67a | 0±0b | 1.33±0.67b | 2.67±1.33b |
| Bt rice 1 | 4.67±1.20a | 7.00±1.00a | 3.67±0.33a | 15.33±1.20a |
| Bt rice 2 | 3.67±0.33a | 6.33±2.08a | 3.63±0.33a | 15.70±3.00a |
| 2nd investigation (abundance) | | | | |
| non-Bt rice | 425.33±107.33b | 4.66±2.33b | 148.00±43.39b | 638.00±69.51b |
| Bt rice 1 | 1291.33±201.17a | 34.33±8.76a | 598.00±150.89a | 1923.67±325.42a |
| Bt rice 2 | 829.67±25.69a | 38.33±10.84a | 521.33±50.11a | 1317.33±63.89a |
| 2nd investigation (number of species) | | | | |
| non-Bt rice | 4.33±0.67a | 2.00±0.58b | 3.00±0.58a | 9.33±0.88b |
| Bt rice 1 | 5.67±0.33a | 7.00±1.53a | 3.67±0.33a | 16.33±1.86a |
| Bt rice 2 | 5.00±0.00a | 6.00±0.00a | 4.00±0.00a | 15.00±0.00a |

Non-Bt and Bt rice plots were sprayed with pesticides when rice pests exceeded the action threshold (Table S1). The first and the second investigation were conducted 5 days and 12 days after the final spray, respectively. The values presented in the table are untransformed means ± SE of each treatment. Means followed by the same letter within the same column in each investigation do not differ significantly at $p = 0.05$ (LSD post hoc test after ANOVA, n = 3).

$F = 34.393$, $p = 0.003$) and measuring time effect (d.f. = 12,52; $F = 59.489$; $p < 0.001$), and no significant block effect (d.f. = 2,4; $F = 1.587$; $p = 0.311$). The survival rate did not differ between the two Bt rice treatments (LSD test, $p = 0.549$), but did between non-Bt rice and Bt rice 1 or Bt rice 2 (LSD test, $p = 0.002$ and $p = 0.003$). Reproduction was observed in the culture medium on the 7th day of the experiment in all treatments, which indicated rice type had no effect on sexual maturity age of *D. magna*. A significant effect of rice type on average reproduction rate was observed (d.f. = 2,4; $F = 8.963$; $p = 0.033$), and LSD tests revealed significant differences between non-Bt rice and Bt rice 1 ($p = 0.018$) or Bt rice 2 ($p = 0.026$) (Fig. 2B). The growth (body length) of *D. magna* was also recorded, but no significant differences were found.

A stock of *P. caudatum* was cultured in the laboratory by traditional methods [29]. The stock culture was diluted to 54.3 cells $ml^{-1}$ in 50 ml culture medium with 10 ml rice straw infusion and approx. 37 ml filtered field water collected from Bt and non-Bt plots. The population growth was inspected daily until the 8th day. The *P. caudatum* expressed lower growth rate in non-Bt rice compared to the two Bt rice treatments (Fig. 3). Repeated-measures ANOVA within RCBD detected significant effects of rice type (d.f. = 2,4; $F = 195.707$; $p < 0.001$), measuring time (d.f. = 8,32; $F = 1790.116$; $p < 0.001$) and the interaction of rice type×measuring time (d.f. = 16,32; $F = 13.109$; $p < 0.001$). The growth rates did not differ between Bt rice 1 and Bt rice 2 (LSD test, $p = 0.072$). The mean density was about 300 cells/ml in non-Bt culture, and 450 and 500 in Bt rice 1 and Bt rice 2 cultures by the end of this experiment.

### Pesticide residue in field-collected water

In our previous study, we revealed that Bt rice (with gene *cry1Ab/1Ac*) in itself has no negative impact on paddy zooplanktons [30]. In order to determine what caused the differences in survival found earlier, we tested pesticide residues in the water collected from the field 10 days after the final spray (see Table S1 for pesticide application dates). We found that residues of pesticides targeting Homopteran and fungal pests were in the same order of magnitude in Bt and non-Bt rice plots, since those pesticides were used both in Bt rice and non-Bt rice (Table 2). The residues of chlorantraniliprole and triazophos, targeting lepidopteran pests, were100 times higher in non-Bt plots than Bt plots (Table 2). Theoretically, pesticides targeting lepidopteran pests would not be detected in Bt plots because they were not used in Bt rice. However, they were detected in Bt plots probably because pesticide spray drifted by wind or/and water although we had made efforts to avoid this. In short, the pesticide residues in non-Bt plots were far higher than Bt plots, which likely accounts for the poor fitness performance of zooplanktons in the non-Bt treatment.

## Discussion

Research on genetic modification has developed rapidly in China. In November 2009, COMA announced that biosafety licenses were issued to two lepidopteran-resistant Bt rice lines in designated areas. However, the decision was the subject of heated scientific and public debates in the following years. The core of these debates is the uncertainty regarding food safety and potential environmental risks.

In this study, we evaluated the environmental risk of two Bt rice lines from a new angle. Firstly, we focused on paddy aquatic organisms as the non-target organisms in the environmental assessment of Bt rice. Previous non-target effect studies of Bt crops mainly focuses on terrestrial invertebrates, such as earthworms, collembolans, mites, woodlice and nematodes [31–35]. Like those terrestrial invertebrates, aquatic organisms also carry out important ecological functions, such as nutrient cycling, decomposition of organic matter and maintaining aquatic food webs [36]. In

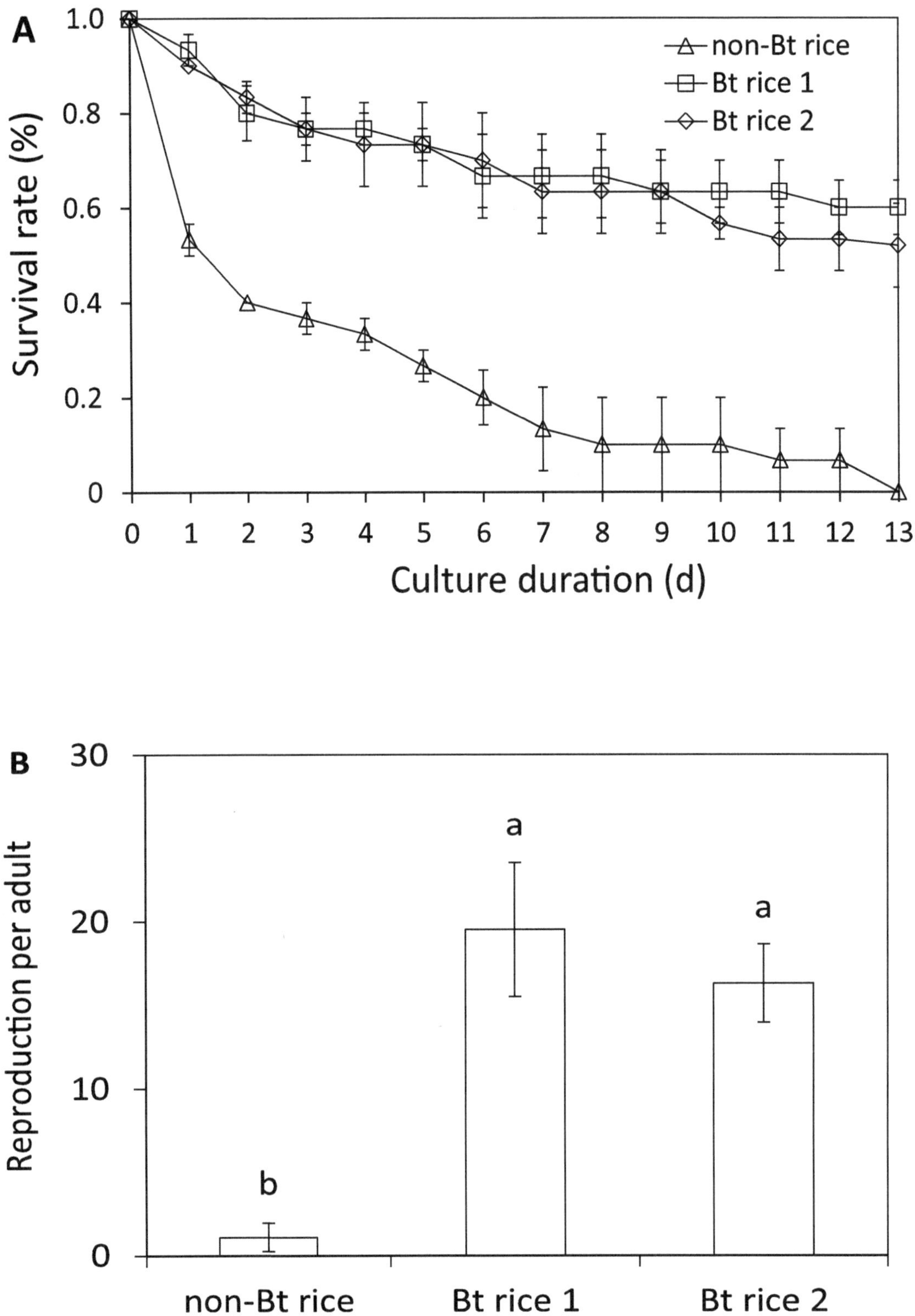


**Figure 2. Survival (A) and reproductive rate (B) of *D. magna* reared in water collected from non-Bt rice and Bt rice plots.** Water was collected 6 days after the final pesticide sprays and the culture medium was renewed every 3 days with field water collected on that very day. Non-Bt and Bt rice plots were sprayed with pesticides when rice pests exceeded the action threshold (see Table S1). Error bars indicated standard error. Different letters capped on the bars indicate significant difference at $p = 0.05$ (LSD post hoc test after ANOVA).

many areas, the water matrix of rice fields is connected to nearby ponds and lakes which are important resources for rice farming farmers for additional income and food by fish production [37]. Secondly, we assessed Bt rice's environmental effects in compar-

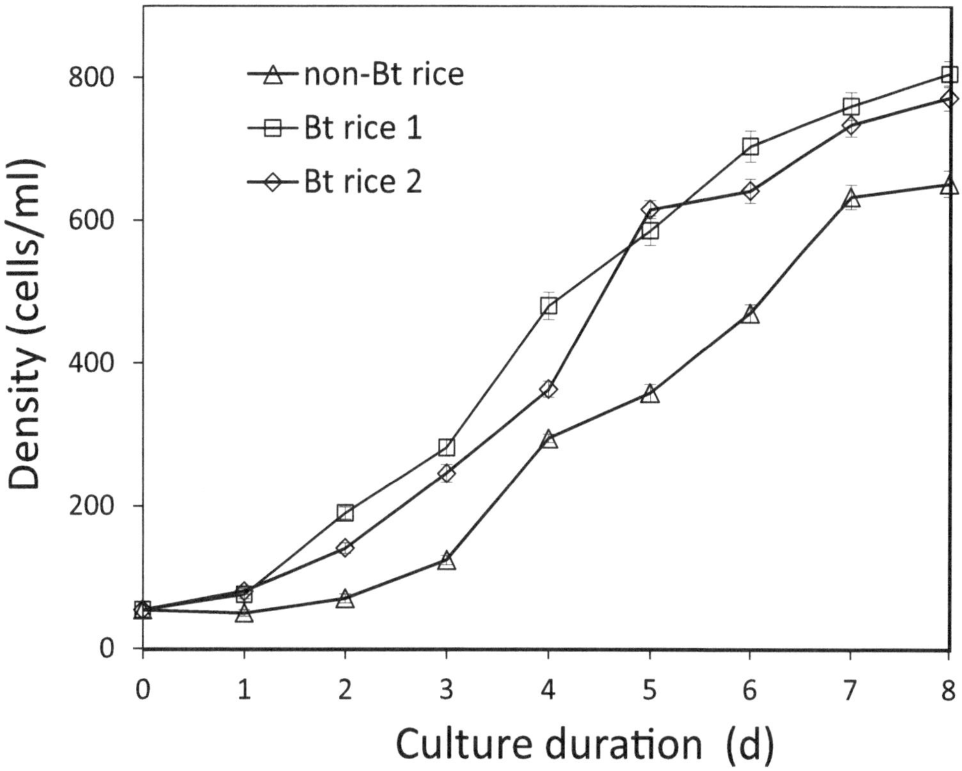


**Figure 3. Population growth of *P. caudatum* in culture medium made of field water.** Field water was collected from non-Bt rice and Bt rice plots 11 days after the final pesticide spray (see Table S1). Error bars indicate the standard error.

ison with non-Bt rice under realistic agricultural conditions in which pesticides were used in both Bt and non-Bt fields as required. Former field studies demonstrated that although Bt rice (*cry1Ab/Ac*) reduced pesticide use by up to 60%, they were still required against fungal and non-lepidopteran pests in order to guarantee no yield loss [14,15]. However, almost all previous studies on Bt rice's non-target effect were conducted without considering the factor of pesticide application [4].

In this study, Bt rice was sprayed twice and non-Bt rice was sprayed 5 times throughout the whole growing season (Table S1), which resulted in equal yield between Bt and non-Bt rice (Fig. 1). The Bt rice expressed good resistance to lepidopteran pests, more specifically rice leaf folder and rice stem borers, and the two sprays only targeted a fungal disease and rice planthopper. The same results were reported in Bt cotton and Bt maize which also greatly reduced insecticide applications [38].

Our first field investigation found that the total abundance of three groups of zooplankton Rotifera, Cladocera and Copepoda in non-Bt field was about 95% lower than that in Bt rice field, and the total number of species was 80% lower (Table 1). The second sampling was conducted 7 days after the first one, and a rise in abundance and diversity was observed, however zooplanktons were still significantly more abundant and diverse in the Bt rice field. In contrast to this result, our previous two-year field study, in which we did not use pesticides, found the total population density of rotifera, cladocera and copepods did not differ significantly between Bt (*cry1Ab/1Ac*) and non-Bt rice fields [30]. The differences we report here are therefore not due to Bt rice itself but from the differences in pesticide application. We tested pesticide residues using LC-MS/MS in water samples collected from rice fields and found that pesticide (targeting lepidopteran pests) residues were around 100 times higher in non-Bt than Bt

**Table 2.** Pesticide residue ± SE (μg $L^{-1}$) in water collected from non-Bt and Bt rice plots.

| Pesticide | Non-Bt rice | Bt rice 1 | Bt rice 2 |
|---|---|---|---|
| chlorantraniliprole | 15.19±1.41 | 0.17±0.04 | 0.17±0.04 |
| difenoconazole | 5.44±1.37 | 1.47±0.89 | 2.57±1.38 |
| propiconazole | 35.21±1.15 | 42.67±14.92 | 26.93±8.89 |
| triazophos | 80.45±11.91 | 0.78±0.23 | 1.19±0.52 |
| buprofezin | 0.73±0.29 | 0.16±0.08 | 0.28±0.21 |

Field water was collected 10 days after the final spray (see table S1 for the spraying date). Pesticide residues were quantified by liquid chromatography-tandem mass spectrometry (LC-MS/MS).

plots (Table 2). It is widely documented that pesticides enter into aquatic environments and accumulate to impact certain aquatic organisms. Some zooplanktons are frequently used in ecotoxicological tests of anthropogenic chemicals [39]. The qualitative and quantitative characteristics of zooplankton population depend on the degree of water pollution [40]. The high abundance of zooplankton in Bt rice fields corresponds with low water pollution from reduced pesticide application, and can be considered a positive effect of Bt rice on aquatic environment.

In addition to field investigation, we conducted laboratory experiment to rear *D. magna* and *P. caudatum* in field-collected water. The survival rate of *D. magna* declined to zero in non-Bt rice field-collected water by the end of the 13-days rearing experiment, while the survival rate was up to 60% in Bt rice field water (Fig. 2A). The amount of reproduction in Bt rice field water was more than 15 times higher than in non-Bt rice field water (Fig. 2B). The population of *P. caudatum* also developed faster in Bt rice field-collected water, and the final population density was about 20% higher in Bt rice field-collected water than non-Bt rice water (Fig. 3). Up to now, only a few studies have assessed non-target effects of Bt crops on aquatic organisms. Rosi-Marshall and colleagues reported that consumption of Bt corn (*cry1Ab*) debris reduced growth and increased mortality of caddisfly larvae in laboratory feeding trials [41]. Bøhn's study demonstrated that *D. magna* fed on Bt maize (*cry1Ab*) showed significantly reduced fitness performance compared to the non-Bt control [42]. Rosi-Marshall's study concluded that Bt δ-endotoxin in Bt maize byproducts harmed aquatic trichopterans insects which are phylogenetically close to Lepidoptera, and Bøhn's study concluded a toxic effect rather than a lower nutritional value of Bt maize accounting for the lower fitness of *D. magna*. Their findings were quite different from the results of our present study. Although Bt protein is sometimes detected in Bt rice field water [43], we did not consider Bt proteins to have negative impact on paddy *Daphnia* because a another water flea species, *Daphnia hyalina*, was shown to perform as well in culture medium with high concentrations of Bt toxin as in the control medium [30]. In this study, we cultured *D. magna* and *P. caudatum* just in water collected from rice field and did not feed with Bt rice materials because water flea and paramecium typically do not feed directly on rice products or by-products [44]. In their studies, the non-target organisms were fed with high doses of Bt maize materials for a longer time than ours. Nevertheless, those aquatic organisms would not use a single food source for that long in the natural environment. Furthermore, they did not consider agricultural chemicals (such as pesticides) that probably existed in their materials, making their conclusions questionable [45]. We found that field-collected water from Bt rice fields was more suitable for survive survival and reproduction of paddy zooplanktons in short-term rearing trials, and we also detected higher pesticide residues in the water from non-Bt rice fields than Bt rice fields. Our results therefore indicate that pesticide residues in field-collected water play a more important role in zooplankton fitness than Bt toxins and Bt rice, which highlight the importance of considering pesticide application in non-target environmental assessments of Bt crops.

## Conclusions

In order to settle the debate on Bt rice and realize its potential benefit as early as possible [46], scientific experiments are needed to reveal the real environmental effects of Bt rice. We assessed the influence of two Bt rice lines on aquatic organisms under normal field management which used pesticide when pests exceeded the action threshold. The results indicated that Bt rice required reduced pesticide application therefore produced lower pesticide residues in the field water body. The abundance and diversity of zooplankton were significantly higher in Bt rice fields than non-Bt rice fields, and *D. magna* and *P. caudatum* displayed far higher survival and fecundity in Bt rice field water than non-Bt rice field water. These results demonstrate that the Bt rice lines are more friendly to the aquatic ecosystem than conventional rice. Pesticide application in rice fields has long been blamed for biodiversity loss in rice-based ecosystems [47] and it is expected that biodiversity will gradually be restored following the release of Bt rice.

## Acknowledgments

We thank Hui Jing for identifying and counting zooplanktons and Mingyuan He for providing *D. magna* samples and Yongjun Lin for providing Bt rice seeds. We also thank James Hereward for valuable comments on grammar and syntax of the manuscript.

## Author Contributions

Conceived and designed the experiments: YW. Performed the experiments: GL. Analyzed the data: YW GL. Contributed to the writing of the manuscript: YW. Provided some advice in experimental design: BL. Performed field management: GZ.

## References


1. Redona ED (2004) Rice Biotechnology for Developing Countries in Asia. In: A Eaglesham editors. Agricultural Biotechnology: Finding Common International Goals. National Agricultural Biotechnology Council (NABC) Report No. 16. Ithaca (New York): National Agricultural Biotechnology Council. 201–232.
2. Khush GS (1997) Origin, dispersal, cultivation and variation of rice. Oryza: From Molecule to Plant, (Springer), 25–34.
3. Pathak MD, Khan ZR (1994) Insect pests of rice Insect pests of rice. International Rice Research Institute, Los Banos, Philippines. 89 p.
4. Chen M, Shelton A, Ye GY (2011) Insect-resistant genetically modified rice in China: from research to commercialization. Annu Rev Entomol 56: 81–101.
5. Lu C (2010) The first approved transgenic rice in China. GM Crops 1(3): 113–115.
6. Niu S, Stanway D (8 Mar 2013) China delays GMO corn and rice to woo the public. The China Post. Available: https://www.chinapost.com.tw/china/national-news/2013/03/08/372455/China-delays.htm Accessed 15 Mar 2014.
7. Tian JC, Liu ZC, Yao HW, Ye GY, Peng YF (2008) Impact of transgenic rice with a cry1Ab gene on parasitoid subcommunity structure and the dominant population dynamics of parasitoid wasps in rice paddy. J Environ Entomol 1: 003.
8. Liu Z, Ye G, Hu C, Swapan KD (2001) Effects of Bt transgenic rice on population dynamics of main non-target insect pests and dominant spider species in rice paddies. Acta Phytophylacica Sinica 29(2): 138–144.
9. Jiang Y, He J, Dong B, Xu H, Fu Q, et al. (2003) Effects of transgenic Bt rice on the biological characteristics of *Apanteles chilonis* (Munakata) (Hymenoptera: Braconidae). Acta Entomol Sinica 47(1): 124–129.
10. Chen M, Zhao JZ, Ye GY, Fu Q, Shelton AM (2006) Impact of insect-resistant transgenic rice on target insect pests and non-target arthropods in China. Insect Sci 13(6): 409–420.
11. Chen M, Ye G, Liu Z, Fang Q, Hu C, et al. (2009) Analysis of Cry1Ab toxin bioaccumulation in a food chain of Bt rice, an herbivore and a predator. Ecotoxicology 18(2): 230–238.
12. Bai Y, Jiang M, Cheng J, Wang D (2006) Effects of Cry1Ab toxin on *Propylea japonica* (Thunberg) (Coleoptera: Coccinellidae) through its prey, *Nilaparvata lugens* Stål (Homoptera: Delphacidae), feeding on transgenic Bt rice. Environ Entomol 35(4): 1130–1136.

13. Bai Y, Jiang M, Cheng J (2005) Effects of transgenic cry1Ab rice pollen on fitness of Propylea japonica (Thunberg). J Pest Sci 78(3): 123–128.
14. Wang Y, Zhang G, Du J, Liu B, Wang M (2010) Influence of transgenic hybrid rice expressing a fused gene derived from cry1Ab and cry1Ac on primary insect pests and rice yield. Crop Prot 29(2): 128–133.
15. Huang J, Hu R, Rozelle S, Pray C (2005) Insect-resistant GM rice in farmers' fields: assessing productivity and health effects in China. Science 308(5722): 688–690.
16. Carstens K, Anderson J, Bachman P, De Schrijver A, Dively G, et al. (2012) Genetically modified crops and aquatic ecosystems: considerations for environmental risk assessment and non-target organism testing. Transgenic Res 21(4): 813–842.
17. Ali AB (1990) Seasonal dynamics of microcrustacean and rotifer communities in Malaysian rice fields used for rice-fish farming. Hydrobiologia 206(2): 139–148.
18. Vanni MJ (2002) Nutrient cycling by animals in freshwater ecosystems. Annu Rev Ecol Syst 33(1): 341–370.
19. Liess M, Ohe PCVD (2005) Analyzing effects of pesticides on invertebrate communities in streams. Environ Toxicol Chem 24(4): 954–965.
20. Biesinger KE, Christensen GM (1972) Effects of various metals on survival, growth, reproduction, and metabolism of Daphnia magna. J Fish Board of Can 29(12): 1691–1700.
21. Shiny K, Remani K, Nirmala E, Jalaja T, Sasidharan V (2005) Biotreatment of wastewater using aquatic invertebrates, *Daphnia magna* and *Paramecium caudatum*. Bioresour Technol 96(1): 55–58.
22. Tu J, Zhang G, Datta K, Xu C, He Y, et al. (2000) Field performance of transgenic elite commercial hybrid rice expressing *Bacillus thuringiensis* δ-endotoxin. Nat Biotechnol 18(10): 1101–1104.
23. Chen H, Tang W, Xu C, Li X, Lin Y, et al. (2005) Transgenic indica rice plants harboring a synthetic cry2A* gene of *Bacillus thuringiensis* exhibit enhanced resistance against lepidopteran rice pests. Theor Appl Genet 111(7): 1330–1337.
24. Emmanuel B, Onyema I (2007) The Plankton and fishes of a tropical creek in South-western Nigeria. Turk. J. Fish. Aquat. Sci 7(2): 105–113.
25. Lovern SB, Strickler JR, Klaper R (2007) Behavioral and physiological changes in *Daphnia magna* when exposed to nanoparticle suspensions (titanium dioxide, nano-C60, and C60HxC70Hx). Environ Sci Technol 41(12): 4465–4470.
26. Stein JR (1973) Handbook of phycological methods: Culture methods and growth measurements (Cambridge University Press, New York). 472 p.
27. Villegas-Navarro A, Gonzalez M, Lopez ER, Aguilar RD, Marcal WS (1999) Evaluation of *Daphnia magna* as an indicator of toxicity and treatment efficacy of textile wastewaters. Environ Int 25(5): 619–624.
28. Miyoshi N, Kawano T, Tanaka M, Kadono T, Kosaka T, et al. (2003) Use of Paramecium Species in Bioassays for Environmental Risk Management: Determination of IC50 Values for Water Pollutants. J Health Sci 49(6): 429–435.
29. Ishikawa T, Hota M (2006) Interaction of two swimming Paramecia. J Exp Biol 209(22): 4452–4463.
30. Wang Y, Huang J, Hu H, Li J, Liu B, et al. (2013) Field and laboratory studies on the impact of two Bt rice lines expressing a fusion protein Cry1Ab/1Ac on aquatic organisms. Ecotox Environ Safe 92: 87–93.
31. Zwahlen C, Hilbeck A, Howald R, Nentwig W (2003) Effects of transgenic Bt corn litter on the earthworm *Lumbricus terrestris*. Mol Ecol 12(4): 1077–1086.
32. Heckmann LH, Griffiths BS, Caul S, Thompson J, Pusztai-Carey M, et al. (2006) Consequences for Protaphorura armata (Collembola: Onychiuridae) following exposure to genetically modified *Bacillus thuringiensis* (Bt) maize and non-Bt maize. Environ Pollut 142(2): 212–216.
33. Griffiths BS, Caul S, Thompson J, Birch ANE, Scrimgeour C, et al. (2006) Soil Microbial and Faunal Community Responses to Maize and Insecticide in Two Soils. J Environ Qual 35(3): 734–741.
34. Clark BW, Prihoda KR, Coats JR (2006) Subacute effects of transgenic Cry1Ab *Bacillus thuringiensis* corn litter on the isopods *Trachelipus rathkii* and *Armadillidium nasatum*. Environ Toxicol Chem 25(10): 2653–2661.
35. Saxena D, Stotzky G (2001) *Bacillus thuringiensis* (Bt) toxin released from root exudates and biomass of Bt corn has no apparent effect on earthworms, nematodes, protozoa, bacteria, and fungi in soil. Soil Biol Biochem 33(9): 1225–1230.
36. Simpson I, Roger PA (1995) The impact of pesticides on nontarget aquatic invertebrates in wetland ricefields: a review. Impact of pesticides on farmer health and the rice environment, (Springer), 249–270.
37. Kangmin L (1988) Rice-fish culture in China: A review. Aquaculture 71(3): 173–186.
38. Benbrook CM (2012) Impacts of genetically engineered crops on pesticide use in the US-the first sixteen years. Environ Sci Eur 24(1): 1–13.
39. Hanazato T (2001) Pesticide effects on freshwater zooplankton: an ecological perspective. Environ Pollut 112(1): 1–10.
40. Vandysh O (2004) Zooplankton as an indicator of the state of lake ecosystems polluted with mining wastewater in the Kola Peninsula. Russ J Ecol 35(2): 110–116.
41. Rosi-Marshall EJ, Tank JL, Royer TV, Whiles MR, Evans-White M, et al. (2007) Toxins in transgenic crop byproducts may affect headwater stream ecosystems. Proc Natl Acad Sci USA 104(41): 16204–16208.
42. Bøhn T, Primicerio R, Hessen DO, Traavik T (2008) Reduced fitness of *Daphnia magna* fed a Bt-transgenic maize variety. Arch Environ Contam Toxicol 55(4): 584–592.
43. Wang Y, Hu H, Huang J, Li J, Liu B, et al. (2013) Determination of the movement and persistence of Cry1Ab/1Ac protein released from Bt transgenic rice under field and hydroponic conditions. Soil Biol Biochem 58: 107–114.
44. O'Sullivan P, Reynolds CS (2008) The lakes handbook: limnology and limnetic ecology, (John Wiley & Sons). 667 p.
45. Beachy RN, Fedoroff NV, Goldberg RB, McHughen A (2008) The burden of proof: A response to Rosi-Marshall, et al. Proc. Natl. Acad. Sci. USA 105: E9.
46. James C (2009) Global Status of Commercialized Biotech/GM Crops, 2009 (ISAAA Ithaca, NY).
47. Halwart M (2008) Biodiversity, nutrition and livelihoods in aquatic rice-based ecosystems. Biodiversity 9(1–2): 36–40.

3

# Spinosad and the Tomato Borer *Tuta absoluta*: A Bioinsecticide, an Invasive Pest Threat, and High Insecticide Resistance

**Mateus R. Campos[1], Agna Rita S. Rodrigues[1], Wellington M. Silva[1], Tadeu Barbosa M. Silva[1], Vitória Regina F. Silva[1], Raul Narciso C. Guedes[2], Herbert Alvaro A. Siqueira[1]***

**1** Departamento de Agronomia (Entomologia), Universidade Federal Rural de Pernambuco, Recife, PE, Brazil, **2** Departamento de Entomologia, Universidade Federal de Viçosa, Viçosa, MG, Brazil

## Abstract

The introduction of an agricultural pest species into a new environment is a potential threat to agroecosystems of the invaded area. The phytosanitary concern is even greater if the introduced pest's phenotype expresses traits that will impair the management of that species. The invasive tomato borer, *Tuta absoluta* (Meyrick) (Lepidoptera: Gelechiidae), is one such species and the characterization of the insecticide resistance prevailing in the area of origin is important to guide management efforts in new areas of introduction. The spinosad is one the main insecticides currently used in Brazil for control of the tomato borer; Brazil is the likely source of the introduction of the tomato borer into Europe. For this reason, spinosad resistance in Brazilian populations of this species was characterized. Spinosad resistance has been reported in Brazilian field populations of this pest species, and one resistant population that was used in this study was subjected to an additional seven generations of selection for spinosad resistance reaching levels over 180,000-fold. Inheritance studies indicated that spinosad resistance is monogenic, incompletely recessive and autosomal with high heritability ($h^2 = 0.71$). Spinosad resistance was unstable without selection pressure with a negative rate of change in the resistance level ($= -0.51$) indicating an associated adaptive cost. Esterases and cytochrome P450-dependent monooxygenases titration decreased with spinosad selection, indicating that these detoxification enzymes are not the underlying resistance mechanism. Furthermore, the cross-resistance spectrum was restricted to the insecticide spinetoram, another spinosyn, suggesting that altered target site may be the mechanism involved. Therefore, the suspension of spinosyn use against the tomato borer would be a useful component in spinosad resistance management for this species. Spinosad use against this species in introduced areas should be carefully monitored to prevent rapid selection of high levels of resistance and the potential for its spread to new areas.

**Editor:** Youjun Zhang, Institute of Vegetables and Flowers, Chinese Academy of Agricultural Science, China

**Funding:** The financial support provided by the National Council of Scientific and Technological Development (CNPq-484240/2011-0) (HAAS), CAPES Foundation (MRC), and the Insecticide Resistance Action Committee – Brazilian Section (IRAC-BR) (HAAS) was greatly appreciated and acknowledged here. The funders had no role in study design, data collection and analysis, decision to publish, or preparation of the manuscript.

* Email: siqueira@depa.ufrpe.br

## Introduction

Invasive agricultural pest species are widely recognized as a major threat to agroecosystems and agricultural production [1–3]. An additional phytosanitary concern is that the introduced pest's phenotype could include inheritable traits that could impose management difficulties, such as resistance to insecticides [4–7]. The invasive species, *Tuta absoluta* (Meyrick) (Lepidoptera: Gelechiidae), the tomato borer or tomato leafminer (also tomato pinworm), is one such species. It is of South American origin but was introduced into Europe as early as 2006. This pest has subsequently spread to North Africa and the Middle East and is now threatening the whole of Asia, particularly China and India, the two leading world tomato producers [8–11].

From its Peruvian origin, the tomato borer has spread in South America. Its eventual introduction into Brazil, the leading neotropical tomato producer [11], led to drastic changes in tomato production in the country with a dramatic increase in insecticide use in the early 1980's [10]. Problems with insecticide resistance in the tomato borer were soon detected in the late 1990's and early 2000's in Chile, Brazil and Argentina for the insecticides initially used against this species, including organophosphates, pyrethroids, abamectin and cartap [12–17]. This resistance led to subsequent registration and large-scale use of new insecticides, particularly in Brazil, including insect growth regulators, indoxacarb, chlorfenapyr, spinosyns, and diamides [10,18,19]. Organically produced tomatoes imposed additional restrictions and challenges for tomato borer control, culminating in the use of bioinsecticides, such as the spinosyn spinosad and

*Bacillus thuringiensis* Berliner, aided by alternative supporting control methods [10,18].

Insecticide registration and use against the tomato borer in South America led to corresponding waves of change in the prevailing patterns of insecticide resistance congruent with the patterns of insecticide use and control failures [13,19,20,21]. The trends, closely followed in Brazil, were intensive use of chitin synthesis inhibitors succeeded the use of abamectin, cartap and pyrethroids against the borer in tomato fields, reaching high levels of resistance (>100-fold), followed by evidences of control failure with this group of insect growth regulators [20,21]. The bioinsecticide spinosad, a compound of natural origin used in neotropical tomato fields (both organic and conventional fields), has become one of the main compounds used against the tomato borer, but reports of resistance have started to appear both in Brazil and Chile [21,22,23].

The appeal of spinosad, a fermentation product of the soil actinomycete *Saccharopolyspora spinosa* (Mertz and Yao), includes its safety profile and acceptable use in organically produced tomatoes [24–26]. However, spinosad resistance counterpoints this appeal and the potential for further use. More seriously, the swift development of insecticide resistance in neotropical field populations of the tomato borer is suggestive of a rapid evolution of spinosad resistance, which remains to be tested [19,21]. The introduction of the tomato borer from South America into Europe, suggests additional problems for managing this destructive species in newly infested areas, further threatening the current world tomato production [19].

The high genetic homogeneity reported among populations of the tomato borer from South America and Europe give credence to the apparent high level of dispersion of the species and a shared origin [27,28]. These findings also support the hypothesis of a single invasive event for the tomato borer in Europe [19,28]. The emerging studies of insecticide resistance in Europe and, particularly, the survey of pyrethroid resistance due to altered target site sensitivity also provides support for the single-introduction event of the tomato borer [29–31]. Additionally, the introduced borer phenotype was likely resistant to at least pyrethroid insecticides [19,31], but may also be capable of rapid development of resistance to other insecticides, including bioinsecticides widely used in traditional and organic tomato production.

In our study, a field population of the tomato borer, already exhibiting spinosad resistance, was subjected to further selection for spinosad resistance to assess the rate of development and level of resistance likely to be achieved with intensive use of this insecticide. The spinosad-selected strain of tomato borer was also subsequently used for the genetic characterization of spinosad resistance and assessment of its stability. This strain was also utilized to evaluate the potential involvement of detoxification by esterases and cytochrome P450-dependent monooxygenases as the underlying resistance mechanism and to assess its cross-resistance spectrum. Based on previous findings in Brazil, a fast response to spinosad selection, reaching high levels of resistance (>100-fold) in few generations (<10) and monogenic resistance was expected. The involvement of cytochrome P450-dependent monooxygenases was previously suggested in Chilean populations of the tomato borer [22]. Evidence of cross-resistance has not yet been detected in the tomato borer, except within pyrethroids and chitin synthesis inhibitors [13,20,31]. Therefore, cross-resistance is more likely among spinosyns than between spinosad and insecticides from other groups, especially if altered target site sensitivity is involved.

**Table 1.** Relative toxicity of spinosad to successive generations of spinosad-selected and -unselected strains of the tomato borer *Tuta absoluta*.

| Population | Generation | Degrees of freedom | Slope ± SE | $LC_{50}$ (95% CI) (μg a.i./mL) | $LC_{99}$ (95% CI) (μg a.i./mL) | $RR_{50}$ (95% CI) | $\chi^2$ |
|---|---|---|---|---|---|---|---|
| Unselected | F1 | 5 | 1.85±0.23 | 0.41 (0.30–0.52) | 7.37 (4.36–17.04) | – | 1.51 |
| Spinosad-selected | F2 | 6 | 1.21±0.18 | 1.41 (0.882–2.14) | 117.77 (42.813–703.52) | 3.43 (3.25–3.61) | 1.51 |
| | F3 | 6 | 1.06±0.20 | 13.13 (4.08–43.74) | 2048.90 (270.67–0.28×$10^7$) | 32.01 (32.44–32.58) | 9.91 |
| | F4 | 5 | 1.14±0.26 | 294.15 (147.85–520.62) | 3166.00 (7419.20–0.13×$10^7$) | 717.14 (716.26–718.01) | 2.66 |
| | F5 | 5 | 1.98±0.31 | 1333.67 (718.99–2109.73) | 19760.00 (9020.30–0.11×$10^6$) | 3256.03 (3255.62–3258.03) | 5.24 |
| | F7 | 6 | 1.16±0.18 | 2200.41 (1051.24–3944.73) | 22.09×$10^7$ (6.0×$10^7$–3.9×$10^9$) | 5380.88 (5379.52–5382.24) | 6.74 |
| | F12 | 5 | 1.70±0.36 | 3706.34 (2055.99–6152.69) | 86640.00 (32808.00–0.82×$10^6$) | 9030.57 (9028.39–9032.74) | 1.98 |
| | F13 | 4 | 1.45±0.27 | 1180.48 (698.73–1756.64) | 47952.00 (18525.00–0.35×$10^6$) | 2880.06 (2878.68–2881.44) | 0.79 |
| | F14 | 4 | 1.62±0.20 | 1637.63 (1240.59–2117.19) | 44902.00 (23312.00–0.13×$10^6$) | 3993.27 (3993.08–3993.37) | 3.16 |
| | F16 | 4 | 1.47±0.19 | 1717.33 (998.16–2764.20) | 65180.00 (21576.00–0.90×$10^6$) | 4191.55 (4192.49–4192.61) | 5.46 |
| | F18 | 5 | 1.08±0.19 | 956.66 (542.62–1537.35) | 0.13×$10^6$ (38335.00–0.10×$10^7$) | 2338.54 (2338.42–2338.65) | 1.78 |
| | F21 | 4 | 1.86±0.23 | 1034.05 (664.63–1462.17) | 18271.00 (8876.20–76320.00) | 2526.35 (2526.20–2526.49) | 4.35 |
| | F22 | 4 | 2.04±0.23 | 1290.22 (1017.12–1601.71) | 17848.00 (11037.00–37045.00) | 3150.06 (3149.90–3150.21) | 1.56 |

All of the concentration-mortality curves followed the probit model based on the $\chi^2$ goodness-of-fit test ($P>0.05$).

**Table 2.** Heritability estimate ($h^2$) of spinosad resistance for a seven-generation spinosad-selected strain of the tomato borer *Tuta absoluta*.

| Parameters | Generations F1–F7 | |
|---|---|---|
| Response estimate | $F_1$ $CL_{50}$ (log) (µg a.i./mL) | 0.41 (−0.39) |
| | $F_7$ $CL_{50}$ (log) (µg a.i./mL) | 2200.00 (3.34) |
| | Response to selection (R) | 0.53 |
| Estimate of differential selection | Selection-surviving Individuals (*p*; %) | 31.25 |
| | Intensity of selection (i) | 1.12 |
| | Initial slope | 1.85 |
| | Final slope | 1.16 |
| | Phenotypic standard deviation (sF) | 0.66 |
| | Differential selection (S) | 0.75 |
| | Generations for 10-fold increase in resistance (G) | 1.88 |
| | Herdability ($h^2$) | 0.71 |

## Materials and Methods

### Ethics Statement

This study did not involve any endangered or protected species. Although the insect species studied is a pest species, permits were secured for the collection of the original field populations. The laboratory colonies were initially established from over 200 field-collected individuals.

### Insects

Populations of the tomato borer were collected from experimental and commercial tomato fields during 2010/2011 in four regions in Brazil. These insect populations were subjected to an initial screening for spinosad resistance and the populations from Iraquara (state of Bahia, Brazil) and Pelotas (state of Rio Grande do Sul, Brazil) were used for our experiments. The insects were laboratory-maintained in wooden cages with anti-aphid mesh. The cages were separate in larvae cage (45×45×45 cm) and adult

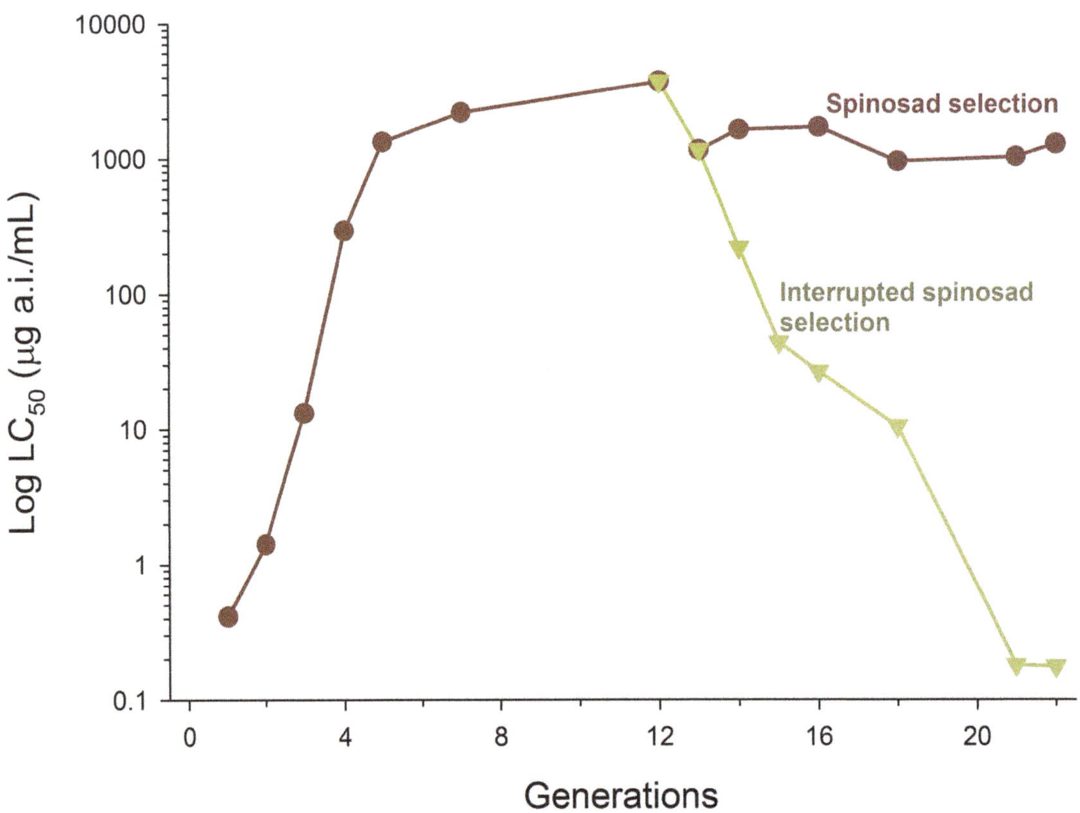


**Figure 1. $LC_{50}$s for spinosad with successive selections for spinosad resistance of the tomato borer *Tuta absoluta*.** After 12 generations of spinosad selections, the selected line was split into two, one line maintaining selection and one line with interrupted selection.

**Table 3.** Relative toxicity of spinosad in spinosad-susceptible and (selected) spinosad-resistant strains, the progeny of reciprocal crosses ($F_1$: ♀R×♂S and ♀S×♂R) and of backcrosses [$F_1$ (pooled)×(selected) spinosad-resistant strain] of the tomato borer *Tuta absoluta*.

| | **No. insects tested** | | **$LC_{50}$ (95% CI)** | | **Degree of dominance (D)** | **$LC_{90}$ (95% CI)** | | **Degree of dominance (D)** | |
|---|---|---|---|---|---|---|---|---|---|
| Strain | (degrees of freedom) | Slope ± SE | (μg a.i./mL) | $RR_{50}$ (95% CI) | at $LC_{50}$ (±SE) | (μg a.i./mL) | $RR_{90}$ (95% CI) | at $LC_{90}$ (±SE) | $\chi^2$ |
| Spinosad susceptible | 320 | 1.18±0.13 | 0.01 | 1.00 | - | 0.113 | 1.00 | - | 5.32 |
| (standard) | (6) | | (0.007–0.01) | (0.81–1.19) | | (0.071–0.222) | (0.81–1.19) | | |
| Spinosad-resistant ($F_{15}$ of | 238 | 1.47±0.19 | 1717.30 | 183122.81 | - | 12730.00 | 1354308.97 | - | 5.46 |
| selected strain) | (4) | | (998.00–2764.00) | (183123.17–182847.10) | | (6481.00–56151.00) | (1354309.17–1354309.25) | | |
| Reciprocal cross ♂ S×♀ R | 216 | 1.51±0.19 | 0.26 | 27.27 | −0.45±0.03 | 1.81 | 15.96 | −0.52±0.06 | 5.32 |
| | (5) | | (0.17–0.39) | (27.65–27.44) | | (1.02–4.98) | (27.65–27.44) | | |
| Reciprocal cross ♂ R×♀ S | 211 | 1.60±0.19 | 0.38 | 40.31 | −0.39±0.03 | 2.38 | 21.03 | −0.47±0.06 | 5.83 |
| | (5) | | (0.25–0.59) | (40.12–40.29) | | (1.30–7.16) | (20.68–21.37) | | |
| Pooled $F_1$ of reciprocal | 427 | 1.55±0.13 | 0.31 | 33.15 | −0.42±0.02 | 2.11 | 18.67 | −0.49±0.05 | 9.18 |
| crosses (RC) | (5) | | (0.21–0.46) | (33.48–33.25) | | (1.22–5.32) | (18.34–18.94) | | |
| Backcross (Pooled $F_1$ RC× | 347 | 0.57±0.05 | 18.15 | 1947.35 | - | 4314.1 | 38374.52 | - | 7.21 |
| Spinosad-resistant) | (9) | | (9.40–35.36) | (1947.03–1947.66) | | (1479.10–18945.00) | (38373.96–38375.12) | | |

All of the concentration-mortality curves followed the probit model based on the $\chi^2$ goodness-of-fit test ($P>0.05$).

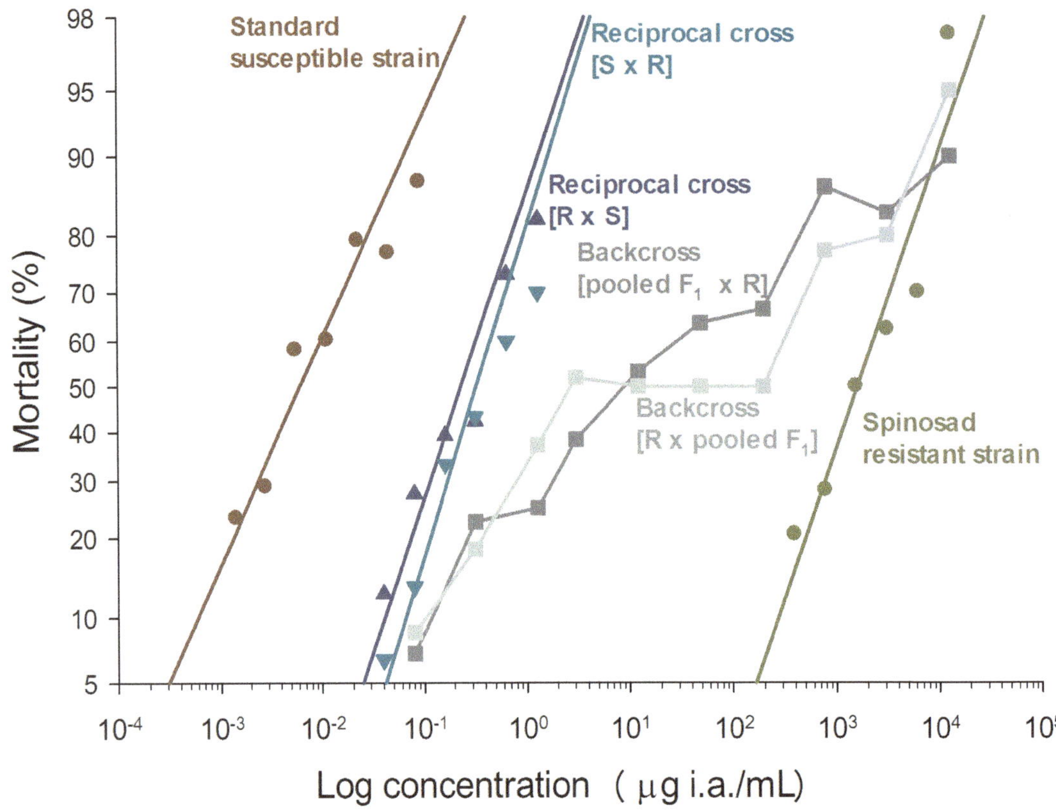


**Figure 2. Spinosad concentration-mortality curves (with observed data as symbols) for the (standard) spinosad susceptible strain, (selected) spinosad resistant strain, the $F_1$ progeny of the reciprocal crosses and the backcross progeny (pooled $F_1$ RC×spinosad-resistant) of the tomato borer *Tuta absoluta*.**

cage (30×30×30 cm). The adult cage was used for oviposition only, where leaves of tomato were provided daily as substrate. Adults of *T. absoluta* were fed with 10% glucose solution (Yoki, 10 Brazil), while the larvae were fed with tomato leaves from Santa Clara tomato cultivar (IC11 5500), cultivated under greenhouse conditions without any insecticide application [20,21]. The insects were maintained under the controlled conditions of 25±1°C temperature, 65±5% relative humidity and 12:12 (L:D) photoperiod.

## Insecticides

The bioinsecticide spinosad was used in its commercial formulation registered for use in tomato fields against the tomato borer (480 g a.i./L, suspension concentrate, Dow AgroSciences, Franco da Rocha, SP, Brazil) [18]. The insecticides used in the cross-resistance bioassays were (the commercial formulations used are indicated between parentheses): abamectin (18 g a.i./L, emulsifiable concentrate, Syngenta Proteção de Cultivos, São Paulo, SP, Brazil), cartap (500 g a.i./Kg, soluble powder, Iharabras, Paulínia, SP, Brazil), chlorantraniliprole (200 g a.i./L, suspension concentrate, DuPont Brazil, Paulínia, SP, Brazil), chlorfenapyr (240 g a.i./L, suspension concentrate, BASF S.A., São Paulo, SP, Brazil), chlorpyrifos (480 g a.i., emulsifiable concentrate, Dow AgroSciences, Santo Amaro, SP, Brazil), indoxacarb (300 g a.i./Kg, water dispersible granule, DuPont Brazil, Paulínia, SP, Brazil), permethrin (384 g a.i./L, emulsifiable concentrate, FMC Química do Brazil, Campinas, SP, Brazil), spinetoram (250 g a.i./Kg, water dispersible granule, Dow AgroSciences, Franco da Rocha, SP, Brazil), and thiamethoxam (250 g a.i./Kg, water dispersible granule, Syngenta Proteção de Cultivos, São Paulo, SP, Brazil). The synergists piperonyl butoxide (PBO-90%) and *S,S,S* – Tributylphosphorotrithioate (DEF-98%) were purchased from Sigma-Aldrich, Milwaukee, WI, EUA).

## Concentration-mortality bioassays

The concentration-mortality bioassays were performed as described previously and validated for the tomato borer, *T. absoluta* [21–23,32]. The insecticide solutions were diluted in water containing 0.01% Triton X-100 and a control treatment without insecticide was used to record natural mortality. Insecticide-treated tomato leaves were placed in Petri dishes (9 cm diameter) with ten 2$^{nd}$ instar larvae of the tomato borer and were maintained under controlled environmental conditions (25±1°C temperature, 65±5% relative humidity and 12:12 (L:D) photoperiod). Larval mortality was assessed after 48 hours of exposure by prodding the insects with a fine hairbrush. Larvae were considered dead if they were unable to move the length of their body.

## Selection for spinosad resistance

The tomato borer population from Iraquara, previously identified as resistant to spinosad [23], was subjected to spinosad selection after four generations under laboratory conditions. The original Iraquara population was split into two lines, one maintained without insecticide exposure and the other maintained under spinosad selection for 22 generations. Between 1,500 and 2,000 2$^{nd}$ instar larvae of the tomato borer surviving exposure to increasing discriminatory concentrations of spinosad (selected based on the concentration-mortality bioassays) were used for

**Table 4.** Dominance of spinosad resistance based on a range of spinosad concentrations including $LC_{50}$s from the susceptible parental strains and pooled $F_1$ progeny of reciprocal crosses estimated for the tomato borer *Tuta absoluta.*

| Concentrations (µg a.i./mL) | Strains | No. insects | Mortality (%) | Survival performance | Estimated dominance (*h*) |
|---|---|---|---|---|---|
| 0.005 | Spinosad-resistant | 19 | 0.00 | 1.00 | - |
| | Spinosad-susceptible | 29 | 3.45 | 0.97 | - |
| | Pooled $F_1$ of reciprocal crosses | 30 | 0.00 | 1.00 | 1.00 |
| 0.05 | Spinosad-resistant | 25 | 0.00 | 1.00 | - |
| | Spinosad-susceptible | 28 | 25.00 | 0.75 | - |
| | Pooled $F_1$ of reciprocal crosses | 29 | 0.00 | 1.00 | 1.00 |
| 0.50 | Spinosad-resistant | 22 | 0.00 | 1.00 | - |
| | Spinosad-susceptible | 30 | 100.00 | 0.00 | - |
| | Pooled $F_1$ of reciprocal crosses | 31 | 58.06 | 0.42 | 0.42 |
| 5.00 | Spinosad-resistant | 21 | 0.00 | 1.00 | - |
| | Spinosad-susceptible | 29 | 100.00 | 0.00 | - |
| | Pooled $F_1$ of reciprocal crosses | 30 | 100.00 | 0.00 | 0.00 |
| 10.00 | Spinosad-resistant | 29 | 13.79 | 0.86 | - |
| | Spinosad-susceptible | 31 | 100.00 | 0.00 | - |
| | Pooled $F_1$ of reciprocal crosses | 30 | 100.00 | 0.00 | 0.00 |

The concentration range used also discriminates for high spinosad resistance, as observed in the spinosad-selected strain. The estimated dominance (*h*) varies from 0 (completely recessive) to 1 (completely dominant), where 0.5 corresponds to co-dominance, $0<h<0.5$ corresponds to incompletely recessive and $0.5<h<1.0$ corresponds to incompletely dominant.

selection in each generation. After the reduction in egg laying by the spinosad-selected population following the 8th generation of selection, the discriminating concentration of 500 µg a.i./mL was maintained. The average mortality of the spinosad-selected population of Iraquara between the $F_2$ and $F_7$ generations was used to estimate the heritability of spinosad resistance. The Pelotas population of the tomato borer, previously identified as susceptible to spinosad, was maintained in the laboratory without insecticide selection as a susceptible standard population.

## Stability of spinosad resistance

The spinosad-resistant population selected for 13 generations was split into two lines, one was maintained under spinosad selection as previously described, and the other was maintained without spinosad selection. Both lines were subjected to spinosad concentration-mortality bioassays during each subsequent generation until the 22nd generation to verify the stability of spinosad resistance without the selection pressure of the bioinsecticide.

**Table 5.** Direct test of monogenic inheritance for spinosad resistance in the tomato borer *Tuta absoluta* by comparing expected and observed mortality of the progeny of the backcrosses between the pooled F1 progeny of the reciprocal crosses and the (selected) spinosad-resistant strain.

| Concentration (mg a.i./L) | Observed mortality (%) | Expected mortality (%) | $\chi^{2}$* | *P* |
|---|---|---|---|---|
| 0.04 | 0.00 | 0.00 | 0.00 | 1.00 |
| 0.08 | 6.90 | 8.62 | 0.11 | 0.74 |
| 0.32 | 22.58 | 18.33 | 0.37 | 0.54 |
| 1.28 | 25.00 | 37.27 | 2.06 | 0.15 |
| 3.05 | 38.46 | 52.00 | 2.86 | 0.09 |
| 12.21 | 53.33 | 50.00 | 0.13 | 0.72 |
| 48.83 | 63.89 | 50.00 | 2.78 | 0.10 |
| 195.31 | 66.67 | 50.00 | 3.33 | 0.07 |
| 781.25 | 86.67 | 77.50 | 1.45 | 0.23 |
| 3125.00 | 83.33 | 80.00 | 0.21 | 0.65 |
| 12500.00 | 90.00 | 95.00 | 1.58 | 0.21 |
| Total | | | $\Sigma\chi^2 = 14.89$ | 0.14 |

*Non-significant at $P>0.05$.

**Table 6.** Relative toxicity of insecticides to the parental spinosad-resistant strain and its derived strain after 15-generations of selection for spinosad resistance.

| Insecticides | No. insects | Slope ± SE | $LC_{50}$ (95% CI) (μg a.i./mL) | $RR_{50}$ (95% CI) | $\chi^2$ (degrees of freedom) |
|---|---|---|---|---|---|
| **Parental spinosad-resistant strain** | | | | | |
| Spinosad | 264 | 1.85±0.23 | 0.410 (0.31–0.51) | - | 1.51 (5) |
| Spinetoram | 276 | 1.72±0.18 | 0.29 (0.23–0.38) | - | 2.91 (5) |
| Abamectin | 282 | 1.56±0.24 | 0.54 (0.31–0.78) | - | 2.07 (5) |
| Chlorantraniliprole | 243 | 2.84±0.38 | 12.18 (9.38–15.10) | - | 2.40 (4) |
| Cartap | 265 | 2.25±0.25 | 173.65 (137.26–214.16) | - | 0.95 (5) |
| Chlorfenapyr | 283 | 1.65±0.21 | 1.08 (0.74–1.44) | - | 0.77 (5) |
| Indoxacarb | 279 | 3.25±0.47 | 0.86 (0.69–1.04) | - | 0.41 (5) |
| Thiamethoxam | 300 | 1.65±0.17 | 1008.86 (717.56–1389.27) | - | 5.49 (5) |
| Permethrin | 281 | 1.87±0.21 | 269.15 (204.91–342.36) | - | 5.49 (6) |
| Chlorpyrifos | 273 | 2.30±0.23 | 509.16 (416.99–623.34) | - | 2.30 (5) |
| **Selected spinosad-resistant strain** | | | | | |
| Spinosad | 238 | 1.47±0.19 | 1717.30 (998.16–2764.20) | 4191.55 (4191.38–4191.72) | 5.46 (4) |
| Spinetoram | 211 | 1.62±0.23 | 195.94 (140.94–261.88) | 656.99 (656.82–657.15) | 2.32 (4) |
| Abamectin | 282 | 1.66±0.18 | 2.85 (2.12–3.66) | 5.25 (5.00–5.50) | 2.15 (6) |
| Chlorantraniliprole | 244 | 1.80±0.22 | 0.42 (0.30–0.55) | 0.03 (0.01–0.20) | 1.26 (5) |
| Cartap | 243 | 1.21±0.20 | 105.34 (72.22–164.21) | 0.61 (0.41–0.80) | 0.32 (4) |
| Chlorfenapyr | 432 | 1.19±0.12 | 3.80 (2.94–4.89) | 3.53 (3.35–3.71) | 3.86 (5) |
| Indoxacarb | 313 | 1.68±0.19 | 1.19 (0.73–1.72) | 1.38 (1.22–1.54) | 6.64 (6) |
| Thiamethoxam | 212 | 1.91±0.23 | 3573.35 (2414.90–5147.74) | 3.54 (3.40–3.69) | 4.40 (4) |
| Permethrin | 238 | 1.62±0.19 | 662.07 (497.68–864.87) | 2.46 (2.30–2.62) | 2.12 (5) |
| Chlorpyrifos | 299 | 1.82±0.19 | 951.97 (758.56–1221.65) | 1.87 (1.73–2.00) | 4.75 (5) |

All of the concentration-mortality curves followed the probit model based on the $\chi^2$ goodness-of-fit test ($P>0.05$).

## Inheritance of spinosad resistance

The inheritance of spinosad resistance was determined through reciprocal crosses between spinosad-selected insects (after 13 generations of selection) and susceptible insects (from Pelotas). Thirty-five crosses were performed for each reciprocal cross with the adults maintained in separate rearing cages for progeny production and concentration-mortality bioassays. The $LC_{50}$ values (and the $LC_{90}$ values) were estimated for both parental strains and reciprocal crosses were used to calculate the degree of dominance (D) of spinosad resistance [38–40]. The estimated dominance ($h$) of spinosad resistance was tested through concentration-mortality bioassays with spinosad for the parental (susceptible and (selected) spinosad-resistant) strains and the $F_1$ progeny from the reciprocal crosses [38]. Five spinosad concentrations (0.005, 0.05, 0.5, 5 and 10 μg a.i./mL), in addition to untreated controls (with the application of only water and adjuvant), were used against individuals of the pooled $F_1$ of reciprocal crosses (n = 180), spinosad-resistant (n = 135), and spinosad-susceptible (n = 177) populations.

The monogenic basis of spinosad resistance was tested using backcrosses with the $F_1$ individuals obtained in the reciprocal crosses between spinosad-selected parental insects. The 2$^{nd}$ instar larvae obtained from such backcrosses were subjected to concentration-mortality bioassays with spinosad and to a direct test of inheritance to recognize their mono or polygenic basis.

## Pattern of cross-resistance

The 2$^{nd}$ instar larvae of the 15$^{th}$ and 16$^{th}$ generations of spinosad selection were used in concentration-mortality bioassays with the insecticides abamectin, cartap, chlorantraniliprole, chlorfenapyr, chlorpyrifos, indoxacarb, permethrin, spinetoram and thiamethoxam to detect the potential spectrum of spinosad cross resistance. The bioassay methods used were those previously described for the concentration-mortality bioassays.

## Synergism of spinosad

The 2$^{nd}$ instar larvae of spinosad susceptible and resistant colonies were used in concentration-mortality bioassays with the insecticides spinosad + PBO and spinosad + DEF to detect whether metabolism is involved in the resistance. The bioassay methods used were those previously described for the concentration-mortality bioassays, but all larvae were topically treated (0.2 μL/larvae) with a concentration of either PBO (1 μg/μL) or DEF (1 μg/μL) before exposure to spinosad.

## Protein extraction and enzyme bioassays

Three batches of ten 3$^{rd}$ instar larvae were collected during each generation of the spinosad selection for triplicate determinations of enzyme activity. The crude insect homogenate was prepared by grinding ten larvae in 0.2 mL of sodium phosphate buffer (0.02 M, pH 7.2). The crude homogenate was filtered through glass-wool and centrifuged at 10,000 $g_{max}$ for 15 min. The pellet was discarded and the supernatant was used for determining

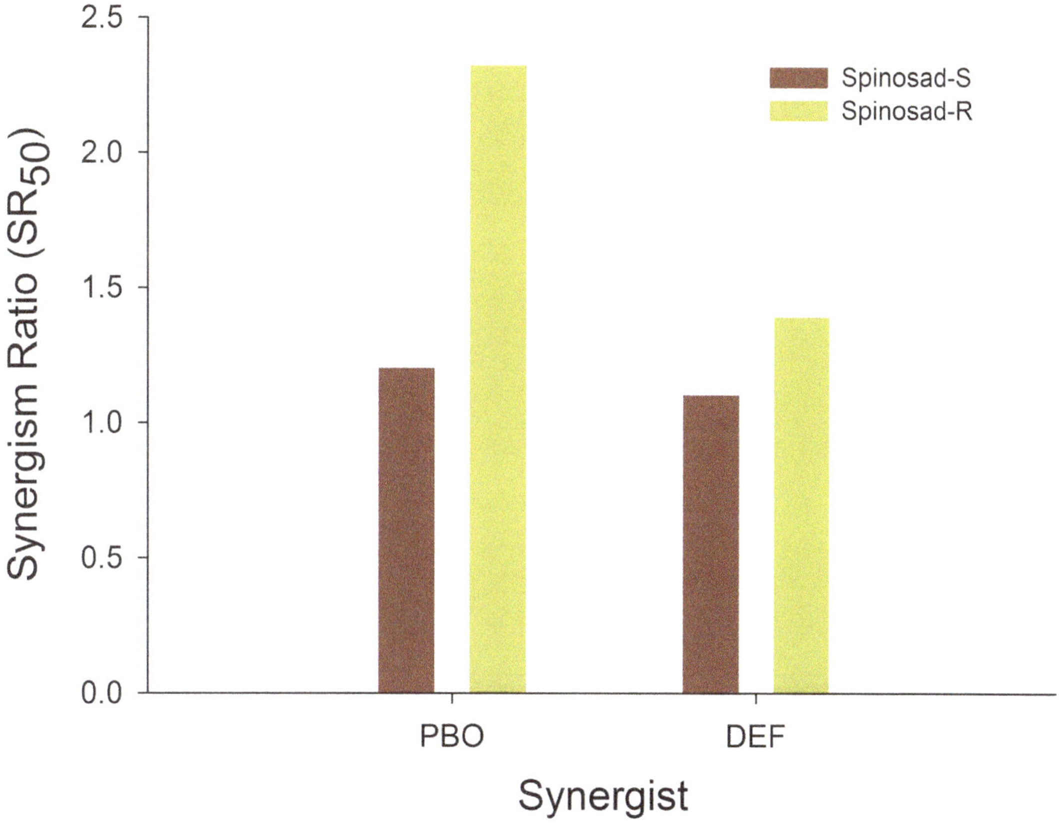


**Figure 3. Synergism of spinosad toxicity in spinosad-susceptible and -resistant strains of the tomato borer *Tuta absoluta*.**

protein content and total esterase activity. The supernatant from the 10,000 $g_{max}$ centrifugation was further centrifuged at 100,000 $g_{max}$ to obtain the microsomal fraction, which was resuspended in 500 µL sodium phosphate [0.1 M, pH 7.5+ glycerol (20%)] and used for the determination of cytochrome P450 *O*-demethylase activity.

Protein concentration was determined following the bicinchoninic acid method using bovine serum albumin as standard [33]. Total esterase activity was determined following the methods of van Asperen [34] using α-naphthyl acetate as substrate and a standard curve of α-naphthol to estimate the esterase activity expressed in nmol α-naphthol/min/mg protein. Cytochrome P450 (*O*-demethylase) activity was determined using *p*-nitroanisole as substrate generating *p*-nitrophenol [35] and enzyme activity was expressed in nmol *p*-nitrophenol/min/mg protein.

## Statistical analyses

The concentration-mortality data were subjected to probit analysis using the software Polo-Plus (LeOra Software Co., Petaluma, CA, USA) with correction for the natural mortality (without insecticide exposure) in the bioassays [36]. The level of resistance was estimated using resistance ratio (RR) estimates, which were considered significant when the 95% confidence interval of the RR did not include the value 1.0 [37].

The stability of spinosad resistance was estimated based on the average response of the spinosad-selected population between the 13th and 18th generations, corresponding to the average rate of change in the absence of the insecticide (RC). Spinosad resistance is unstable if the rate of change in the absence of spinosad is negative (RC<0). The number of generations required for a 10-fold reduction in spinosad resistance (G) can also be used to estimate the average rate of change in resistance without insecticide exposure using the formula $G = RC^{-1}$ [38,39].

The degree of dominance (D) of spinosad resistance was calculated according to the method of Hartl [38] and Stone [40], using the formula $D = 2\ (2.L_2 - L_1 - L_3)/(L_1 - L_3)$, where $L_1$, $L_2$, and $L_3$ are the log values of the $LC_{50}$s (or of $LC_{90}$s) of the spinosad-selected, $F_1$ (between resistant and selected strains), and spinosad-susceptible strains, respectively. The values of D may range from −1 to +1, with the former corresponding to complete recessive inheritance and the later to complete dominance [40].

The estimated dominance (*h*) was calculated for each concentration: $h = (w12 - w22)/(w11 - w22)$; where $w11$, $w12$, and $w22$ represent fitness values determined for resistant homozygotes, heterozygotes, and susceptible homozygotes, respectively [38]. The fitness value of resistant homozygotes was considered 1.0, while the fitness values of heterozygotes and susceptible homozygotes were calculated as the ratio between the observed survival rate of the pooled $F_1$ progeny of the reciprocal crosses and the survival rate of the (selected) spinosad-resistant strain. The *h*-values varies from 0 (completely recessive) to 1 (completely dominant), where 0.5 corresponds to co-dominance, $0<h<0.5$ corresponds to incompletely recessive and $0.5<h<1.0$ corresponds to incompletely dominant.

The monogenic or polygenic basis of spinosad resistance in the tomato borer was initially estimated by comparing the slopes of the concentration-mortality curves of the $F_1$ reciprocal crosses (between resistant and susceptible lines) and their backcrosses. The results from the backcrosses were compared with the monogenic expectation using the $\chi^2$ test [41,42]. The minimal

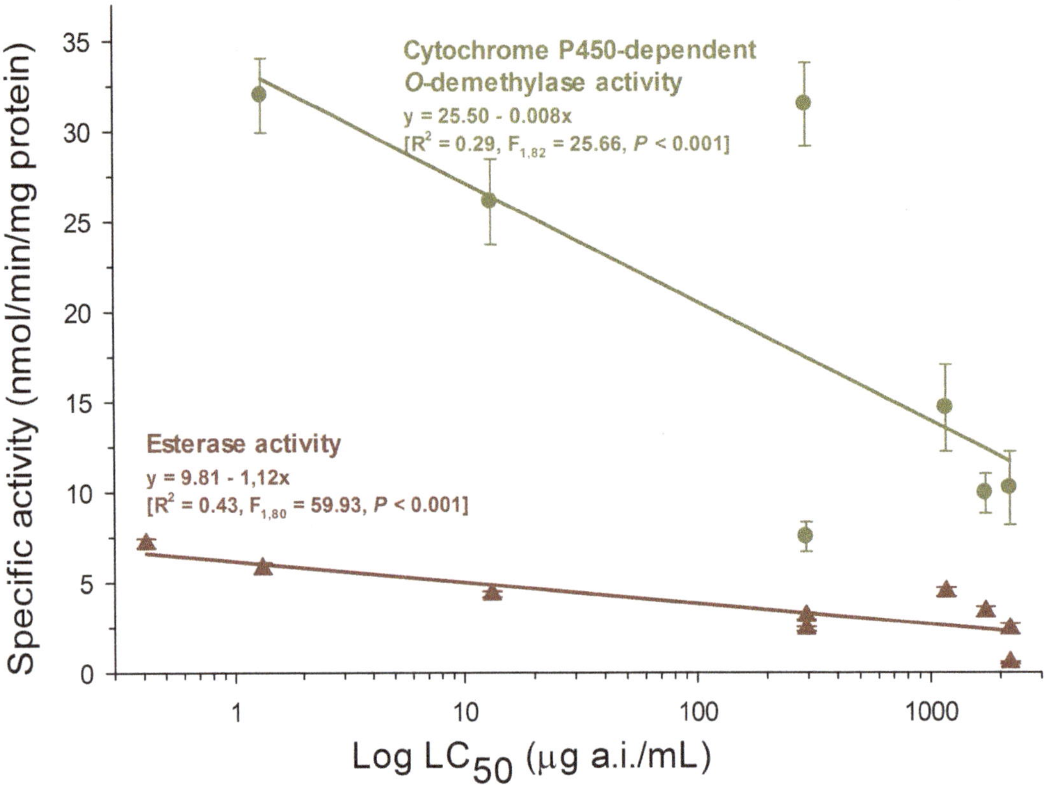


**Figure 4. Relationship between detoxification enzyme activity and $LC_{50}$s for spinosad in spinosad-selected generations of the tomato borer *Tuta absoluta*.**

number of effective genes ($n_E$) was estimated with the formula $n_E = (L_2{-}L_1)^2/8\sigma_s^2$ [43], where $L_2$ and $L_1$ are the log of $LC_{50}$ for the spinosad-selected strain and for the susceptible strain, respectively. The phenotypic variance ($\sigma_s^2$) was estimated with the formula $\sigma_s^2 = \sigma_{B1}^2+\sigma_{B2}^2-(\sigma_{F1}^2+½.\sigma_{P1}^2+½.\sigma_{P2}^2)$, where $\sigma_{F1}^2$, $\sigma_{B1}^2$, $\sigma_{P1}^2$ and $\sigma_{P2}^2$ are the phenotypic variances of the $F_1$ progeny, of the $F_1$-spinosad resistance backcross progeny, of the spinosad resistant strains, and of the spinosad susceptible strain, respectively. The $F_1$-spinosad susceptible backcross was not carried out and, therefore, we considered $\sigma_{B2}^2 = 0$.

The heritability ($h^2$) of spinosad resistance was estimated using the formula $h^2 = R/S$, where R is the response to selection and S is the differential selection; the 10-fold increase in spinosad resistance was estimated using the formula $G = R^{-1}$ [38,39]. The response to selection (R) was calculated by the formula $R = (L_f{-}L_i)/n$, where $L_f$ and $L_i$ are the log $LC_{50}$ of the 2nd and 7th generations and $n$ is the number of generations under selection. The differential selection (S) was estimated with the formula $S = i.\sigma_F$, where $i$ is the selection intensity and $\sigma_F$ is the phenotypic standard deviation [39]. The selection intensity ($i$) was estimated for $p$, which is the percentage of individuals surviving the selection [40]. The phenotypic standard deviation ($\sigma_F$) was estimated using the formula $\sigma_F = ½.(\beta_i+\beta_f)^{-1}$, where $\beta_i$ is the initial slope and $\beta_f$ is the final slope of the concentration-mortality curve.

The synergism ratio was calculated dividing the $LC_{50}$ unsynergized by the $LC_{50}$ synergized for each colony and synergist. The results of enzyme activity were subjected to analyses of variance and Tukey's HSD test ($P<0.05$) when appropriate, after ascertaining normality and homoscedasticity assumptions (PROC GLM and PROC UNIVARIATE) [44]. Linear regression analyses between enzyme activity and $LC_{50}$s at each generation of spinosad selection were performed using the procedure PROC REG in SAS [44].

## Results

### Selection for spinosad resistance

The selection for spinosad resistance at each generation of the initial Iraquara population led to a steady increase of the level of spinosad resistance until the 7th generation of selection, reaching a 5,000-fold increase in the level of resistance (>180,000-fold based on standard susceptible stain) (Table 1, Fig. 1). The selection response (R) was 0.53 and the differential selection (S) was estimated at 0.75, leading to a high heritability of spinosad resistance ($h^2 = 0.71$) and representing a 10-fold increase in the level of resistance at each 1.88 generations (Table 2).

### Stability of spinosad resistance

Although the selection for spinosad resistance was rapid, it reached a plateau after the 7th generation of selection with no further increases in the level of resistance with additional selection maintained until the 22nd generation (Fig. 1). After reaching a plateau in the selection for spinosad resistance, the selection was interrupted in a line of selected insects exhibiting high spinosad resistance (>100,000-fold based on standard-susceptible strain) and such high resistance levels quickly eroded with a negative rate of change in subsequent generations without selection (RC = −1.06). Spinosad resistance was, therefore, unstable without spinosad exposure and resulted in a 10-fold reduction in the level of spinosad resistance at each 1.57 generations with a return to susceptibility resembling the original strains after eight generations without selection (Fig. 1).

## Inheritance of spinosad resistance

The spinosad-selected strain exhibited very high levels of spinosad resistance (180,000-fold) compared with the standard susceptible strain, both of which exhibited similar variability of responses based on the overlapping standard errors of the slope from the concentration-mortality curves of both strains, indicating their relatively similar homogeneity of responses to spinosad (i.e., they exhibit similar levels of homozygosis). The $F_1$ progeny of the reciprocal crosses between spinosad-susceptible and -resistant strains exhibited intermediate levels of spinosad resistance (ca. 27- and 40-fold) (Table 3; Fig. 2), which were not significantly different based on the Polo-Plus $\chi^2$ test of equality of the concentration-mortality curves ($\chi^2 = 3.53$; df = 2, $P>0.05$). Therefore, spinosad resistance is an autosomal trait (i.e., not sex-linked) for the tomato borer.

The degree of dominance of spinosad resistance was estimated for the $F_1$ progeny of both reciprocal crosses between spinosad-susceptible and -resistant strains, and also for the pooled data from both progenies, providing values ranging from −0.39 to −0.45 (at the $LC_{50}$). The estimates of degree of dominance at the $LC_{90}$ for the same progenies were similar ranging from −0.47 and −0.52. These findings indicate that spinosad resistance is incompletely recessive, which was further confirmed by estimating the dominance using a range of five concentrations against the spinosad-susceptible and -resistant strains and their $F_1$ progeny (pooled together from both reciprocal crosses strain) (Table 4). At high concentrations, full recessiveness prevailed, while at low concentrations full dominance prevailed and the incompletely recessive pattern prevailed at intermediate spinosad concentrations as would be expected for an incompletely recessive pattern of inheritance.

The direct test for monogenic inheritance of spinosad resistance provided non-significant variation between expected and observed frequencies at increasing spinosad concentrations (Table 5). The overall $\chi^2$ test for the 11 spinosad concentrations tested was not significant ($\chi^2 = 14.88$, df = 10, $P>0.05$) (Table 5). The minimal number of effective genes ($n_E$) involved in spinosad resistance was 0.63 indicating, again, a monogenic trait.

## Spinosad cross-resistance spectrum

The concentration-mortality curves for nine different insecticides (of different groups) used against the tomato borer were estimated for the parental spinosad-resistant strain (Iraquara) and its spinosad-selected strain after 15 generations of selection to allow the recognition of potential patterns of cross-resistance (i.e., a single resistance mechanism leading to resistance to two or more insecticides). An eventual increase in resistance with the increase of spinosad resistance by selection indicates cross-resistance. However, among the insecticides tested, only spinetoram exhibited a significant increase in resistance with selection for spinosad resistance (Table 6). Therefore, cross-resistance was observed only between spinosad and spinetoram, another spinosyn insecticide.

## Synergism of spinosad

The synergism ratios with PBO and DEF were 1.2- and 1.1-fold, respectively (Fig. 3) for susceptible colony, while it was 2.32- and 1.39-fold in the resistant colony for PBO and DEF, respectively. Therefore, the synergisms caused by PBO and DEF against spinosad for the resistant colony were 1.93- and 1.26-fold greater compared to the susceptible colony, respectively. This suggests that such enzymes play a minimal role in the spinosad resistance.

## Activity of detoxification enzymes

The activity of esterases and cytochrome P450-dependent monooxygenases was determined for the spinosad-selected strain after different generations of selection to identify a potential increase in detoxification activity with the increase in spinosad resistance. However, the activity of both detoxification enzymes significantly decreased with selection for spinosad resistance indicating that they are not the underlying mechanism of this phenomenon (Fig. 4).

## Discussion

In countries managing invasive species and maintaining records of insecticide resistance development, insecticide use against the tomato borer can be described as waves of use of different (insecticide) groups. This is the case in Chile and Brazil, where the initial use of organophosphates, pyrethroids, and cartap was replaced by abamectin and subsequently by insect growth regulators as the main insecticide groups under use [10,12–16,19–21]. The main determinant in the replacement of an insecticide is the sequential development of insecticide resistance to the main insecticides being utilized at a given time, this phenomenon has evolved quickly in the tomato borer leading to control failures and relatively fast changes in the patterns of insecticide use [10,12–17,19–21]. Three insecticides are the main compounds currently being used against the tomato borer in Brazil, the spinosyn spinosad, the pyrrole chlorfenapyr, and the diamide chlorantraniliprole [18–21].

Quick reselection for resistance to insecticides has limited the range of available compounds for managing the tomato borer, increasing reliance on few molecules for this objective [20,21]. The concern of spreading insecticide resistant phenotypes of the tomato borer justifies the examination of the occurrence of insecticide resistance in this species. The new focus is on the few compounds under effective use, particularly in the likely centers of spread of the species. The emergence of spinosad resistance in South America is cause for concern, and the levels of resistance seemed to have increased quickly, but there is little information available beyond an initial survey [19–23]. The quick development of spinosad resistance in the region is suggestive of a highly inheritable (monogenic) trait, which was confirmed in our study.

Very high levels of spinosad resistance (>180,000-fold) were achieved within seven generations of selection from a field population already exhibiting resistance to spinosad [23]. The heritability of spinosad resistance proved high, with enough field variability to allow a quick selection for resistance, which has also been observed for spinosad resistance in the diamondback moth *Plutella xylostella* (L.) and the American serpentine leafminer *Liriomyza trifolii* (Burgess) [45,46]. A monogenic pattern of inheritance is consistent with the fast selection and evolution of spinosad resistance and such autosomal monogenic inheritance was observed in the tomato borer, which was incompletely recessive. This inheritance seems to be the general pattern for spinosad resistance in insect pest species [47–49]. The simple inheritance and high heritability of spinosad resistance in the tomato borer reinforces the phytosanitary concerns of the quick dispersion of this pest species and the spread of insecticide resistant phenotypes or populations amenable to fast local selection for spinosad resistance [19,31].

The potential cross-resistance to other insecticidal compounds is another issue of concern because it may further limit the management tools available against a pest species, particularly an invasive and very destructive species that is already difficult to control, such as the tomato borer [8–10,19]. Cross-resistance in spinosad-resistant insect populations seems limited to related compounds [50,51], as observed for the tomato borer where only cross-resistance to spinetoram, another spinosyn, was observed. This pattern of cross-resistance among spinosyns has been associated with altered target site sensitivity in the insect strains resistant to these compounds and their site of action [50]. This is consistent with our findings in the tomato borer, despite an earlier suggestion of the potential involvement of enhanced cytochrome P450-dependent monooxygenase activity in spinosad resistance in this species [22]. Initial correlational evidence of the potential involvement of enhanced esterase activity in spinosad resistance also suggested enhanced detoxification as a potential mechanism [23]. However, these earlier suggestions did not provide substantiated evidence for this possibility and our results did not support this hypothesis. Our results demonstrated altered target site sensitivity as the underlying mechanism of spinosad (and spinetoram) resistance in the tomato borer [50,52].

Spinosad resistance evolved quickly in the tomato borer under spinosad pressure reaching a high threshold of selection with over a level of 180,000-fold level of resistance, maintained with continued selection. However, the interruption of spinosad selection led to a quick erosion of spinosad resistance reestablishing the initial (reduced) levels of resistance after eight generations. This finding indicates that there is a fitness cost associated with insecticide resistance, which has been reported in different insect species and with different insecticides [46,53–55]. The fitness disadvantage of spinosad resistant populations of the tomato borer without selection pressure by spinosyn applications allows for the potential of moderation as an insecticide resistance management strategy. In this case, the reduction of spinosad use for a few generations of the tomato borer (>10) will allow the eventual reestablishment of susceptibility to spinosad and the subsequent reuse of this insecticide in the area. Reselection for spinosad resistance is likely to be rapid based on what has been observed for other insecticides in the tomato borer and in other species [4–7,20,21], but insecticide rotation with compounds of different modes of action and detoxification should extend the field use of spinosyns against the tomato borer.

In summary, very high levels of spinosad resistance were quickly selected for in the tomato borer with a monogenic autosomal pattern of inheritance that was incompletely recessive. A cross-resistance spectrum to spinetoram, another spinosyn, was observed, which suggested that the likely resistance mechanism involved is an altered target site sensitivity, given that the activity of esterases and cytochrome P450-dependent monooxygenases were not associated with spinosad resistance in the tomato borer. Spinosad resistance was unstable without spinosad selection, suggesting that the suspension of spinosyn use against the tomato borer would be a useful component in spinosad resistance management for this species.

Spinosad use against this species in introduced areas should be carefully monitored to prevent rapid selection for high levels of resistance and the potential for its spread to new areas.

## Acknowledgments

We would like to acknowledge the IRAC-BR and the affiliated agrochemical companies, who assisted in field collection of the insect populations and provided the insecticide formulations.

## Author Contributions

Conceived and designed the experiments: MRC ARSR HAAS. Performed the experiments: MRC TBMS WMS VRFS. Analyzed the data: MRC ARSR TBMS WMS HAAS. Contributed reagents/materials/analysis tools: RNCG HAAS. Contributed to the writing of the manuscript: MRC ARSR RNCG HAAS.

## References


1. Olson LJ (2006) The economics of terrestrial invasive species: a review of the literature. Agric Resour Econ Rev 35: 178–194.
2. Haack RA, Herard F, Sun JH, Turgeon JJ (2010) Managing invasive populations of Asian longhorned beetle and citrus longhorned beetle: a worldwide perspective. Annu Rev Entomol 55: 521–546.
3. Ragsdale DW, Landis DA, Brodeur J, Heimpel GE, Desneux N (2011) Ecology and management of the soybean aphid in North America. Annu Rev Entomol 56: 375–399.
4. Georghiou GP (1972) The evolution of resistance to pesticides. Annu Rev Ecol System 3: 133–168.
5. Brattsten LB, Holyoke CW, Leeper JR, Raffa KF (1986) Insecticide resistance: challenge to pest management and basic research. Science 231: 1255–1260.
6. Whalon ME, Mota-Sanchez D, Hollingworth RM (2008) Global Pesticide Resistance in Arthropods. CABI: Cambridge, MA, USA. 208.
7. Heckel DG (2012) Insecticide resistance after *Silent Spring*. Science 337: 1612–1614.
8. Desneux N, Wajnberg E, Wyckhuys AG, Burgio G, Arpaia S, et al. (2010) Biological invasion of European tomato crops by *Tuta absoluta*: ecology, geographic expansion and prospects for biological control. J Pest Sci 83: 197–215.
9. Desneux N, Luna MG, Guillemaud T, Urbaneja A (2011) The invasive South American tomato pinworm, *Tuta absoluta*, continues to spread in Afro-Eurasia and beyond: the new threat to tomato world production. J Pest Sci 84: 403–408.
10. Guedes RNC, Picanço MC (2012) The tomato borer *Tuta absoluta* in South America: pest status, management and insecticide resistance. Bull OEPP/EPPO Bull 42: 211–216.
11. FAO [Food and Agriculture Organization of the United Nations] (2014) FAOSTAT. Available: http://faostat3.fao.org/faostat-gateway/go/to/download/Q/QC/E. Accessed 2014 Feb 5.
12. Salazar ER, Araya JE (1997) Detección de resistência a insecticidas em La polilla del tomate. Simiente 67: 8–22.
13. Siqueira HAA, Guedes RNC, Picanço MC (2000) Insecticide resistance in populations of *Tuta absoluta* (Lepidoptera: Gelechiidae). Agric For Entomol 2: 147–153.
14. Siqueira HAA, Guedes RNC, Picanço MC (2000) Cartap resistance na synergism in populations of *Tuta absoluta* (Lep., Gelechiidae). J Appl Entomol 124: 233–238.
15. Salazar ER, Araya JE (2001) Respuesta de La polilla del tomate, *Tuta absoluta* (Meyrick), a insecticides em Arica. Agric Tecn 61: 429–435.
16. Siqueira HAA, Guedes RNC, Fragoso DB, Magalhães LC (2001) Abamectin resistance and synergism in Brazilian populations of *Tuta absoluta* (Meyrick) (Lepidoptera: Gelechiidae). Int J Pest Manag 47: 247–251.
17. Lietti MMM, Botto E, Alzogaray RA (2005) Insecticide resistance in Argentine populations of *Tuta absoluta* (Meyrick) (Lepidoptera: Gelechiidae). Neotrop Entomol 34: 113–119.
18. MAPA [Ministério da Agricultura, Agropecuária e Abastecimento] (2014) AGROFIT: Sistema de Agrotóxicos Fitossanitários. Available: http://extranet.agricultura.gov.br/agrofit_cons/principal_agrofit_cons. Accessed 2014 Feb 6.
19. Guedes RNC, Siqueira HAA (2012) The tomato borer *Tuta absoluta*: insecticide resistance and control failure. CAB Rev 7, no. 055. doi:10.1079/PAVSNNR20127055.
20. Silva GA, Picanço MC, Bacci L, Crespo ALB, Rosado JF, et al. (2011) Control failure likelihood and spatial dependence of insecticide resistance in the tomato pinworm, *Tuta absoluta*. Pest Manag Sci 67: 913–920.
21. Gontijo PC, Picanço MC, Pereira EJG, Martins JC, Chediak M, et al. (2012) Spatial and temporal variation in the control failure likelihood of the tomato leaf miner, *Tuta absoluta*. Ann Appl Biol 162: 50–59.
22. Reyes M, Rocha K, Alarcón L, Siegwart M, Sauphanor B (2012) Metabolic mechanisms involved in the resistance of field populations of *Tuta absoluta* (Meyrick) (Lepidoptera: Gelechiidae) to spinosad. Pestic Biochem Physiol 102: 45–50.
23. Campos MR (2013) Bases para o manejo da resistência de *Tuta absoluta* (Meyrick) (Lepidoptera: Gelechiidae) a diamidas e espinosinas [Thesis]. Recife: Universidade Federal Rural de Pernambuco - UFRPE. 128 p.
24. Racke KD (2006) A reduced risk insecticide for organic agriculture: spinosad case study. In: Felsot AS, Racke KD, editors. Crop protection products for organic agriculture: environmental, health and efficacy assessment, vol. 947. Washington: ACS. 92–108.
25. Biondi A, Zappalà L, Stark JD, Desneux N (2013) Do biopesticides affect the demographic traits of a parasitoid wasp and its biocontrol services through sublethal effects? PLoS ONE 8(9): e76548. doi:10.1371/journal.pone.0076548.
26. Puinean AM, Lansdell SJ, Collins T, Bielza P, Millar NS (2013) A nicotinic acetylcholine receptor transmembrane point mutation (G275E) associated with resistance to spinosad in *Frankliniella occidentalis*. J Neurochem 124: 590–601.

27. Suinaga FA, Casali VWD, Picanço M, Foster J (2004) Genetic divergence among tomato leafminer populations based on AFL analysis. Pesq Agropec Bras 39: 645–651.
28. Cifuentes D, Chynoweth R, Bielza P (2011) Genetic study of Mediterranean and South American populations of tomato leafminer *Tuta absoluta* (Povolny, 1994) (Lepidoptera: Gelechiidae) using ribosomal and mitochondrial markers. Pest Manag Sci 67: 1155–1162.
29. Roditakis R, Skarmoutsou C, Staurakaki M, Martínez-Aguirre M del R, García-Vidal L, et al. (2013) Determination of baseline susceptibility of European populations of *Tuta absoluta* (Meyrick) to indoxacarb and chlorantraniliprole using a novel dip bioassay method. Pest Manag Sci 69: 217–227.
30. Roditakis R, Skarmoutsou C, Staurakaki M (2013) Toxicity of insecticides to populations of tomato borer *Tuta absoluta* (Meyrick) from Greece. Pest Manag Sci 69: 834–849.
31. Haddi K, Berger M, Bielza P, Cifuentes D, Field LM, et al. (2012) Identification of mutations associated with pyrethroid resistance in the voltage-gated sodium channel of the tomatoleaf miner (*Tuta absoluta*). Insect Biochem Mol Biol 42: 506–513.
32. IRAC [Insecticide Resistance Action Committee] (2014) IRAC method no. 022: insecticide bioassay for *Tuta absoluta*. Available: http://www.irac-online.org/wp-content/uploads/2009/09/Method_022_Tuta_.pdf. Accessed 2014 Feb 6.
33. Smith PK, Krohn RI, Hermanson GT, Mallia AK, Gartner FH, et al. (1985) Measurement of protein using bicinchoninic acid. Anal Biochem 150: 76–85.
34. van Asperen K (1962) A study of housefly esterases by means of a sensitive colorimetric method. J Insect Physiol 8: 401–416.
35. Netter KJ, Seidel G (1964) An adaptively stimulated *O*-demethylating system in rat liver microssomes and its kinetic properties. J Pharmacol Exp Ther 146: 61–65.
36. LeOra-Software (2005) POLO-Plus, POLO for Windows computer program, ver. 2.0. Petaluma, CA: LeOra-Software.
37. Robertson JL, Russell RM, Preisler HK, Savin NE (2007) Bioassays with Arthropods, 2$^{nd}$ ed. Boca Raton: CRC. 224 p.
38. Hartl DL (2000) A primer of population genetics, 3rd ed. Sunderland: Sinauer. 221 p.
39. Falconer DS, Mackay TFC (1996) Introduction to quantitative genetics, 4th ed. Longman: New York. 480 p.
40. Stone BF (1968) A formula for determining degree of dominance in cases of moonofactorial inheritance of resistance to chemicals. Bull WHO 38: 325–326.
41. Tabashnik BE, Cushing NL, Finson N, Johnson MW (1990) Field development of resistance to *Bacillus thunringiensis* in diamondback moth (Lepidoptera: Plutellidae). J Econ Entomol 83: 1671–1676.
42. Sokal R, Rohlf F (2012) Biometry, 4$^{th}$ ed. New York: Freeman. 937 p.
43. Lande R (1981) The minimum number of genes contributing to quantitative variation between and within populations. Genetics 99: 541–553.
44. SAS Institute (2008) SAS/STAT user's version. Cary, NC: SAS Institute.
45. Ferguson JS (2004) Development and stability of insecticide resistance in the leafminer *Liriomyza trifolii* (Diptera: Agromyzidae) to cyromazine, abamectin, and spinosad. J Econ Entomol 97: 112–119.
46. Sayyed AH, Ferre J, Wright DJ (2000) Mode of inheritance and stability of resistance to *Bacillus thunringiensis* var *kurstaki* in a diamondback moth (*Plutella xylostella*) population from Malaysia. Pest Manag Sci 56: 743–748.
47. Zhao JZ, Li YX, Collins HL, Gusukuma-Minuto L, Mau RF, et al. (2002) Monitoring and characterization of diamondback moth (Lepidoptera: Plutellidae) resistance to spinosad. J Econ Entomol 95: 430–436.
48. Sayyed AH, Saeed S, Noor-ul-ane M, Crickmore N (2008) Genetic, biochemical, and physiological characterization of sponisad resistance in *Plutella xylostella* (Lepidoptera: Plutellidae). J Econ Entomol 101: 1658–1666.
49. Sparks TC, Dripps JE, Watson GB, Paroonagian D (2012) Resistance and cross-resistance to the spinosyns – a review and analysis. Pestic Biochem Physiol 102: 1–10.
50. Salgado VL, Sparks TC (2005) The spinosyns: chemistry, biochemistry, mode of action, and resistance. In: Lawrence IG, Kostas I, Sarjeet SG, editors. Comprehenisve molecular insect science, vol 6. Amsterdam: Elsevier. 137–173.
51. Watson GB, Chouinard SW, Cook KR, Geng C, Gifford JM, et al. (2010) A spinosyn-sensitive *Drosophila melanogaster* nicotinic acetylcholine receptor identified through chemically induced target site resistance, resistance gene identification, and heterologous expression. Insect Biochem Mol Biol 40: 376–384.
52. Salgado VL, Saar R (2004) Desensitizing and non-desensitizing subtypes of alpha-bungarotoxin-sensitive nicotinic acetylcholine receptors in cockroach neurons. J Insect Physiol 50: 867–879.
53. Wang D, Qiu X, Wang H, Qiao K, Wang K (2010) Reduced fitness associated with spinosad resistance in *Helicoverpa armigera*. Phytoparasitica 38: 103–110.
54. Coustau C, Chevillon C, ffrench-Constant R (2000) Resistance to xenobiotics and parasites: can we count the cost? Trends Ecol Evol 15: 378–383.
55. Guedes RNC, Oliveira EE, Guedes NMP, Ribeiro B, Serrão JE (2006) Cost and mitigation of insecticide resistance in the maize weevil, *Sitophilus zeamais*. Physiol Entomol 31: 30–38.

# 4

# Impact of Chronic Neonicotinoid Exposure on Honeybee Colony Performance and Queen Supersedure

**Christoph Sandrock[1]*, Matteo Tanadini[2], Lorenzo G. Tanadini[3], Aline Fauser-Misslin[1,4], Simon G. Potts[5], Peter Neumann[1,4]**

**1** Agroscope, Swiss Bee Research Centre, Berne, Switzerland, **2** Seminar for Statistics, ETH Zurich, Zurich, Switzerland, **3** Division of Biostatistics, Institute for Social and Preventive Medicine, University of Zurich, Zurich, Switzerland, **4** Institute of Bee Health, Vetsuisse Faculty, University of Berne, Berne, Switzerland, **5** School of Agriculture, Policy and Development, University of Reading, Reading, United Kingdom

## Abstract

***Background:*** Honeybees provide economically and ecologically vital pollination services to crops and wild plants. During the last decade elevated colony losses have been documented in Europe and North America. Despite growing consensus on the involvement of multiple causal factors, the underlying interactions impacting on honeybee health and colony failure are not fully resolved. Parasites and pathogens are among the main candidates, but sublethal exposure to widespread agricultural pesticides may also affect bees.

***Methodology/Principal Findings:*** To investigate effects of sublethal dietary neonicotinoid exposure on honeybee colony performance, a fully crossed experimental design was implemented using 24 colonies, including sister-queens from two different strains, and experimental in-hive pollen feeding with or without environmentally relevant concentrations of thiamethoxam and clothianidin. Honeybee colonies chronically exposed to both neonicotinoids over two brood cycles exhibited decreased performance in the short-term resulting in declining numbers of adult bees (−28%) and brood (−13%), as well as a reduction in honey production (−29%) and pollen collections (−19%), but colonies recovered in the medium-term and overwintered successfully. However, significantly decelerated growth of neonicotinoid-exposed colonies during the following spring was associated with queen failure, revealing previously undocumented long-term impacts of neonicotinoids: queen supersedure was observed for 60% of the neonicotinoid-exposed colonies within a one year period, but not for control colonies. Linked to this, neonicotinoid exposure was significantly associated with a reduced propensity to swarm during the next spring. Both short-term and long-term effects of neonicotinoids on colony performance were significantly influenced by the honeybees' genetic background.

***Conclusions/Significance:*** Sublethal neonicotinoid exposure did not provoke increased winter losses. Yet, significant detrimental short and long-term impacts on colony performance and queen fate suggest that neonicotinoids may contribute to colony weakening in a complex manner. Further, we highlight the importance of the genetic basis of neonicotinoid susceptibility in honeybees which can vary substantially.

**Editor:** Nicolas Desneux, French National Institute for Agricultural Research (INRA), France

**Funding:** This research received funding from the European Union Framework 7 under grant agreement no. 244090 (CP-FP) STEP (Status and Trends in European Pollinators, www.STEP-project.net). The funders had no role in study design, data collection and analysis, decision to publish, or preparation of the manuscript.

**Competing Interests:** The authors have declared that no competing interests exist.

* Email: ch.sandrock@gmail.com

## Introduction

Ecosystem services provided by pollinating insects are vital for the maintenance of biodiversity [1,2] and food security through agricultural productivity [3,4,5,6]. Recent evidence for globally paralleling declines of various pollinators [7,8,9,10], however, stands in contrast to prospects of continuously increasing demands for pollination services [4,11,12].

The Western honeybee, *Apis mellifera*, is the predominant managed pollinator worldwide [4,8]. Although global stocks of domestic honeybees have increased during the last half century (except in Europe [13] and in the USA [14]), present and predicted agricultural demands for insect pollination far exceed currently available capacities [15]. Repeated, massive declines of managed honeybee colonies during the last decade in North America, Europe and the middle East [8,16,17,18,19,20,21,22] raised substantial concerns about safeguarding future honeybee pollination services. Despite comprehensive recent research, conclusive evidence for common causal drivers is lacking, which points at multiple interacting factors [8,23,24]. The invasive ectoparasitic mite *Varroa destructor* represents a severe problem for honeybees almost worldwide [25,26,27], in particular due to its vital role as a virus vector [28,29]. Some observations of elevated colony losses were also influenced by the widespread gut parasites *Nosema ceranae* [30,31,32]. Microbes and *V. destructor*-associated viruses are prevalent almost globally and commonly co-occurring, yet there is no uniform evidence so far that even the interactions between some of these pathogenic stressors necessarily result in colony failure [33,34,35,36]. Instead, it appears as if more complex interactions with additional environmental stressors and at multiple levels are of key importance [8,24]. Likely candidates, commonly encountered by honeybees, are routinely applied agricultural pesticides [37,38]. For instance, the use of systemic neonicotinoid insecticides has strongly increased on a global scale during the last decade [39,40,41,42]. Mainly acting as specific agonists of the insect acetylcholine receptors, neonicotinoids disrupt neuromuscular signalling pathways and are thus efficiently used for controlling insect pests [39,43]. Systemic compounds like neonicotinoids can be particularly problematic for pollinating insects through exposure to residues in nectar and pollen of treated crops [41,44,45]. Although field-realistic neonicotinoid residue levels in pollen and nectar are generally assumed to result in sublethal dietary exposure [46], sublethal effects on honeybees include various negative impacts, such as impairment of physiology, cognitive abilities like memory and learning, and foraging and homing behaviour [45,47,48,49,50,51,52,53,54,55,56,57,58,59]. In addition, combined exposure to multiple pesticides can have additive or synergistic adverse effects in bees [60,61,62,63] with eventually underestimated consequences [64]. Similarly, multifactorial impact arising through combined chronic exposure to pathogens and pesticides may trigger detrimental feedbacks of supposedly sublethal individual stressors at the colony level that could explain otherwise enigmatic negative impacts [65]. In this regard, an important but yet poorly understood aspect represents the adverse influence of neonicotinoids on the honeybee's immune system [66,67]. There is growing evidence for detrimental interactions and compromised immunity in honeybees upon combined exposure to neonicotinoids and pathogens, including the prevalent gut-parasite *Nosema* spp. [36,68,69,70,71] and near-ubiquitous viruses, such as the typically *V. destructor*-associated deformed wing virus [72] or the black queen cell virus [36]. To date it remains unclear to what extent these findings can be extrapolated to field settings where, for example, immunity-related patterns in honeybees may be strongly influenced by many more environmental factors, such as the overall nutritional status [24,73,74,75,76]. While several unplanned field exposures resulted in massive effects of neonicotinoids under certain circumstances [41,77,78], there is no compelling evidence so far that field-realistic neonicotinoid exposure resulting from standard agronomic implementations of neonicotinoids pose a serious threat to whole colonies [79,80,81,82,83]. Nevertheless, some regulatory bodies reacted recently [84,85] in order to clarify whether the overall contrasting results may have been influenced by: overestimating individual predictors while dynamics at multiple levels were neglected [65]; by the lack of statistical power for individual studies [44]; or by the possibility that especially sensitive endpoints have simply been missed. For instance, the long-term impacts of neonicotinoids, which might only become evident during sensitive phases like overwintering, have received little attention to date. Moreover, information on how neonicotinoids could impact on queens is virtually lacking [64]. It is unknown whether queens are relatively protected from agricultural pesticides through receiving processed food from hive bees only, or whether trophallactic interactions with the usually most long-lived honeybee in a hive could indeed represent a sink for trace residues of such systemic compounds. Sublethal pesticide exposure could have important consequences for colony fate through compromising the queen's cognitive abilities or immune system and thereby reducing her performance. For instance, it is known that replacement of failing queens by the worker bees, i.e. queen supersedure, can be triggered by reduced oviposition of old or insufficiently mated queens [86,87]. Colony fitness is another sensitive but largely neglected aspect of honeybee colony performance. In bumblebees, for example, it has repeatedly been shown that the negative impact of chronic neonicotinoid exposure on individuals and colony performance was less pronounced compared to queen production [88,89,90]. Similarly alarming fitness effects upon chronic sublethal neonicotinoid exposure are indicated in solitary bees [91]. Reproductive success, however, is vital for inferring long-term population level consequences. Compared to the assessment of much more general traits of colony performance and productivity in honeybees, the quantification of fitness in the true sense is very difficult because of the complex socio-biology of reproduction in honeybees. In managed honeybee colonies, besides male mating success, swarming can be considered as a tangible proxy of fitness [92], which is in practice primarily linked to beekeeping management decisions though.

Here we experimentally assessed the effects of chronic dietary neonicotinoid exposure on honeybee colony performance and fitness on a temporal scale and in relation to the honeybees' genetic background. In a fully crossed experimental design 24 freely flying honeybee colonies, including two groups of sister-queens from different strains (14 and 10 colonies, respectively), were placed at a single apiary and were either exposed to control pollen or pollen that has been spiked with a combination of the two neonicotinoids thiamethoxam and clothianidin (on average 5.31 µg/kg and 2.05 µg/kg, respectively) via in-hive feeding over two brood cycles (see Figure S1). All colonies were then identically maintained and controlled against *V. destructor* throughout a one year monitoring period. During the study four detailed colony assessments, including estimates of numbers of adult bees and the amounts of brood and stores, were conducted to evaluate potential effects on colony performance and productivity in the short- (1.5 months), medium- (3.5 months) and long-term (1 year). We used this experimental design in order to contribute to better understand three currently poorly resolved aspects. First, while the majority of experiments at the colony level applied sublethal chronic neonicotinoid exposure through sucrose solution, we hypothesized that neonicotinoid-contaminated pollen provides stronger exposure of larval stages and nurse bees that may express sublethal effects when performing more complex tasks in later life cycle stages [53,93], thereby potentially resulting in delayed effects on colony performance. Second, contaminated pollen could result in sublethal exposure of honeybee queens that may affect their performance and subsequently cause failure. Therefore, we assessed the fate of queens one year subsequent to experimental feeding of neonicotinoid-spiked pollen, as well as the colonies' propensity to swarm as an indicator for colony fitness. Third, we addressed the question of whether neonicotinoid susceptibility at the honeybee colony level has a genetic basis.

## Materials and Methods

### Experimental setup and colony maintenance

Twenty-four honeybee colonies were established using artificial swarms (1.5 kg of bees) in summer 2010. Two groups of sister-queens originating from different, locally adapted breeding populations were introduced in order to control for the honeybees' genetic background and maternal effects: one group of 14 queens from a region in eastern Germany that is characterized by intense agriculture, and one group of 10 queens from an alpine region in central Switzerland. All queens were freely mated at apiaries in corresponding geographic regions during early summer 2010 and then individually tagged and clipped one forewing. The former group of colonies represented *A. m. carnica*, whereas the latter represented predominantly *A. m. mellifera*, and in the following they are referred to as strain A and B, respectively.

As permitted by the veterinary agency of the canton Zurich, we established an apiary on the land of the research station Agroscope Reckenholz-Tänniken in a rural area near the city of Zurich, Switzerland (47°25′38N 8°31′11E). Two groups of each 12 hives were placed in a single row, with all hive-entrances pointing in the same direction. To meet a fully-crossed experimental design, artificial swarms containing queens of the different strains were randomly allocated to the two experimental groups (N = 7 from strain A and N = 5 from strain B each). Hives containing queens of the two different strains A and B were ordered identically in each experimental group (see Figure S1 for details of the setup). Hives within groups were separated by 1 m, and both groups of hives were separated by approximately 20 m distance, including shrubs as landmarks to minimise forager drift between groups [94]. All colonies were equipped with brand new hive material, including polystyrene hive bodies, wooden frames (200×350 mm in size) and organically certified, pesticide-free wax foundations. Commercial pollen traps (Wienold, Lauterbach, Germany), painted in different colours, were installed at all hive entrances (identical colour sequence in each group), but only activated during the exposure phase (see below). All colonies were identically treated against *V. destructor* using oxalic acid (40 ml sucrose solution containing 3.5% w/w oxalic acid per colony) five days after establishment of artificial swarms (in the absence of capped brood), and fed with commercial sugar syrup (sugar beet based, 73% sugar content, containing equal proportions of glucose, fructose and sucrose; Hostettler's, Zurich, Switzerland) during summer 2010 to promote colony growth. Pollen was assumed to be available in sufficient quantities. All colonies overwintered on eleven combs, and oxalic acid treatment against *V. destructor* was repeated in December 2010.

During spring and summer 2011 colonies were not fed but left to freely collect nectar and pollen. All colonies were simultaneously provided with a second and third hive body containing 11 frames with wax-foundations for comb building in early April 2011 and in mid-May 2011, respectively. The upper hive body provided last was separated by a queen excluder to ensure honey storage only, and on the same day it was provided the experimental treatment was initiated and lasted until end of June 2011 (see below). In mid-July 2011 colonies were taken off their honey combs and subsequently maintained on 22 combs. They received 12.5 kg of sugar syrup during late July and late August 2011 (25 kg in total). After each feeding phase, colonies were simultaneously treated against *V. destructor* using 130 ml of formic acid (70% w/w) evaporating from commercial dispensers (Andermatt Biocontrol AG, Grossdietwil, Switzerland) for about one week each during early August and early September 2011. Colonies were then overwintered and treated with oxalic acid in December 2011 (see above). At no point during the study synthetic acaricides were used for varroa mite management. Subsequent to overwintering colonies were monitored until June 2012 without further intervention. No honey supers were provided in 2012 in order to increase the propensity for swarming, which served as a proxy of fitness one year after the treatment (see below).

## Treatment procedures and residue analyses

In mid-May 2011 pollen traps were activated to vastly prevent pollen inflow, and in-hive pollen feeding was initiated. Pollen patties consisted of 55% honeybee pollen (common stock of commercial pollen with mixed floral content of at least 19 plants; Sonnentracht Imkerei, Bremen, Germany), 5% brewer's yeast and approximately 40% sucrose (two thirds 73% sugar syrup and one third powder sugar). Three times per week (each Monday, Wednesday and Friday) all colonies were provided with two 200 g pollen patties loosely packed in cellophane paper and placed between the two lower hive bodies (i.e. within the brood nest). Disturbance of the colonies was thus reduced to a minimum. The bees easily corroded the cellophane paper to access the content. Two pollen patties of 200 g were generally consumed completely within 48 hours by each colony. Prior to feeding, pollen was gamma ray irradiated (Leoni Studer Hard AG, Däniken, Switzerland) to prevent putative pathogen spill over (e.g., see [95]). One group of 12 colonies (see above) received plain pollen while the other received patties containing environmentally relevant residues of the two neonicotinoids thiamethoxam and clothianidin (see below). Chronic neonicotinoid exposure through in-hive pollen feeding was performed for 46 days (1.5 months) in order to cover two brood cycles, thereby resulting in total provisions of 8 kg of pollen patties per colony. Pollen traps were then deactivated to no longer prevent colonies from storing pollen collected outside.

For both neonicotinoids pure analytical standards (PESTANAL, Fluka; with purities of 99.7 and 99.9% for thiamethoxam and clothianidin, respectively) were purchased (Sigma-Aldrich, Seelze, Germany), dissolved in distilled water (1 mg/L) and then stored at room temperature and protected from light. Aliquots of a single stock solution for each parent compound were added to the sucrose solution, which was then thoroughly mixed into the pollen and yeast. A commercial kneader was used to produce a homogenous paste to be portioned in cellophane paper (200 g) and kept frozen until usage. In total, 20 mixtures of plain and neonicotinoid-spiked pollen were prepared and fed batch-wise over time. A subsample of each of these batches of pollen patty preparations was subjected to residue analyses performed by the United States Department of Agriculture, Agricultural Marketing Service, Science and Technology Laboratory Approval and Testing Division of the National Science Laboratories in Gastonia, North Carolina. All samples were extracted for analysis of agrochemicals using a refined methodology for the determination of pesticides using an approach of the official pesticide extraction method (AOAC OMA 2007.01) using an acetonitrile:water solution and analysed by liquid chromatography coupled with tandem mass spectrometry detection (LC/MS-MS) utilising the parent and confirmatory ions of thiamethoxam and clothianidin. Samples were analysed using certified standard reference materials for the presence of both compounds with a limit of detection of 4.0 ppb for thiamethoxam and 1.0 ppb for clothianidin. Our target concentrations were 5.0 and 2.0 ppb for thiamethoxam and clothianidin, respectively. Since clothianidin is the major metabolite of thiamethoxam [96,97], both bioactive compounds will co-occur in the pollen and nectar of thiamethoxam-treated crops and were therefore applied in combination. The concentrations used here match field-realistic levels of both compounds previously found in pollen of treated crops [41,42,46,80,82,98]. In order to confirm the absence of unexpected additional exposure, we also subjected six random samples of the sugar syrup used for late season feeding (2010 and 2011), six samples of the original pollen stock, as well as four pollen patty samples of the control group to residue analyses. Moreover, pollen trap contents collected during the experimental feeding were pooled across colonies of each experimental group and samples from five collection dates throughout the treatment were taken for residue analyses. Additional matrix endpoints at the end of the experimental pollen feeding phase for both treatment groups were forager bees (1 sample pooled across all colonies) and pupae close to adult emergence (3 samples pooled across 4 colonies each). Furthermore, in both treatments we sampled wax (2 samples pooled across 6 colonies each) and bee bread (4 samples pooled across 3 colonies each) 3 weeks after the pollen treatment when honey supers were removed, as well as one sample of honey from each colony of the neonicotinoid-exposed group (pooled from at least 3 different combs). All samples for residue analyses were stored frozen and shipped on dry ice.

## Population estimates and data collection

Estimates of colony strength were performed using the 'Liebefelder Method' [99,100]. Specifically, we visually estimated the number of adult bees, the amount of capped brood (pupae) and un-capped brood (eggs and larvae), and the amount of honey and pollen stores in $dm^2$. Colony assessments were consistently performed by the same person and alternated between treatment groups during the day. During each colony assessment the presence of the original queen was checked. Successful queen replacement was counted as the presence of a non-tagged, egg-laying queen which possessed two intact forewings, either in the presence or absence of the originally tagged mother-queen. Based on 1 $dm^2$ comb containing 400 cells on average, the estimated proportion of comb area comprising of brood was converted into numbers of individuals in order to treat corresponding response variables as counts, whereas estimates of honey stores were converted into total weight (based on an average weight of 2 kg for fully filled honey combs of the used frame format).

In spring 2011 (mid-May) the first colony assessment was performed and three days later the experimental treatment was initiated. During the pollen feeding phase, contents of all pollen traps were collected and weighed each time when new pollen patties were provided, resulting in 20 pollen collection records per colony in total. In summer 2011 (beginning of July), two days after the last pollen patties were fed, the second colony assessment was conducted to evaluate short-term effects on colony performance. After the exposure phase the control and the neonicotinoid-treated colonies were maintained identically and the third colony assessment was performed 3.5 months after the exposure in autumn 2011 (mid-October), to evaluate medium-term effects colony performance before overwintering. Overwintering success was assessed end of March 2012 and thereafter surviving colonies were inspected on a weekly basis. Finally, the fourth colony assessment for all colonies that survived winter was performed one year after the treatment in spring 2012 (late April) to evaluate long-term effects on colony growth. Afterwards colonies were maintained until June 2012 and inspected for queen cells on a weekly basis and for swarming events at least every second day. Since original queens had one clipped forewing, swarms remained nearby the hive and could thus be easily recognized.

## Statistical analyses

To investigate the effects of exposure to thiamethoxam and clothianidin on honeybee colony performance over time, we analyzed a set of sensitive endpoints within the framework of mixed models.

**Model formulation and selection.** The response variables (endpoints) were numbers of adult honeybees, pupae and eggs and larvae. These were modelled including the explanatory variables (factors) treatment (control and neonicotinoids), honeybee strain (A and B), and assessment date (spring 2011, summer 2011, autumn 2011 and spring 2012) as fixed effects, and colony as a random effect. Residual analysis of all response variables indicated the need for variance stabilization and variables were transformed accordingly. The data for the endpoints number of adult honeybees and pupae were square-root transformed. Residuals for eggs and larvae displayed a more complex pattern of a bow-shaped variance being largest for medium fitted values (~8000) and decreasing for larger and smaller fitted values. This pattern resembles that of a binomially distributed variable, which conforms well to the upper bounded egg-laying rate of a honeybee queen. In such cases effective counts of a given response variable can be divided by the expected maximum number to obtain binomially distributed variables. Based on Khoury *et al.* [101], we set the limit to a daily egg-laying rate of 2000 and, according to the honeybee life-cycle, 16000 eggs and larvae present at any time. Ratios of actually present and maximum possible numbers of eggs and larvae were arcsine square-root transformed to stabilize variances for further analyses, as commonly performed for binomially distributed variables.

Numbers of adult honeybees exhibited an increased variance at the fourth assessment date in spring 2012, which prompted us to use a weighted linear mixed model for this response variable. Weights were set as the inverse of the residual variances of the two groups (for the assessments between spring and autumn 2011 and the assessment in spring 2012, respectively).

For each endpoint, complete models were fitted based on the threefold interaction term of the explanatory variables (fixed effects) plus the random effect. Model simplification was evaluated by hierarchically removing interaction terms based on likelihood ratio tests. The goodness of corresponding Chi-squared approximations was confirmed by model based parametric bootstrapping. During all steps of the model selection, model assumptions and serial correlations of the residuals of colonies were inspected.

**Hypotheses testing with contrasts.** To test for effects of neonicotinoid exposure on colony performance and the influence of the honeybees' genetic background on responses to the neonicotinoid-treatment, one-sided contrasts and corresponding *P*-values (adjusted for multiple testing) were computed for the overall treatment effect at each individual assessment date and, when the threefold interaction significantly contributed to model a given response, also for treatment nested within honeybee strain. Since seasonal effects were expected, contrasts including assessment date were not performed.

**Further statistical analyses.** To investigate effects of chronic exposure to thiamethoxam and clothianidin on honey production, the difference of the log transformed total weights before and after the experimental pollen feeding was analysed using linear regression. The full model was fitted with neonicotinoid exposure and honeybee strain as fixed factors and the interaction between them. In the same way we compared comb areas in $dm^2$ comprising of pollen stores (bee bread) in the hives before and after the experimental pollen feeding in order to evaluate pollen consumptions during the treatment.

The time series of twenty pollen collections per colony sampled during the 1.5 months of experimental in-hive pollen feeding were converted into respective Areas Under the Curve (AUC) for further analysis. The AUC represents the overall colony-specific pollen foraging activity, with higher AUC values corresponding to higher collection performances. We analyzed AUC with a one-sided Mann-Whitney test for a difference between neonicotinoid-exposed and control colonies.

Supersedure of original queens (at any time) and swarming events in spring 2012 yield a yes/no value for each colony. The associations of these variables with neonicotinoid treatment were investigated using Fisher's exact tests for contingency tables of small sample sizes.

All statistical analysis were performed using R [102]. Mixed models were fitted using the lmer function of the *lme4* package [103], and contrasts were performed using the glht function of the *multcomp* package [104].

## Results

### Residue analyses

Original stocks of honeybee pollen and sugar syrup used to prepare pollen patties in our experiment did not contain traceable amounts of the two neonicotinoids thiamethoxam and clothianidin, even when the limit of detection was reduced to 0.1 ppb for both compounds. Thus, thiamethoxam and clothianidin are considered being absent in the pollen patties fed to the control colonies. In contrast, in all samples from the pollen patties that have been spiked with the two neonicotinoids both parent compounds were detected in the range of the target concentrations: The effective mean concentrations ($\pm$ SD across different pollen patty batches) used during the neonicotinoid treatment were determined to be $5.31 \pm 0.75$ ppb for thiamethoxam and $2.05 \pm 1.18$ ppb for clothianidin. The residue analyses document constant chronic exposure to thiamethoxam and clothianidin at field-realistic levels in pollen over a period of two honeybee brood cycles. There was no unexpected additional exposure to thiamethoxam or clothianidin from the outside, as indicated by the lack of traceable residues of both compounds in the pollen collected from pollen traps of both treatment groups during the in-hive pollen feeding phase. In none of the forager bee and pupae samples collected directly after the treatment and in none of the hive samples collected 3 weeks after the treatment (i.e. bee bread, honey and wax) residues of either compound above respective limits of detection could be found in both the control and neonicotinoid treatment. The absence of residues in bees is not surprising given the low concentrations used here, yet the absence of in-hive residues 3 weeks after feeding contaminated pollen over two brood cycles indicates that low level residues may quickly disappear in the hive matrix (see also [80,82]) through consumption, dilution or degradation.

### Colony growth

In the control group and in the group exposed to thiamethoxam and clothianidin each 1 colony of strain A lost their queen after the pollen feeding phase during the formic acid treatment against varroa mites in 2011. Moreover, in each experimental group 1 colony became queenless during winter (originating from strain B and A in the control and in the neonicotinoid-exposed group, respectively). Thus, mixed model analyses are based on 12 colonies per experimental group for the first and the second colony assessment (in both treatment groups 7 colonies originating from strain A and 5 from strain B), while 11 colonies per group were available for the third colony assessment (in both treatment groups 6 colonies originating from strain A and 5 from strain B), and 10 colonies were available for the fourth colony assessment (in the control 6 and 4 from strain A and B, respectively, and for the neonicotinoid treatment each 5 from strain A and B). Queenless colonies were removed immediately after queen loss was recognised.

The data for the three endpoints adult bees, pupae and eggs and larvae across assessment dates are summarized in Fig. 1A–C. Dynamics of colony strength and brood curves displayed the expected general pattern of seasonal variation. However, we detected strong effects of neonicotinoid exposure, as well as interactions of the honeybees' genetic background with neonicotinoid exposure. The number of adult bees and the number of eggs and larvae were each best explained by the threefold interaction term model, i.e. neonicotinoid exposure, honeybee strain and assessment date, while for the number of pupae the retained model included the twofold interactions of assessment date with honeybee strain and neonicotinoid exposure, respectively, but no threefold interaction. Model-based estimates of contrasts and corresponding significance levels are summarized in Table 1. After 1.5 months of experimental pollen feeding, there was a significantly negative influence of the exposure to thiamethoxam and clothianidin on the number of adult bees for both honeybee strains, whereas this effect was much stronger for honeybee strain B than for strain A (Fig. 1A). Overall, average worker populations were 28% smaller in the neonicotinoid treatment compared to the control. There was also a significant overall decrease of the number of eggs and larvae in the neonicotinoid treatment, yet, this effect was not significant when tested within either honeybee strain A or B (Fig. 1B). However, there was no significant effect of nenicotinoid treatment or honeybee strain on the amount of pupae after the experimental pollen feeding (Fig. 1C). Compared to the control, the average amount of total brood had declined by 13% in the colonies exposed to thiamethoxam and clothianidin. No effects of the previous neonicotinoid exposure on the amount of adult honeybees or honeybee brood were detected 3.5 months after the experimental pollen feeding (Fig. 1A–C). Interestingly though, one year after the experimental pollen feeding, the negative impact of the previous neonicotinoid treatment on the number of adult bees was even stronger than directly after exposure to thiamethoxam and clothianidin (Fig. 1A). These effects were significant within honeybee strains A and B, but again much more pronounced in strain B. Similarly, when the overall significant decrease of the amount of eggs and larvae one year after the neonicotinoid treatment was tested within strains, a significant effect was only detected for honeybee strain B (Fig. 1B). Moreover, contrasting to the finding directly after exposure to thiamethoxam and clothianidin, there was also a significantly detrimental effect of the neonicotinoid treatment in the previous year on the amount of pupae (independent of honeybee strain), see Fig. 1C and Table 1.

### Honey production

At treatment initiation all colonies already harboured considerable honey stores due to comparatively early spring flowering in 2011. Strain A and B, respectively, had on average $24.8 \pm 2.9$ and $27.0 \pm 2.9$ kg in the control group, and $25.4 \pm 2.8$ and $23.6 \pm 0.7$ kg in the group subsequently exposed to thiamethoxam and clothianidin. During the experimental pollen feeding, honey stores in the control group increased by $7.7 \pm 4.4$ kg per colony on average ($10.3 \pm 3.6$ and $4.1 \pm 2.2$ kg for colonies of strain A and B, respectively). During the same period honey stores in the neonicotinoid-exposed group also slightly increased for colonies of strain A ($1.8 \pm 2.1$ kg on average), but decreased for neonicotinoid-exposed colonies of strain B ($-4.9 \pm 1.0$ kg on average), resulting in an overall decrease of $-1.0 \pm 3.8$ kg per colony on average. Honey production during the treatment was significantly influenced by both neonicotinoid exposure and honeybee strain: While strain B was significantly less productive than strain A independent of the treatment ($F_{1,21} = 40.40$, $P < 0.001$), neonicotinoid exposure negatively affected honey production in both strains ($F_{1,21} = 68.18$, $P < 0.001$). The interaction between both predictors was not significant ($F_{1,20} = 1.45$, $P = 0.24$) and was thus removed prior to testing the main effects. Overall, the mean honey production over the entire season (including honey production during the pre-treatment phase) remained 29% lower in the neonicotinoid-exposed colonies ($23.7 \pm 2.5$ kg) compared to the control ($33.4 \pm 5.1$ kg).

### Pollen consumption

At treatment initiation, pollen stores (bee bread) in the control group comprised of $17.9 \pm 10.7$ $dm^2$ and those in the group subsequently exposed to thiamethoam and clothianidin comprised

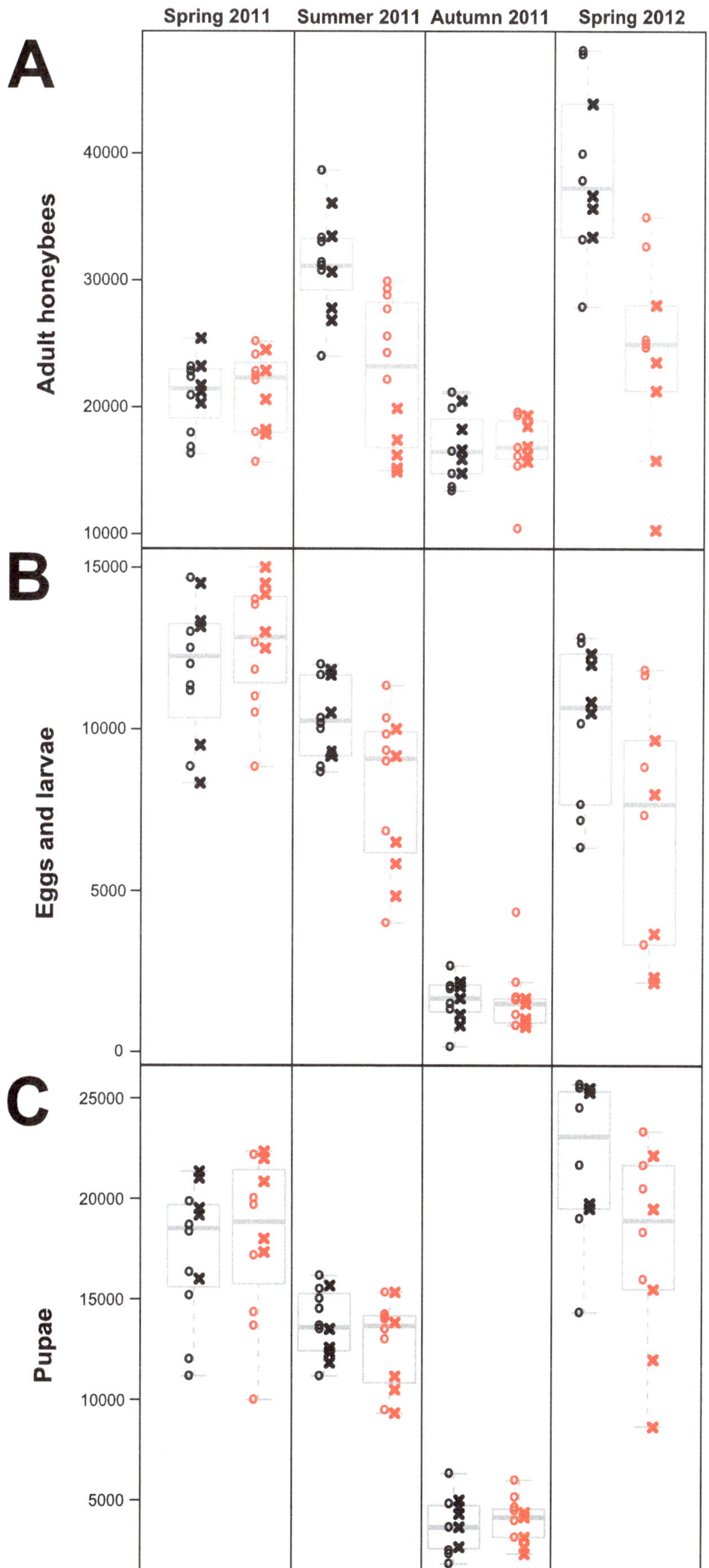
Spring 2011
Summer 2011
Autumn 2011
Spring 2012
A
B
C
Adult honeybees
Eggs and larvae
Pupae
40000
30000
20000
10000
15000
10000
5000
0
25000
20000
15000
10000
5000

**Figure 1. Dynamics of honeybee colony performance.** Data of all three endpoints number of adult bees (A), eggs and larvae (B) and pupae (C) for the different pollen feeding treatments (black = control; red = neonicotinoids) and honeybee strains (circles = strain A; crosses = strain B). The data were obtained at four successive colony assessment dates (X-axis subpanels within figures) performed before (Spring 2011) and directly after the 1.5 months of experimental pollen feeding (Summer 2011), 3.5 months after the treatment (Autumn 2011) and one year later (Spring 2012). Estimated numbers on the Y-axes are truncated for adult bees and pupae for better overview.

of 20.8±3.8 $dm^2$. After the experimental pollen feeding pollen stores within hives comprised of 29.8±8.9 $dm^2$ and 24.9±11.7 $dm^2$ in the control and neonicotinoid-exposed group, respectively. There was no indication that pollen storing and pollen consumption, respectively, during the experimental pollen feeding was influenced by neonicotinoid exposure ($F_{1,21} = 2.63$, $P = 0.12$) or honeybee strain ($F_{1,21} = 2.52$, $P = 0.13$). The interaction between the two predictors was not significant ($F_{1,21} = 0.74$, $P = 0.40$) and was thus removed prior to testing the main effects.

## Pollen collections

Total mean pollen harvests (±SD) per colony, as inferred from pollen trap contents, were 4.4±0.47 kg and 3.58±0.43 kg for the control and neonicotinoid-exposed group, respectively. The colonies exposed to thiamethoxam and clothianidin had 19% lower total pollen collections on average. Pollen collections measured as AUCs were found to be significantly lower in neonicotinoid-exposed colonies ($P<0.001$). While both treatment groups collected similar amounts of pollen during the first 3 weeks of the experimental pollen feeding, colonies exposed to thiamethoxam and clothianidin consistently collected less pollen later on, with mean pollen collections barely reaching more than 50% of the control group during the last week of exposure (Fig. 2).

## Supersedure of queens and tendency to swarm

We found a significant association of neonicotinoid exposure and queen supersedure (in the absence of swarming) ($P = 0.01$): while all ten queens of the control group survived until the end of the experiment (~2 years or swarmed, see below), 6 out of 10 queens of the colonies experimentally exposed to thiamethoxam and clothianidin over 1.5 months were replaced within one year after treatment. The result remained significant when overall queen loss was assessed, i.e. also including the two colonies (one per treatment group) that lost their queen during winter ($P = 0.02$). A negative association of neonicotinoid exposure and swarming events during spring following experimental treatment was found ($P = 0.005$): in the control group 9 out of 10 colonies swarmed until end of May 2012 (5 out of 6 colonies of strain A and all 4 colonies of strain B), while only 2 colonies (one of strain A and B each) of the group that was exposed to thiamethoxam and clothianidin in the previous season swarmed.

# Discussion

The major findings of this study using sublethal chronic exposure of honeybee colonies to thiamethoxam and clothianidin through feeding contaminated pollen were: (*i*) significant short-term (1.5 months) impacts at the colony level resulting in decreased colony performance and productivity; (*ii*) no negative influence in the medium-term (3.5 months) and on colony overwintering; (*iii*) significantly decelerated colony growth in the long-term (1 year) that was associated with higher queen supersedure rates and a reduced tendency to swarm; and (*iv*) significant interactions of the honeybee genetics with the observed effects of neonicotinoids on most parameters. In the following, these findings are discussed in context.

## Short-term impact

At the colony level, honeybee foraging efficiency was negatively influenced during chronic exposure to pollen containing environmentally relevant concentrations of the two neonicotinoids thiamethoxam and clothianidin over two brood cycles (1.5 months). The detected decrease in pollen collection and honey production upon sublethal neonicotinoid exposure are in line with earlier findings of impaired honeybee foraging through impacts on neurophysiological traits and cognitive abilities, including sucrose responsiveness, foraging rates, waggle dancing and memory and learning [47,48,49,50,51,52,53,54,55,56,57,58,59]. In addition, increased forager losses resulting from decreased homing success [48] through compromised navigation memory [58] may have contributed to both reduced foraging efficiency and significantly smaller worker populations of the colonies exposed to thiamethoxam and clothianidin compared to controls (Fig. 1A, Table 1). Indeed, there were no significant effects on the numbers of pupae present at the end of the exposure phase (Fig. 1C, Table 1), which points at higher forager losses rather than decreased worker production of neonicotinoid-exposed colonies during the treatment. As opposed to known effects of neonicotinoid ingestion of foragers through nectar-substitutes [47,48,49,50,51,56,57,58], the overall reduced numbers of adult bees through pollen exposure is intriguing. Pollen is generally not stored extensively within hives but consumed quickly. There was no evidence for differences in pollen storing between treatment groups directly after the experimental feeding. Therefore, it can be assumed that control pollen and pollen spiked with thiamethoxam and clothianidin were similarly consumed within experimental colonies. Contaminated pollen may affect various life stages of honeybees [105], yet pollen is predominantly consumed by nurse bees and larvae, but not foragers. In this regard, the recent finding that sublethal neonicotinoid exposure of honeybee larvae impacts post-emergence olfactory associative behaviour of adults [53] is relevant, and might be one of the largely unresolved effects arising from altered larval gene expression profiles upon sublethal neonicotinoid exposure [93] that may help to explain the strong impediment of colony growth documented here. Delayed sublethal effects extending to later life-cycle stages could similarly apply to larvae and nurse bees consuming contaminated pollen [64] and, depending on colony exposure duration, be able to reinforce decreased foraging efficiency. Our observed pattern of increasing divergence of pollen collections between treatment groups after 2–3 weeks, which roughly corresponds to the adult in-hive phase after emergence [106], may be indicative of the recruitment of less efficient foragers, which have encountered pollen contaminated with thiamethoxam and clothianidin during their nursing phase. The successively greater decline of pollen collections of neonicotinoid-exposed colonies after 5 weeks, in turn coincides with the time frame during which foragers could be expected to be exposed to contaminated pollen from the young larva stage onwards and throughout their entire development. Further research is needed to evaluate the potential for delayed impact on implementing complex foraging tasks through altered metabolic networks caused by sublethal dietary neonicotinoid exposure to honeybee brood [53,93] and eventually also nurse bees [64], the developmental

**Table 1.** Model-based estimates of contrasts and corresponding significance levels of the treatment effect (neonicotinoid *versus* control) and honeybee genetics (strain A *vs.* strain B).

| | Adult bees | | | Eggs and larvae | | | Pupae | | |
|---|---|---|---|---|---|---|---|---|---|
| | Summer 2011 | Autumn 2011 | Spring 2012 | Summer 2011 | Autumn 2011 | Spring 2012 | Summer 2011 | Autumn 2011 | Spring 2012 |
| Neonicotinoids *vs* Control | −60.56*** | −0.73 | −82.96*** | −0.31* | −0.01 | −0.49*** | −4.36 | 1.84 | −15.31** |
| Strain A *vs* strain B | | | | | | | 5.31 | 2.65 | 7.91 |
| Treatment within strain A | −14.07* | −1.33 | −28.59* | −0.10 | 0.03 | −0.06 | | | |
| Treatment within strain B | −46.49*** | 0.59 | −54.37*** | −0.21· | −0.05 | −0.42*** | | | |

Results are shown in the transformed scale for the three response variables adult bees, eggs and larvae and pupae assessed directly after the 1.5 months of treatment (Summer 2011), 3.5 months later (Autumn 2011) and 1 year later (Spring 2012). For adult bees and eggs and larvae (the models that included a significant threefold interaction between treatment, honeybee strain and assessment date) contrasts for treatment effects were also computed within individual honeybee strains at each assessment date. *P* values are adjusted for multiple testing. ***$P<0.001$; **$P<0.01$; *$P<0.05$; · $0.05<P<0.1$.

stages during which fundamental physiological processes required for adult olfaction and learning performance are settled [107].

The observation of strongly reduced pollen collection of the colonies exposed to thiamethoxam and clothianidin during the last week of the treatment could also be interpreted as a response to significantly reduced numbers of larvae (Fig. 1B, Table 1). However, a general decrease in foraging efficiency is also indicated by significantly reduced honey production. Because experimental pollen was provided constantly in sufficient amounts, significantly reduced investment in rearing larvae after 1.5 months of combined exposure to thiamethoxam and clothianidin is best explained by the declining numbers of workers: higher forager losses trigger premature forager recruitment, which results in fewer nurse bees available for brood rearing [101,108]. Although physiological changes in nurse bees may have played an additional role [54,55,93,109], reduced investment in brood rearing was probably reinforced by the reallocation of worker resources at the colony level towards the end of the exposure period rather than by adverse effects on the larvae themselves. Although there is some indication for negative effects of neonicotinoids and other pesticides on honeybee larvae [62,93,110], there is so far no evidence that field-realistic neonicotinoid exposure in the absence of additional stressors results in increased larval mortality [36,64]. The latter would also be difficult to reconcile with the observation that the numbers of pupae did not differ between treatments, even after 1.5 months of neonicotinoid exposure. Interestingly, short-term effects of the combined exposure to thiamethoxam and clothianidin on the number of adult bees were influenced by the honeybees' genetic background. While effects on colonies of strain A were only marginally significant, those on colonies of strain B were highly significant (Fig. 1, Table 1).

## Medium-term impact

The colony assessment during autumn 2011, 3.5 months after the experimental pollen feeding, revealed that there were neither effects of the combined exposure to thiamethoxam and clothianidin nor of the honeybees' genetic background on colony strength and the amount of brood (Fig. 1A–C, Table 1). All colonies were strong and well-fed, and thus overwintered successfully, except one colony of each treatment which lost their queens during winter. This reiterates the general view that a sustainable varroa mite management is a major aspect of honeybee colony health, thereby limiting colony losses [25,26]. Our comprehensive varroa mite management, comprising of a combination of integrated actions using organic acids [27] but avoiding potentially detrimental synthetic acaricides [47,52,111,112], was apparently sufficient to limit *V. destructor*-associated damage during the sensitive over-wintering period, and throughout the entire study. A thorough varroa mite control pre-supposed, the overcoming of the previously observed short-term effects of neonicotinoids on colony strength and brood rearing 4–5 brood cycles later shows that chronic sublethal neonicotinoid-exposure alone does not trigger elevated honeybee colony winter losses [17,20], although contrasting results have been found in other studies [113,114].

There were no obvious clinical symptoms of infections with widespread honeybee pathogens such as the ectoparasitic mite *V. destructor*, microsporidian gut parasites *Nosema* ssp., chalkbrood (the fungal pathogen *Ascophaera apis*) or European and American foulbrood (bacterial diseases caused by *Melissococcus plutonius* and *Paenibacillus larvae*, respectively) at any time during the study. Nevertheless, it cannot be excluded that some of the above described short-term effects on colony performance may have been influenced by detrimental neonicotinoid-pathogen interactions at the level of individual bees [36,68,69,70,71,72]. Higher

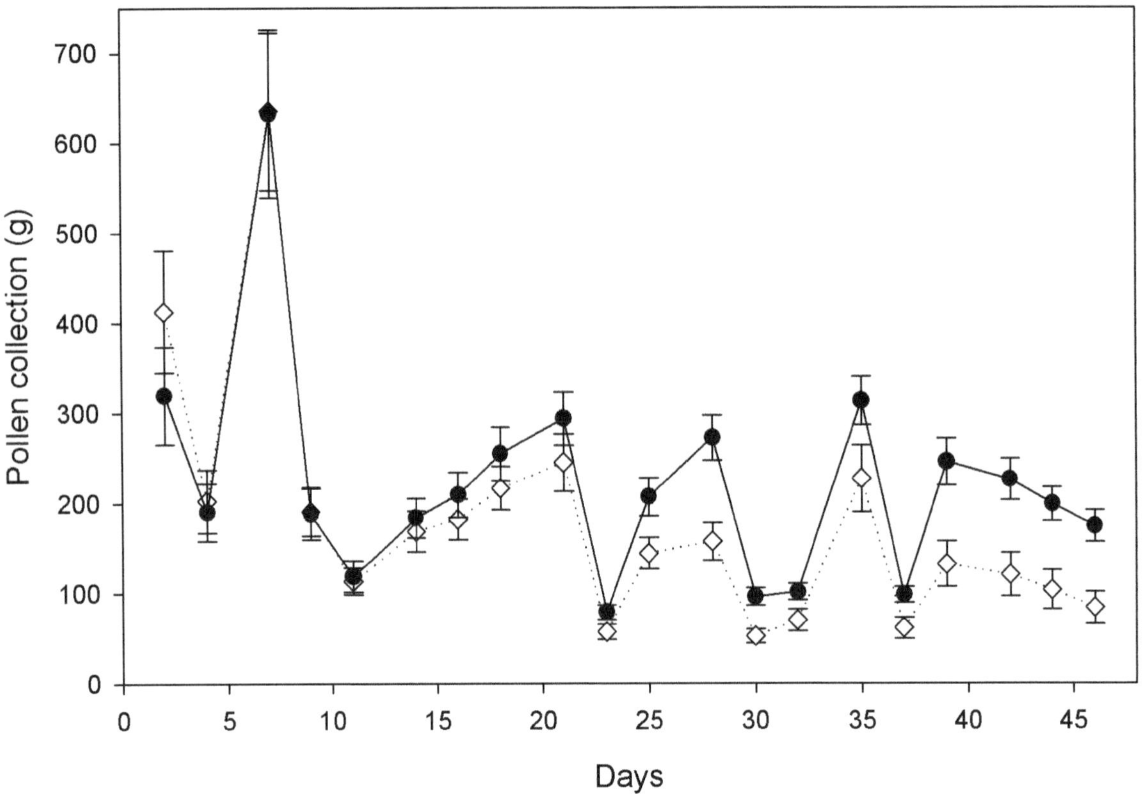


**Figure 2. Pollen collections.** Mean (±SD) fresh weights of pollen collections for control (black) and neonicotinoid-exposed (white) colonies over the course of the treatment period (pollen-trap contents were weighed in 2-2-3 days intervals throughout the study).

forager losses associated with overall shifts in forager age structure may also cause generally decreased immunological competence at the colony level [115]. However, these differentiations between individual drivers and their interactions are only of limited practical importance because free-flying honeybees will almost inevitably carry ubiquitous pathogens and thus encounter combined pressures. In fact, it is very likely that latent infections with the most common *V. destructor*-associated honeybee viruses and *Nosema* spp. have also been present in the experimental apiary. The common garden approach to maintain all colonies at the same isolated field site for almost one year prior to the experiments actually aimed at 'synchronizing' viral and microbial landscapes across colonies, thereby resulting in potential errors of systematic conservative nature. If multiple pressures were indeed present in our experiment, the observed recovery from decreased performance of previously neonicotinoid-exposed colonies after several brood cycles conforms to model predictions, showing that even detrimental interactions presumably can be tolerated to some extent and do not necessarily result in disrupting colony function [65].

## Long-term impact

Given that significantly negative short-term impacts of combined exposure to thiamethoxam and clothianidin during summer 2011 had faded with respect to colony strength during autumn (medium-term) and overwintering success, the patterns observed during the subsequent spring were unexpected. Colonies that received pollen spiked with thiamethoxam and clothianidin during the previous spring exhibited significantly lower numbers of adult bees compared to controls one year later, whereas the effects were much stronger for strain B than for strain A (Fig. 1A, Table 1). In addition, there were significantly negative overall effects of last year's neonicotinoid exposure on the amount of pupae (Fig. 1C, Table 1) and eggs and larvae (Fig. 1B, Table 1). For the latter endpoint, however, significant effects were only found for strain B (Table 1). These patterns of decelerated colony growth are intriguing, because all hive samples collected 3 weeks after the experimental feeding of neonicotinoid-spiked pollen patties did not contain traceable residues of thiamethoxam and clothianidin. Although trace residues below corresponding limits of detection could have still been present within hives one year after the treatment, it is highly unlikely that this explains detrimental effects that far exceeded those observed directly after the treatment (Table 1). Instead, reduced performance of previously neonicotinoid-exposed queens appears to be a more plausible explanation for long-term effects. While all original, individually tagged and wing-clipped queens were recognized in autumn 2011, the assessment in spring 2012 revealed that there were non-tagged egg-laying queens with intact wings in some colonies previously exposed to thiamethoxam and clothianidin. This clearly shows queen supersedure during the previous autumn as a response to poor queen performance [86,87], which could pose a risk for successful overwintering. Yet, supersedure during autumn bears a considerable risk for colony failure as well, because virgin queens may not be sufficiently mated during their mating flights due to a comparatively lower number of available drones. Queen replacement in the neonicotinoid treatment group in the presence of original queens, and without swarming, was also observed during spring 2012. This indication of impaired performance of previously neonicotinoid-exposed queens, together with a likely reduced mating success of replacement queens from the previous autumn, is thus considered to be responsible for the overall decelerated colony growth of the neonicotinoid treatment group one year after exposure. The causal patterns of queen failure could be manifold, including compromised immunity that has been shown to result in increased pathogen loads in honeybee workers [36,68,69,72]. Taking into account that honeybee colony fate essentially depends on the queen, the long-term impact of neonicotinoids on queens we observed, resulting in 60% supersedure within one year, represents a novel finding of major importance that clearly deserves further research. Increased queen supersedure rates were shown upon sublethal pyrethroid exposure [116] and detrimental effects on queen development and performance have also been reported from exposure to commonly encountered in-hive acaricides [117,118,119,120], some of which are known to have similarly negative influence like neonicotinoids on worker bees [47,52,108,121]. General effects of neurotoxic insecticides on queens may thus not be fundamentally surprising. Queen failure has repeatedly been indicated to be an important aspect shaping the present enigma of widespread colony losses [22,25,122], which may also be influenced by neonicotinoids. While direct contact exposure to agrochemicals can be generally considered much more relaxed for queens compared to workers, this might be different regarding oral exposure to systemic compounds, such as neonicotinoids. Here, ingestion of contaminated food represents a principal risk for all developmental stages and different castes of honeybees. Moreover, the queen's life span as well as her level of trophallactic interactions far exceed that of workers [123], which may predestine her as a potential sink of sublethal exposure to systemic insecticides through pollen-based queen nutrition. Our results suggest that sublethal neonicotinoid effects on queens can jeopardize colony fate in the long-term, and should thus not be ignored.

Compared to the need for maintaining productive queens within colonies, swarming is less relevant for practical beekeeping, especially commercial operations. Nevertheless, apart from drone mating success, colony splitting through swarming represents a true fitness estimate in honeybees. According to overall reduced colony strength, colonies that were exposed to thiamethoxam and clothianidin in the previous season exhibited a significantly lower swarming success.

## The role of honeybee genetics

The detection of significant interactions of the honeybees' genetic background with the effects of chronic neonicotinoid exposure on both colony performance in the short-term and queen performance in the long-term was a major insight of our study. Honeybee strain B, *A. m. mellifera* sister queens originating from a Swiss alpine region, was much more susceptible to the combination of thiamethoxam and clothianidin compared to strain A, *A. m. carnica* sister queens originating from a German region characterized by intense agriculture. For example, neonicotinoid-exposed colonies of strain B had the lowest numbers of adult bees directly after exposure (Fig. 1A), and none of their queens survived throughout the one year post-exposure period. This indicates that honeybee susceptibility to the here applied neonicotinoids may also include a genetic or epigenetic component, as has been previously suggested for neonicotinoids and other pesticides [17,93,111,124,125,126]. It remains unresolved to what extent the outcome of our study was influenced by the usage of different honeybee ecotypes and/or by their distinct breeding histories linked to different environments. It could be possible that *A. m. mellifera* tends to be more susceptible to oral neonicotinoid exposure than *A. m. carnica* bees, as similarly found in another recent study [126]. Yet, compared to strain B, the particular breeding regime of strain A prior to our study may have simply included an unintended selection for higher neonicotinoid tolerance as a non-lineage-specific trait. Further research is required to explore the potential genetic basis underlying variable responses to sublethal neonicotinoid exposure in honeybees. Causal patterns could include detoxification genes and corresponding expression profiles, e.g., cytochrome P450 monooxygenases [93], yet, compared to many other insects, honeybees are equipped with a limited set of detoxification genes [127] and specifically the pathways for nitroguanidine neonicotinoid metabolism in honeybees remain less resolved compared to other insecticidal compounds [61,111,128,129,130,131].

## Conclusions

In line with a recent meta-analysis [44], our results clearly indicate that neonicotinoids negatively impact on honeybee colony performance after chronic sublethal exposure throughout two brood cycles. Virtually all tested contrasts (18 out of 21) produced negative estimates for the effects of exposure to thiamethoxam and clothianidin (Table 1). It is supposed that sublethal neonicotinoid exposure through pollen has a stronger impact at the honeybee colony level compared to nectar-substitute feeding (e.g., [132]). Therefore more studies focussing on effects of sublethal exposure of larvae or nurse bees extending to the performance of complex tasks of adult forager honeybees are needed. Similarly, sublethal effects of neonicotinoids on honeybee queens clearly deserve in depth investigation.

It remains uncertain whether the observed colony level responses were stronger influenced by either thiamethoxam or clothianidin, or by the possibility for interactive effects of the combination of these neonicotinoids, both being ranked as having high risk potential to honeybees [64]. Yet, since clothianidin is the major metabolite of thiamethoxam, both bioactive compounds will be present in pollen and nectar of crops treated with thiamethoxam [80,82,96,97,98]. Thus, the combined exposure as well as the residue levels administered in pollen patties in this study represent a biologically relevant exposure scenario.

Exposure to almost any sublethal dosage of these highly potent insecticides could trigger adverse chronic effects in a time-dependant context [133,134]. Our experimental exposure to contaminated pollen over 1.5 months can be considered as a worst-case scenario for agricultural settings, where honeybee colonies may encounter neonicotinoid-treated crops repeatedly throughout the season, but likely for shorter individual exposure periods. In bumblebees, it has recently been shown, that microcolonies can recover after short periods of sublethal exposure [135] to imidacloprid, which may well be similar, or even more pronounced in honeybee colonies. In a similar way, the supposedly exclusive exposure to pollen contaminated with neonicotinoids in our setup can be regarded as a worst case because in the field honeybee colonies generally exploit multiple available pollen resources at any time, some of which may only occasionally be contaminated with agricultural pesticides [38]. Nonetheless, it remains unclear whether repeated pulsed exposures over the entire season, either through consecutively available neonicotinoid-treated honeybee-attractive crops providing pollen and nectar (e.g., oilseed rape, sunflowers and maize that may each flower for 2–3 weeks) or through the general accumulation of neonicotinoid residues in the environment [38,41,42,82], could result in detrimental effects as found here after 1.5 months of continuous chronic exposure. Moreover, although distinguishing between continuous and shorter but repeated periods of neonicotinoid exposure could represent different risk potentials for the honeybee worker population succumbed to generally high turn-over rates [101], the suspected buffering capacities at the colony level may deplete in the long-term, if the queen was still affected through pulsed exposures. Thus, a worst case scenario that applies realistic exposure levels while settling at the upper possible range of overall exposure duration deserves consideration. Currently there is no evidence that field-realistic neonicotinoid exposure could be directly involved in colony losses [79,80,81,82,83]. However, available studies do not fully allow to address the question of whether neonicotinoids are really contributing to colony weakening, because several aspects in corresponding experimental designs remain unsettled, such as lacking sufficient statistical power [44], the putative influence of honeybee genetics (see above), too small distances between test fields and other factors that challenge the prerequisite of adequate control fields, such as no pesticide treatment at all [41,79,80,81,82], queen effects that remain elusive [25] or queen rotation practices that impede uncovering potential effects on queen performance [80]. Interestingly, this long-term impact detected in our study, notably supersedure of failing queens, compounds recent criticism that the presence of a given predictor may not necessarily correlate with impaired colony function [65] by showing that delayed effects can emerge also in the absence of causal stressors, in this case the lack of traceable amounts of the applied neonicotinoids within hives several weeks after exposure. There is an urgent need for more thoroughly designed studies to clarify the threats of neonicotinoids to honeybees, and pollinators in general [42]. The growing body of scientific awareness on sublethal side-effects of pesticides on non-target pollinators, ranging from gene expression profiles of individual developmental stages to entire life-time fitness performances of different species of bees, represents a unique opportunity to benefit the current framework of pesticide risk assessment [64,84,136,137].

Finally, the here detected interactions of honeybee genetics and impacts of chronic neonicotinoid exposure on colony performance suggest that there is genetic variability for neonicotinoid susceptibility, and thus potential to partly counteract negative effects through selective breeding of more tolerant honeybee strains.

## Supporting Information

**Figure S1 Colony set up at the experimental apiary.** Two groups of honeybee colonies each comprising of 7 colonies originating from strain A (blue squares) and 5 colonies originating from strain B (yellow squares) were placed in a row in front of a small forest with entrances pointing in the same direction. All colonies were maintained approximately one year before and one year after the experimental feeding with either control or neonicotinoid-spiked pollen patties over two brood cycles.

## Acknowledgments

We thank Jennifer Brandenburg, Rebecca Basile and Mario Waldburger for assistance and Nicolas Desneux, Werner Stahel and three anonymous reviewers for valuable comments. The infrastructure was courtesy provided by Agroscope Reckenholz-Tänniken ART, Switzerland.

## Author Contributions

Conceived and designed the experiments: CS. Performed the experiments: CS AFM. Analyzed the data: MT LGT CS. Contributed reagents/materials/analysis tools: PN CS. Wrote the paper: CS PN SGP.

## References

1. Bascompte J, Jordano P, Olesen JM (2006) Asymmetric coevolutionary networks facilitate biodiversity maintenance. Science 312: 431–433.
2. Fontaine C, Dajoz I, Meriguet J, Loreau M (2006) Functional diversity of plant-pollinator interaction webs enhances the persistence of plant communities. PLOS Biol 4: e1.
3. Garibaldi LA, Aizen MA, Klein AM, Cunningham SA, Harder LD (2011) Global growth and stability of agricultural yield decrease with pollinator dependence. Proc Natl Acad Sci USA 108: 5909–5914.
4. Klein AM, Vaissière BE, Cane JH, Steffan-Dewenter I, Cunningham SA, et al. (2007) Importance of pollinators in changing landscapes for world crops. Proc R Soc Lond B Biol Sci 274: 303–313.
5. Gallai N, Salles JM, Settele J, Vaissière BE (2009) Economic valuation of the vulnerability of world agriculture confronted with pollinator decline. Ecol Econ 68: 810–821.
6. Lautenbach S, Seppelt R, Liebscher J, Dormann CF (2012) Spatial and temporal trends of global pollination benefit. PLOS ONE 7: e35954.
7. Biesmeijer JC, Roberts SPM, Reemer M, Ohlemuller R, Edwards M, et al. (2006) Parallel declines in pollinators and insect-pollinated plants in Britain and the Netherlands. Science 313: 351–354.
8. Potts SG, Biesmeijer JC, Kremen C, Neumann P, Schweiger O, et al. (2010) Global pollinator declines: trends, impacts and drivers. Trends Ecol Evol 25: 345–353.
9. Cameron SA, Lozier JD, Strange JP, Koch JB, Cordes N, et al. (2011) Patterns of widespread decline in North American bumble bees. Proc Natl Acad Sci USA 108: 662–667.
10. Goulson D, Lye GC, Darvill B (2008) Decline and conservation of bumble bees. Annu Rev Entomol 53: 191–208.
11. Aizen MA, Garibaldi LA, Cunningham SA, Klein AM (2008) Long-term global trends in crop yield and production reveal no current pollination shortage but increasing pollinator dependency. Curr Biol 18: 1572–1575.
12. Aizen MA, Garibaldi LA, Cunningham SA, Klein AM (2009) How much does agriculture depend on pollinators? Lessons from long-term trends in crop production. Ann Bot 103: 1579–1588.
13. Potts SG, Roberts SPM, Dean R, Marris G, Brown MA, et al. (2010) Declines of managed honey bees and beekeepers in Europe. J Apic Res 49: 15–22.
14. Ellis JD, Evans JD, Pettis JS (2010) Colony losses, managed colony population decline and Colony Collapse Disorder in the United States. J Apic Res 49: 134–136.
15. Aizen MA, Harder LD (2009) The global stock of domesticated honey bees is growing slower than agricultural demand for pollination. Curr Biol 19: 915–918.
16. vanEngelsdorp D, Hayes J Jr., Underwood RM, Pettis J (2008) A survey of honey bee colony losses in the U.S., fall 2007 to spring 2008. PLOS ONE 3: e4071.
17. vanEngelsdorp D, Evans JD, Saegerman C, Mullin C, Haubruge E, et al. (2009) Colony collapse disorder: A descriptive study. PLOS ONE 4: e6481.
18. Cox-Foster DL, Conlan S, Holmes EC, Palacios G, Evans JD, et al. (2007) A metagenomic survey of microbes in honey bee colony collapse disorder. Science 318: 283–287.
19. Anderson D, East IJ (2008) The latest buzz about colony collapse disorder. Science 319: 724–725.
20. van der Zee R, Pisa L, Andronov S, Brodschneider R, Charrière J-D, et al. (2012) Managed honey bee colony losses in Canada, China, Europe, Israel and Turkey, for the winters of 2008–9 and 2009–10. J Apic Res 51: 100–114.
21. Neumann P, Carreck N (2010) Honey bee colony losses. J Apic Res 49: 1–6.
22. Brodschneider R, Moosbeckhofer R, Crailsheim K (2010) Surveys as a tool to record winter losses of honey bee colonies: a two-year case study in Austria and South Tyrol. J Apic Res 49: 23–30.
23. Johnson RM, Evans JD, Robinson GE, Berenbaum MR (2009) Changes in transcript abundance relating to colony collapse disorder in honey bees (*Apis mellifera*). Proc Natl Acad Sci USA 106: 14790–14795.
24. Vanbergen AJ, the Insect Pollinators Initiative (2013) Threats to an ecosystem service: pressures on pollinators. Front Ecol Environ 11: 251–259.
25. Genersch E, von der Ohe W, Kaatz H, Schroeder A, Otten C, et al. (2010) The German bee monitoring project: a long term study to understand periodically high winter losses of honey bee colonies. Apidologie 41: 332–352.
26. Le Conte Y, Ellis M, Ritter W (2010) *Varroa* mites and honey bee health: can Varroa explain part of the colony losses? Apidologie 41: 353–363.
27. Rosenkranz P, Aumeier P, Ziegelmann B (2010) Biology and control of *Varroa destructor*. J Invertebr Pathol 103 Suppl 1: S96–119.
28. Nazzi F, Brown SP, Annoscia D, Del Piccolo F, Di Prisco G, et al. (2012) Synergistic parasite-pathogen interactions mediated by host immunity can drive the collapse of honeybee colonies. PLOS Pathog 8: e1002735.
29. Martin SJ, Highfield AC, Brettell L, Villalobos EM, Budge GE, et al. (2012) Global honey bee viral landscape altered by a parasitic mite. Science 336: 1304–1306.

30. Higes M, Martin-Hernandez R, Garrido-Bailon E, Gonzalez-Porto AV, Garcia-Palencia P, et al. (2009) Honeybee colony collapse due to *Nosema ceranae* in professional apiaries. Environ Microbiol Rep 1: 110–113.
31. Antunez K, Martin-Hernandez R, Prieto L, Meana A, Zunino P, et al. (2009) Immune suppression in the honey bee (*Apis mellifera*) following infection by *Nosema ceranae* (Microsporidia). Environ Microbiol 11: 2284–2290.
32. Higes M, Meana A, Bartolome C, Botias C, Martin-Hernandez R (2013) *Nosema ceranae* (Microsporidia), a controversial 21st century honey bee pathogen. Environ Microbiol Rep 5: 17–29.
33. Evans JD, Schwarz RS (2011) Bees brought to their knees: microbes affecting honey bee health. Trends Microbiol 19: 614–620.
34. Bromenshenk JJ, Henderson CB, Wick CH, Stanford MF, Zulich AW, et al. (2010) Iridovirus and microsporidian linked to honey bee colony decline. PLOS ONE 5: e13181.
35. Cornman RS, Tarpy DR, Chen Y, Jeffreys L, Lopez D, et al. (2012) Pathogen webs in collapsing honey bee colonies. PLOS ONE 7: e43562.
36. Doublet V, Labarussias M, De Miranda JR, Moritz RFA, Paxton JR (2014) Bees under stress: sublethal doses of a neonicotinoid pesticide and pathogens interact to elevate honey bee mortality across the life cycle. Environ Microbiol. (in press): DOI:10.1111/1462-2920.12426.
37. Mullin CA, Frazier M, Frazier JL, Ashcraft S, Simonds R, et al. (2010) High levels of miticides and agrochemicals in North American apiaries: implications for honey bee health. PLOS ONE 5: e9754.
38. Krupke CH, Hunt GJ, Eitzer BD, Andino G, Given K (2012) Multiple routes of pesticide exposure for honey bees living near agricultural fields. PLOS ONE 7: e29268.
39. Elbert A, Haas M, Springer B, Thielert W, Nauen R (2008) Applied aspects of neonicotinoid uses in crop protection. Pest Manag Sci 64: 1099–1105.
40. Jeschke P, Nauen R, Schindler M, Elbert A (2011) Overview of the status and global strategy for neonicotinoids. J Agric Food Chem 59: 2897–2908.
41. van der Slujs JP, Simon-Delso N, Goulson D, Maxim L, Bonmatin JM, et al. (2013) Neonicotinoids, bee disorders and the sustainability of pollinator services. Curr Opin Environ Sustain 5: 293–305.
42. Goulson D (2013) An overview of the environmental risks posed by neonicotinoid insecticides. J Appl Ecol 50: 977–987.
43. Matsuda K, Buckingham SD, Kleier D, Rauh JJ, Grauso M, et al. (2001) Neonicotinoids: insecticides acting on insect nicotinic acetylcholine receptors. Trends Pharmacol Sci 22: 573–580.
44. Cresswell JE (2011) A meta-analysis of experiments testing the effects of a neonicotinoid insecticide (imidacloprid) on honey bees. Ecotoxicology 20: 149–157.
45. Desneux N, Decourtye A, Delpuech JM (2007) The sublethal effects of pesticides on beneficial arthropods. Annu Rev Entomol 52: 81–106.
46. Blacquière T, Smagghe G, van Gestel CA, Mommaerts V (2012) Neonicotinoids in bees: a review on concentrations, side-effects and risk assessment. Ecotoxicology 21: 973–992.
47. Williamson SM, Wright GA (2013) Exposure to multiple cholinergic pesticides impairs olfactory learning and memory in honeybees. J Exp Biol 216: 1799–1807.
48. Henry M, Beguin M, Requier F, Rollin O, Odoux JF, et al. (2012) A common pesticide decreases foraging success and survival in honey bees. Science 336: 348–350.
49. Belzunces LP, Tchamitchan S, Brunet J (2012) Neural effects of insecticides in the honey bee. Apidologie 43: 348–370.
50. Decourtye A, Devillers J (2010) Ecotoxicity of neonicotinoid insecticides to bees. In: Thany SH, editor. Insect nicotinic acetylcholine receptors. New York: Springer. 85–95.
51. Schneider CW, Tautz J, Grünewald B, Fuchs S (2012) RFID tracking of sublethal effects of two neonicotinoid insecticides on the foraging behavior of *Apis mellifera*. PLOS ONE 7: e30023.
52. Palmer MJ, Moffat C, Saranzewa N, Harvey J, Wright GA, et al. (2013) Cholinergic pesticides cause mushroom body neuronal inactivation in honeybees. Nat Commun 4: 1634.
53. Yang EC, Chang HC, Wu WY, Chen YW (2012) Impaired olfactory associative behavior of honeybee workers due to contamination of imidacloprid in the larval stage. PLOS ONE 7: e49472.
54. Hatjina F, Papaefthimiou C, Charistos L, Dogaroglu T, Bouga M, et al. (2013) Sublethal doses of imidacloprid decreased size of hypopharyngeal glands and respiratory rhythm of honeybees in vivo. Apidologie 44: 467–480.
55. Oliveira RA, Roat TC, Carvalho SM, Malaspina O (2013) Side-effects of thiamethoxam on the brain and midgut of the africanized honeybee *Apis mellifera* (Hymenoptera: Apidae). Environ Toxicol. (in press): DOI:10.1002/tox.21842.
56. Eiri DM, Nieh JC (2012) A nicotinic acetylcholine receptor agonist affects honey bee sucrose responsiveness and decreases waggle dancing. J Exp Biol 215: 2022–2029.
57. Ramirez-Romero R, Chaufaux J, Pham-Delègue MH (2005) Effects of Cry1Ab protoxin, deltamethrin and imidacloprid on the foraging activity and the learning performances of the honeybee *Apis mellifera*, a comparative approach. Apidologie 36: 601–611.
58. Fischer J, Müller T, Spatz AK, Greggers U, Grünewald B, et al. (2014) Neonicotinoids interfere with specific components of navigation in honeybees. PLOS ONE 9: e91364.
59. Han P, Niu CY, Lei CL, Cui JJ, Desneux N (2010) Use of an innovative T-tube maze assay and the proboscis extension response assay to assess sublethal effects of GM products and pesticides on learning capacity of the honey bee *Apis mellifera* L. Ecotoxicology 19: 1612–1619.
60. Gill RJ, Ramos-Rodriguez O, Raine NE (2012) Combined pesticide exposure severely affects individual- and colony-level traits in bees. Nature 491: 105–108.
61. Iwasa T, Motoyama N, Ambrose JT, Roe RM (2004) Mechanism for the differential toxicity of neonicotinoid insecticides in the honey bee, *Apis mellifera*. Crop Prot 23: 371–378.
62. Zhu W, Schmehl DR, Mullin CA, Frazier JL (2014) Four common pesticides, their mixtures and a formulation solvent in the hive environment have high oral toxicity to honey bee larvae. PLOS ONE 9: e77547.
63. Thompson HM, Fryday SL, Harkin S, Milner S (2014) Potential impacts of synergism in honeybees (*Apis mellifera*) of exposure to neonicotinoids and sprayed fungicides in crops. Apidologie (in press): DOI:10.1007/s13592-014-0273-6.
64. Sànchez-Bayo F, Goka K (2014) Pesticide residues and bees - a risk assessment. PLOS ONE 9: e94482.
65. Bryden J, Gill RJ, Mitton RA, Raine NE, Jansen VA (2013) Chronic sublethal stress causes bee colony failure. Ecol Lett. 16: 1463–1469.
66. James RR, Xu J (2012) Mechanisms by which pesticides affect insect immunity. J Invertebr Pathol 109: 175–182.
67. Mason R, Tennekes H, Sànchez-Bayo F, Jepsen PU (2013) Immune suppression by neonicotinoid insecticides at the root of global wildlife declines. J Environ Immunol Toxicol 1: 3–12.
68. Vidau C, Diogon M, Aufauvre J, Fontbonne R, Vigues B, et al. (2011) Exposure to sublethal doses of fipronil and thiacloprid highly increases mortality of honeybees previously infected by *Nosema ceranae*. PLOS ONE 6: e21550.
69. Pettis JS, vanEngelsdorp D, Johnson J, Dively G (2012) Pesticide exposure in honey bees results in increased levels of the gut pathogen *Nosema*. Naturwissenschaften 99: 153–158.
70. Alaux C, Brunet JL, Dussaubat C, Mondet F, Tchamitchan S, et al. (2010) Interactions between *Nosema* microspores and a neonicotinoid weaken honeybees (*Apis mellifera*). Environ Microbiol 12: 774–782.
71. Aufauvre J, Misme-Aucouturier B, Viguès B, Texier C, Delbac F, et al. (2014) Transcriptome analyses of the honeybee response to *Nosema ceranae* and insecticides. PLOS ONE 9: e91686.
72. Di Prisco G, Cavaliere V, Annoscia D, Varricchio P, Caprio E, et al. (2013) Neonicotinoid clothianidin adversely affects insect immunity and promotes replication of a viral pathogen in honey bees. Proc Natl Acad Sci USA 110: 18466–18471.
73. Naug D (2009) Nutritional stress due to habitat loss may explain recent honeybee colony collapses. Biol Conserv 142: 2369–2372.
74. Alaux C, Dantec C, Parrinello H, Le Conte Y (2011) Nutrigenomics in honey bees: digital gene expression analysis of pollen's nutritive effects on healthy and varroa-parasitized bees. BMC Genomics 12: 496.
75. Alaux C, Ducloz F, Crauser D, Le Conte Y (2010) Diet effects on honeybee immunocompetence. Biol Lett 6: 562–565.
76. Mao W, Schuler MA, Berenbaum MR (2013) Honey constituents up-regulate detoxification and immunity genes in the western honey bee *Apis mellifera*. Proc Natl Acad Sci USA 110: 8842–8846.
77. Chauzat MP, Martel AC, Blanchard P, Clément MC, Schurr F, et al. (2010) A case report of a honey bee colony poisoning incident in France. J Apic Res 49: 113–115.
78. Tremolada P, Mazzoleni M, Saliu F, Colombo M, Vighi M (2010) Field trial for evaluating the effects on honeybees of corn sown using Cruiser and Celest xl treated seeds. Bull Environ Contam Toxicol 85: 229–234.
79. Cutler GC, Scott-Dupree CD (2007) Exposure to clothianidin seed-treated canola has no long-term impact on honey bees. J Econ Entomol 100: 765–772.
80. Pilling E, Campbell P, Coulson M, Ruddle N, Tornier I (2013) A four-year field program investigating long-term effects of repeated exposure of honey bee colonies to flowering crops treated with thiamethoxam. PLOS ONE 8: e77193.
81. Nguyen BK, Saegerman C, Pirard C, Mignon J, Widart J, et al. (2009) Does imidacloprid seed-treated maize have an impact on honey bee mortality? J Econ Entomol 102: 616–623.
82. Pohorecka K, Skubida P, Miszczak A, Semkiw P, Sikorski P, et al. (2012) Residues of neonicotinoid insecticides in bee collected plant materials from oilseed rape crops and their effect on bee colonies. Journal of Apicultural Science 56: 115–134.
83. Cresswell JE, Desneux N, vanEngelsdorp D (2012) Dietary traces of neonicotinoid pesticides as a cause of population declines in honey bees: an evaluation by Hill's epidemiological criteria. Pest Manag Sci 68: 819–827.
84. Cressey D (2013) Europe debates risk to bees. Nature 496: 408.
85. Stokstad E (2013) Pesticides under fire for risks to pollinators. Science 340: 674–676.
86. Butler CG (1957) The process of queen supersedure in colonies of honeybees (*Apis mellifera* Linn.). Insectes Soc 4: 211–223.
87. Pettis JS, Higo H, Winston ML, Pankiw T (1997) Queen rearing suppression in the honey bee - evidence for a fecundity signal. Insectes Soc 44: 311–322.
88. Whitehorn PR, O'Connor S, Wackers FL, Goulson D (2012) Neonicotinoid pesticide reduces bumble bee colony growth and queen production. Science 336: 351–352.

89. Larson JL, Redmond CT, Potter DA (2013) Assessing insecticide hazard to bumble bees foraging on flowering weeds in treated lawns. PLOS ONE 8: e66375.
90. Fauser-Misslin A, Sadd BM, Neumann P, Sandrock C (2014) Influence of combined pesticide and parasite exposure on bumblebee colony traits in the laboratory. J Appl Ecol 51: 450–459.
91. Sandrock C, Tanadini LG, Pettis J, Biesmeijer JC, Potts SG, et al. (2014) Sublethal neonicotinoid insecticide exposure reduces solitary bee reproductive success. Agric For Entomol 16: 119–128.
92. Moritz RFA, Southwick EE (1992) Bees as superorganisms: an evolutionary reality. Berlin: Springer-Verlag. 395p.
93. Derecka K, Blythe MJ, Malla S, Genereux DP, Guffanti A, et al. (2013) Transient exposure to low levels of insecticide affects metabolic networks of honeybee larvae. PLOS ONE 8: e68191.
94. Neumann P, Moritz RFA, Mautz D (2000) Colony evaluation is not affected by drifting of drone and worker honeybees (*Apis mellifera* L.) at a performance testing apiary. Apidologie 31: 67–79.
95. Higes M, Martin-Hernandez R, Garrido-Bailon E, Garcia-Palencia P, Meana A (2008) Detection of infective *Nosema ceranae* (Microsporidia) spores in corbicular pollen of forager honeybees. J Invertebr Pathol 97: 76–78.
96. Maienfisch P, Angst M, Brandl F, Fischer W, Hofer D, et al. (2001) Chemistry and biology of thiamethoxam: a second generation neonicotinoid. Pest Manag Sci 57: 906–913.
97. Nauen R, Ebbinghaus-Kintscher U, Salgado VL, Kaussmann M (2003) Thiamethoxam is a neonicotinoid precursor converted to clothianidin in insects and plants. Pestic Biochem Phys 76: 55–69.
98. Dively GP, Kamel A (2012) Insecticide residues in pollen and nectar of a cucurbit crop and their potential exposure to pollinators. J Agric Food Chem 60: 4449–4456.
99. Imdorf A, Buehlmann G, Gerig L, Kilchenmann V, Wille H (1987) A test of the method of estimation of brood areas and number of worker bees in free-flying colonies. Apidologie 18: 137–146.
100. Delaplane KS, van der Steen J, Guzman E (2013) Standard methods for estimating strength parameters of *Apis mellifera* colonies. In: Dietemann V, Ellis JD, Neumann P, editors. The COLOSS BEEBOOK, Volume 1: standard methods for *Apis mellifera* research.
101. Khoury DS, Myerscough MR, Barron AB (2011) A quantitative model of honey bee colony population dynamics. PLOS ONE 6: e18491.
102. R Core Development_Team (2012) R: A language and environment for statistical computing. R Foundation for Statistical computing, Vienna, Austria. Available: http://cran.r-project.org. Accessed 27 April 2014.
103. Bates D, Maechler M, Bolker B, Walker S (2013) lme4: Linear mixed-effects models using Eigen and S4 classes. R package version 10-4. Available: http://CRANR-projectorg/package = lme4.
104. Hothorn T, Bretz F, Westfall P (2008) Simultaneous inference in general parametric models. Biometrical Journal 50: 346–363.
105. Rortais A, Arnold G, Halm MP, Touffet-Briens F (2005) Modes of honeybees exposure to systemic insecticides: estimated amounts of contaminated pollen and nectar consumed by different categories of bees. Apidologie 36: 71–83.
106. Seeley TD (1982) Adaptive significance of the age polyethism schedule in honeybee colonies. Behav Ecol Sociobiol 11: 287–293.
107. Masson C, Pham-Delègue MH, Fonta C, Gascuel J, Arnold G, et al. (1993) Recent advances in the concept of adaptation to natural odour signals in the honeybee, *Apis mellifera* L. Apidologie 24: 169–194.
108. Wu JY, Anelli CM, Sheppard WS (2011) Sub-lethal effects of pesticide residues in brood comb on worker honey bee (*Apis mellifera*) development and longevity. PLOS ONE 6: e14720.
109. Smodis Skerl MI, Gregorc A (2010) Heat shock proteins and cell death in situ localisation in hypopharyngeal glands of honeybee (*Apis mellifera carnica*) workers after imidacloprid or coumaphos treatment. Apidologie 41: 73–86.
110. Gregorc A, Ellis JD (2011) Cell death localization *in situ* in laboratory reared honey bee (*Apis mellifera* L.) larvae treated with pesticides. Pesticide Biochemistry and Physiology 99: 200–207.
111. Johnson RM, Dahlgren L, Siegfried BD, Ellis MD (2013) Acaricide, fungicide and drug interactions in honey bees (*Apis mellifera*). PLOS ONE 8: e54092.
112. Hawthorne DJ, Dively GP (2011) Killing them with kindness? In-hive medications may inhibit xenobiotic efflux transporters and endanger honey bees. PLOS ONE 6: e26796.
113. Lu C, Warchol KM, Callahan RA (2012) *In situ* replication of honey bee colony collapse disorder. Bull Insectology 65: 99–106.
114. Lu C, Warchol KM, Callahan RA (2014) Sub-lethal exposure to neonicotinoids impaired honey bees winterization before proceeding to colony collapse disorder. Bull Insectology 67: 125–130.
115. Wilson-Rich N, Dres ST, Starks PT (2008) The ontogeny of immunity: development of innate immune strength in the honey bee (*Apis mellifera*). J Insect Physiol 54: 1392–1399.
116. Bendahou N, Fleche C, Bounias M (1999) Biological and biochemical effects of chronic exposure to very low levels of dietary cypermethrin (Cymbush) on honeybee colonies (Hymenoptera: Apidae). Ecotoxicol Environ Saf 44: 147–153.
117. Pettis JS, Collins AM, Wilbanks R, Feldlaufer M (2004) Effects of coumaphos on queen rearing in the honey bee, *Apis mellifera*. Apidologie 35: 605–610.
118. Haarmann T, Spivak M, Weaver D, Weaver B, Glenn T (2002) Effects of fluvalinate and coumaphos on queen honey bees (Hymenoptera: Apidae) in two commercial queen rearing operations. J Econ Entomol 95: 28–35.
119. DeGrandi-Hoffman G, Chen Y, Simonds R (2013) The effects of pesticides on queen rearing and virus titers in honey bees (*Apis mellifera* L.). Insects 4: 71–89.
120. Collins AM, Pettis JS, Wilbanks R, Feldlaufer M (2004) Performance of honey bee (*Apis mellifera*) queens reared in beeswax cells impregnated with coumaphos. J Apic Res 43: 128–134.
121. Garrido PM, Antunez K, Martin M, Porrini MP, Zunino P, et al. (2013) Immune-related gene expression in nurse honey bees (*Apis mellifera*) exposed to synthetic acaricides. J Insect Physiol 59: 113–119.
122. vanEngelsdorp D, Tarpy DR, Lengerich EJ, Pettis JS (2013) Idiopathic brood disease syndrome and queen events as precursors of colony mortality in migratory beekeeping operations in the eastern United States. Prev Vet Med 108: 225–233.
123. Crailsheim K (1998) Trophallactic interactions in the adult honeybee (*Apis mellifera* L.). Apidologie 29: 97–112.
124. Elzen PJ, Elzen GW, Rubink W (2003) Comparative susceptibility of European and Africanized honey bee (Hymenoptera: Apidae) ecotypes to several insecticide classes. Southwest Entomol 28: 255–260.
125. Suchail S, Guez D, Belzunces LP (2000) Characteristics of imidacloprid toxicity in two *Apis mellifera* subspecies. Environ Toxicol Chem 19: 1901–1905.
126. Laurino D, Manino A, Patetta A, Porporato M (2013) Toxicity of neonicotinoid insecticides on different honey bee genotypes. Bull Insectology 66: 119–126.
127. Claudianos C, Ranson H, Johnson RM, Biswas S, Schuler MA, et al. (2006) A deficit of detoxification enzymes: pesticide sensitivity and environmental response in the honeybee. Insect Mol Biol 15: 615–636.
128. Johnson RM, Mao W, Pollock HS, Niu G, Schuler MA, et al. (2012) Ecologically appropriate xenobiotics induce cytochrome P450s in *Apis mellifera*. PLOS ONE 7: e31051.
129. Mao W, Schuler MA, Berenbaum MR (2011) CYP9Q-mediated detoxification of acaricides in the honey bee (*Apis mellifera*). Proc Natl Acad Sci USA 108: 12657–12662.
130. Suchail S, Debrauwer L, Belzunces LP (2004) Metabolism of imidacloprid in *Apis mellifera*. Pest Manag Sci 60: 291–296.
131. Johnson RM, Pollock HS, Berenbaum MR (2009) Synergistic interactions between in-hive miticides in *Apis mellifera*. J Econ Entomol 102: 474–479.
132. Faucon JP, Aurieres C, Drajnudel P, Mathieu L, Ribière M, et al. (2005) Experimental study on the toxicity of imidacloprid given in syrup to honey bee (*Apis mellifera*) colonies. Pest Manag Sci 61: 111–125.
133. Tennekes H, Sànchez-Bayo F (2012) Time-dependent toxicity of neonicotinoids and other toxicants: implications for a new approach to risk assessment. J Environ Anal Toxicol S4: S4-001.
134. Tennekes HA, Sànchez-Bayo F (2013) The molecular basis of simple relationships between exposure concentration and toxic effects with time. Toxicology 309: 39–51.
135. Laycock I, Cresswell JE (2013) Repression and recuperation of brood production in *Bombus terrestris* bumble bees exposed to a pulse of the neonicotinoid pesticide imidacloprid. PLOS ONE 8: e79872.
136. Decourtye A, Henry M, Desneux N (2013) Environment: Overhaul pesticide testing on bees. Nature 497: 188.
137. Osborne JL (2012) Ecology: Bumblebees and pesticides. Nature 491: 43–45.

# 5

# Use of Bed Nets and Factors That Influence Bed Net Use among Jinuo Ethnic Minority in Southern China

**Jian-wei Xu[1], Yuan-mei Liao[2], Hui Liu[1]*, Ren-hua Nie[1], Joshua Havumaki[3]**

**1** Yunnan Institute of Parasitic Diseases, Puer, China, **2** Health School of Ganzhou, Jiangxi Province, Ganzhou, China, **3** Foundation for Innovative New Diagnostics, Cointrin/Geneva, Switzerland

## Abstract

***Background:*** Insecticide-treated nets (ITNs) are an integral part of vector control recommendations for malaria elimination in China. This study investigated the extent to which bed nets were used and which factors influence bed net use among Jinuo Ethnic Minority in China-Myanmar-Laos border areas.

***Methods and Findings:*** This study combined a quantitative household questionnaire survey and qualitative semi-structured in-depth interviews (SDI). Questionnaires were administered to 352 heads of households. SDIs were given to 20 key informants. The bed net to person ratio was 1:2.1 (i.e., nearly one net for every two people), however only 169 (48.0%) households owned at least one net and 623 (47.2%) residents slept under bed nets the prior night. The percentages of residents who regularly slept under nets (RSUN) and slept under nets the prior night (SUNPN) were similar (48.0% vs. 47.2%, $P>0.05$), however the percentage correct use of nets (CUN) was significantly lower (34.5%, $P<0.0001$). The annual cash income per person (ACIP) was an independent factor that influenced bed net use ($P<0.0001$), where families with an ACIP of CNY10000 or more were much more likely to use nets. House type was strongly associated with bed net use (OR: 4.71, 95% CI: 2.81, 7.91; $P<0.0001$), where those with traditional wood walls and terracotta roofs were significantly more likely to use nets, and the head of household's knowledge was an independent factor (OR: 5.04, 95% CI: 2.72, 9.35; $P<0.0001$), where those who knew bed nets prevent malaria were significantly more likely to use nets too.

***Conclusions:*** High bed net availability does not necessarily mean higher coverage or bed net use. Household income, house type and knowledge of the ability of bed nets to prevent malaria are all independent factors that influence bed net use among Jinuo Ethnic Minority.

**Editor:** Roly D. Gosling, University of California, San Francisco, United States of America

**Funding:** The authors thank the fifth grant to China of the Global Fund to fight AIDS, Tuberculosis and Malaria for financial sponsorship. The funder had no role in study design, data collection and analysis, decision to publish, or preparation of the manuscript.

**Competing Interests:** The authors have declared that no competing interests exist.

* Email: liubible@126.com

## Introduction

Malaria is a global disease. The World Health Organization (WHO) estimated that between 2000 and 2010, global malaria incidence decreased by 17% and malaria-specific mortality rates decreased by 26%. Reported malaria cases have reduced by more than 50% in 34 of the 99 malaria endemic countries [1]. Currently, malaria prevalence is decreasing and 32 of the 99 countries have either declared a national policy for malaria elimination or are pursuing spatially progressive elimination within their borders [2,3]. Countries in the Asia Pacific region are making substantial progress towards eliminating malaria [4]. China has seen a great reduction in malaria burden, and is aiming to eliminate malaria by 2020 [5]. Vector control is one strategy that China is employing to eliminate malaria. This method requires that more than 90% of the population in ongoing transmission areas has at least one method of protection against mosquitoes. Therefore insecticide-treated nets (ITNs) are an integral component of the recommendations for vector control [6].

In China, Yunnan province has the heaviest burden of malaria among the country's 31 Provinces/Municipalities/Autonomous Regions, 1522 (34.0%) of 4479 malaria cases reported in the country in 2011 were from Yunnan [7]. China's border areas originally belonged to hyper-endemic malaria zones. The Jinuo People, whose language belongs to the Tibetan-Burman group of the Sino-Tibetan family [8], are an ethnic minority living in the China-Myanmar-Laos border area of the Yunnan Province in southern China. The burden of malaria is substantial amongst Jinuo People [9]. The Jinuo people used slash-and-burn agriculture for rice and maize before 1958. They then started small-scale rice cultivation in irrigated paddy fields. Rubber growing was introduced into the Jinuo Mountain Area in the 1990s. Currently, rubber typping is the main source of economic income in the region. Rubber tappers may bring malaria into the community. Workers go into the forest at 3:00–4:00 am to tap rubber. They may contract malaria in the forest and bring it into the community. The results of a survey conducted in 56 villages of Jinuo Mountain Township showed that 153 (10.3%) of 1500 residents had a malaria episode between March to November 1993. There was a 5.8% (30/513) parasite rate amongst febrile patients in Jinuo Mountain Township in 1993 [10]. This was significantly higher ($x^2 = 11.48$, $P = 0.0007$) when compared with the 3.1% (19489/620778) rate in Yunnan province [11] in 1993. The WHO Global Malaria Program has recommended full coverage of long lasting insecticidal nets (LLINs) in areas targeted for malaria prevention [12]. However, the use of bed nets among ethnic minorities is not well researched. This study investigated the extent bed nets including untreated nets, ITNs and LLINs are used and also which factors influence bed net use among the Jinuo Ethnic Minority from July to September, the high risk season of malaria, 2011.

## Methods

### Study site and population

This study was conducted on China-Myanmar-Laos border, Xishuangbanna prefecture in southern Yunnan, China (Figure 1). Jinuo tribe was recognized as an ethnic minority by the Chinese government in June 1979. It was the last of the 55 ethnic minority groups to be recognized in China. There are about 20,000 members of the Jinuo ethnic minority and they live in 56 villages on the Jinuo Mountain. The Jinuo Mountain has a tropical rain forest climate. Therefore, malaria is a substantial public health problem among the Jinuo people. Both *Plasmodium falciparum* and *P. vivax* are prevalent [7]. Malaria vectors are complex, and the primary vector *is Anopheles minimusis*. Peak malaria transmission occurs during the rainy seasons from May to November each year [13,14].

### Household survey

The study combined a quantitative household questionnaire and qualitative semi-structured in-depth interviews (SDIs). The data collection tools (questionnaires and SDI guidelines) were developed in Chinese because the Jinuo language is only a spoken language and most heads of households understand Chinese. When a respondent did not understand Chinese, a researcher from Jinuo Township Hospital who understands both Jinuo and Chinese conducted the interview in the Jinuo language and then filled the questionnaire in Chinese. Households were the units of sampling. The list of households for sampling was obtained through the three Villager Committee Offices in Jinuo Mountain. Ten out of the 56 villages were sampled by a simple computer randomization, and then every household in each selected village was surveyed. A sample size of 350 households was required for the 5% precision around a 35% point estimate for the proportion of community residents sleeping under nets the prior night and 95% of confidence limits. A household was defined as all those eating from the same cooking pot. The interviewers first introduced the purpose of the project, the topic and type of questions. Then an oral informed consent was obtained, and finally the questionnaire was administered to heads of households. When permission was obtained, all bed nets in each household were checked to determine how much they had been used and their textile integrity.

### Semi-structured in-depth interviews

Structure in-depth Interviews (SDIs) were administered to 20 key informants to explore potential factors which influenced bed net use. The key informants consisted of two staff members that were responsible for control of communicable diseases in the Jinuo Township Hospital, seven health workers who worked for the ten selected villages, six village leaders who were appointed by the local government and five representatives of the villagers who were selected by recommendations from their fellow villagers during the household survey. The key informants discussed local health problems, malaria incidence, perception of the villagers, symptoms and treatment seeking behavior, transmission and prevention of malaria, and finally factors which influence the use of bed nets.

### Concept definitions

The use of bed nets was categorized as regularly sleeping under bed nets (RSUN), sleeping under bed nets the prior night (SUNPN) and correct use of bed nets (CUN). RSUN was defined as a person who reported habitually using nets on a daily basis. CUN was defined as an individual who closes the net and puts the hemline under the mattress before sleeping. Annual cash income per person (ACIP) was used as a surrogate for socioeconomic status because ACIP is more easily obtained. Every household owns a rubber plantation, and rubber tapping is the main source of income in the region. Nets were classified in three ways; untreated nets, ITNs and LLINs. An untreated net was either never treated or had been treated with insecticide 12 or more months prior to the survey. An ITN was treated less than 12 months prior to the survey. There is a gradual loss of insecticide over time in LLINs, reducing their protective effect. LLINs therefore expire three or more years after their production; however the dates of LLIN production could not be identified in the survey, so the LLIN were therefore defined as having been owned for three years or less. Personal characteristics of family members (PCFM) and characteristics of households and household heads (CHHH) were examined separately to determine factors that influence bed net use. PCFM are characteristics of a person in each household. CHHH are characteristics of a household and head of household who responded during the interview survey.

### Data analysis

Double data entry and cleaning of quantitative data was done in Epidata 3.1. The dataset was then analyzed in EpiInfo 7. Household ownership, coverage and use of untreated nets, ITNs and LLINs were all analyzed. Chi-squared test compared the percentage of the bed net use across different demographic groups. Sleeping under nets the prior night (SUNPN) was used as the outcome variable to decrease recall bias. Personal characteristics of family members were independent variables; a multivariate logistic model was used to assess the association of personal variables and the use of bed nets. Meanwhile, SUNPN for at least one family member was the outcome variable and the characteristics of households and their heads of households (respondents) were independent variables. The multivariate logistic analysis was conducted to analyze these difference factors and their impact on the use of nets. Qualitative data analysis was carried out to explore potential factors which influence the use of bed nets. Qualitative data analysis was conducted with support of TAMS 3.0 software. The data were encoded on the basis of emerging themes and a codebook was created. Trends in the data were identified by producing matrices which combined and compared information from the different key informants.

### Ethical approval

According to the Helsinki Declaration, ethical approval for the study was granted by the Ethics Committee of Yunnan Institute of Parasitic Diseases, China. The Ethics Committee approved a verbal consent procedure as sufficient because the study was interview-based, and did not include any human specimens. The purpose and procedures of the study were explained and disclosed to all participants before obtaining informed consent. They could choose whether or not to participate in the study, and could also refuse a response to any question at any time. The participants were for oral consent at the start of the survey and were advised that they could skip questions or end the interview at any time.

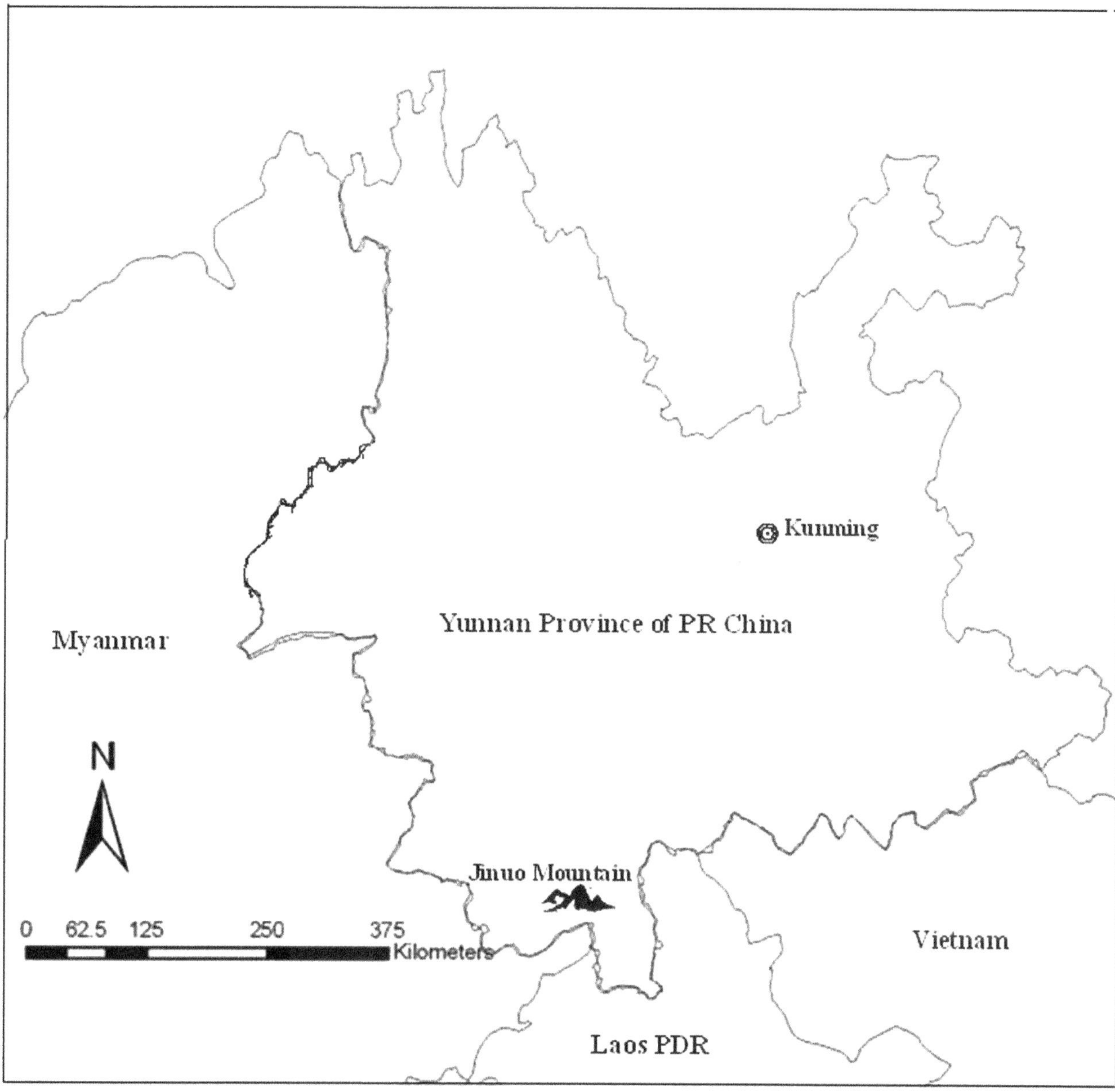


**Figure 1. The study site: Jinuo Mountain, Yunnan Province, China.**

Their consent was assumed if they did not refuse to answer questions. No one was coerced into participation in the study and if individuals wanted to withdraw from the study, they were allowed to do so without any issue.

## Results

### Characteristics of respondents

The total number of households was 482 across the ten sampled villages. 370 households were available at the time of sampling and 18 of them chose not to participate in the study. Three hundred and fifty two households with a total 1319 individuals participated in the questionnaire survey. The median of annual cash income per person (ACIP) of 352 households was CNY6667 (range CNY375–37500) or about US$1092 (range US$ 62–6148). Questionnaires were administered to 287 (81.5%) male and 65 (18.5%) female heads of households. The mean age of the 352 respondents was 41.3 (95% CI, 40.0, 42.5) years old. Three hundred and twenty one (91.2%) of the respondents were married. One hundred and eighty two (51.7%) of the respondents completed secondary school or higher, 154 (43.8%) had completed primary school and 16 (4.5%) were illiterate. The median number of family members was four (range 1–9) persons. The twenty key informants who were interviewed were composed of 10 males and 10 females ranging from 26 to 52 years old.

### Ownership and types of the bed nets

Ownership of bed nets was low among the Jinuo ethnic minority. One hundred and sixty nine (48.0%) households owned at least one net, reporting a total of 633 nets amongst all households. Three hundred and fifty eight (56.6%) nets were untreated nets s, 126(19.9%) were ITNs and 149 (23.5%) were

LLINs. The net to person ratio was 1:2.1(nearly one net for every two people). According to villager-self report in the survey, 482 (76.1%) of the nets were obtained by the villagers themselves commercially. Based on records of LLIN distribution in Jinuo Township Hospital, only 149 (23.5%) of free LLINs were delivered by the national malaria elimination program. One hundred and forty five (41.2%) of households owned at least one ITN or one LLIN. Eighty four (13.3%) nets were cotton textile and 549 (86.7%) were polyester-based. One hundred and thirty three (21.0%) nets had at least one hole (Figure 2).

## Use of the bed nets

The rate of bed net use was low; the key informants estimated that only half of residents consistently slept under the nets. Results from the survey revealed that 686 (52.0%) residents had never slept under nets, 633 (48.0%) residents regularly slept under the nets (RSUN), and 623(47.2%) residents slept under the nets the prior night (SUNPN). The SUNPN group included 353 (26.8%) residents who used untreated nets, 124 (9.4%) used ITNs and 146 (11.1%) used LLINs. Total RSUN and SUNPN use was similar ($P>0.05$), however the proportion (34.5%)of correct use of the nets (CUN) was significantly lower ($P<0.0001$) than RSUN and SUNPN (table 1).

Bed net use was then categorized by age, educational status and annual cash income per person (ACIP). The percentage of CUN was significantly lower than RSUN and SUNPN for all categories except for those who had completed senior middle or higher education. Rates of bed net use including RSUN, SUNPN and CUN were approximately equal between males and females ($P>0.05$). All five villager representatives said that "The oldest and the ill residents commonly slept under the nets more". Additionally, children under five years old had a slightly higher rate of use than other age categories ($P<0.05$). It was observed that children under five years old slept under the nets more correctly than those aged 50 years or older, older people usually paid less attention to CUN and the maintenance of bed nets. Education status did not significantly predict differences in the bed net use ($P>0.05$). However, bed net use rates were significantly different in the four categories of economic status. Families with an ACIP of CNY10000 or more were much more likely to use nets, their rates of RSUN, SUNPN and CUN were 88.89%, 87.58% and 72.55% respectively (table 1).

## Influence factors on use of the bed nets

In summary of the qualitative study, only people of Balaxia Village considered it important to use nets; however people in nine out of the ten selected villages did not feel it was important to sleep under bed nets. Reasons include that they felt it was too hot to sleep under bed nets, they did not feel at risk for malaria infection, they could burn mosquito coils and spray insecticides instead of bed net use, they found hanging bed nets to be troublesome, and others said that they didn't have enough money to buy bed nets. The results of the multivariate logistic regression analysis (MVLRA) showed that economic status was strongly associated with bed net use. The three poorest categories were significantly less likely to use nets than those with more than CNY10000 ACIP ($P<0.0001$). Residents who slept outside at night were slightly less likely to use nets than those that didn't (OR: 0.58, 95% CI: 0.32, 0.94; $P = 0.047$) in the hot season (table 2).

The qualitative information from the SDIs and household visits indicated that people living in modern houses used nets less because there are fewer holes reducing vector entry in modern houses. Additionally, they tended to use other measures such as window screens and mosquito coils. MVLRA of household variables and respondents identified that house type was strongly associated with use of the bed nets (OR: 4.71, 95% CI: 2.81, 7.91; $P<0.0001$), where those with traditional wood walls and terracotta roofs were significantly more likely to sleep under nets the prior night. Additionally, respondents living in houses with windows and screen doors were less likely to use nets (OR: 0.54, 95% CI: 0.29, 0.99; $P = 0.045$) (table 3).

One villager head commented, "Providing information on the risk of mosquitoes such as transmission of malaria and other communicable diseases, communicating on the benefits of the bed nets, especially ITNs and LLINs and providing free bed nets will promote the use of nets". Results of MVLRA showed that bed net use was not associated with head of household's knowledge about malaria transmission, however it was closely related to the knowledge that the bed nets prevent malaria infection (OR: 5.04, 95% CI: 2.72, 9.35; $P<0.0001$), where those who knew bed nets preventing malaria were significantly more likely to use nets

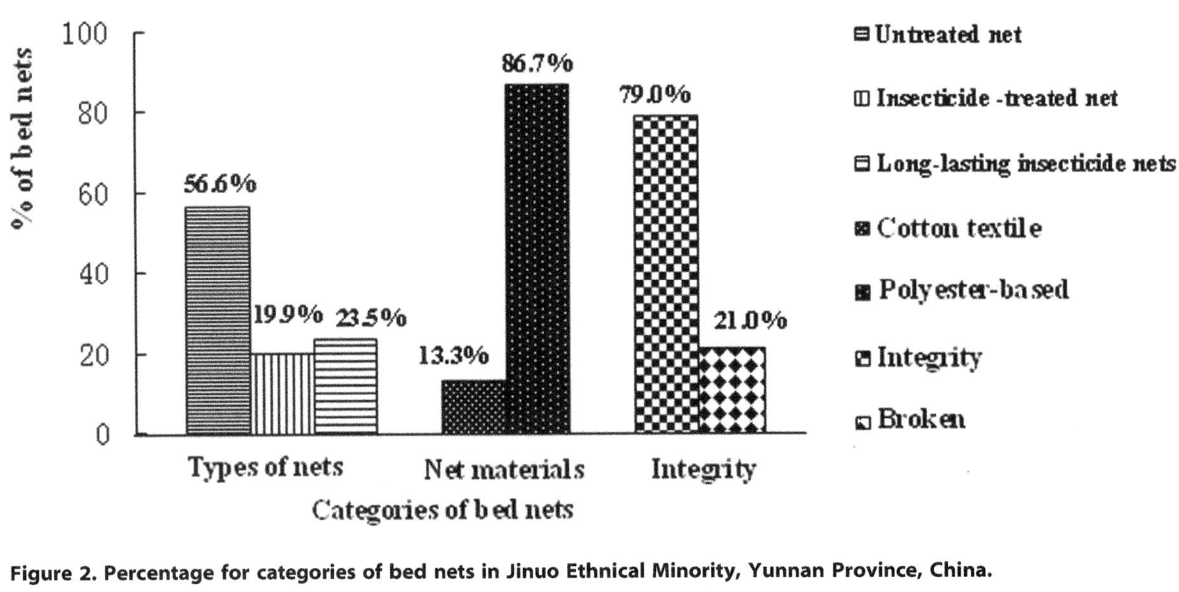


**Figure 2. Percentage for categories of bed nets in Jinuo Ethnical Minority, Yunnan Province, China.**

**Table 1.** Bed net use among different groups of Jinuo People, Yunnan, China.

| | Regular | Prior night | Correct | $x^2$ value | P-value |
|---|---|---|---|---|---|
| **Sex** | | | | | |
| Male (n = 671) | 311 (46.6%) | 305 (45.5%) | 224 (33.3%) | 28.94 | <0.0001 |
| Female (n = 648) | 322 (49.7%) | 318 (49.1%) | 231 (35.7%) | 33.00 | <0.0001 |
| $x^2$ value | 1.48 | 1.73 | 0.75 | | |
| P-value | 0.2244 | 0.1881 | 0.3869 | | |
| **Age** | | | | | |
| ≤5 (n = 98) | 55 (56.4%) | 55 (56.4%) | 41 (41.8%) | 5.34 | 0.0455 |
| 5–14(n = 91) | 42 (46.2%) | 42 (46.2%) | 27 (29.7%) | 6.83 | 0.0219 |
| 15–49(n = 851) | 390 (48.8%) | 379 (44.5%) | 300 (35.3%) | 23.27 | <0.0001 |
| ≥50(n = 279) | 146 (52.3%) | 147 (52.7%) | 87 (31.2%) | 34.13 | <0.0001 |
| $x^2$ value | 6.42 | 8.96 | 4.85 | | |
| P-value | 0.0930 | 0.0298 | 0.1834 | | |
| **Educational status** | | | | | |
| Illiterate (n = 266) | 137 (51.5%) | 139 (52.3%) | 90 (33.8%) | 23.29 | <0.0001 |
| Primary (n = 425) | 208 (48.9%) | 206 (48.5%) | 135 (31.8%) | 33.19 | <0.0001 |
| Junior mid (n = 514) | 242 (47.1%) | 234 (45.5%) | 190 (37.0%) | 12.43 | 0.0010 |
| Senior mid or higher (n = 114) | 46 (40.5%) | 44 (38.6%) | 40 (35.1%) | 0.69 | 0.4123 |
| $x^2$ value | 4.31 | 6.97 | 2.86 | | |
| P-value | 0.2303 | 0.0730 | 0.4139 | | |
| **Annual cash income per person (CNY)** | | | | | |
| ≤6000 (n = 241) | 78 (32.4%) | 74 (30.7%) | 49 (20.3%) | 10.21 | 0.0027 |
| 6001–8000 (n = 453) | 229 (50.6%) | 228 (50.3%) | 165 (36.4%) | 23.91 | <0.0001 |
| 8001–10000 (n = 472) | 190 (40.3%) | 187 (39.6%) | 130 (27.5%) | 21.07 | <0.0001 |
| >10000 (n = 153) | 136 (88.9%) | 134 (87.6%) | 111 (72.6%) | 17.89 | 0.0001 |
| $x^2$ value | 138.62 | 139.08 | 130.29 | | |
| P-value | <0.0001 | <0.0001 | <0.0001 | | |
| **Total (n = 1319)** | **633 (48.0)** | **623 (47.2%)** | **455 (34.5%)** | | **<0.0001** |

(table 3). A village health worker in Balaxia, the site of a malaria outbreak in 2000, said, "In the malaria outbreak, many people contracted malaria. The outbreak improved awareness and knowledge of malaria among the villagers. They know bed nets are effective in preventing malaria. Currently, in our village, bed nets are available in every house and most people use them. Furthermore, bed net use has become part of our normal living habits. Most of the villagers, including those living in modern houses, always sleep under the nets". The village head of Balaxia commented "Not only do nets prevent mosquito bites, but they also prevent nuisances caused by other pests such as cockroaches, bedbugs and fleas. Additionally, the nets are dustproof, windproof and preserve warmth in the winter".

## Discussion

Yunnan province has one of greatest burdens of malaria in China, and Jinuo Mountain is one of the highest endemic areas in China. Of 352 households visited, 169 of them had any nets. Together these 169 houses owned 633 nets. Only 47.2% of 1319 residents surveyed had slept under them the previous night. Two hundred and fifty (19.0%) of the nets were ITNs or LLINs. One hundred and forty five (41.2%) households owned at least one LLIN or ITN. The results of a survey for the Fifth Round of China Global Fund Malaria Project in 2007 showed that 65.8% (25797/39234) of households owned at least one net and 8.68%(3404/39234) owned at least one ITN in Yunnan province [15]. With the implementation of The National Malaria Control Program, coverage and use of ITNs and LLINs increased in Yunnan. 89.7% (278/310) of households owned at least one LLIN or ITN and 30.6% (789/2582) had slept under LLINs or ITNs the previous night by 2010 [16]. This shows that the ownership of general nets, ITNs and LLINs, and the use of ITNs and LLINs among Jinuo ethnic minority are lower than the mean in the rest of Yunnan. The results of the study also indicated that Jinuo people did not have equal to access bed nets because 482 (76.1%) of the nets used were procured by villagers themselves commercially, and only 149 (23.5%) of LLINs were free from the national malaria elimination program.

More than 80% of the residents in Tanzanian households which have a net to person ratio better than 1:4 reported using a net the previous night [17]. While a net to person ratio of 1:2.1 among the Jinuo people only ensured that 47.2% of people slept under bed nets the previous night. MVLRA identified that annual cash income per person (ACIP) was significantly associated with the bed net use ($p<0.0001$). Overall, free distribution of bed nets can increase their coverage and equity. When commercial sectors are responsible for bed net availability, the ownership and use of nets is concentrated among the richest families [18]. For instance, the

**Table 2.** Personal Variables related to use of bed nets in Jinuo people, Yunnan, China.

| | SUNPN*(%) | Univariate OR (95% CI) | Adjusted OR (95% CI) | P values |
|---|---|---|---|---|
| **Sex** | | | | |
| Male (n = 671) | 305 (45.5%) | 0.86(0.69–1.08) | 0.79 (0.54–1.99) | 0.337 |
| Female (n = 648) | 318 (49.1%) | 1 | 1 | |
| **Age (years)** | | | | |
| <15 (n = 189) | 97 (51.3%) | 0.95(0.64–1.39) | 0.87(0.60–1.55) | 0.915 |
| 15–49 (n = 851) | 390 (48.8%) | 0.76(0.57–1.01) | 0.68(0.47–1.42) | 0.169 |
| ≥50 (n = 279) | 147 (52.3%) | 1 | 1 | |
| **Educational status** | | | | |
| Illiterate & prim (n = 691) | 345 (49.9) | 1.26 (1.00–1.57) | 1.58(0.91–1.68) | 0.145 |
| Mid or higher (n = 628) | 278 (44.3) | 1 | 1 | |
| **Annual cash income per person (CNY)** | | | | |
| ≤6000 (n = 241) | 74 (30.7%) | 0.06 (0.03–0.11) | 0.10(0.08–0.58) | <0.0001 |
| 6001–8000 (n = 453) | 228 (50.3%) | 0.14(0. 08–0.25) | 0.18(0.12–0.59) | <0.0001 |
| 8001–10000 (n = 472) | 187 (39.6%) | 0.09 (0.05–0.16) | 0.12 (0.07–0.46) | <0.0001 |
| >10000 (n = 153) | 134 (87.6%) | 1 | 1 | |
| **Rest outdoor at night of hot season** | | | | |
| Yes (n = 307) | 123(40.2) | 0.66(0.50–0.86) | 0.58(0.32–0.94) | 0.047 |
| No (n = 1012) | 510(50.4) | 1 | 1 | |

*SUNPN = sleeping under a net the prior night.

Wa people in a focus group discussion in Yunnan said "if nets were provided bed nets freely, we would use them" [19]. On other hand, more than other vector control methods, ITN programs largely depend on the acceptance and active involvement of communities. Involvement of communities in promoting use of bed nets also depends on their knowledge, perception, attitude and behavior towards nets [20,21]. Knowledge, information and distribution of free bed nets are established as the first step towards encouraging the use of nets [22]. The MVLRA did not identify the association between the net use and head of household's knowledge of the cause of malaria and its transmission. However, the knowledge that bed nets prevent malaria was an independent factor for bed net use. This indicated that among the Jinuo ethnic minority, direct knowledge worked better than more technical knowledge such as the cause of malaria and the life cycle of parasites. When considering strategies to encourage behavioral changes among the Jinuo people, special attention should be given to what points should be communicated. On other hand, knowledge itself was not sufficient to guarantee bed net use. This is because knowledge does not always confer a certain behavior or action. Individual perception, awareness, the availability of nets and an environment which enables net use are necessary for behavioral change [23–25]. House type and availability of windows and screen doors was strongly associated with bed net use among the Jinuo Ethnical Minority. Additionally, residents who slept outdoors at night during the hot season were less likely to use nets than others. They said "sleeping under nets is uncomfortable and hanging a net is a burden" during the SDIs. The Wa people also said "sleeping under a net is too hot, so villagers prefer not to use it" [19]. These results indicate that human, socio-economic and the environmental factors can all affect the use of bed nets. Additionally, the nuisance of mosquitoes and perceived threat of malaria were the main determinants of bed net use [23]. In Balaxia village, most people have been sleeping under bed nets since the outbreak of malaria in 2000. Whereas in other nine villages, people did not sleep under bed nets because they felt like the risk of malaria was very low. The villager representatives commented, "Malaria was common ten years ago, but now it is difficult to find or has disappeared entirely".

The epidemiological trends of malaria have changed in countries where it is targeted for elimination which are often predominantly adult men, with shared social, behavioral, and geographical risk characteristics [26]. Most Jinuo families earn their income from rubber tapping. Adults go the forest at 3:00–4:00 am to tap rubber. Therefore bed nets are unlikely to prevent malaria transmission for these rubber tappers since they are exposed at night. They might then bring malaria from the forest into the community. Bed net use within this community can prevent malaria transmission from the malaria infected workers to others. Therefore, bed net use within this community is very important for both these workers and other villagers. On the other hand, other measures, such as topical and spatial repellents for rubber tappers are needed. Additionally, among villagers, long-lasting insecticidal hammocks and mosquito traps should be used before sleeping and throughout the night for those that sleep outside [27,28].

The study had several limitations. Firstly, considering the low prevalence of pregnant women and low malaria incidence in general, information on pregnant women was not specifically gathered. Therefore, the study has not identified whether pregnant women were more likely to sleep under nets. Secondly, if demographic groups are not similar, bias might be introduced [29]. Selection bias and information bias may have influenced the study results. One hundred and twelve (23.2%) households were not available at the time of sampling, and 18 (3.7%) households chose not to participate in the study. However it is not known why residents were away from their homes or why they opted out of participating in the study. This might have led to selection bias.

**Table 3.** Household variables and respondents (Head of Household) related to use of nets in Jinuo ethnic minority, Yunnan, China.

| | SUNPN1*(%) | Univariate OR (95% CI) | Adjusted OR (95% CI) | P values |
|---|---|---|---|---|
| **Sex of respondents** | | | | |
| Male (n = 287) | 142(49.5) | 0.97(0.61–1.55) | 0.99(0.54–1.74) | 0.934 |
| Female(n = 65) | 27(41.5) | 1 | 1 | |
| **Age (years) of respondents** | | | | |
| **<35 (n = 101)** | 48(47.5) | 1.38(0.80–2.38) | 1.27(0.78–2.44) | 0.457 |
| **≥35 (n = 251)** | 121(48.2) | 1 | 1 | |
| **Marriage status of respondents** | | | | |
| **Married (n = 321)** | 156(48.6) | 1.31(0.62–2.76) | 1.06 (0.47–2.88) | 0.696 |
| **Single (n = 31)** | 13(41.9) | 1 | 1 | |
| **Family size** | | | | |
| **≤5 (n = 197)** | 91(46.2) | 0.85(0.56–1.29) | 0.88(0.54–1.36) | 0.508 |
| **>6 (n = 155)** | 78(50.3) | 1 | 1 | |
| **Family decision** | | | | |
| **Husband (n = 175)** | 88(50.3) | 1.12(0.79–1.82) | 1.20(0.77–1.89) | 0.458 |
| **Wife or co-decision (n = 177)** | 81(45.8) | 1 | 1 | |
| **House types** | | | | |
| **Traditional (n = 173)** | 118(68.2) | 5.38(3.41–8.50) | 4.71(2.81–7.91) | <0.0001 |
| **Modern (n = 104)** | 51(28.5) | 1 | 1 | |
| **Window and door screen** | | | | |
| **Yes (n = 91)** | 25(27.5) | 0.31(0.18–0.52) | 0.54(0.29–0.99) | 0.045 |
| **No (n = 261)** | 144(55.2) | 1 | 1 | |
| **Household heads knew malaria cause** | | | | |
| **Yes (n = 185)** | 98(53.0) | 1.53(1.00–2.32) | 0.83(0.42–1.67) | 0.607 |
| **No (n = 167)** | 71(42.5) | 1 | 1 | |
| **Household heads knew malaria transmission** | | | | |
| **Yes (n = 202)** | 108(53.5) | 1.68(1.09–2.57) | 1.32(0.64–2.69) | 0.452 |
| **No (n = 150)** | 61(40.7) | 1 | 1 | |
| **Household heads knew bed nets against malaria infection** | | | | |
| **Yes (n = 244)** | 108(53.5) | 4.47(2.68–7.43) | 5.04(2.72–9.35) | <0.0001 |
| **No (n = 108)** | 61(40.7) | 1 | 1 | |

*SUNPN1 = sleeping under a net the prior night at least one family member.

However, the selection bias would be limited because 73.1% of households were involved in the study. Despite the analysis adjusting for sex and age to identify independent influence factors, a higher proportion (81.5%) of male respondents might have also caused selection bias. Data on relatively sensitive topics, such as annual cash income per person (ACIP) was obtained by self-reporting. This is a potential source information bias. In the interviews, the validity of ACIP was confirmed through the number of rubber trees owned by each household.

In Conclusion, high bed net availability does not necessarily mean higher coverage or bed net use. The household income, house type and knowledge of the ability of bed nets to prevent malaria are independent influence factors which influence bed net use among Jinuo Ethnic Minority.

## Acknowledgments

We are grateful to Dr Kang Zhu from Jinuo Township Hospital for his interviewing the respondents who could not understand Chinese; and Dr Qiang Huang from Jinghong Center for Disease Control and Prevention for his coordination of the field survey. The study would not have been possible without the generous and willing cooperation of the study population.

## Author Contributions

Conceived and designed the experiments: JWX YML HL. Performed the experiments: JWX YML HL RHN. Analyzed the data: JWX YML JH HL. Wrote the paper: JWX YML JH HL RHN.

## References

1. WHO (2011) World malaria report 2011. Geneva: World Health Organization.
2. Feachem RG, Phillips AA, Hwang J, Cotter C, Wielgosz B, et al. (2010) Shrinking the malaria map: progress and prospects. Lancet 376: 1566–1578.
3. Tatem AJ, Smith DL, Gething PW, Kabaria CW, Snow RW, et al. (2010) Ranking of elimination feasibility between malaria-endemic countries. Lancet published online Oct 29. DOI: 10.1016/S0140-6736(10)61301-3.
4. Gosling RD, Whittaker M, Gueye CS, Fullman N, Baquilod M, et al. (2012) Malaria elimination gaining ground in the Asia Pacific. Malaria Journal 11:346.

5. Ministry of Health, People's Republic of China (2010) Malaria elimination action plan 2010–2020 (in Chinese). Beijing: The Ministry of Health.
6. China Centre for Diseases Control and Prevention (2011) Technical guidelines for malaria elimination (in Chinese). Beijing: China Centre for Diseases Control and Prevention.
7. Xia ZG, Yang MN, Zhou SE (2012) Malaria Situation in the People's Republic of China in 2011. Chin J Parasitol Parasit Dis (abstract in English) 30: 419–422.
8. Gai X (1986) Jinuo-yu Jianzhi (Introduction of the Jinuo Language). Kunming: Minzu Press.
9. Xu JW, Liu H (1997) Border malaria in Yunnan, China. Southeast Asian J Trop Med Public Health 28:456–459.
10. Zhou HN, Dong LM (1994) An epidemiological survey on malaria in Jinuo Mountain. J practical parasit dis (in Chinese) 2:44–45
11. Zhu DF (1997) The implementation of the eighth malaria control program and situation analysis in Yunnan Province. J practical parasit dis (in Chinese) 5:45–46
12. WHO (2007) WHO Global Malaria Programme: Position Statement on ITNs. Geneva: World Health Organization.
13. Dong XS (2000) The malaria vectors and their ecology in Yunnan Province (abstract in English). Chin J Parasit Dis 13: 144–147.
14. Dong XS (2010) The mosquito fauna of Yunnan Province. Kunming: Yunnan Science and Technology Press (abstract in English): 49–84.
15. Zhou S, Yang R, Lv Q, Yang ZH, Du LF et al. (2009) Situation analysis of insecticide treatment bed net in Yunnan malaria endemic areas and its generalization strategy. Chin J Vector Bio & Control (abstract in English) 20: 326–328.
16. Li BF, Yang YM, Xu JW, Chen GW, Zhou S et al. (2013) Baseline investigation of the national malaria elimination program and projects funded by the Global Malaria Fund in Yunnan Province. J Pathogen Bio (abstract in English) 8:448–472.
17. Tsuang A, Lines J, Hanson K (2010) Which family members use the best nets? An analysis of the condition of mosquito nets and their distribution within households in Tanzania. Malaria Journal 9:211
18. Noor A M, Amin AA, Akhwale WS, Snow WS (2007) Increasing coverage and decreasing inequity in insecticide-treated bed net use among rural Kenyan children. PLoS Medicine 4(8/e255): 1341–1348.
19. Xu JW, Xia M, Tao H, Zhong YC, Wu XH et al. (2004) Current situation, attitudes and scaling up obstacles toward usage of bednets in Wa ethnic. China Trop Med 4:319–332.
20. Stewart T and Marchand RP (2002) Factors that affect the success and failure of insecticide treated net programs for malaria control in SE Asia and the Western Pacific. Manila: WHO/WPRO.
21. Xu JW, Wu XH, Zhong YC (2004) Bednet use and influential factors of two ethnical minorities at China-Myanmar border (abstract in English). Chin J Schisto Control 16:51–53.
22. Mishra RN, Singh SP, Vanlerberghe V, Sundar S, Boelaert M, et al. (2010) Lay perceptions of kala-azar, mosquitoes and bed nets in Bihar India. Trop Med Int Health 15 (suppl. 2): 36–41.
23. Atkinson AJ, BobogareA, Fitzgerald L, Boaz L, Appleyard B, et al. (2009) A qualitative study on the acceptability and preference of three types of long-lasting insecticide-treated bed nets in Solomon Islands: implications for malaria elimination. Malaria Journal 8:119.
24. Xu JW, Wu XH, WEI C, Petlueng P, Tao H, et al. (2010) A pilot study on strengthening malaria control for ethnic minorities (abstract in English). Chin J Vector Biol Control 21:527–530.
25. Xu JW, Liu H (2012) The Challenges of malaria elimination in Yunnan Province, People's Republic of China. Southeast Asian J Trop Med Public Health 43: 819–824.
26. Cotter C, Sturrock HJW, Hsiang MS, Liu J, Phillips AA, et al. (2013) The changing epidemiology of malaria elimination: new strategies for new challenges. Lancet, published online Apr 15. http://dx.doi.org/10.1016/S0140-6736(13)60310-4.
27. Sochantha T, Van Bortel W, Savonnaroth S, Marcotty T, Speybroeck N, et al. (2010) Personal protection by long-lasting insecticidal hammocks against the bites of forest malaria vectors. Trop Med Int Health 15: 336–341.
28. Thang ND, Erhart A, Speybroeck N, Xa NX, Thanh NN, et al. (2009) Long-lasting insecticidal hammocks for controlling forest malaria: a community-based trial in a rural area of Central Vietnam. PLoS One 4: e7369.
29. Ward KA, Spokes PJ, McAnulty JM (2011) Case-control study of risk factors for hospitalization caused by pandemic (HIN1) 2009. Emerging Infectious Diseases 17: 1409–1416.

# 6

# Quantifying Synergy: A Systematic Review of Mixture Toxicity Studies within Environmental Toxicology

**Nina Cedergreen***

Department of Plant and Environmental Sciences, University of Copenhagen, Frederiksberg, Denmark

## Abstract

Cocktail effects and synergistic interactions of chemicals in mixtures are an area of great concern to both the public and regulatory authorities. The main concern is whether some chemicals can enhance the effect of other chemicals, so that they jointly exert a larger effect than predicted. This phenomenon is called synergy. Here we present a review of the scientific literature on three main groups of environmentally relevant chemical toxicants: pesticides, metal ions and antifouling compounds. The aim of the review is to determine 1) the frequency of synergy, 2) the extent of synergy, 3) whether any particular groups or classes of chemicals tend to induce synergy, and 4) which physiological mechanisms might be responsible for this synergy. Synergy is here defined as mixtures with minimum two-fold difference between observed and predicted effect concentrations using Concentration Addition (CA) as a reference model and including both lethal and sub-lethal endpoints. The results showed that synergy occurred in 7%, 3% and 26% of the 194, 21 and 136 binary pesticide, metal and antifoulants mixtures included in the data compilation on frequency. The difference between observed and predicted effect concentrations was rarely more than 10-fold. For pesticides, synergistic mixtures included cholinesterase inhibitors or azole fungicides in 95% of 69 described cases. Both groups of pesticides are known to interfere with metabolic degradation of other xenobiotics. For the four synergistic metal and 47 synergistic antifoulant mixtures the pattern in terms of chemical groups inducing synergy was less clear. Hypotheses in terms of mechanisms governing these interactions are discussed. It was concluded that true synergistic interactions between chemicals are rare and often occur at high concentrations. Addressing the cumulative rather than synergistic effect of co-occurring chemicals, using standard models as CA, is therefore regarded as the most important step in the risk assessment of chemical cocktails.

**Editor:** Aamir Nazir, CSIR-Central Drug Research Institute, India

**Funding:** No funding provided apart from my own salary from the University of Copenhagen. The funders had no role in study design, data collection and analysis, decision to publish, or preparation of the manuscript.

**Competing Interests:** The author has declared that no competing interests exist.

* E-mail: ncf@plen.ku.dk

## Introduction

### Background

Cocktail effects and synergistic interactions of chemicals in mixtures are an area of great concern to both the public [1,2] and regulatory authorities in the US and Europe ([3] and references therein). There are two general aspects underlying this concern: The first is the uncertainty as to whether we are monitoring and regulating the most harmful chemicals? The second concerns whether the chemicals we regulate on a single compound basis, and deem "safe", potentiate or are being potentiated by other chemicals so that they jointly exert a larger effect than predicted? The latter is called synergy, and is one of the factors that create uncertainty around models proposed for the implementation in chemical risk assessment of mixtures. For those legislations where mixtures are considered, which are few, dose- or concentration additivity is proposed as the default model [3]. But are synergistic interactions really an area that should concern us? Earlier reviews have shown that synergistic interactions, at least within pesticide mixtures and realistic low-dose chemical mixtures in mammals, are a rather rare phenomenon, constituting approximately 5% of the tested mixture combinations [4–7]. This percentage is rather low given the fact that experiments are often designed to search for synergistic interactions, thereby biasing the databases towards synergistic interactions. If, however, these 5% are combinations that often co-occur in humans and the environment, they might nonetheless be of quantitative importance. Hence, if we could identify the groups of chemicals that are likely to induce synergistic interactions, special precautions could be taken in the risk assessment of these chemicals. Identifying the potential synergists would reduce the uncertainty of using the models proposed for risk assessment of mixtures of the remaining 95% of antagonistic or non-interacting chemicals [3].

The aim of this review is therefore to define which groups of chemicals are involved in well documented synergistic interactions, and if possible, to identify the mechanisms behind their synergistic effects. This will be done within three large groups of chemicals that often co-occur in the environment at measurable concentrations: The first group consists of pesticides, which is probably the most well studied chemical group within ecotoxicological mixtures studies. This is not only due to the use of chemical mixtures in pesticide formulations and tank mixtures and the resulting co-occurance in agricultural areas, but just as much because of the in depth knowledge of their physiological mode of action. This makes them ideal candidates for testing mixture models based on chemical mode of action and understanding the physiological mechanisms behind possible interactions [8,9]. The second group of chemicals are metals. Metals typically co-occur in potentially toxic concentration in relation to mining, smelting and other industrial activities and a substantial body of literature on metal mixtures is available [10]. The last group are antifouling biocides, which consist both of traditional organic biocides, organo-metals and metal ions [11], making this group a mixture of the two above thereby opening the possibility of finding other synergistic mechanisms. Antifoulants co-occur in harbour areas and marine and freshwater areas with substantial boat traffic [12,13].

Chemical mixtures from waste water treatment plants, oil spills, industrial effluents and other sources yielding very complex mixtures have not been included for two reasons: The first is that they are often chemically very poorly described; hence, we do often not know which chemicals cause the majority of the toxicity [14,15]. The second is that the probability for severe interactions decrease when the number of pollutants adding to the joint toxicity increase [16,17], hence, severe interactions are more likely to occur when a few chemicals dominate the overall toxicity, as is more often seen for e.g. pesticide pollution, in comparison to effluent pollution [18].

## The Definition of Synergy

Defining synergy as two or more chemicals exerting a larger effect than predicted implies that we can predict joint effects of chemicals under certain assumptions. The aim of being able to do so, has been a research topic for more than a century [19], and the two major concepts underlying all valid assessments of joint chemical effects were framed already in the first part of the twentieth century by Loewe and Muischnek (1926) and Bliss (1939), respectively [20,21]. Loewe and Muischnek (1926) based their concept on the assumption that all chemicals in a mixture acted on the same biological target site and therefore could be viewed as being dilutions of each other, each having a different chemical potency. Hence, if the chemical potency of chemical A and B in a binary mixture was based on the Effect Concentration (*EC*) of each chemical causing *x*% effect on any endpoint in a test-system ($1/EC_{xA}$ and $1/EC_{xB}$), then the sum of the concentration of chemicals ($c_A$ and $c_B$) multiplied with their respective potency in a mixture provoking x% effect would be equal to 1 [20]:

$$\frac{c_A}{EC_xA} + \frac{c_B}{EC_xB} = 1 \qquad (1)$$

The concept has been re-invented several times since 1926 and has received many names such as Loewe Additivity, Dose Addition, The Additive Dose Model or Concentration Addition, depending on whether it has been used within pharmaceutical, agricultural, toxicological or ecotoxicological sciences [22]. In this review we will use the term Concentration Addition (CA). Bliss (1939) worked with test-systems where mortality was the endpoint, and added another way of looking at mixtures, in the cases where the tested chemicals obviously did not affect the organisms through a similar molecular target. Different target sites were by Bliss defined by their concentration-response curves having different shapes [21]. Bliss viewed death by a chemical as a stochastic event. The probability of surviving or dying due to exposure to several chemicals acting on independent targets in the organism could therefore be calculated based on probabilities of surviving or dying from exposure to the individual chemicals [21]. Hence, the probability of surviving two independently acting chemicals ($R_{mix}$) would be equal to the probability of surviving the first chemical ($R_1$) multiplied by the probability of surviving the second chemical ($R_2$). Or, if assessing the probability of dying from two independently acting chemicals ($E_{mix}$), this is equal to the probability of dying from the first chemical ($E_1$) plus the probability of dying from the second chemical ($E_2$), minus the probability of dying from both chemicals ($E_1 \times E_2$) [21].

$$E_{mix} = E_A + E_B - E_A E_B \qquad (2)$$

This concept has likewise been re-invented several times and has been named Bliss Independence, Response Multiplication, Response Addition, Effect Addition, Independent Action a.o. depending on the inventor and context [22]. In this review we will use the term Independent Action (IA). Both concepts can be extended to an infinite number of chemicals and can be used to predict mixture toxicity effects of all mixture ratios and effect levels, providing that entire dose- or concentration response relationships for the single chemicals in the desired test-system are available. Often such data are not available and reduced approaches must be used. A recent review of mixture models and their uses can be found in Cedergreen et al. (2013) [22]. How they are proposed to be used in different chemical legislation is reviewed by Backhaus and Faust (2010) [3]. Common for both concepts is also the assumption that the chemicals do not interact chemically or affect the toxicity of each other [20,21]. If the chemicals do interact, the joint effects might deviate from the predictions resulting either in the before mentioned synergistic effects or in antagonistic effects, which are defined as smaller effects than predicted [22].

Synergy can therefore be defined in relation to two basic concepts: CA and IA. Empirical evidence, however, shows that even mixture toxicity of dissimilarly acting compounds can be described with a high level of accuracy with CA, as well as with IA, despite their different underlying assumptions [6,23]. CA generally generate slightly more conservative predictions (predicting larger effects than IA), and as databases on chemicals often only provides $EC_x$ data or No Observable Effect Concentrations (NOECs) or Lowest Observable Effect Concentrations (LOECs) which only makes CA predictions possible and not IA, CA is most often the recommended model for risk assessment purposes [3]. In this review, synergy is therefore defined in relation to CA predictions.

Experimental data are always determined with variance. For mixture studies this applies both to the toxicity data of the individual compounds used to make the model prediction, and to the tested mixture toxicity data. The consequence of this is that small deviations from the reference models can be difficult to detect statistically and repeat experimentally [24]. Biologically significant and reproducible synergy is therefore here defined as a more than two fold deviation from CA, as was also proposed by Belden et al (2007) [5]. That is, the concentration predicted to yield a certain effect is more than twice the concentration actually observed giving the proposed effect [5]. Belden calls the ratio of predicted versus observed effect concentrations for the Model Deviation Ratio (MDR) [5]. Many of the mixtures showing MDRs slightly below two, most likely also include true synergists. But to exclude false positives and to focus on combinations where the size of the synergistic interactions might be of quantitative importance, we have chosen to set the MDR limit defining synergy at two.

# Materials and Methods

## Identification of Experiments

A flow chart of the record selection for each of the three toxicant groups is presented in Figure 1. To evaluate the frequency of chemicals, chemical mixtures and species groups involved in synergistic ($MDR>2$), additive ($0.5 \leq MDR \leq 2$) and antagonistic

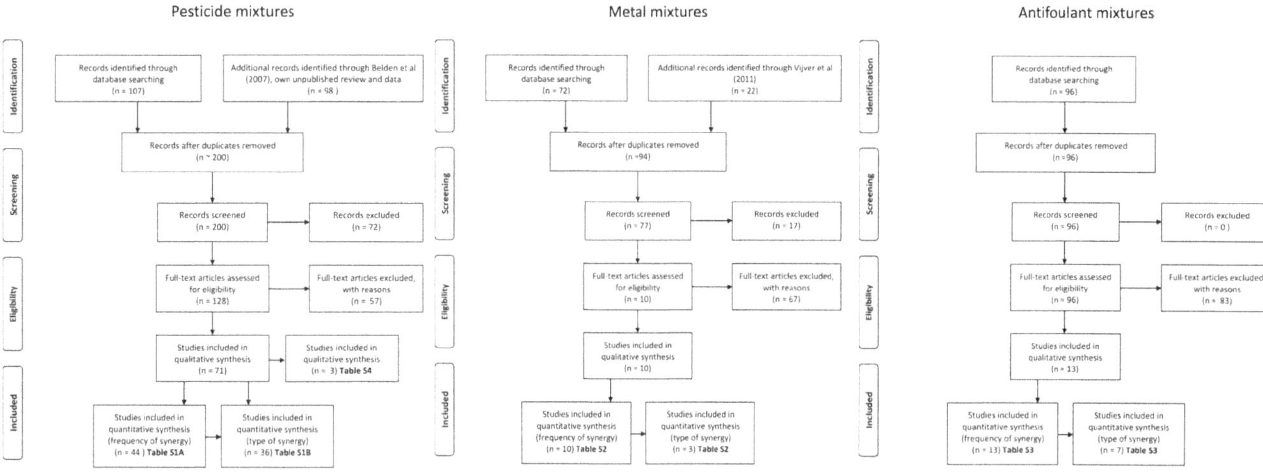


**Figure 1. PRISMA 2009 Flow Diagram** [90]**.** A flow diagram depicting the process of selection of records used in the review for the three main groups of toxicants: Pesticides, metals and antifoulants. Data selection has, for pesticides and metals, been built on previous reviews and data compilations, given in the top right text-box, supplemented with database searched using ISI Web of Science. Search criteria and criteria for selecting eligible records are given in the Material and Methods section. For each toxicant the search resulted in two types of databases: One to determine the frequency of synergy in a randomly selected number of mixtures studies, and another focussing only on defined synergistic mixtures. It should be noted that many records contain data on several independent mixtures studies; hence the number of records given in the figure does not match the number of selected studies reported in the results section. References to tables in supporting material giving the raw data on specific chemical mixtures, test species, endpoint and timecourse of the experiment, and the record providing the information are given in the figure.

(MDR<0.5) mixture experiments, the database of Belden et al. (2007) was used for the pesticides, the one of Vijver et al (2011) was used as a starting point for the metals, while our own data-collection was used for the antifoulants (Figure 1, Supporting information: Table 1A, 2 and 3).

For the pesticide mixtures, Belden et al (2007 and our own database on synergistic interactions was expanded with more recent studies screening the database ISI Web of Science using the search words "pesticide*", "mixture*" and "synerg*" in the period 2008–2013. For the metals, the review by Vijver et al (2011) [10] was supplemented by newer studies using ISI Web of Science and the search words "metal*", "mixture*", "synerg*" and "toxic*" for the period 2009–2013. The antifoulant mixture compilation using ISI Web of Science and the search words "antifoul*" and "mixture*" for the time period 1990–2013 to be able to detect the frequency of synergy in a similar way as had been done in the study by Belden et al. (2007).

Only studies complying with the criteria developed by Belden et al (2007) were used: Mixture studies should be conducted using only pure substances. Hence, studies using formulated pesticide or formulated antifouling biocides were excluded, as the formulation products could affect the results. Studies using metals in the form of nano-particles were likewise excluded. To avoid biasing the database with similar experiments, duplicated experiments using the same mixture and species presented in the same manuscript were entered in the database as one study, but giving the MDRs of each individual replicate. If multiple mixture ratios were tested in the same experiment, the MDR from the mixture ratio closest to the ratio where both chemicals contributed equally to the toxicity (equipotent ratio) was used in the cases of isobole designs, where several mixture ratios were tested. Otherwise the numerically larger MDR was used. Finally, the experiment had to be conducted in a way that an MDR could be calculated. That is, comparable $ECx$ values or Toxic Units ($1/EC_x$) from individual compounds and their mixtures should be available either directly or from reading off graphs. From each study, the following information was collected: The chemicals involved, the species tested, the higher taxonomic group of the species, monitored endpoint and duration of the toxicity test, and the original reference where the raw data were reported. Studies on species communities were not included.

It should be noted that the published data does not represent a random selection of chemical mixtures tested on representative ecological species, but rather represent mixtures selected because of co-occurrence or suspicion of synergy tested on standard laboratory species. The choice of chemicals biases the database towards detecting synergies, while the choice of robust laboratory organisms, on the other hand, might give conservative estimates on synergies as they might not represent the most susceptible species.

All data treatments were done in excel.

## Results

### The Frequency of Synergy

Figure 1 presents the selection process of record for the study. Several of the records reported more than one mixture toxicity experiment. In the following the individual mixture toxicity experiments will be discussed. The records from where data has been retrieved can be found in the tables S1–S4 in File S1 in the supporting information. A PRISMA Checklist for reviews is given in Checklist S1.

**Pesticide mixtures.** The database of Belden et al (2007) provided data on 207 pesticide mixtures of which 194 were binary and another 13 consisted of more than two pesticides [5].

**Metal mixtures.** Evaluating the meta-analysis of Vijver et al (2011) on metal mixtures according to the criteria set by Belden et al (2007), reduced the number of usable studies from 22 to 6 studies reporting 10 experiments where MDR could be calculated and another 7 experiments, where data shown on graphs could be evaluated as being over or under-predicted by CA (Table S2 in File S1). Of the remaining 15 studies, five studies only allowed IA predictions and 10 studies reported metal tissue accumulations, but not effects. Since there is not always a straight forward

**Table 1.** The overall group, name and proposed Modes of Action (MoA) of the antifouling compounds.

| Group | Name | IUPAC name | Mode of Action |
|---|---|---|---|
| Photosystem II inhibitors | Atrazin | 6-chloro-$N^2$-ethyl-$N^4$-isopropyl-1,3,5-triazine-2,4-diamine | Inhibits the electron transport in photosystem II |
| | Irgarol1051 | 2-tert-butylamino)-4-(cyclopropylamino)-6-(methylthio)-1,3,5-triazine | Inhibits the electron transport in photosystem II |
| | Seanine211 | 4,5-dichloro-2-n-octyl-4-isothiazoline-3-one | Inhibits the electron transport in photosystem II |
| | Diuron | 3-(3,4.dichlorophenyl)-1,1-dimethylurea | Inhibits the electron transport in photosystem II |
| Metals and organometals | Cd | Cadmium ion | General toxicant, interacts with enzymes[a] |
| | Cu | Copper ion | General toxicant, interacts with enzymes[a] |
| | CuPT | Copper 2-pyridinethiol-1-oxide | General toxicant, interacts with enzymes[a] |
| | Zn | Zink ion | General toxicant, interacts with enzymes[a] |
| | ZnPT | Zinc 2-pyridinethiol-oxide | General toxicant, interacts with enzymes[a] |
| | Ziram | Zinc bis(N,N'-dimethyl)-dithiocarbamate | Dimethyldithiocarbamate fungicide with Zn. Inhibitor of enzymes containing copper ions or sulfhydryl groups, including P450 monooxygenases of the CYP 2A6 group[b] |
| | TBT | tri-butyl-tin-chloride | PSII inhibitor (with tin), endocrine disruptor[c] |
| Fungicides | Chlorothalonil | Tetrachloroisophthalonitrile | Conjugation with, and depletion of, thiols (particularly glutathione) from germinating fungal cells, leading to disruption of glycolysis and energy production, fungistasis and fungicidal action. |
| | Dichlofluanid | *N*-dichlorofluoromethylthio-*N'*,*N'*-dimethyl-*N*-phenylsulfamide | Multi-site mode of action, non-specific thiol reactant, inhibiting respiration. |
| | IPBC | 3-iodo-2-propynyl butylcarbamate | AChE inhibitor and fungicide and bactericide[d] |
| | PTPB | Pyridine triphenylboron | Fungicide[d] |
| | TCMTB | 2-thio cyano methyl thio benzothiazole | Fungicide, Inhibitor of mitochondrial electron transport[c] |
| | Tolylfluanid | *N*-dichlorofluoromethylthio-*N'*,*N'*-dimethyl-*N*-*p*-tolylsulfamide | Multi-site mode of action, non-specific thiol reactant, inhibiting respiration. |

[a]Altenburger, 2011 [88].
[b]Walker, 2009 [49].
[c]Fernandez-Alba et al, 2002 [89].
[d]Zhou et al, 2006 [40].
Particularly for the fungicides, which have multiple and often undefined modes of action, different target sites are given in different references. For herbicides and fungicides used as pesticides we use the definition of Tomlin 2002 [29]. For the remaining compounds, the source of the MoA are given as footnotes.

correlation between tissue accumulation and toxic effect [25,26] these were disregarded. A recent paper of Xu et al (2011) [27] added another 11 metal mixtures where MDR-values could be calculated, making it a total of 28 mixtures from 8 studies tested on 7 species. Of these, 21 mixtures were binary while the remaining 7 mixtures consisted of more than two metals.

**Antifoulants mixtures.** For antifoulants 136 mixtures where MDR-values could be calculated were found. These were presented in 14 studies comprising mixtures of 20 chemicals tested on 15 different species (Table S3 in File S1). There were 103 binary mixtures and 33 mixtures with more than two chemicals. The frequencies of synergy in the binary mixtures were 7%, 3% and 26% for pesticides, metals and antifoulants, respectively, while 88%, 86% and 64% was within two fold of the CA prediction (Figure 2). For the 13 pesticide mixtures where more than two chemicals were included, only one was synergistic [5], while for the 33 antifoulant mixtures with more than two chemicals 61% showed severe synergy (Table S3 in File S1).

## Types of Synergy

**Pesticide mixtures.** In addition to Belden et al (2007) [5] and the review by Cedergreen et al (2008) [28] another 84 papers were reviewed for synergy where the MDR ratios were >2. This resulted in a database on synergistic interactions including 73 cases of synergy from both Belden et al (2007) and the data search compiled from 36 studies. These studies tested the effect of combinations of 54 pesticides on 27 different species. Of all the mixture combinations, 69 were binary mixtures while the remaining four mixtures consisted of combinations of three or five organophosphate insecticides or eight chloroacetamide herbicide safeners (Table S1B in File S1). Dividing the pesticides into groups with common modes of action according to Tomlin

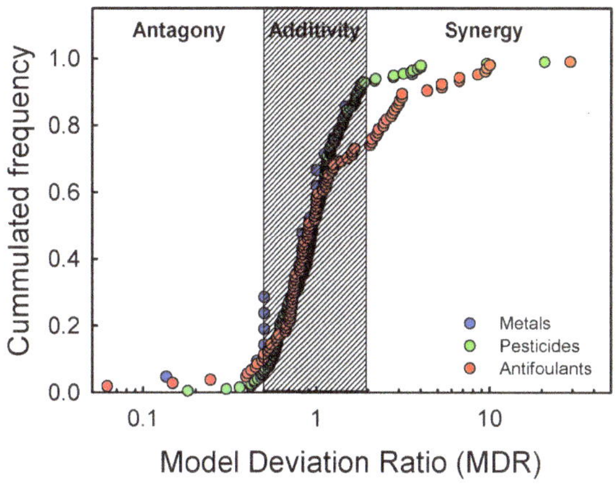


**Figure 2. Cummulated frequency of Model Deviation Ratios.** Cummulated frequency of Model Deviation Ratios. (MDR) of binary mixtures of pesticides ($n=195$), metals ($n=20$), and antifoulants ($n=103$). The hatched interval where $0.5 \leq \text{MDR} \leq 2$ defines the mixtures that deviates less than two-fold from a Concentration Addition predictions. Mixtures having MDR values<0.5 are termed antagonistic, while mixtures with MDR values>2 are synergistic.

(2002) [29] showed that particularly five groups of pesticides were overrepresented in the synergistic mixtures. These were the organophosphate and carbamate insecticides (Cholinesterase inhibitors), azole fungicides (Ergosterol biosynthesis inhibitors), triazine herbicides (Photosystem II inhibitors) and pyrethroid insecticides (interferes with sodium channels in nerve cells) (Figure 3A). Grouping the cholinesterase inhibitors together and looking at which of the binary combinations of the above pesticide groups induced synergy in auto-trophic organisms (plants and algae) and hetero-trophic organisms (microorganisms and animals) showed no cases of synergy within the autotrophic organisms (Figure 3B). In the group of hetero-trophic organisms 69 of the 73 synergistic mixtures (95%) contained either cholinesterase inhibitors (organophosphates or carbamates) or azole fungicides (Figure 3C). The remaining four mixtures were the before mentioned mixture of 8 herbicide safeners, a mixture of a pyrethroid with an organochloride insecticide, a pyrethroid insecticide and a piperidine fungicide and a photosystem II (PSII) inhibiting herbicide and a growth regulator (Table S1B in File S1). Of the 69 binary mixtures 76% contained a cholinesterase inhibitor and another 24% an azole fungicide (Figure 3C). The triazines only entered in synergistic mixtures together with either chlorpyriphos, diazinon, malathion, methidathion, methyl-parathion, which belong to the phosphorothioate and phosphorodithioates class of organophosphates, or trichlorfon, a phosphate class organophosphate. Pyrethroids, on the other hand, only entered in synergistic mixtures together with azole fungicides.

An evaluation of which types of the pesticides from the review of Belden et al (2007) were dominant in the antagonistic mixtures and those conforming to CA, showed that cholinesterase inhibitors and azole fungicides made up 29% of the antagonistic mixtures and 48% of the mixtures conforming to CA (Figure 3B and C), which is considerably less than the 95% of the synergistic mixtures. Hence, though these modes of action were present in all types of mixtures, they were clearly overrepresented in the mixtures displaying synergistic interactions. The triazines occurred in 1% of the antagonistic mixtures, 22% of the concentration additive mixtures and in 12% of the synergistic mixtures. Hence, triazines did not seem to occur particularly frequently in the synergistic mixtures, and when they did, only in mixtures with the before

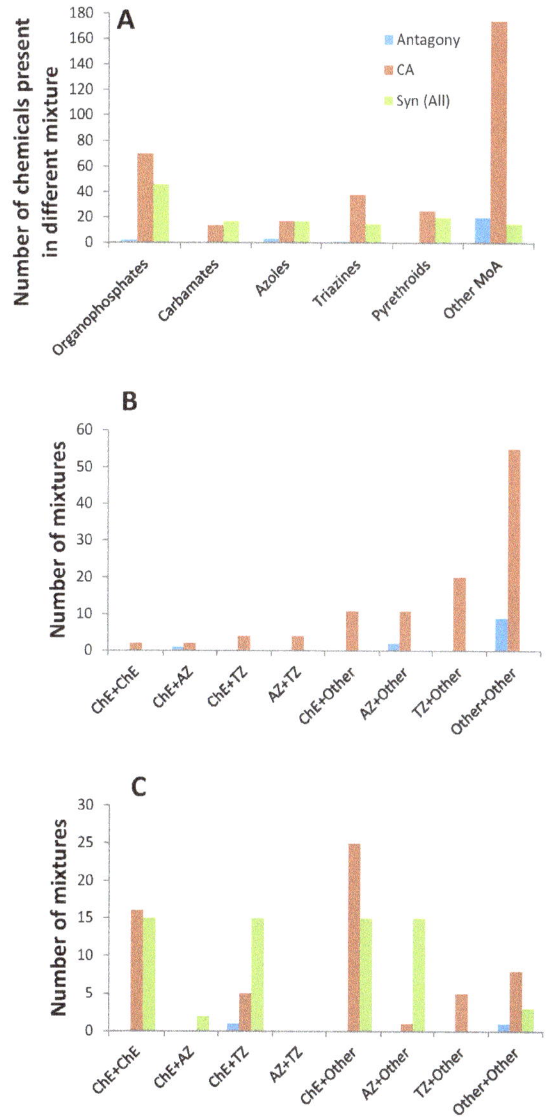


**Figure 3. Frequency of pesticide antagony, additivity and synergy.** Figure 2A shows the number of times a pesticide belonging to the group of organophosphates, carbamates, azoles, triazines, pyrethroids or some other Mode of Action (other MoA) occur in a binary mixture resulting in antagony (blue bars), concentration additivity (CA) (red bars) or synergy (green bars). In figure B and C, the number of binary combinations of cholinesterase inhibitors (ChE) (The organophosphates and carbamates), azoles (AZ), triazines (TZ) and other Modes of Action (Other) resulting in either antagony, concentration additivity or synergy are shown for mixtures tested on B) autotropic organisms (plants and algae, $n=120$) or C) heterotrophic organisms (microorganisms and animals, $n=128$).

mentioned organophosphates. The 19 triazine mixtures with an MDR<1 were dominated by Auxin transport inhibitors, branched chain- and aromatic amino acid synthesis inhibitors, while the 19 triazine mixtures with MDR values between 1 and 2 were dominated by organophosphates, PSII inhibitors and cell division inhibiting herbicides. All 22 additive mixtures including pyrethroids, were mixtures with organophosphates, carbamates or other pyrethroids (Table S1A in File S1).

**Metal mixtures.** Going through the 55 selected potential papers found on ISI Web-of-Science using the key-words given

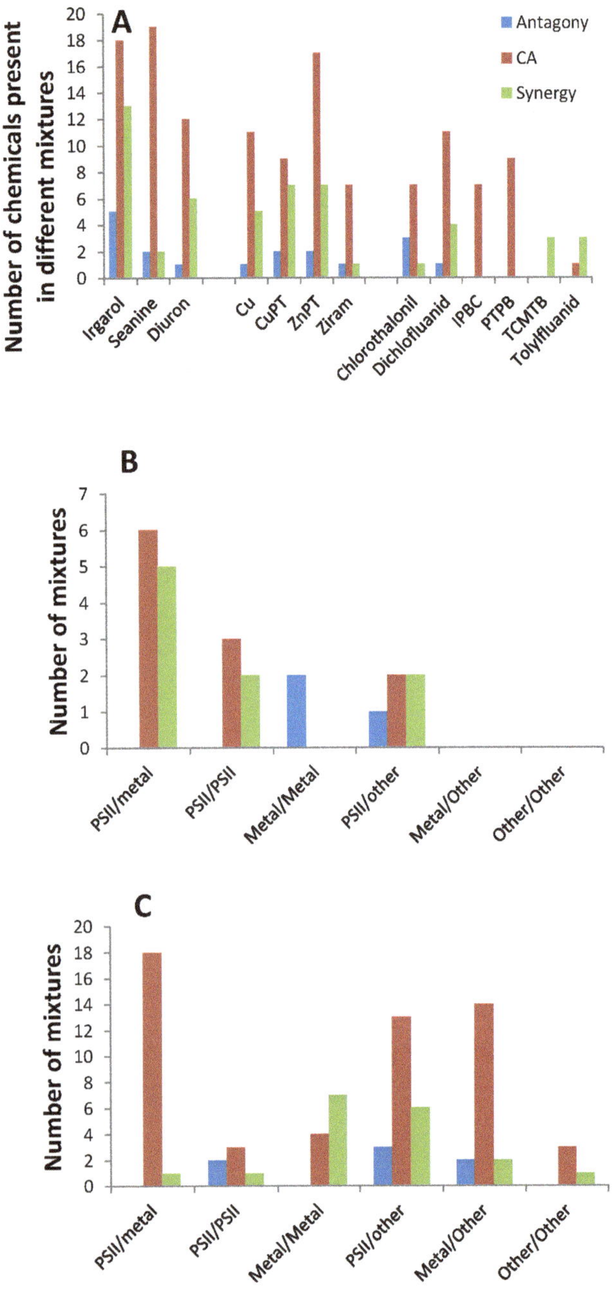


**Figure 4. Frequency of antifoulant antagony, additivity and synergy.** Figure 3A shows the number of times each of the antifoulants occur in a binary mixture resulting in antagony (blue bars), concentration additivity (CA) (red bars) or synergy (green bars). Antifoulants occurring in less than 1% of the mixtures were excluded. In figure B and C, the number of binary combinations of photosystem II herbicides (PSII) metal ions or metal containing compounds (Metal) and other organic compounds (Other) resulting in either antagony, concentration additivity or synergy are shown for mixtures tested on B) auto-tropic organisms (plants and algae, $n = 23$) or C) heterotrophic organisms (microorganisms and animals, $n = 80$).

above only revealed two additional studies with three experiments where MDR>2 could be estimated from figures (Table S2 in File S1). Hence, despite the large numbers of studies made on metal mixtures, well documented severe synergistic metal-metal interactions seem to be rare. The four binary mixtures giving synergy were, Cd+Zn, Cu+Zn, Cu+Cd and Cd+As tested on the shrimp *Penaeus setiferus*, the fish *Gobiocypris rarus* and the water-flee *Daphnia magna* (Table S2 in File S1).

**Antifoulants mixtures.** In the antifoulants database (Table S3 in File S1), we found 47 cases of synergy from 8 studies, testing the effect of mixtures of 12 chemicals on 9 organisms. Another 7 chemicals were tested that did not occur in any of the synergistic mixtures. The antifoulants were more difficult to categorise according to physiological mode of action compared to the pesticides, as this information is not required for registration. While pesticides are often developed to act physiologically very specifically in specific target organisms, antifoulants are selected to be toxic to the wide range of organisms settling on ship hulls. Hence, their physiological mode of action is more likely to be general, targeting physiological pathways important for a broad range of species. Hence, the analysis of the frequency of chemicals in synergistic, additive and antagonistic mixtures were done on the individual chemicals roughly divided into three groups: Herbicides (2,4-D, atrazine, irgarol 1051, Seanine 211, and diuron), metals and metal containing organic compounds (Cd, Cu, Cu Pyrethrione (PT), Zn, ZnPT, Ziram and Tributhyltin (TBT)), and other organic compounds (chlorothalonil, dichlorfluanid, 3-iodo-2-propynyl butylcarbamate (IPBC), pyridine triphenylboron (PTPB), 2-thio cyano methyl thio benzothiazole (TCMTB) and tolylfluanid). Chemical class and proposed modes of action are given in Table 1. For the 103 binary mixtures the frequency of synergy was markedly higher than the frequency of antagony for mixtures containing either irgarol or diuron, Cu, CuPT or ZnPT, TCMTB, dichlorofluanid or tolyfluanid (Mixtures including Cd, Zn, or TBT were excluded in this analysis as they were included in <3 binary mixtures each) (Figure 4A). Analysing the frequency of binary mixtures combined of the above defined three overall groups for the 23 binary studies on plants and algae and the 80 studies on animals and microorganisms separately, showed that all synergistic mixtures tested on plants or algae contained a PSII inhibiting herbicide either in combination with another PSII inhibiting herbicide, or metal or an organic antifoulant (Figure 4B). This is contrary to the pesticide study, where no synergy was found in studies on auto-trophic organisms at all (Figure 3B and C, Table S1B in File S1). For the autotrophic organisms the PSII inhibitors were present in slightly more additive than synergistic mixtures (Figure 4B). For the group of heterothrophs, PSII inhibiting herbicides also caused synergy, particularly together with dichlorfluanid, tolyfluanid and TCMTB (Figure 4C, Table S3 in File S1). Combinations of two metal containing compounds induced synergy in seven of 11 cases for this group (64%). Hence, contrary to mixtures of metal ions (Table S2 in File S1), mixtures of organically bound metals seem to be much more potent in inducing synergy.

Of the 23 ternary mixtures of antifoulants and the ten quarternary mixtures, four mixtures were antagonistic, nine additive and the remaining 20 mixtures had an MDR>2 (Table S2 in File S1). The frequency of occurrence of the different antifoulants in antagonistic, additive and synergistic mixtures is shown in figure 5, confirming that particularly irgarol, Seanine, CuPT, dichlofluanid and tolylfluanid often occur in synergistic mixtures.

**Synergistic mixtures of metals and organic compounds.** In the search for synergistic mixtures, some mixtures showed up that could not be categorised as either pesticides or metals, as they contained both. Synergistic interactions between metals and pesticides seemed to be quite frequent compared to synergistic mixtures of metals alone, but since a comprehensive database on mixture experiments of metals and

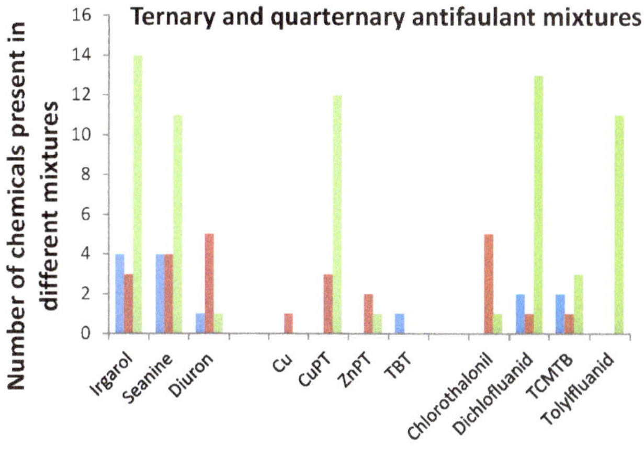


**Figure 5. Frequency of antifoulant interactions in ternary and quaternary mixtures.** The number of times each of the antifoulants occur in a ternary ($n = 23$) or quaternary ($n = 10$) mixture resulting in antagony (blue bars), concentration additivity (CA) (red bars) or synergy (green bars).

pesticides has not been made, this cannot be tested. Table S4 in File S1 in the Supplementary material show 11 mixtures of metals and pesticides from three studies, of which eight had a MDR value >2.

## Discussion

### Which are the Chemicals Causing Synergy?

The review showed that for pesticides, the combinations causing synergy were not random but included either cholinesterase inhibitors or azole fungicides in 95% of the described cases. The proposed mechanisms behind these synergies are relatively well investigated, as discussed below. The synergy frequency for metal ion mixtures was very low, hence no general conclusion in terms of which compounds caused synergy could be made. When metal ion synergy occurred, it was in the mg $L^{-1}$ concentration range for three of the four cases [30,31]. These concentrations are high, compared to the concentrations normally found in metal polluted waters being in the lower ng to µg $L^{-1}$ range [32,33]. For the antifouling compounds synergistic interactions were also related to specific chemical groups, though more synergistic combinations of different chemical groups were involved than seen for the pesticides (Figure 3, 4). The high frequency of synergistic interactions observed for the antifoulants, particularly in the mixtures with more than two active ingredients, is most likely due to the selection for compounds able to induce synergy in antifouling products, which most often are composed of more than one active ingredient [34]. The mechanisms behind the synergistic interactions of the antifoulants are, contrary to what is seen for the pesticides, rarely investigated. In the following, the proposed mechanisms behind the synergistic interactions of pesticide, metal ions and antifouling mixtures are discussed.

### Mechanisms Causing Synergistic Interactions

Interactions between chemicals can basically affect six processes that are important for the resultant toxicity of a chemical towards an organism: bioavailability, uptake, internal transportation, metabolization, binding at the target site and excretion. The synergistic interactions identified in the present study are most likely caused by interactions around one or more of these processes. In the following, known mechanisms behind the identified synergistic interactions will be discussed in terms of which processes are most likely affected by the interactions.

**Bioavailability.** Interactions between chemicals can take place outside the organism, with one chemical affecting the availability of the other. This is commonly seen for metal ions, where ion speciation and competition for binding sites to organic matter in soil, sediments and the water phase can change free ion availability and composition [35–37]. If a less toxic ion replaces a bound or chelated ion with a higher toxicity, this will lead to apparent synergistic interactions, if the toxicities are estimated based on total metal concentrations rather than bioavailable concentrations. These types of interactions, however, most often occur when binding sites are limited. Hence, either the metal ion concentrations are high, or the binding site density low, as would for example be the case with ions in water with low concentrations of dissolved organic matter or mineral ions as calcium carbonate and other salts [37]. None of the four metal-metal ion synergies found in this study (Table S4 in File S1) allow for an assessment of whether the synergistic interactions occurred due to changed ion availability, as only total metal concentrations were given. Changes in speciation outside the organism as a cause of synergistic interaction has, however, been well documented for mixtures of pyrithione antifoulants [38]. When ZnPT and Cu ions are mixed together the more toxic CuPT complex is formed, making the mixture more toxic than predicted from the toxicities of ZnPT and Cu alone [38–40]. As the affinity of pyrithione for Cu is higher than for Zn, then the equilibrium between the metal-pyrithione complexes and free pyrithione will be shifted in favour of CuPT [41]. If there is a metal ion surplus in the ZnPT and CuPT synergistic mixture observed by Koutsaftis and Aoyama (2006) [42], a shift towards a larger proportion of CuPT might be taking place. It could therefore be hypothesised that changes in speciation outside the organism is a main mechanism behind all the reported synergistic interactions of metal/metal mixtures among the antifoulants (Figure 4C), apart from the mixture of Cu and Ziram [40].

**Uptake rates and transport to the target site.** One chemical can affect the uptake rate of the other by for example competition at biological ligands or competitive inhibition of transport proteins, as is often observed for interactions on metal uptake ([35,36] and references herein); though not all studies explicitly describe external ion availabilities, making it difficult to determine whether the interactions measured on internal concentration stem from interactions on bioavailability or on uptake.

Interactions on uptake rates have, however, also been measured for combinations of organic contaminants. When Belden and Lydy (2000) investigated the synergistic interactions between the herbicide atrazine and the organophosphate insecticide chlorpyriphos, they found that the addition of atrazine increased chlorpyriphos uptake by 40% [43]. This increase in contaminant uptake was proposed to be caused by an increased oxygen consumption, leading to higher ventilation rates and thereby higher uptake rates of a contaminant as chlorpyriphos, which is predominantly taken up over the gills. Increased ventilation alone could not explain the observed four-fold increase in toxicity. Hence, though it is likely that many contaminants will increase ventilation rates when the organisms start spending energy metabolizing them, thereby increasing uptake of other contaminants taken up over gills, lungs and tracheid's, the quantitative importance of this extra uptake is most likely of little importance for the more severe synergistic cases reported in the literature.

Potential important effects on uptake was also proposed by Kennaugh et al (1993) in a study on the effect of the known synergist Piperonyl Butoxide (PBO) on the cytochrome P450 mediated metabolic rate of the pyrethroid insecticide Permethrin in permethrin resistant and wildtype *Helicoverpa armigera* [44]. The ability of PBO to break the 20-fold resistance could not be explained by differences in P450 monooxygenase mediated permethrin detoxification rates, since they were identical for the resistant and non-resistant genotypes. Hence, it was proposed that PBO instead affected a P450 mediated "penetration resistance" developed by the resistant strain, making the resistant strain take up less pyrethroid. The proposed effect on uptake rates was, however, never confirmed by actual studies of Permethrin uptake. Hence, the P450 mediated "penetration resistance" is still a hypothesis.

Many of the synergists known to enhance uptake belong to the large group of surfactants and other additives added to formulated pesticides with the exact purpose of enhancing the uptake of the active compounds [45]. As this review has excluded all studies with formulated compounds and surfactants, the database does not include examples on surfactant synergies, despite of their frequent use. Though there is a proven effect of the surfactants on uptake of active compounds when hitting their target at high concentrations, it is likely that most lose their "uptake enhancing" potency when diluted in environmental matrices, even though they might still act as dilute pollutants adding to the overall toxicity according to concentration addition. This is supported by a toxicity study on formulated versus technical herbicides on aquatic plants and algae showing no difference in potency for nine out of ten herbicides [46].

The transport rate of one chemical towards its molecular target can be affected by the presence of another chemical, as is for example the proposed mechanisms behind the strong antagonistic responses often seen in plants when a rapidly acting photosynthetic inhibitor is mixed with a slower acting systemic herbicide [47]. No studies have, to my knowledge, shown that one chemical can actively increase the transport of another chemical to their target.Thist is, nonetheless, the proposed mechanism behind many hypotheses regarding nano-particle facilitated increase in chemical toxicity [48], which we will not touch upon here, and therefore cannot be excluded either for chemical/chemical interactions.

**Metabolic enzyme activities.** Alternations of metabolic activity that are the most well investigated mechanisms behind observed synergistic patterns. A chemical can either increase or decrease the metabolization rate of another chemical. Decreased metabolization will typically lead to a higher toxicity than expected, when the toxic effect is caused by the unchanged parent compound. In contrast, increased metabolization will increase the toxicity of chemicals which are metabolically activated.

Synergistic interactions involving azole fungicides are most likely all examples of cases where the metabolization of the pesticides is inhibited by the azole. Azole fungicides are known inhibitors of a wide range of P450 monooxygenases, which are enzymes responsible for the phase I metabolization of lipophilic compounds [49], together with a range of biosynthesis processes in both plants and animals [50,51]. Hence, the toxicity of lipophilic insecticides such as pyrethroids are often severely enhanced when mixed with azole fungicides [49,52–55].

The synergistic cases involving cholinesterase inhibitors, which made up 76% of all the synergistic pesticide mixtures, most likely all also involve interactions on metabolism. The dominant mechanisms are, however, different depending on which compounds are involved. Basically three mechanisms can be involved:

First, besides the target enzyme acethylcholinesterase (AChE), organophosphate and carbamate insecticides can also inhibit esterases, which are responsible for phase II metabolization of other xenobiotics, including organophosphates and carbamates themselves [49]. Although having the same mode of action and therefore supposedly following concentration addition, mixtures of some organophosphates and carbamates do act synergistically [56–61] (Figure 3).

Second, organophosphates, from the phosphorothioate and phosphorodithioates class of organophosphates, must be metabolically activated to their more active oxon form in order to inhibit the target site AChE [43]. This means that compounds that can induce the production of P450 monooxygenases, will increase the rate of oxon formation and hence increase the toxicity of the organophosphates. This mechanism has been proposed as being the main mechanism responsible for the cases of synergy between triazine herbicides and organophosphates [43,62]. Belden and Lydy (2000) elegantly showed how the amount of polar metabolites of chlorpyriphos increased in *Chironomus tentans* in the presence of atrazine [43], explaining the majority of the observed synergy. Triazine herbicides have also been shown to induce P450 activity in fish [63,64]. The fact that all the cases of pesticide synergy between triazines and organophosphates include organophosphates belonging to the class of phosphorothioate and phosphorodithioates (chlorpyriphos, diazinon, malathion, methidathion, methyl-parathion) [43,62,62,65–68], or being transformed into one [69,29], indicate that triazine induced P450 induction is the main cause of this synergistic interaction. However, not only triazines induce P450 activity. Many xenobiotics, ranging from polyaromatic hydrocarbons, dioxins and ethanol [49,70] to natural substances in honey and metal ions [71,72] are proven P450 inducers. The synergistic interactions between organo(thio-)phosphates and neonicotenoids in the nematode *Caenorhabditis elegans* were also proposed to stem from P450 induction of neonicotenoids [73]. It even seems as if compounds that inhibit P450 activity at high concentrations induce activity at low concentrations or on a longer time-scale. Azole fungicides have, for example, shown to give protective effects against pyrethroid toxicity at low doses in bees [52] and the aquatic invertebrate *Daphnia magna* (pers.obs.), and have been shown to induce synergy together with organophosphates in birds pre-treated with prochloraz [74–76]. In the bird studies increased metabolism of the organophosphates was measured, strongly indicating P450 induction [75,76].

Third and finally, phosphorothioate organophosphates are known to inhibit some types of P450 monooxygenases, thereby not only affecting phase II but also affecting phase I metabolism of xenobiotics [49]. New studies have shown that the inhibitions and activations of different P450 genes are compound specific [77], as are the xenobiotics affinities for the different monooxygenases [49]. Hence, it is likely that the majority of severe synergistic interactions can be explained by interactions on metabolism. Which types of interactions plays the largest role for specific chemical combinations, and at which concentrations and time-scales the interactions are most severe for different species, is, however, still largely unexplored.

**Excretion.** As the ability of an organism to excrete a compound is mainly related to its ability to transform xenobiotis to an excretable form, excretion is closely related to metabolization. One exception is active excretion of essential metals as Cu and other ions, for which specific transporters or other excretion systems exist, aiding in keeping internal concentrations within a non-toxic range [78,79]. It could be hypothesised that interactions on these excretion processes could lead to synergistic interactions if they were in some way inhibited, though none of the synergistic mixtures included in this study have proven these mechanisms to be important.

## Synergistic Interactions where the Mechanisms are Unknown

Apart from the pesticide and metal examples given above, where mechanisms causing synergistic interactions are, if not proven, then at least suggested, the review also revealed synergistic chemical combinations where the mechanisms are unknown. These were mainly the interactions between PSII herbicides with other PSII herbicides, metals or non-azole fungicides in the antifouling mixtures together with the mixtures of metals and organo-metals (Table S3 in File S1) or simply metals and organic pesticides (Table S4 in File S1).

Photosystem II herbicides did not induce synergy in any of the 33 mixtures performed on plants or algae in the pesticide database (Table S1 in File S1). Hence, it was surprising to find that significant synergy was found in nine of the 21 antifoulant mixtures including PSII herbicides when tested on plants or algae. Five of these nine mixtures were with the metals Cd, Cu and Zn, which were not part of any of the PSII mixture in the pesticide database. A proposed synergistic mechanism between metals and PSII inhibitors in autotrophs could be that metals might prevent the repair of not only damaged PSII complexes, which are constantly repaired during photosynthesis [80], but also the damage caused by the reactive oxygen species (ROS) created by the PSII inhibition and the metals themselves, by interacting with enzymes responsible for the repair. The two synergistic PSII/PSII mixtures were between irgarol and diuron, while the remaining two were between irgarol and chorothalonil or TCMTB. The synergies between irgarol and the two general fungicides, chorothalonil and TCMTB, could be similar to the mechanism proposed for the PSII/metal interactions, as both fungicides create ROS [81] and additionally chlorothalonil conjugates with gluthatione [29], an important ROS scavenger. These hypotheses, however, need to be tested.

The mode of action of PSII inhibitors in heterotrophs is largely unknown, as these organisms lack photosystems. The studies on pesticides revealed that triazines, such as irgarol, can induce P450 activity in heterotrophs, thereby enhancing the effect of the organophosphates which needed to be metabolically activated. A study by Suzuki et al (2004) show that also dichlofluanid and chlorothalonil need activation by P450 monooxygenases to reach their full lipid-oxidation potential, with dichlofluanid being far more potent than chlorothalonil [81]. As tolylfluanid is chemically related to dichlofluanid, it might also have to be oxidised by P450 to be fully activated. A study on fish have shown that also the metal ions $Cu^{+}$ and $Pb^{+}$ can induce P450 activity [72]. Five of the seven synergistic mixtures involving PSII inhibitors or metals together with organics are mixtures of the triazine irgarol or Cu and either dichlofluanid or tolylfluanid. In addition, for the twenty synergistic mixtures with more than two antifoulants, these combinations were present in all but three mixtures. It could therefore be hypothesised that the main mechanism behind the synergy between irgarol or Cu (or CuPT), and the fungicides dichlofluanid and tolylfluanid were irgarol and Cu mediated P450 induction leading to faster activation of dichlofluanid and tolylfluanid. But this hypothesis would have to be tested.diuron and TCMTB also induced synergy together with irgarol in two cases each, in the heterotrophic organisms, but information in terms of possible P450 induced activation of these two compounds has not been found.

Metal induced P450 activity could possibly also play a role in the synergies with the phosphorothioate organophosphates malathion, chlorpyriphos and dimethoate [56,82](Sejerøe 2011) (Table S4 in File S1), but as synergies between metal ions an phosphate organophosphates as dichlorvos, the carbamate carbofuran and the azole penconazole [56]Sejerøe 2011) was also found, other mechanisms are most likely also of importance. Lister et al (2011) found an increased uptake and metabolization rate of chlorpyriphos in the presence of Ni, but data were two variable to say anything definite [83]. However, as both P450 monooxygenases and esterases are important enzymes in many biochemical processes, changes in their activity could also affect uptake, excretion or possibly some of the mechanisms used for inactivation of metals in different organisms, though these hypotheses must be subjected to experimental scrutiny.

## Is Synergy of any Importance in Nature?

For synergistic interactions to take place in the environment, interacting chemicals have both to co-occur and to be present at levels high enough to induce the synergy. Co-occurrence does happen, as has been shown for both pesticides and antifoulants [11,59,84]. Looking at the cases presented in this review, however, most experiments showing significant synergy use chemical concentrations in the high $\mu g\ L^{-1}$ to mg $L^{-1}$ range, which is considerably above the concentrations most often monitored in the environment (pg $L^{-1}$ to the low $\mu g\ L^{-1}$ range) [11,59,84]. Very few studies though use realistic concentration ranges, Laetz et al (2009) being such an exception. It is, however, likely that a threshold for synergistic interactions exists for most synergists, and that only a few proven synergists will act as synergists at any endpoint when diluted down to realistic environmental levels. This loss of efficiency as a synergist has for example been shown for piperonyl butoxide (PBO), a known P450 inhibitor, when used to formulate pyrethroid insecticides for mosquito control [85]. In this case, adding the synergist to the aquatic environment did not increase the efficacy of the insecticide towards an aquatic crustacean. Another case, however, showed PBO to enhance pyrethroid toxicity down to concentrations as low as 25 $\mu g\ L^{-1}$ [86]. Hence, more data is needed to determine if a lower threshold for synergists interfering with metabolic processes do exist. In these studies it will be important to include sub-lethal endpoints such as growth and reproduction so that true long term effects on population growth can be estimated.

## Conclusion

From the present review of possible mechanisms causing the observed synergies, it can be concluded that interactions on metabolic processes affecting the transformation of xenobiotics seem to be far the most common mechanism of synergy, though interactions on availability and uptake might play an important role for metal/metal synergies. For the synergistic interactions between pesticides, with cholinesterase inhibitors and azole fungicides being present in 95% of the described synergistic cases, the chemical groups causing synergy can be well defined. For the antifoulants the pattern was less clear, primarily due to the lack of knowledge on the interference of the compounds with metabolic processes. However, knowing that most synergistic interactions most likely stem from interactions on metabolic processes, it would be possible to screen for potential synergists using either *in vitro* assays on P450 monooxygenase or esterase inhibition potential, or by investigating metabolization kinetics *in vivo* in representative test species; though the latter is quite labour intensive.

In the introduction it was stated that if we could identify the groups of chemicals that are likely to induce synergistic interactions, special precautions could be taken in the risk assessment of these chemicals. The present review shows that some groups of potential synergists can indeed be identified, while others need more research to be specifically defined as synergists.

That said, considering the generally high chemical concentrations needed to induce synergistic interactions, their importance as synergists within naturally occurring exposure scenarios is most likely of a relatively small importance compared to the additive effect of many co-occuring pollutants. Even if one compound enhances the effect of another compound four-fold, it only takes another three compounds of a similar strength to arrive at the same joint toxicity. And considering the complex pollution patterns monitored [13,33,87], the additive effect of the many co-occurring pollutants might likely project a larger hazard than those of the presence of a few synergist. Hence, in a regulatory perspective addressing the cumulative effect of co-occurring chemicals is the first and most important step in providing a more realistic hazard assessment of chemical cocktails in both man and environment.

## Supporting Information

**Checklist S1 PRISMA 2009 checklist for systematic reviews and meta-analyses.**

**File S1** Table S1A. Antagonistic and additive pesticide mixtures. All binary antagonistic and concentration additive pesticide mixtures from Belden et al (2007) sorted with increasing Model Deviation ration (MDR). The synergistic mixtures from Belden et al (2007) are included in Table S1B. For information on species tested, endpoint and original references, please see Belden et al (2007), Supplementary material, Table 1. Table S1B. Synergistic pesticide mixtures. The mixtures are sorted with increasing MDR and including information on the test species, its phylum, sub-phylum or class, the endpoint tested and the reference of the original study. The synergistic mixtures also included in Belden et al (2007) are given in bold. In the cases where the same mixtures were repeated on the same organism in independent experiments, MDR-values are given for all experiments and are sorted according to the highest MDR value. One full ray-design is defined as one experiment, even though several mixture ratios were tested. Table S2. Metal mitures. Antagonistic and concentration additive mixtures of metal ions from Vijvers et al (2011) and Xu et al (2011) from which MDR-values could be calculated, sorted with the binary mixtures first and then with increasing MDR. Below are the four synergistic mixtures found of which one mixture, given in bold, was obtained from Vijvers et al (2011). The table includes information on the test species, its phylum, sub-phylum or class, the endpoint tested and the reference of the original study. The last three entries are the three extra synergistic mixtures found by the additional database study. Table S3. Mixtures of antifoulants. All mixtures of antifoulants (Antif) from which MDR-values could be calculated, sorted with the binary mixtures first and then with increasing MDR. The table includes information on the test species, its phylum, sub-phylum or class, the endpoint tested and the reference of the original study. For full chemical names and chemical class and mode of action of the antifaulants, please consult Table 1 in the manuscript. The following names are abbreviated: Irgarol1051 (Irgarol), Sea-nine211 (Seanine), Chlorothalonil (Chlorot.), Dichlofluanid (Dichlo), Tolylfuanid (Tolyl). Table S4. Additional synergistic mixtures. Synergistic mixtures between metals and organic compounds which did not fit into any of the three categories; pesticides, metals or antifoulants, sorted with increasing MDR. The table includes information on the test species, its phylum, sub-phylum or class, the endpoint tested and the reference of the original study.

## Acknowledgments

I wish to thank Christian Neerstrøm and Maj-Britt A. Bjergager for starting the compilation of literature on pesticides and Andreas Kretschmann, Michele Gottardi, Claus Svendsen, David Spurgeon and Helle Marcussen for fruitful discussions on modes of synergistic interactions, and comments on an earlier version of the manuscript. No additional funding was received for this study apart from my salary provided by the University of Copenhagen.

## Author Contributions

Conceived and designed the experiments: NC. Wrote the paper: NC.

## References


1. Colborn T, Dumandski D, and Myers JP (1996) Our Stolen Future: Are We Threatening Our Fertility, Intelligence, and Survival? A Scientific Detective STory. New York: Dutton. 306 p.
2. Kepner J (2004) Synergy: The Big Unknowns of Pesticide Exposure. Pesticides and You 23: 17–20.
3. Backhaus T, Blanck H, Faust M (2010) Hazard and Risk Assessment of Chemical Mixtures under REACH: State of the Art, Gaps and Options for Improvement. 3/10: 1–75.
4. Deneer JW (2000) Toxicity of mixtures of pesticides in aquatic systems. Pest Manage Sci 56: 520.
5. Belden JB, Gilliom RJ, Lydy MJ (2007) How well can we predict the toxicity of pesticide mixtures to aquatic life? Int Environ Assess Manage 3: 364–372.
6. Cedergreen N, Christensen AM, Kamper A, Kudsk P, Matthiasen S, et al.(2008) A review of independent action as a reference model for binary mixtures of compounds with different molecular target sites. Environ Toxicol Chem 27: 1621–1632.
7. Boobis A, Budinsky R, Collie S, Crofton K, Embry M, et al.(2011) Critical analysis of literature on low-dose synergy for use in screening chemical mixtures for risk assessment. Crit Rev Toxicol 41: 369–383.
8. Altenburger R, Backhaus T, Boedeker W, Faust M, Scholze M, et al. (2000) Predictability of the toxicity of multiple chemical mixtures to Vibrio fischeri: Mixtures composed of similarly acting chemicals. Environ Toxicol Chem19: 2341–2347.
9. Backhaus T, Altenburger R, Boedeker W, Faust M, Scholze M, et al. (2000) Predictability of the toxicity of multiple mixture of dissimilarly acting chemicals to *Vibrio Fischeri*. Environ Toxicol Chem19: 2348–2356.
10. Vijver MG, Elliott EG, Peijnenburg WJGM, De Snoo GR (2011) Response Predictions for Organisms Water-Exposed to Metal Mixtures: A Meta-Analysis. Environ Toxicol Chem30: 1482–1487.
11. Readman JW (2006) Development, occurance and regulation of antifouling paint biocides: Historical review and future trends. In: Antifouling Paint Biocides. Berlin: Springer. 1–15.
12. Mohr S, Schroder H, Feibicke M, Berghahn R, Arp W, et al. (2008) Long-term effects of the antifouling booster biocide Irgarol 1051 on periphyton, plankton and ecosystem function in freshwater pond mesocosms. Aquat Toxicol 90: 109–120.
13. Cassi R, Tolosa I, de Mora S (2008) A survey of antifoulants in sediments from Ports and Marinas along the French Mediterranean coast. Mar Pollut Bull 56: 1943–1948.
14. Vermeirssen ELM, Hollender J, Bramaz N, van der Voet J, Escher BI (2010) Linking Toxicity in Algal and Bacterial Assays with Chemical Analysis in Passive Samplers Deployed in 21 Treated Sewage Effluents. Environ Toxicol Chem29: 2575–2582.
15. Grung M, Naes K, Fogelberg O, Nilsen AJ, Brack W, et al.(2011) Effects-Directed Analysis of Sediments From Polluted Marine Sites in Norway. J Toxicol Environ Health-Part A-Current Issues 74: 439–454.
16. Warne MSJ, Hawker DW (1995) The Number of Components in A Mixture Determines Whether Synergistic and Antagonistic Or Additive Toxicity Predominate - the Funnel Hypothesis. Ecotoxicol Environ Saf 31: 23–28.
17. Cedergreen N, Sørensen H, Svendsen C (2012) Can the effect of ternary mixtures be predicted from binary mixture toxicity results? Sci Tot Environ 427–428: 229–237.
18. Alabaster JS, Lloyd R (1980) Water Quality Criteria for Freshwater Fish. London: Butterworth. 200 p.
19. Fraser TR (1892) Lecture on The Antagonism Between the Actions of Active Substances. Britt Med J Nov. 2: 485–487.

20. Loewe S, Muischnek H (1926) Über kombinationswirkungen. 1. Mitteilung: Hilfsmittel der Fragestellung. Naunyn-Schmiedebergs Arch Exp Pathol Pharmakol 114: 313–326.
21. Bliss CI (1939) The toxicity of poisons applied jointly. Ann App Biol 26: 585–615.
22. Cedergreen N, Svendsen C, Backhaus T (2013) Chemical Mixtures: Concepts for Predicting Toxicity. In: Jorgensen SE, editors. Encyclopedia of Environmental Management. New York: Taylor & Francis. 2601–2610.
23. Altenburger R, Boedeker W, Faust M, Grimme LH (1996) Regulations for combined effects of pollutants: Consequences from risk assessment in aquatic toxicology. Food Chem Toxicol 34: 1155–1157.
24. Cedergreen N, Kudsk P, Mathiassen SK, Sørensen H, Streibig JC (2007) The reproducibility of binary mixture toxicity studies. Environ Toxicol Chem26: 149–156.
25. Lock K, Janssen CR (2001) Zinc and cadmium body burdens in terrestrial oligochaetes: Use and significance in environmental risk assessment. Environ Toxicol Chem 20: 2067–2072.
26. Cedergreen N, Johansson HKL, Nielsen JK, Nørhave N, Spurgeon D, et al. (2013) Low temperatures enhance the chronic toxicity of Cd and Cu through different mechanisms. Environ Toxicol Chem. 32: 2274–2283.
27. Xu X, Li Y, Wang YA, Wang YH (2011) Assessment of toxic interactions of heavy metals in multi-component mixtures using sea urchin embryo-larval bioassay. Toxicol in Vitro 25: 294–300.
28. Cedergreen N, Streibig JC (2004) Independent action and apparent synergistic effects between toxicants. SETAC Globe 5: 33–34.
29. Tomlin CDS (2002) The e-Pesticide Manual.
30. Vanegas C, Espina S, Botello AV, Villanueva S (1997) Acute toxicity and synergism of cadmium and zinc in white shrimp, Penaeus setiferus, juveniles. Bull Environ Contam Toxicol 58: 87–92.
31. Zhu B, Wu ZF, Li J, Wang GX (2011) Single and joint action toxicity of heavy metals on early developmental stages of Chinese rare minnow (Gobiocypris rarus). Ecotoxicol Environ Saf 74: 2193–2202.
32. Rahman MS, Saha N, Molla AH, Al-Reza SM (2014) Assessment of Anthropogenic Influence on Heavy Metals Contamination in the Aquatic Ecosystem Components: Water, Sediment, and Fish. Soil Sediment Contam 23: 353–373.
33. Nriagu JO, Wong HKT, Lawson G, Daniel P (1998) Saturation of ecosystems with toxic metals in Sudbury basin, Ontario, Canada. Sci Tot Environ 223: 99–117.
34. Okamura H, Mieno H (2006) Present status of antifouling systems in Japan: tributyltin substitutes in Japan. In: Antifouling paint biocides. Berlin: Springer. 201–212.
35. Posthuma L, Baerselman R, Van Veen RPM, Dirven-Van Breemen EM (1997) Single and joint toxic effects of copper and zinc on reproduction of Enchytraeus crypticus in relation to sorption of metals in soils. Ecotoxicol Environ Saf 38: 108–121.
36. Elliott NG, Swain R, Ritz DA (1986) Metal Interaction During Accumulation by the Mussel Mytilus-Edulis-Planulatus. Mar Biol 93: 395–399.
37. Meyer JS, Ranville JF, Santore RC, Ryan AC, Gorsuch JW, et al.(2013) Toxicity of metal mixtures to *Daphnia magna*: Competitive binding to dissolved organic matter and biotic ligand can explain non-additive interactions. SETAC Europe 23rd Annual Meeting, Glasgow, UK, 12–16 May 2013.
38. Mochida K, Ito K, Harino H, Kakuno A, Fujii K (2006) Acute toxicity of pyrithione antifouling biocides and joint toxicity with copper to red sea bream (Pagrus major) and toy shrimp (Heptacarpus futilirostris). Environ Toxicol Chem25: 3058–3064.
39. Bao VWW, Leung KMY, Kwok KWH, Zhang AQ, Lui GCS (2008) Synergistic toxic effects of zinc pyrithione and copper to three marine species: Implications on setting appropriate water quality criteria. Mar Pollut Bull 57: 616–623.
40. Zhou XJ, Okamura H, Nagata S (2006) Remarkable synergistic effects in antifouling chemicals against Vibrio fischeri in a bioluminescent assay. J Health Sci 52: 243–251.
41. Nakajima K, Yasuda T, Nakazawa H (1990) High-Performance Liquid-Chromatographic Determination of Zinc Pyrithione in Antidandruff Preparations Based on Copper Chelate Formation. J Chromatography 502: 379–384.
42. Koutsaftis A, Aoyama I (2006) The interactive effects of binary mixtures of three antifouling biocides and three heavy metals against the marine algae Chaetoceros gracilis. Environ Toxicol 21: 432–439.
43. Belden JB, Lydy MJ (2000) Impact of atrazine on organophosphate insecticide toxicity. Environ Toxicol Chem19: 2266–2274.
44. Kennaugh L, Pearce D, Daly JC, Hobbs AA (1993) A piperonyl butoxide synergizable resistance to permethrin in *Helocoverpa armigera* which is not due to increased detoxification by cytochrome P450. Pest Biochem Physiol 45: 234–241.
45. Stephenson GR, Solomon KR (2007) Pesticide selection, formulation and application. In: Stephenson GR, Solomon KR, editors. Pesticides and the Environment. Guelph: Canadian Network of Toxicology Centres Press. 57–68.
46. Cedergreen N, Streibig JC (2005) The toxicity of herbicides to non-target aquatic plants and algae: Assessment of predictive factors and hazard. Pest Manage Sci 61: 1152–1160.
47. Cedergreen N, Kudsk P, Matthiasen S, Streibig JC (2007) Combination effects of herbicides: Do species and test system matter? Pest Manage Sci 63: 282–295.
48. Hartmann NB, Von der Kammer F, Hofmann T, Baalousha M, Ottofuelling S, Baun A (2010) Algal testing of titanium dioxide nanoparticles-Testing considerations, inhibitory effects and modification of cadmium bioavailability. Toxicol 269: 190–197.
49. Walker CH (2009) Factors determining the toxicity of organic pollutants to animals and plants. In: Organic Pollutants. London: CRC Press. 17–66.
50. Bolwell GP, Bozak K, Zimmerlin A (1994) Plant Cytochrome-P450. Phytochemistry 37: 1491–1506.
51. Guengerich FP (2008) Cytochrome P450 and chemical toxicology. Chem Res Toxicol 21: 70–83.
52. Johnson RM, Dahlgren L, Siegfried BD, Ellis MD (2013) Acaricide, Fungicide and Drug Interactions in Honey Bees (Apis mellifera). Plos One: 8.
53. Nørgaard KB, Cedergreen N (2010) Pesticide cocktails can interact synergistically on aquatic crustaceans. Environ Sci Pollut Res 17: 957–967.
54. Pilling ED, Jepson PC (1993) Synergism Between EBI Fungicides and A Pyrethroid Insecticide in the Honeybee (Apis-Mellifera). Pest Sci 39: 293–297.
55. Pilling ED, Bromley-Challanor KAC, Walker CH, Jepson PC (1995) Mechanism of synergism between the pyrethroid insecticide lamda-cyhalothrin and the imidazole fungicide prochloraz in the honeybee (*Apis mellifera* L.). Pest Biochem Physiol 51: 1–11.
56. Forget J, Pavillon JF, Beliaeff B, Bocquene G (1999) Joint action of pollutant combinations (pesticides and metals) on survival (LC50 values) and acetylcholinesterase activity of Tigriopus brevicornis (Copepoda, Harpacticoida). Environ Toxicol Chem18: 912–918.
57. Marking LL, Dawson VK (1975) Method for Assessment of Toxicity Or Efficacy of Mixtures of Chemicals. US Fish and Wildlife Service Investigations in Fish Control 67: 1–8.
58. Cacciatore LC, Kristoff G, Guerrero NRV, Cochon AC (2012) Binary mixtures of azinphos-methyl oxon and chlorpyrifos oxon produce in vitro synergistic cholinesterase inhibition in Planorbarius corneus. Chemosphere 88: 450–458.
59. Laetz CA, Baldwin DH, Collier TK, Hebert V, Stark JD, et al. (2009) The Synergistic Toxicity of Pesticide Mixtures: Implications for Risk Assessment and the Conservation of Endangered Pacific Salmon. Environ Health Perspec 117: 348–353.
60. Mcleese DW, Metcalfe CD (1979) Toxicity of Mixtures of Phosphamidon and Methidathion to Lobsters (Homarus-Americanus). Chemosphere 8: 59–62.
61. Woods M, Kumar A, Correll R (2002) Acute toxicity of mixtures of chlorpyrifos, profenofos, and endosulfan to Ceriodaphnia dubia. Bull Environ Contam Toxicol 68: 801–808.
62. Schuler LJ, Trimble AJ, Belden JB, Lydy MJ (2005) Joint toxicity of triazine herbicides and organophosphate insecticides to the midge Chironomus tentans. Arch Environ Contam Toxicol 49: 173–177.
63. Fu Y, Li M, Liu C, Qu J, Zhu W, et al. (2013) Effect of atrazine and chlorpyrifos exposure on ctochrome P450 contents and enzyme activities in common carp gills. Ecotoxicol Environ Saf 94: 28–36.
64. Dong XL, Zhu LS, Wang JH, Wang J, Xie H, et al. (2009) Effects of atrazine on cytochrome P450 enzymes of zebrafish (Danio rerio). Chemosphere 77: 404–412.
65. Jin-Clark Y, Lydy MJ, Zhu KY (2002) Effects of atrazine and cyanazine on chlorpyrifos toxicity in Chironomus tentans (Diptera: Chironomidae). Environ Toxicol Chem21: 598–603.
66. Lydy MJ, Austin KR (2004) Toxicity assessment of pesticide mixtures typical of the Sacramento-San Joaquin Delta using Chironomus tentans. Arch Environ Contam Toxicol 48: 49–55.
67. Anderson TD, Lydy MJ (2002) Increased toxicity to invertebrates associated with a mixture of atrazine and organophosphate insecticides. Environ Toxicol Chem21: 1507–1514.
68. Lydy MJ, Linck SL (2003) Assessing the impact of triazine herbicides on organophosphate insecticide toxicity to the earthworm Eisenia fetida. Arch Environ Contam Toxicol 45: 343–349.
69. Pape-Lindstrom PA, Lydy MJ (1997) Synergistic toxicity of atrazine and organophosphate insecticides contravenes the response addition mixture model. Environ Toxicol Chem16: 2415–2420.
70. Vanderweiden MEJ, Tibosch HJH, Bleumink R, Sinnige TL, Vandeguchte C, et al. (1993) Cytochrome-P450-1A Induction in the Common Carp (Cyprinus-Carpio) Following Exposure to Contaminated Sediments with Halogenated Polyaromatics. Chemosphere 27: 1297–1309.
71. Johnson RM, Mao WF, Pollock HS, Niu GD, Schuler MA, et al. (2012) Ecologically Appropriate Xenobiotics Induce Cytochrome P450s in Apis mellifera. Plos One 7.
72. Henczova M, Deer AK, Filla A, Komlosi V, Mink J (2008) Effects of Cu2+ and Pb2+ on different fish species: Liver cytochrome P450-dependent monooxygenase activities and FTIR spectra. Comparative Biochemistry and Physiology C-Toxicology & Pharmacology 148: 53–60.
73. Svendsen C, Siang P, Lister LJ, Rice A, Spurgeon DJ (2010) Similarity, Independence, Or Interaction for Binary Mixture Effects of Nerve Toxicants for the Nematode Caenorhabditis Elegans. Environ Toxicol Chem29: 1182–1191.
74. Snegaroff J, Bach J (1989) Effects of the fungicide prochloraz on xenobiotic metabolism in rainbow trout - Inhibition invitro and time course of induction invivo. Xenobiotica 19: 255–267.
75. Johnston G, Walker CH, Dawson A (1994) Interactive Effects Between Ebi Fungicides (Prochloraz, Propiconazole and Penconazole) and Op Insecticides (Dimethoate, Chlorpyrifos, Diazinon and Malathion) in the Hybrid Red-Legged Partridge. Environ Toxicol Chem13: 615–620.

76. Johnston G, Walker CH, Dawson A (1994) Interactive Effects of Prochloraz and Malathion in Pigeon, Starling and Hybrid Red-Legged Partridge. Environ Toxicol Chem13: 115–120.
77. Mao WF, Schuler MA, Berenbaum MR (2011) CYP9Q-mediated detoxification of acaricides in the honey bee (Apis mellifera). Proceedings of the National Academy of Sciences of the United States of America 108: 12657–12662.
78. Burkhead JL, Reynolds KAG, Abdel-Ghany SE, Cohu CM, Pilon M (2009) Copper homeostasis. New Phytologist 182: 799–816.
79. Buchwalter DB, Cain DJ, Martin CA, Xie L, Luoma SN, et al. (2008) Aquatic insect ecophysiological traits reveal phylogenetically based differences in dissolved cadmium susceptibility. Proc Natl Acad Sci U S A 105: 8321–8326.
80. Taiz L, Zeman F (2010) Photosynthesis: The light reaction. In: Plant Physiology. Sunderland, USA: Sinauer Associates Inc. 163–198.
81. Suzuki T, Nojiri H, Isono H, Ochi T (2004) Oxidative damages in isolated rat hepatocytes treated with the organochlorine fungicides captan, dichlofluanid and chlorothalonil. Toxicol 204: 97–107.
82. Amorim MJB, Pereira C, Menezes-Oliveira VB, Campos B, Soares AMVM, et al.(2012) Assessing single and joint effects of chemicals on the survival and reproduction of Folsomia candida (Collembola) in soil. Environ Pollut 160: 145–152.
83. Lister LJ, Svendsen C, Wright J, Hooper HL, Spurgeon DJ (2011) Modelling the joint effects of a metal and a pesticide on reproduction and toxicokinetics in Lumbricid earthworms. Environ International 37: 663–670.
84. Wang H, Li Y, Huang HH, Xu X, Wang YH (2011) Toxicity Evaluation of Single and Mixed Antifouling Biocides Using the Strongylocentrotus Intermedius Sea Urchin Embryo Test. Environ Toxicol Chem30: 692–703.
85. Lawler SP, Dritz DA, Johnson CS, Wolder M (2008) Does synergized pyrethrin applied over wetlands for mosquito control affect Daphnia magna zooplankton or Callibaetis californicus mayflies? Pest Manage Sci 64: 843–847.
86. Brander SM, Werner I, White JW, Deanovic LA (2009) Toxicity of a dissolved pyrethroid mixture to *Hyalella azteca* at environmentally relevant concentrations. Environmental Toxicology and Chemistry 28: 1493–1499.
87. Martin JD, Crawford CG, Larson SJ (2003) Pesticides in streams. 1–5.
88. Altenburger R (2011) Understanding combined effects for metal co-exposure in ecotoxicology. In: Sigel A, Sigel H, Sigel KO, editors. Metal Ions in Life Sciences. London: Royal Society of Chemistry. 1–26.
89. Fernandez-Alba AR, Hernando MD, Piedra L, Chisti Y (2002) Toxicity evaluation of single and mixed antifouling biocides measured with acute toxicity bioassays. Anal Chim Acta 456: 303–312.
90. Moher D, Liberati A, Tetzlaff J, Altman DG (2009) Preferred Reporting Items for Systematic Reviews and Meta-Analyses: The PRISMA Statement. Plos Medicine 6: 1–6.

# 7

# Multiple Routes of Pesticide Exposure for Honey Bees Living Near Agricultural Fields

**Christian H. Krupke[1]*, Greg J. Hunt[1], Brian D. Eitzer[2], Gladys Andino[1], Krispn Given[1]**

**1** Department of Entomology, Purdue University, West Lafayette, Indiana, United States of America, **2** Department of Analytical Chemistry, The Connecticut Agricultural Experiment Station, New Haven, Connecticut, United States of America

## Abstract

Populations of honey bees and other pollinators have declined worldwide in recent years. A variety of stressors have been implicated as potential causes, including agricultural pesticides. Neonicotinoid insecticides, which are widely used and highly toxic to honey bees, have been found in previous analyses of honey bee pollen and comb material. However, the routes of exposure have remained largely undefined. We used LC/MS-MS to analyze samples of honey bees, pollen stored in the hive and several potential exposure routes associated with plantings of neonicotinoid treated maize. Our results demonstrate that bees are exposed to these compounds and several other agricultural pesticides in several ways throughout the foraging period. During spring, extremely high levels of clothianidin and thiamethoxam were found in planter exhaust material produced during the planting of treated maize seed. We also found neonicotinoids in the soil of each field we sampled, including unplanted fields. Plants visited by foraging bees (dandelions) growing near these fields were found to contain neonicotinoids as well. This indicates deposition of neonicotinoids on the flowers, uptake by the root system, or both. Dead bees collected near hive entrances during the spring sampling period were found to contain clothianidin as well, although whether exposure was oral (consuming pollen) or by contact (soil/planter dust) is unclear. We also detected the insecticide clothianidin in pollen collected by bees and stored in the hive. When maize plants in our field reached anthesis, maize pollen from treated seed was found to contain clothianidin and other pesticides; and honey bees in our study readily collected maize pollen. These findings clarify some of the mechanisms by which honey bees may be exposed to agricultural pesticides throughout the growing season. These results have implications for a wide range of large-scale annual cropping systems that utilize neonicotinoid seed treatments.

**Editor:** Guy Smagghe, Ghent University, Belgium

**Funding:** This research was funded by an award from the North American Pollinator Protection Campaign to CK, GH and BE and by an award from the Managed Pollinator Coordinated Agricultural Project (United States Department of Agriculture National Institute of Food and Agriculture 2009-8511805718) to GH. The funders had no role in study design, data collection or analysis, decision to publish or manuscript preparation.

**Competing Interests:** CK has received funding in the past from agrochemical companies that manufacture and distribute a wide range of pesticides, including many of those mentioned in this manuscript. These entities include the Bayer Corporation, BASF and Syngenta. None of these entities were involved with this work or are aware of it in any way, financial or otherwise.

* E-mail: ckrupke@purdue.edu

## Introduction

Pollinator health is receiving increased attention as both managed pollinators (i.e. honey bees) and native pollinator populations decline worldwide [1–3]. Several causal mechanisms (including viral pathogens, parasitic mites and pesticides) have been proposed and investigated as contributing causes [4]. Pesticide exposure has received significant attention and recently-published analyses of pollen from managed bees located near agricultural environments demonstrated that many agricultural chemicals (including insecticides, miticides, fungicides and herbicides) are detectable in honey bee wax and pollen samples [5,6]. Of the many compounds detected, the neo-nicotinoid group has arguably received the most attention. These compounds act as nicotinic acetylcholine receptor agonists in insects, causing persistent excitation of these receptors and eventually death [7]. As a group, neonicotinoids possess several key attributes that have facilitated heavy adoption in both agricultural and urban environments, including low vertebrate toxicity [8] and the ability to be translocated by plants. Neonicotinoids are also persistent, offering the potential for a large window of activity. The half-lives of these compounds in aerobic soil conditions can vary widely, but are best measured in months (148–1,155 days for clothianidin) [8]. Among the largest single uses of these compounds is application to maize seed; production of maize for food, feed and ethanol production represents the largest single use of arable land in North America. Maize planting reached unprecedented levels in the US in 2010 (35.7 million hectares [9], and is expected to increase. Virtually all if the maize seed planted in North America (the lone exception being organic production = 0.2% of total acreage [9]) is coated with neonicotinoid insecticides. There are 2 major compounds used: clothianidin and thiamethoxam; the latter is metabolized to clothianidin in the insect. The current application rates for these compounds range from 0.25 to 1.25 mg/kernel. These compounds are highly toxic to honey bees: a single kernel contains several orders of magnitude of active ingredient more than the published LD50 values for honey bees (defined as the amount of material that will kill 50% of exposed individuals), which ranges from 22–44 ng/bee for clothianidin (contact toxicity) [8,10]. Maize seeds are typically planted at a rate ca. 12,500 kernels/hectare, making it essential that any potential routes for honey bee exposure to these compounds be quantified as thoroughly as possible. This study was initiated in response to reports of bee kills at Indiana apiaries in spring of 2010. These reports coincided with the peak period of maize planting in the area [11]. Results of analyses of these bees and pollen from the hives revealed that both clothianidin and thiamethoxam were present on dead bees and in pollen collected from a single hive. These compounds were also present in dead bees from other hives but not in bees from hives that did not show mortality. Also found was atrazine, a herbicide that is commonly used in maize production and is relatively non-toxic to honey bees [12].

These preliminary results prompted additional experiments to determine how honey bees may be gaining exposure to clothianidin and other pesticides commonly applied to either maize seed or to plants later in the season. We collected samples from a variety of potential exposure routes near agricultural fields and analyzed them to determine whether pesticides were present. We sampled soils, pollen (both collected by honey bees and directly from plants), dandelion flowers, and dead and healthy bees. We also examined waste products produced during the planting of treated seed. Sowing of maize seed in North America is accomplished using tractor-drawn planters that employ a forced air/vacuum system and a perforated disc to pick up individual seeds and drop them into the planting furrow at the selected spacing. Because maize kernels are treated with neonicotinoids and other compounds (usually fungicides) they do not flow readily and may stick to one another, causing uneven plant spacing. To remedy this, talc (a common mineral composed of hydrated magnesium silicate) is typically added to seed boxes to reduce friction and stickiness and ensure smooth flow of seed during planting. Much of the talc is exhausted during planting, either down with the seed or behind the planter and into the air using an exhaust fan. We sampled waste talc following planting to determine whether this material was contaminated with pesticides abraded from treated seeds. This waste is a mixture of the talc that has been in contact with treated maize kernels and minute pieces of the seeds themselves.

## Results

Soil collected from areas near our test site revealed that neonicotinoid insecticide residues were present in all samples tested (Table 1), with clothianidin occurring in each field sampled. Herbicide residues were also found in these samples. Sampling of the waste talc from planting activities revealed that extremely high concentrations of clothianidin were found in talc exposed to treated seed (Table 2). Fungicides applied to the seed were also found. Analysis of talc used to plant untreated seed contained low quantities of the same pesticides, this is likely due to contamination and reflects the difficulties associated with thorough cleaning of equipment between plantings. Direct sampling of anthers revealed that many of the same compounds were present in maize grown from treated seed, albeit in far lower concentrations (Table 3). Collection of pollen from traps in the field demonstrated that thiamethoxam was present in 3 of 20 samples, while pollen containing clothianidin was present in 10 of 20 samples (Table 4). Fungicides were also frequently detected: azoxystrobin and propiconazole were found in all pollen samples, while trifloxystrobin was found in 12 of the 20 samples analyzed. Maize pollen was frequently collected by foraging honey bees while it was available: maize pollen comprised over 50% of the pollen collected by bees, by volume, in 10 of 20 samples.

The samples collected in 2011 revealed some similar trends (Table 5); clothianidin was found on all the dead and dying bees we sampled, while the apparently healthy bees we sampled from the same locations did not contain detectable levels of clothianidin. Atrazine and metolachlor were also found, providing further evidence that these bees were foraging near agricultural fields; as these herbicides are commonly applied prior to or during maize planting. When we sampled the contents of wax combs removed from two hives at the same apiary, we found both clothianidin and thiamethoxam in pollen removed from both hives. Nectar did not contain either compound. The miticide coumaphos was found at low levels in each nectar and pollen sample as well. Both soil and dandelion flowers obtained from the fields closest to the affected apiary (soybeans in 2010, unplanted when sampled in 2011) contained clothianidin (Table 6), therefore clothianidin in/on the dandelions could have resulted from translocation from the soil to the flower, from surface contamination of the flowers from dust, or a combination of these two mechanisms. Dandelion flowers growing far from agricultural areas served as controls; no neonicotinoids were detected.

## Discussion

The results we present here more clearly define some potential intersections between foraging honey bees and some of the seed treatments used during planting of maize. These results demonstrate that honey bees living and foraging near agricultural fields are exposed to neonicotinoids and other pesticides through multiple mechanisms throughout the spring and summer. The potential for greatest exposure (and the period when mortality was noted), occurs during planting time when there is potential for exposure to extremely high concentrations of neonicotinoids in waste talc that is exhausted to the environment during and after planting. Furthermore, we show that bees living in these environments will forage for maize pollen and transport pollen

**Table 1.** Pesticide concentrations found in soil samples taken from production fields surrounding study area, all concentrations shown are expressed as parts per billion.[1,2]

| FIELD HISTORY | THIAMETHOXAM LOD = 1.0 | CLOTHIANIDIN LOD = 2.0 | IMIDACLOPRID LOD = 1.0 | METOLACHLOR LOD = 2.0 | ATRAZINE LOD = 0.5 |
|---|---|---|---|---|---|
| MAIZE-MAIZE* | ND | **6.3** | **2.9** | **5.9** | **52.0** |
| SOY-SOY | ND | **9.6** | **7.3** | **11.1** | **7.8** |
| MAIZE-SOY | ND | **4.9** | **ND** | **6.1** | **8.5** |
| SOY-MAIZE | ND | **2.1** | **ND** | **ND** | **22** |

[1]ND = Not detected.
[2]* = Experimental field where hives were placed in 2010.

**Table 2.** Pesticide concentrations found in talc samples removed from planter following planting with treated and untreated maize seed, all concentrations shown are expressed as parts per million.[1,2]

| SEED TYPE | THIAMETHOXAM LOD = 0.02 | CLOTHIANIDIN LOD = 0.04 | METALAXYL LOD = 0.04 | TRIFLOXYSTROBIN LOD = 0.04 |
|---|---|---|---|---|
| TALC ONLY | ND | ND | ND | ND |
| COMMERCIALLY TREATED MAIZE SEED 1 | 735 | 3400 | 116 | 66 |
| COMMERCIALLY TREATED MAIZE SEED 2** | 68 | 10000 | 92 | 3.8 |
| COMMERCIALLY TREATED MAIZE SEED 3 | 13240 | 4900 | 263 | 503 |
| COMMERCIALLY TREATED MAIZE SEED 4 | 70 | 15030 | 131 | 4.4 |
| COMMERCIALLY TREATED MAIZE SEED 5 | 588 | 11413 | 116 | 189 |
| UNTREATED MAIZE SEED*** | 12 | 12 | 4 | 4 |

[1]ND = Not detected.
[2]**, *** = Talc samples taken following planting of treated and untreated sections of experimental fields, respectively, in 2010.

containing neonicotinoids to the hive. Pollen contaminated with levels of neonicotinoids similar to those shown in our results has been known to impair pollinator health [13,14,15]. Although we anticipated that planter dust may cause higher pesticide concentrations in samples taken immediately after planting our plots, we found the opposite trend: Pollen collected just prior to our planting period contained the highest levels of neonicotinoids detected (and not detected) in samples from both the treated and untreated fields. This may reflect the high variability in the types of pollen being brought back to the hive. Most of the maize in our study area was shedding pollen before and during our atypically late planting period (mid-July). After sampling anthers directly and identifying maize pollen in our samples, we know that pollen originating from treated seed does contain clothianidin, although not at the levels found in some of the bee-collected pollen samples, indicating the likelihood of additional pathways or sources. The levels of clothianidin in bee-collected pollen that we found are approximately 10-fold higher than reported from experiments conducted in canola grown from clothianidin-treated seed [16].

Detection of agricultural pesticides (and neonicotinoids in particular) in hives (including honey, pollen and wax) has been documented in the past: Bee-collected pollen was found to contain the neonicotinoid imidacloprid [17] in one study, although no adverse effects upon adults or brood were found [18]. However, a more recent study found that rearing brood in comb contaminated with pesticides (including the neonicotinoids found in our study, thiamethoxam and clothianidin) led to delayed worker development [6]. A field study examining the effects upon honey bees of clothianidin-treated canola found low levels of clothianidin in both pollen and nectar (0.93ppb and 2.59 ppb, respectively), but also found no significant effects upon honey bee populations [16]. In studies in maize, guttation droplets produced by plants grown from neonicotinoid treated seed were shown to have from 10–100 mg/L of the pesticides and were found to cause paralysis and eventual death when fed to honey bees [19], while other studies have found traces of the seed treatment imidacloprid on vegetation near maize plantings and have hypothesized that sowing treated seed can cause dispersal of dust containing insecticide [20]. Further evidence of detrimental effects of planting treated maize seed was noted by researchers in Italy, who found that honey bee mortality increased on the day seeds were planted and that numbers of foragers declined in the days following planting [21]. A subsequent study demonstrated that bees that were induced to fly near a maize planter in Europe showed up to 100 ng of clothianidin/bee upon analysis. Interestingly, however, these bees did not die unless they were kept in conditions of high humidity [22].

Detection of clothianidin in pollen, both in stored pollen in cells and in pollen traps is a critical finding because clothianidin is even more toxic when administered to bees orally, with an LD50 of 2.8–3.7 ng/bee [23,24]. Given an average weight of 80–100 mg/bee, some of our pollen sample concentrations exceed the oral LD50. This, combined with the result that our samples of dead and dying honey bees consistently demonstrated the presence of clothianidin, suggests that the levels of both clothianidin and thiamethoxam found in our sampling of stored pollen in May of 2011 may have contributed to the deaths of the bees we analyzed. However, our analytical methods do not allow us to determine what fraction of the pesticide is on the surface of bees (contact toxicity, due to drift of soil or planter exhaust) vs. inside the body (oral toxicity, due to ingestion of contaminated pollen or guttation droplets). A combination of these exposure modalities is not unlikely.

Our results also demonstrate that clothianidin is present in the surface soil of agricultural fields long after treated seed has been

**Table 3.** Pesticide concentrations found in pollen removed from maize anthers at anthesis. Samples were taken from the experimental field where hives were placed. All concentrations shown are expressed as parts per billion.[1]

| | THIAMETHOXAM LOD = 0.2 | CLOTHIANIDIN LOD = 0.5 | METALAXYL LOD = 0.5 | TRIFLOXYSTROBIN LOD = 0.5 |
|---|---|---|---|---|
| TREATED MAIZE SEED 2 | 1.7 | 3.9 | 3.1 | 1.7 |
| UNTREATED MAIZE SEED | ND | 0.3 | 4.0 | 5.5 |

[1]ND = Not detected.

**Table 4.** Pesticide concentrations found in pollen samples removed from returning foragers of hives placed adjacent to maize fields at planting, all concentrations shown are expressed as parts per billion.[1]

| HIVE NUMBER | DATE | % MAIZE POLLEN | THIAMETHOXAM LOD=0.5 | CLOTHIANIDIN LOD=1.0 | TRIFLOXYSTROBIN LOD=0.4 | AZOXYSTROBIN LOD=0.5 | PROPICONAZOLE LOD=2.0 |
|---|---|---|---|---|---|---|---|
| HIVE #5: Untreated | 2 d pre-planting | 82.7 | ND | ND | 1.6 | 22 | 9.3 |
| | 1 d pre-planting | 78.5 | ND | 20.1 | 3.9 | 23 | 12.5 |
| | At planting | 54.7 | ND | ND | 17.9 | 9.6 | 15.0 |
| | 1 d post planting | 16.4 | ND | ND | 18.0 | 9.2 | 9.6 |
| | 2 d post planting | 56.5 | ND | ND | 1.6 | 82 | 41.0 |
| HIVE #7: Untreated | 2 d pre-planting | 58.8 | ND | 1.1 | ND | 1.4 | 7.3 |
| | 1 d pre-planting | 73.4 | ND | 13.1 | ND | 17.9 | 10.9 |
| | At planting | 47.9 | ND | ND | 0.4 | 18.9 | 5.6 |
| | 1 d post planting | 8.8 | ND | 6.7 | 0.9 | 12.9 | 14.3 |
| | 2 d post planting | 52.9 | ND | 3.4 | 0.4 | 26.9 | 15.8 |
| HIVE #6: Treated | 2 d pre-planting | 69.1 | ND | ND | 0.5 | 12.4 | 5.6 |
| | 1 d pre-planting | 76.3 | ND | ND | ND | 5.4 | 7.6 |
| | At planting | 59.5 | ND | ND | 1.8 | 7.1 | 4.8 |
| | 1 d post planting | 18.2 | ND | ND | ND | 66 | 23.8 |
| | 2 d post planting | 18.7 | ND | ND | 1.2 | 7.5 | 6.5 |
| HIVE #8: Treated | 2 d pre-planting | 43.0 | 7.4 | 88 | 9.8 | 30.5 | 9.2 |
| | 1 d pre-planting | 24.0 | 2.3 | 25 | 3.3 | 11.3 | 9.5 |
| | At planting | 35.4 | 1.2 | 10 | 2.6 | 6.7 | 6.6 |
| | 1 d post planting | 2.6 | ND | 12 | 2.1 | 7.5 | 9.1 |
| | 2 d post planting | 13.5 | ND | 3.9 | 2.6 | 4.3 | 3.2 |

[1]ND = Not detected.

planted in that field. All soil samples we collected contained clothianidin, even in cases where no treated seed had been planted for 2 growing seasons. During the spring planting period, dust that arises from this soil may land on flowers frequented by bees, or possibly on the insects themselves. Of potentially greater concern are the very high levels of neonicotinoids (and fungicides) found in the talc that has been exposed to treated seed, since part of this highly mobile material is exhausted to the outside environment during planting and after planting. The large areas being planted with neonicotinoid treated seeds, combined with the high persistence of these materials and the mobility of disturbed soil and talc dust, carry potential for effects over an area that may exceed the boundaries of the production fields themselves. A key mechanism for honey bee exposure may occur during the period

**Table 5.** Pesticide concentrations found in/near apiary colonies during planting period in 2011, all concentrations shown are expressed as parts per billion.[1]

| SAMPLE TYPE | Sample wt. (g) | THIAMETHOXAM LOD=0.5 | CLOTHIANIDIN LOD=1.0 | METOLACHLOR LOD=2.0 | ATRAZINE LOD=0.5 | AZOXYSTROBIN LOD=0.5 | COUMAPHOS LOD=1.0 |
|---|---|---|---|---|---|---|---|
| Dead/dying bees | 2.96 | ND | 6.9 | 3.0 | 6.5 | ND | ND |
| Dead/dying bees | 2.47 | ND | 10.8 | 1.7 | 3.9 | ND | ND |
| Dead/dying bees | 1.32 | ND | 3.8 | 5.5 | 9.5 | ND | ND |
| Dead/dying bees | 2.57 | ND | 4.9 | 0.8 | 4.6 | ND | ND |
| Dead/dying bees | 1.62 | ND | 13.3 | 1.1 | 3.9 | ND | ND |
| Healthy bees | 0.59 | ND | ND | ND | 5.9 | ND | ND |
| Nectar hive 1 (healthy) | 5.78 | ND | ND | ND | 0.5 | 0.6 | 1.1 |
| Nectar hive 2 (sick) | 5.72 | ND | ND | ND | ND | 0.3 | 4.7 |
| Pollen hive 1 (healthy, 2 samples) | 5.05 | 6.2±4.9 | 2.9±1.3 | 28.5±3.5 | 16±1.4 | 28.5±3.5 | 1.3±0.4 |
| Pollen hive 2 (sick, 2 samples) | 5.08 | 20.4±2.7 | 10.7±2.3 | 81.5±0.7 | 36.5±3.5 | 0.8±0.3 | 2.7±0.3 |

When two aliquots of the same sample were analyzed the results are expressed as ± the standard deviation of the two analyses.
[1]ND = Not detected.

**Table 6.** Pesticide concentrations found in unplanted fields near apiary during planting period in 2011, all concentrations shown are expressed as parts per billion.[1]

| SAMPLE TYPE | Sample wt. (g) | THIAMETHOXAM LOD = 1.0 | CLOTHIANIDIN LOD = 1.0 | METOLACHLOR LOD = 0.5 | ATRAZINE LOD = 0.2 | AZOXYSTROBIN LOD = 0.2 | COUMAPHOS LOD = 1.0 |
|---|---|---|---|---|---|---|---|
| Soil, unplanted field 1, Soybeans 2010 (2 samples) | 5.15, 5.01 | **ND** | **6.0±0.3** | **1014±14** | **771±170** | **0.2±0.1** | **ND** |
| Soil, unplanted field 2, Soybeans 2010 (2 samples) | 5.28, 5.43 | **ND** | **8.9±0.1** | **8.3±0.7** | **160±15** | **26±17** | **ND** |
| Dandelions near maize field | 2.96 | **ND** | **1.4** | **49** | **677** | **ND** | **ND** |
| Dandelions near maize field | 3.81 | **1.6** | **5.9** | **64** | **1133** | **ND** | **ND** |
| Dandelions near maize field | 4.51 | **1.3** | **3.1** | **28** | **522** | **ND** | **ND** |
| Dandelions near maize field | 4.05 | **2.9** | **1.1** | **60** | **269** | **ND** | **ND** |
| Dandelions near maize field | 3.10 | **1.1** | **1.6** | **5.7** | **125** | **ND** | **ND** |
| Dandelions near maize field | 3.44 | **ND** | **9.4** | **295** | **1004** | **ND** | **ND** |
| Dandelion, CAES (non-agricultural area) | 3.93 | **ND** | **ND** | **ND** | **0.3** | **ND** | **ND** |

When two aliquots of the same sample were analyzed the results are expressed as ± the standard deviation of the two analyses.
[1]ND = Not detected.

when maize is typically planted across much of the Midwest (mid-April through early May). At this time, the energetic requirements of honey bee colonies are increasing rapidly and pollen and nectar resources are being gathered for colony growth. Talc and soil dusts from planting are mobile and have the potential to contaminate any flowering plants that are commonly found in or near agricultural fields and are visited by honey bees, including dandelion (*Taraxacum officinale*), which has been shown to be a preferred pollen and nectar source for honey bees during this period, when floral resources are relatively limited [25].

Later in the season, when planting is largely complete, we found that honey bees will collect maize pollen that contains translocated neonicotinoids and other pesticides from seed. Translocation of neonicotinoids into pollen has previously been reported for maize grown from imidacloprid-treated seed [26], although the degree to which honey bees in our study gathered maize pollen was surprising. The finding that bee-collected pollen contained neonicotinoids is of particular concern because of the risks to newly-emerged nurse bees, which must feed upon pollen reserves in the hive immediately following emergence. Pollen is the primary source of protein for honey bees, and is fed to larvae by nurse bees in the form of royal jelly. A bee will consume 65 mg of pollen during the 10 day period it spends as a nurse bee [27], therefore a concentration of 20 ng/g (ppb) in pollen would correspond to a dose of 1.3 ng (65 mg×20 ng/g) or almost 50% of the oral LD50 of ca. 2.8 ng/bee [23]. Some of our pollen concentrations were even higher, although it is important to note that LD50 is measured as a one-time dose, while exposure through contaminated pollen would be spread out over the 10 d period and that there is likely substantial metabolic decay of the compounds during this time. Lethal levels of insecticides in pollen are an obvious concern, but sublethal levels are also worthy of study as even slight behavioral effects may impact how affected bees carry out important tasks such as brood rearing, orientation and communication.

Also potentially important are the three fungicides found in bee-collected pollen samples (trifloxystrobin and azoxystrobin and propiconazole). Azoxystrobin and trifloxystrobin are frequently used in maize seed treatments as protectants and all three of these compounds are also widely applied to maize in North America, even in the absence of disease symptoms [28]. These compounds are typically applied using aerial application during anthesis. Propiconazole has been shown to synergize toxicity of some neonicotinoids (thiacloprid and acetamiprid) to honey bees in the laboratory, although the same results have not been shown in field studies [10,29]. Although these fungicides are not acutely toxic to honey bees [5], the fact that they are routinely applied to areas that bees will frequent (i.e. maize plants at anthesis) coupled with the difficulties and uncertainties in assessing the toxicity of pesticide mixtures [30], indicate that they should be considered in future work.

In evaluating our results, it is important to bear in mind that toxicity is only one variable in addressing pesticide risks to pollinators – the intersection between toxicity and exposure is key in determining how much risk is posed by a toxicant to a given organism. These components are assessed by regulators in developing a "risk cup" which combines these parameters to assess the cumulative risks of a given toxicant to an organism [31]. In the case of honey bees, the toxicity of the neonicotinoid seed treatments used for large acreage field crops has been well-established [14,31], although when assessing the overall threat to posed to honey bee populations, calculations are complicated even further by the observation that sublethal doses of insecticides can weaken bees and increase susceptibility to key parasites or pathogens [32].

Because we found these compounds in pollen, oral LD50 is a relevant parameter in discussing toxicity to honey bees. In terms of acute toxicity (based on the oral LD50 of 2.8 ng/bee [23]), the amount of clothianidin on a single maize seed at the rate of 0.5 mg/kernel contains enough active ingredient to kill over 80,000 honey bees. However, the overall level of risk has been more difficult to quantify, as there has not been a clear mechanism whereby honey bees could be exposed to high levels of these compounds – once the treated seed is planted, opportunities for honey bee exposure to concentrations of neonicotinoids over a wide area should drop dramatically (although see [19]). Our results suggest that of the factors we quantified in this study, used talc exhausted during and after planting (the latter would occur during routine cleaning of planting equipment) stands out as potential routes for exposure that should be prioritized for further quantification and remediation. A recently published review of the risks posed by planting treated seeds in the E.U. estimates that measures taken there may reduce the dust generated during planting by 99% [33]. In North America, different planting equipment is used and there are currently no guidelines for disposal of waste talc, nor are there devices for filtering exhaust material from the vacuum planting systems. Producers may be largely unaware that this material is highly toxic to pollinators. However, given the unprecedented levels of maize production across the United States, coupled with the increasing adoption of neonicotinoid seed treatments in other annual crops covering a wide area, including soybeans (31.3 million ha), wheat (24.7 million ha), and cotton (4.4 million ha, all figures 2010 planting) [9], it is clear that this material presents a risk that is worthy of further investigation and possibly corrective action.

Our findings have implications both for honey bees located near these crops year-round, but also for migratory colonies (bees used to pollinate winter-flowering specialty crops in western North America, such as almonds and other fruit and nut crops). Many of these colonies reside in areas where treated seed is used extensively (i.e. the upper Midwestern United States) during the period from early spring through late fall. During this period, these bees forage on a variety of crops that may be planted using neonicotinoid treated seed, including maize, soybeans and canola. Although our study was confined to honey bees, these results are relevant for any pollinators that forage in or near agricultural fields, both in the crop itself or on other flowering plants (i.e. weeds) that are present in or near the field.

## Materials and Methods

### Ethics statement

No specific permits were required for the described field studies. All studies were performed at a Purdue Agricultural Center, which is comprised of agricultural land that is owned and maintained by Purdue University specifically for research trials. All products used during maize planting (seed, herbicides, fertilizer etc.) were registered for legal use and were applied in accordance with label guidelines. These field studies did not involve endangered or protected species.

We based our study design upon planting practices, which helped us to identify several risk factors: during maize planting, tractor movement and wind generate large quantities of dust (i.e. disturbed soil) that may land directly on bees and/or the flowers that they visit, making this dust one potential route of exposure for foraging bees.

### Sample preparation and chemical analysis

All samples were analyzed using a modified version of the QuEChERs protocol [34]. The pesticides were extracted from the matrix (i.e. soil, talc or pollen) by agitation with water, acetonitrile and the salts magnesium sulfate and sodium acetate. After centrifuging, a portion of the supernatant acetonitrile layer containing the pesticides was further cleaned by the addition of solid phase dispersants (primary secondary amine, magnesium sulfate, C-18 silica). An aliquot of solvent was then concentrated and analyzed by liquid chromatography/mass spectrometry utilizing an Agilent 1200 LC interfaced to a Thermo -LTQ mass spectrometer. The MS was operated in a positive electrospray mode with a separate MS/MS scan function utilized for each pesticide monitored. This analytical technique provides both qualitative (retention time and mass spectra) and quantitative information. The combined techniques allow the various pesticides to be unambiguously identified at the parts per billion concentration level.

### Pollen identification

A portion of each pollen sample from 2010 experiments was used to count and identify the types of pollen collected by the bees. The identification of maize pollen was accomplished using comparisons to reference slides prepared from maize pollen in our study area. The other pollen types were identified using comparisons to online images (www.polleninfo.org). The four pollen types identified were maize (*Zea mays*), common dandelion (*Taraxacum officinale*), plantain (*Plantago* spp.) and goldenrod (*Solidago* spp.). To estimate the proportion of maize pollen present, a subsample of pollen suspended in water was pipetted onto a disposable hemacytometer (DRM-700 Cell-VU CBC) and the pollen grains were counted. Fresh maize pollen grains are prolate spheroidal in shape [35]. The maize pollen volume was estimated measuring the principal diameters of five individual grains from each subsample, resulting in an average volume for each maize pollen grain of $3.69\times10^{-4}$ mm$^3$.

### Field experiment design, planting and sample collection methods

Our experimental field was surrounded by fields that had been planted with maize and soybeans. We sampled soil collected from the experimental field and each of the adjacent fields. Soil samples of 500 cc were taken at four equidistant locations within each field on July 7, 2010. Because we were most interested in the content of dust that may arise from disturbed soil, we confined our sampling to the top 2–3 cm of soil. The soil samples from each field were pooled and mixed prior to analysis. A 500 cc sample of soil was taken from each pooled sample for chemical analysis. On July 8 2010, we placed 8 hives surrounding an unplanted field in an area of long-term maize and soybean planting. The field dimensions were approximately 142 m wide by 148 m long. This area was divided in half, and the hives were then arranged such that each field half had a hive in the center of each border. Hive entrances faced the field interior. Pollen traps (Model #M00682, Dadant and Sons, Inc. M. Hamilton, IL) were placed in the entrances of each hive and pollen samples collected daily both before and after the field was planted. Samples were labeled and stored at −10°C prior to analysis. Hives were selected for approximately equal populations. Each hive contained about 20,000 to 30,000 bees. This level of stocking is much lower than other apiaries that are kept closer to Purdue University, and not expected to result in excessive competition for resources. However, honey bees will forage up to several kilometers if necessary and individual hives will vary in their foraging needs and resource preferences. We chose this number of hives to give an adequate sampling of the agricultural fields in which they were placed, without the expectation that they would forage primarily in the experimental field.

On July 12 2010, we planted half of the field with commercial maize seed treated with 1.25 mg/kernel of clothianidin, adding talc to each seed box at the recommended rate (approx. 240 cc talc/ 75 kg of maize seed) [36]. Because untreated maize seed was not available commercially, we used maize harvested the previous year to plant the other half of the field. Talc was added to this seed at the same rate as above. Fields were planted using a 6-row John Deere 7200 MaxEmerge planter. Collection of waste talc for analysis was performed following planting by manually removing approximately 50 g of talc from the manifold of the planter vacuum system. The planter and vacuum system was exhausted thoroughly and cleaned with compressed air prior to each planting and following each collection at a location far from the experimental fields. Samples of fresh pollen from maize anthers were taken by removing the entire tassel from approximately 100 plants in the field while pollen was being shed. Samples were vigorously shaken into paper bags. The resulting mixture of pollen and anthers was spread over paper towels. Anthers and other large debris were removed so that only pollen grains remained for analysis.

## 2011 sampling

During the spring of 2011, we again received reports of dead and dying bees at a local apiary, located in a small wooded area near maize and soybean production fields in northwestern Indiana. As in 2010, these reports coincided with local planting and tillage activities. We collected bees from the entrances of several hives on May 10$^{th}$ and 12$^{th}$, 2011. We also collected apparently healthy returning foragers from hives at the same apiary. We removed frames containing nectar and pollen from two colonies at this location: one frame was taken from a hive with dead bees near its entrance, the second frame was removed from a nearby hive without any dead bees visible. Pollen and nectar from these frames was removed from cells for analysis, and two separate analyses of pollen samples were conducted. Finally, we collected samples of surface soil (using the methods outlined above) and dandelion flowers (multiple areas sampled, approximately 7–10 flowers were collected/sample) from maize fields within 2 km of this apiary that were being planted at the time of our bee collections.

We also collected additional waste talc samples in 2011, using commercially available neonicotinoid treated maize seed from several different manufacturers. Because our goal was to develop a representative sample from a variety of maize hybrids used in our research area, all hybrids were selected based upon agronomic suitability for local planting. Both clothianidin and thiamethoxam treated seed was used, at application rates ranging from 0.25 mg/kernel to 1.25 mg/kernel. Talc was added to each seed box at the recommended rate (approx. 240 cc talc/75 kg of maize seed) (36). Fields were planted using a 6-row John Deere 7200 MaxEmerge planter. Collection of waste talc for analysis was performed following planting by manually removing approximately 50 g of talc from the manifold of the planter vacuum system using a scoopula. The planter and vacuum system was exhausted thoroughly and cleaned with compressed air prior to each planting and following each collection.

## Acknowledgments

We appreciate the comments of anonymous reviewers whose thoughtful and thorough reviews helped improve the manuscript. We thank Alexzandra Murphy, who assisted with collation of background literature. Caitlin Race, Madeline Spigler and Nicole Parker assisted with colony maintenance and sample collection. We are also grateful for the assistance of Peter Illingworth and Larry Bledsoe who aided in planting of fields.

## Author Contributions

Conceived and designed the experiments: CK GH BE. Performed the experiments: CK BE GA. Analyzed the data: CK BE. Contributed reagents/materials/analysis tools: BE CK. Wrote the paper: CK GH BE. Maintained and transported honeybee colonies used for these experiments: KG.

## References


1. vanEngelsdorp D, Underwood R, Caron D, Hayes J (2007) An estimate of managed colony losses in the winter of 2006–2007: A report commissioned by the Apiary Inspectors of America. Am Bee J 147: 599–603.
2. vanEngelsdorp D, Hayes J, Underwood RM, Pettis J (2008) A survey of honey bee colony losses in the U.S., fall 2007 to spring 2008. PLoS ONE 3: 1–6 e4071.
3. Biesmeijer JC, Roberts SPM, Reemer M, Ohlemuller R, Edwards M, et al. (2006) Parallel declines in pollinators and insect-pollinated plants in Britain and the Netherlands. Science 313: 351–354.
4. Alaux C, Brunet JT, Dussaubat C, Mondet F, Tchamitchan S, et al. (2009) Interactions between Nosema microspores and a neonicotinoid weaken honey bees. Env Microbiol 12: 774–782.
5. Mullin CA, Frazier M, Frazier JL, Ashcraft S, Simonds R, et al. (2010) High levels of miticides and agrochemicals in North American apiaries: Implications for honey bee health. PLoS ONE 5: e9754.
6. Wu J, Anelli C, Sheppard W (2011) Sub-lethal effects of pesticide residues in brood comb on worker honey bee (*Apis mellifera*) development and longevity. PLoS ONE 6: e14720.
7. Jeschke P, Nauen R (2008) Neonicotinoids – from zero to hero in insecticide chemistry, Pest Mgt Sci 64: 1084–1098.
8. US Environmental Protection Agency, Office of Pesticide Programs. Fact-sheet Clothianidin (2003) EPA Publication 7501C; www.epa.gov/opprd001/factsheets/clothianidin.pdf.
9. National Agriculture Statistics Service (NASS), online database: Crop production 2010 summary (2010) Available: http://usda.mannlib.cornell.edu/MannUsda/viewDocumentInfo.do?documentID=10471. Accessed 2011 July 10.
10. Iwasa T, Motoyama N, Ambrose JT, Roe RM (2004) Mechanism for the differential toxicity of neonicotinoid insecticides in the honey bee, *Apis mellifera*. Crop Prot 23: 371–378.
11. National Agriculture Statistics Service (NASS) online database (2010) Indiana crop and weather report, report for week ending April 25, 2010. Available: www.nass.usda.gov. Accessed 2011 April 15.
12. US Environmental Protection Agency, Office of Pesticide Programs. Decision documents for atrazine (2003) Available: http://www.epa.gov/oppsrrd1/REDs/atrazine_combined_docs.pdf. Accessed 2011 April 15.
13. Decourtye A, Lacassie E, Pham-Delegue M (2003) Learning performances of honey bees (*Apis mellifera* L.) are differentially affected by imidacloprid according to the season. Pest Mgt Sci 59: 269–278.
14. Desneux N, Decourtye A, Delpuech JM (2007) The sublethal effects of pesticides on beneficial arthopods. Annu Rev Entomol 52: 81–106.
15. Mommaerts V, Reynders S, Boulet J, Besard L, Sterk G (2010) Risk assessment for side-effects of neonicotinoids against bumblebees with and without impairing foraging behavior. Ecotoxicol 19: 207–215.
16. Cutler GC, Scott-Dupree C (2007) Exposure to clothianidin seed-treated canola has no long-term impact on honey bees. J Econ Entomol 100: 765–772.
17. Chauzat MP, Faucon JP, Martel AC, Lachaize J, Cougoule N, et al. (2006) A survey of pesticide residues in pollen loads collected by honey bees in France. J Econ Entomol 99: 253–262.
18. Chauzat MP, Carpentier P, Martel AC, Bougeard S, Cougoule N, et al. (2009) Influence of pesticide residues on honey bee (Hymentoptera: Apidae) colony health. Environ Entomol 38: 514–523.
19. Girolami V, Mazzon L, Squartini A, Mori N, Marzaro M, et al. (2009) Translocation of neonicotinoid insecticides from coated seeds to guttation drops: a novel way of intoxication for bees. J Econ Entomol 102: 1808–1815.
20. Greatti M, Barbattini R, Stravisi A, Sabatini AG, Rossi S (2006) Presence of the a.i. imidacloprid on vegetation near corn fields sown with Gaucho dressed seeds. Bull Insectol 59: 99–103.
21. Tremolada P, Mazzoleni M, Saliu F, Colombo M, Vighi M (2010) Field trial for evaluating the effects on honey bees of corn sown using Cruiser and Celest xl treated seeds. Bull Environ Contam Toxicol 85: 229–234.
22. Girolami V, Marzaro M, Vivan L, Mazzon L, Greatti M, et al. (2011) Fatal powdering of bees in flight with particulates of neonicotinoids seed coating and humidity implication. doi: 10.1111/j.1439-0418.2011.01648.x.
23. Laurino D, Porporato M, Patetta A, Manino A (2011) Toxicity of neonicotinoid insecticides to honey bees: laboratory tests. Bull Insectol 64: 107–113.
24. US Environmental Protection Agency, Office of Chemical Safety and Pollution Prevention (2011) PC Code: 044309. Memorandum: Clothianidin registration of Prosper T400 seed treatment on mustard seed and Poncho/Votivo seed treatment on cotton. (2010). Available: www.epa.gov/pesticides/chem_search/cleared_reviews/csr_PC-044309_2-Nov-10_b.pdf Accessed 2011 Aug 1.
25. Ginsberg HS (1974) Ecology of bees in an old field. Ecology 64: 165–175.
26. Bonmatin JM, Marchand PA, Charvet R, Moineau I, Bengsch ER, et al. (2005) Quantification of imidacloprid update in maize crops. J Agri Food Chem 53: 5336–5341.
27. Rortais A, Arnold G, Hahm MP, Touffet-Briens F (2005) Modes of honey bee exposure to systemic insecticides: estimated amounts of contaminated pollen and nectar consumed by different categories of bees. Apidologie 36: 71–83.
28. Paul PA, Madden LV, Bradley CA, Robertson AE, Munkvold GP, et al. (2011) Meta-analysis of yield response of hybrid field corn to foliar fungicides in the US corn belt. ;doi: 10.1094/phyto-03-11-0091.
29. Schmuck R, Stadler T, Schmidt HW (2003) Field relevance of a synergistic effect observed in the laboratory between an EBI fungicide and a chloronicotinyl insecticide in the honey bee (*Apis mellifera* L, Hymenoptera). Pest Mgt Sci 59: 279–286.
30. Lydy M, Belden J, Wheelock C, Hammock B, Denton D (2004) Challenges in regulating pesticide mixtures. Ecology and Society 9(6): 1.

31. Decourtye A, Devillers J In: Steeve Herve Thany, editor. Insect Nicotinic Acetylcholine Receptors. Springer Press, New York. pp 85–96.
32. Vidau C, Diogon M, Aufauvre J, Fontbonne R, Vigues B, et al. (2011) Exposure to sublethal doses of fipronil and thiacloprid highly increases mortality of honey bees previously infected by *Nosema ceranae*. PLoS One 6: e21550.
33. Thompson HM (2010) Risk assessment for honey bees and pesticides: recent developments and 'new issues'. Pest Mgt Sci 66: 1157–1162.
34. Anastassides M, Lehotay SJ, Stajnbaher D, Schenk FJ (2003) Fast easy multiresidue method employing acetonitrile partitioning and 'dispersive solid phase extraction' for determination of pesticide residues in produce. J AOAC Int 86: 412–431.
35. Kiesselbach TA (1949) The structure and reproduction of corn. Research Bulletin 161. LincolnNE: Univ. of Nebraska, College of Agriculture.
36. Deere & Company, Publication A83046. Ready to plant guide. Available: http://www.deere.com/en_US/ag/media/pdf/ready_to_plant/rtp_guide.pdf. Accessed 2011 Jun 20.

# 8

# Expression of Xenobiotic Metabolizing Cytochrome P450 Genes in a Spinosad-Resistant *Musca domestica* L. Strain

**Dorte H. Højland, Karl-Martin Vagn Jensen, Michael Kristensen***

Department of Agroecology, Aarhus University, Slagelse, Denmark

## Abstract

***Background:*** Spinosad is important in pest management strategies of multiple insect pests. However, spinosad resistance is emerging in various pest species. Resistance has in some species been associated with alterations of the target-site receptor, but in others P450s seems to be involved. We test the possible importance of nine cytochrome P450 genes in the spinosad-resistant housefly strain 791spin and investigate the influence of spinosad on P450 expression in four other housefly strains.

***Results:*** Significant differences in P450 expression of the nine P450 genes in the four strains after spinosad treatment were identified in 40% of cases, most of these as induction. The highly expressed *CYP4G2* was induced 6.6-fold in the insecticide susceptible WHO-SRS females, but decreased 2-fold in resistant 791spin males. *CYP6G4* was constitutively higher expressed in the resistant strain compared to the susceptible strain. Furthermore, *CYP6G4* gene expression was increased in susceptible WHO-SRS flies by spinosad while the expression level did not alter significantly in resistant fly strains. Expression of *CYP6A1* and male *CYP6D3* was constitutively higher in the resistant strain compared to the susceptible. However, in both cases male expression was higher than female expression.

***Conclusion:*** *CYP4G2*, *CYP6A1*, *CYP6D3* and *CYP6G4* have expressions patterns approaching the expectations of a hypothesized sex specific spinosad resistance gene. *CYP4G2* fit requirements of a spinosad resistance gene best, making it the most likely candidate. The overall high expression level of *CYP4G2* throughout the strains also indicates importance of this gene. However, the data on 791spin are not conclusive concerning spinosad resistance and small contributions from multiple P450s with different enzymatic capabilities could be speculated to do the job in 791spin. Differential expression of P450s between sexes is more a rule than an exception. Noteworthy differences between spinosad influenced expression of P450 genes between a field population and established laboratory strains were shown.

**Editor:** John Vontas, University of Crete, Greece

**Funding:** This work was funded by a mobility stipend to Dorte H. Højland from the Danish Agency for Science Technology and Innovation. The funders had no role in study design, data collection and analysis, decision to publish, or preparation of the manuscript.

**Competing Interests:** The authors have declared that no competing interests exist.

* Email: michael.kristensen@agrsci.dk

## Introduction

Spinosad is a mixture of two macrocyclic lactones; spinosyn A and spinosyn D isolated from the actinomycete bacteria, *Saccharopolyspora spinosa* and has been developed as a commercial insecticide [1]. Its mode of action is unique as its primary target site appears to be a subtype of the nicotinic acetylcholine receptors (nAChRs) with also a secondary target site suggested to be the γ-aminobutyric acid (GABA)-gated chloride channel [2]. Spinosad has been introduced for insect control initially in 1997 and plays an essential role for the control of Lepidoptera and Diptera pests as well as planthoppers, spider mites, ticks, fleas and head lice. There are more than 30 examples of resistance to spinosad and half of these were selected in the field [3]. The mechanism of spinosad resistance has been implied to involve changes of target site and metabolism [4].

A key element in preventing development of resistance as well as resistance management is the understanding of the molecular mechanisms potentially responsible for resistance. Prior to the introduction a decade ago of bait-formulated spinosad for housefly control in Denmark we established a baseline for spinosad toxicity of Danish houseflies. Field populations of houseflies collected from livestock farms throughout Denmark showed up to 7.5-fold resistances (compared to the susceptible reference strain WHO-SRS) to spinosad in feeding bioassays, which was considered to reflect the natural variation of spinosad toxicity in field populations [5]. Additional studies with insecticide-resistant laboratory strains showed that there was no cross-resistance to the major insecticide classes in the housefly [5].

Spinosad resistance in the laboratory housefly strain *rspin* [6] selected for high spinosad resistance, is inherited as a recessive trait linked to autosome 1. The resistance in the *rspin* strain does apparently not involve P450-mediated metabolism as resistance level is unchanged upon pretreatment with the xenobiotic metabolism inhibitor piperonyl butoxide (PBO) [6]. Likewise, spinosad resistance in a Chinese housefly strain was autosomal, although incompletely dominant, controlled by multiple genes and not influenced by PBO [7].

Spinosad resistance in the multi-resistant Danish field population (791a) has been shown to be associated with P450 activity in a study implementing PBO based bioassays as well as gene expression studies [8]. In the spinosad-selected strain 791spin strain, originating from 791a, the male determining factor are located on autosome III [9]. The housefly *CYP4G2* is also located on chromosome III [10]. Insect CYP4G is remarkable by having orthologs distributed across Insecta [11]. The CYP4G enzymes function as oxidative decarbonylases, catalysing the terminal step in insect hydrocarbon production synthesizing alkanes making up a large part of the cuticular hydrocarbons [12].

If spinosad influence expression of xenobiotic genes including P450 genes, we hypothesize that variation of this response is a selectable trait in adaptation to an environment influenced by spinosad, thus having a resistance potential. Additionally, the coordination of P450 gene expression in response to spinosad can elucidate important new insights into the general response to a xenobiotic compound. This paper thus investigates the constitutive and spinosad-induced expression levels of nine resistance associated housefly P450 genes. CYP6A1 assists in the metabolism of organochlorine- and organophosphate-insecticides [13,14] and is constitutively over-expressed in an organophosphate-resistant strain [15,16] as well as in neonicotinoid-resistant strains [17]. *CYP6A36* is constitutively overexpressed in a permethrin-resistant strain, whereas no difference was found for *CYP6A37* [18]. *CYP6D1* is constitutively over-expressed in pyrethroid-resistant houseflies [19,20] and in neonicotinoid-resistant strains [17]. *CYP6D3* is constitutively over-expressed in pyrethroid-resistant houseflies and previously reported involved in insecticide resistance in the housefly [20]. *CYP6G4* is the potential housefly orthologue of *Drosophila melanogaster CYP6G1*, where over-expression is correlated with DDT and neonicotinoid resistance [21,22]. The mitochondrial CYP12A1 and CYP12A2 metabolize a variety of insecticides and xenobiotics and are constitutively over-expressed in the diazinon-resistant Rutgers strain [23].

## Methods

### Houseflies

Housefly breeding followed standard laboratory conditions. Egg laying was performed on crumpled filter paper soaked in whole milk. Breeding jars (5 L plastic buckets) containing 4 L of medium were seeded with 200 mg of eggs, corresponding to 2700 eggs. The breeding medium consisted of wheat bran 400 g, lucerne meal 200 g, bakers yeast 10 g, malt extract 15 mL, whole milk 500 mL and water 500 mL. For adult feeding, cube sugar and water were given continuously. Feeding started after emergence with whole-milk powder mixed with icing sugar (1:1 w/w) [24].

The insecticide-susceptible standard reference strain WHO-SRS was received in 1988 from the Department of Animal Biology, University of Pavia, Italy.

The unselected neonicotinoid resistant laboratory population 766b was collected in 2005 in Denmark. Female and male 766b flies were 140-fold and 130-fold resistant to imidacloprid at $LC_{50}$, respectively [25]. The 766b strain is susceptible to spinosad [8].

The multi-resistant 791a laboratory population was collected in the context of a resistance survey in 1997 in Denmark. The strain has never been selected after collection. The strain was highly resistant to pyrethroid, anti-cholinesterase and showed some resistance to the chitin synthesis disrupting larvicides as well as fipronil [26–28]. Female and male 791a flies were 20-fold and 22-fold resistant to imidacloprid at $LC_{50}$, respectively [17]. 791a females were 27-fold spinosad-resistant at $LC_{50}$, whereas 791a male houseflies were susceptible (5-fold resistant) [8].

The spinosad selected 791spin strain was created by selection of the spinosad resistant field strain 791a. The initial selection of 791spin was made by 24 h non-choice feeding sugar impregnated with 71 μg spinosad $g^{-1}$ sugar; males ($N = 573$, 9% survival) and females ($N = 406$, 32% survival). Selection was repeated in generations 2, 5, 7, 10, 13, 18 and 22 after the initial selection, with increasing concentrations of spinosad. 791spin females were 21-fold spinosad-resistant at $LC_{50}$, whereas 791spin male houseflies were 6-fold resistant which is considered to be within the natural variation in spinosad toxicity in susceptible Danish field populations [5]. Female and male 791spin flies were considered susceptible to imidacloprid having 3-fold and 4-fold resistance factors at $LC_{50}$, respectively [8]. The strain is retained by regular annually selections with spinosad-impregnated sugar: male flies 0.4 mg spinosad per g sugar and female flies 3.2 mg spinosad per g sugar, for a maximum 72 hours.

The field population 845b (unselected) was collected August 23. 2011 at a dairy farm located at Nykøbing Mors, Denmark (56°53′51.07″N, 8°48′42.81″E). The flies were collected on private land with consent of the owner. The field collection did not involve endangered or protected species. The strain could be characterized as a good representative for Danish field populations with no or a low level of resistance towards neonicotinoids and pyrethroids [29]. $F_1$ and $F_2$ flies from the 845b field population were used for experiments testing resistance level as well as the influence of spinosad on expression on P450 genes.

### Treatment of houseflies for gene expression analysis

Five to seven days old, adult male and female flies were subjected to a non-choice feeding test with spinosad (88%, 76.1% spinosyn A and 11.9% spinosyn D, DOW AgroSciences). The insecticide were diluted with analytic-grade acetone and impregnated on. Susceptible male and female flies were given spinosad in dose 0.063 mg per g sugar and male and female flies from the field populations 766b and 845b were given spinosad in dose 0.25 mg and 0.5 mg per g sugar, respectively. Males and females of multi-resistant strain 791a were given respectively 0.5 mg and 2 mg spinosad per g sugar [17].

All flies had access to water, milk and sugar *ad libitum* before trials. Flies, which were alive at day 5 (130–500 flies) were placed in cages with full access to water and were given excess of spinosad-treated granular sugar in a small petri-dish as the only food. The feeding tests were carried out at 25–26°C, 60–65% RH in continuous light. Twenty-four hours upon test start living and fresh looking flies were collected by vacuum suction, immediately sedated by cold and killed at −20°C. The flies were hereafter kept on −80°C until RNA extraction.

Parallel experimental set-ups were performed with resistant and susceptible flies, with the one exception that the sugar offered were left acetone-coated. These flies served as reference for examinations of constitutive gene expression levels.

### Primer designs for quantitative real-time PCR

Gene specific primer pairs were designed based on sequences obtained from the NCBI GenBank with the exception of *CYP4G2*, which was extracted from Zhu *et al.* [10]. Gao *et al.* [30] mentioned *CYP6G4* (*Musca domestica*) to be an ortholog to *CYP6G1* (*Drosophila melanogaster*). The *CYP6G1* gene was blasted against the GenBank database (blastn, http://blast.ncbi.nlm.nih.gov/) with 65% identity.

*CYP4G2*, F: 5′-cgaggaggatgatgaaataagcaagc-3′, R: 5′-ttggacatggccatcatggcatct-3′; *CYP6A1* (GenBank: M25367), F: 5′-aattttgccaatcgtggtctg-3′, R: 5′-tccaccattaccaagtggcc-3; *CYP6A36* (DQ642009), F: 5′-aaaggcatggccgttgttat-3′, R: 5′-acttgagaagcggcaaaatg-3′; *CYP6A37* (DQ642010), F: 5′-atgcaaatcctcatccccg-3′, R: 5′-ccgtgactttgtcatgggaga-3′; *CYP6D1* (U22366), F: 5′-gcaaatgcactcaggatttcc-3′, R: 5′-tgcccaagagggagatgataa-3′; *CYP6D3* (AF200191), F: 5′-tgccccataagggaggct-3′, R: 5′-agaccattgactggtactaaaaccg-3′; *CYP6G4* (FJ911556), F: 5′-gctgcaaagcaaattggg-3′, R: 5′-actacgcaccacattcag-3′; *CYP12A1* (U86618), F: 5′-atccgttgaccttgggaaatg-3′, R: 5′-tcatcctgcagcaaacctgtt-3′; *CYP12A2* (U94698), F: 5′-cctatttgagggcctgcat-3′, R: 5′-tgggaacacgatagccact-3′; *GAPDH* (DQ386609) F: 5′-ccggtatctccctcaacg-3′, R: 5′-tgacacggttggagtaaccga-3′.

The primer pairs used were designed not to span introns since the present study used gDNA for external standards in real-time PCR runs. To avoid non-specific amplification all RNA samples were routinely treated with DNase before use. Upon optimization forward and reverse primers were used in optimal concentration 150 nM. Amplicon sequence specificity was verified dissociation curves giving rise to single peaks at the specific melting temperature of the products. A full summary on primer design, optimization and validation is given by Markussen and Kristensen [17].

## RNA and DNA extraction

Total RNA from whole bodies of houseflies was extracted using the RNeasy Maxi kit (Qiagen, Ballerup, Denmark). Pools of flies (*approx.* 1.2 g equivalent to 60 flies) were thoroughly grinded with liquid nitrogen, a mortar and pestle and homogenized with buffer added β-mercaptoethanol supplied by the RNeasy kit according to the manufacturer's protocol. Isolated RNA was DNase-treated and concentrated using the RNeasy MinElute kit (Qiagen). Gel electrophoresis and spectrophotometry (Nanodrop; NanoDrop Technologies, Wilmington, USA) was performed to assess the integrity and the concentration of each RNA sample, which was dissolved in RNase-free water and stored at −20°C until use.

Extraction of gDNA used for external standards was performed according to the manufacturer's protocol for the DNeasy kit (Qiagen). Genomic DNA was stored as stocks of 125 ng $\mu L^{-1}$ at −20°C corresponding to ~120,000 copies of a single-copy gene. The mass of the haploid housefly genome (the C-value; http://www.genomesize.com) is ~1.04 pg therefore 1 ng of gDNA from *M. domestica* contains *approx.* 962 copies of a single-copy gene. A fresh 10-fold serial dilution at five quantities ranging from 125 ng (~120000 gene copies) to 0.0125 ng (~12 gene copies) was prepared for each real-time PCR run.

## RT reaction and real-time PCR

First-strand cDNA was synthesized from RNA using the High Capacity Reverse Transcription kit (ABI). In a single RT reaction, 1 μg tRNA was mixed with 10 μL RT buffer, 10 μL random primers [10X], 4 μL dNTPs [25X], 5 μL Multiscribe Reverse Transcriptase [5 U $\mu L^{-1}$] and RNase free water (Qiagen) to a final volume of 100 μL. The reaction was incubated at 25°C for 10 min for primer-RNA binding, followed by reverse transcription at 37°C for 120 min and 85°C for 5 min.

PCR of 20 μL reactions were performed using 20 ng of the cDNA samples, SYBR Green PCR master mix (ABI) and 150 nM of primers specific for the *CYP4G2*, *CYP6A1*, *CYP6A36*, *CYP6A37*, *CYP6D1*, *CYP6D3*, *CYP6G4*, *CYP12A1*, *CYP12A2* and *GAPDH* genes. All samples and the external standards were run in four technical replicates. The strain variance is accounted for by randomization of the flies selected for RNA purification as well as the number of flies used; approx. 60 per sample in addition to 2–4 biological replicas depending on the availability of houseflies. Absence of gDNA was confirmed through PCR runs of no-RT controls in replicates of two.

The PCR runs were performed on ABI PRISM 7500 HT Sequence Detection Systems with Sequence Detection system software version 1.4 (ABI) initiated by a 2 min activation step at 50°C followed by a polymerase activation step for 10 min at 95°C. Amplification was obtained by 40 cycles of 15 s at 95°C with a 1 min anneal and extending step at 60°C. A final dissociation stage at 95°C for 15 sec, 60°C for 15 sec and 95°C for 15 sec was added to generate a melting curve for verification of amplification product specificity. The PCR data in Table 1 and 2 are presented as the mean copy number per 20 ng of RNA ± standard deviation of 2–4 biological replicates. Statistical analysis was undertaken using pairwise Wilcoxon Ranking test, where a P-value less than 0.05 was considered to be statistically significant [31].

## Results and Discussion

The xenobiotic response of P450s is known to play an important role in the development of insecticide resistance. Enhancement of insecticide detoxification has been associated with increased P450 expression of genes leading to resistance [11]. The present experimental setup was selected in order to elucidate which, if any, of the nine P450 genes associated with resistance in the housefly could be responsible or partly responsible for spinosad resistance in the 791spin strain.

The P450s in this study were selected based on the documentation of their involvement in resistance and/or degradation of xenobiotics. Additionally *CYP4G2* were selected because spinosad resistance possibly is linked to chromosome III [9], where *CYP4G2* is located. The house-keeping gene *GAPDH* was added as a control of expression levels.

The four additional housefly strains of this study comes with different histories; an insecticide-susceptible reference strain (WHO-SRS) kept in laboratory breeding for decades, a neonicotinoid-resistant and spinosad-susceptible field population (766b) kept in breeding for 8 years without selection, a multi-resistant (including spinosad) field population (791a) kept in breeding without selection for 15 years and a newly collected spinosad-susceptible field population (845b).

### Gene expression of a house-keeping gene in Danish houseflies

The house-keeping gene *GAPDH* was included in this study to assess the influence of spinosad on general gene expression. For the most part, *GAPDH* gene expression remained constant within resistant laboratory strains (Table 2). The newly-collected 845b strain had a lower level of *GAPDH* gene expression than the other four strains. Short-term spinosad exposure did not alter expression significantly in any of the five strains tested here ($P>0.15$), with the exception of the insecticide-susceptible WHO-SRS strain. Here expression was increased 1.8- and 1.9-fold in males and females, respectively ($P<0.0006$).

### 791spin P450 gene expression

A comparison of P450 expression with or without spinosad exposure of 791spin and WHO-SRS showed multiple deviations from neutrality (Figure 1 and 2). Differential expression is mostly up-regulation for both strains, but a few cases of down-regulation were also observed for 791spin (Figure 2). Spinosad exposure in 791spin males caused minor effects (2-fold or less) in both directions. In females, no decreasing effect was observed for spinosad exposure, while the *CYP6A36*, *CYP6G4* and *CYP12A1* were up-regulated 4-, 8.3- and 3.4-fold, respectively (Table 1).

Most of the genes tested here had a higher constitutive expression in 791spin than in the susceptible WHO-SRS strain for both males and females (Table 1). *CYP6A36* had the lowest level of expression of the P450s investigated, while the highest expression was that of *CYP4G2* for both sexes.

**Table 1.** Constitutive and spinosad induced P450 gene expression of a spinosad-susceptible housefly strain (WHO-SRS) and spinosad -resistant housefly strain (791spin) measured by quantitative real-time PCR.

| Strain | Gene | Male | | | | Female | | | | Constitutive | Spinosad |
|---|---|---|---|---|---|---|---|---|---|---|---|
| | | Constitutive | Spinosad | Fold- change | P-value[a] | Constitutive | Spinosad | Fold- change | P-value[a] | P-value[b] | P-value[b] |
| **WHO-SRS** | *CYP4G2* | 947±517 | 1,810±611 | | 0.1014 | 2,540±1,858 | 16,680±508 | 6.6 | 0.0050 | 0.1797 | 0.0034 |
| | *CYP6A1* | 8.9±4.3 | 25.9±3.1 | 2.9 | 0.0006 | 1.9±1.2 | 3.2±0.3 | 1.7 | 0.0411 | 0.0012 | 0.0012 |
| | *CYP6A36* | 6.2±1.1 | 7.2±1.8 | | 0.3829 | 3.9±1.8 | 8.4±5.0 | | 0.0593 | 0.0734 | 0.8665 |
| | *CYP6A37* | 118±44.9 | 310±22.9 | 2.6 | 0.0034 | 53.3±5.2 | 141±18.1 | 2.6 | 0.0007 | 0.0034 | 0.0007 |
| | *CYP6D1* | 1,090±331 | 1,730±1007 | | 0.9999 | 284±58.3 | 1,320±174 | 4.7 | 0.0003 | 0.0012 | 0.9999 |
| | *CYP6D3* | 69.4±11.0 | 206±45.9 | 3.0 | 0.0003 | 13.5±4.7 | 57.9±19.7 | 4.3 | 0.0006 | 0.0006 | 0.0003 |
| | *CYP6G4* | 154±35.5 | 394±128 | 2.6 | 0.0012 | 50.8±22.9 | 348±133 | 6.9 | 0.0043 | 0.0022 | 0.4318 |
| | *CYP12A1* | 21.6±10.2 | 12.7±9.8 | | 0.1807 | 7.0±0.8 | 13.2±1.8 | 1.9 | 0.0003 | 0.0003 | 0.9452 |
| | *CYP12A2* | 36.4±7.5 | 41.5±5.5 | | 0.2786 | 10.9±3.6 | 22.5±5.0 | 2.1 | 0.0006 | 0.0002 | 0.0002 |
| | *GAPDH* | 4,860±680 | 8,570±1508 | 1.8 | 0.0006 | 4,550±319 | 8,700±676 | 1.9 | 0.0003 | 0.5358 | 0.3829 |
| **791spin** | *CYP4G2* | 2,100±497 | 1,010±318 | 0.5 | 0.0021 | 11,790±7106 | 7,800±2400 | | 0.6943 | 0.0014 | 0.0003 |
| | *CYP6A1* | 24.3±8.8 | 42.0±14.4 | 1.7 | 0.0201 | 6.2±4.0 | 6.8±3.6 | | 0.5358 | 0.0008 | 0.0002 |
| | *CYP6A36* | 2.4±0.7 | 3.8±1.1 | | 0.3717 | 0.8±0.5 | 3.2±0.7 | 4.0 | 0.0148 | 0.0003 | 0.8745 |
| | *CYP6A37* | 327±71.7 | 288±19.7 | 0.9 | 0.0019 | 166±77.1 | 128±14.1 | | 0.9999 | 0.0007 | 0.0003 |
| | *CYP6D1* | 797±144 | 599±274 | | 0.2359 | 662±569 | 472±57 | | 0.7104 | 0.6806 | 0.9999 |
| | *CYP6D3* | 405±49.8 | 467±83.5 | | 0.2198 | 51.5±28.1 | 428±191 | 8.3 | 0.0012 | 0.0001 | 0.9372 |
| | *CYP6G4* | 363±95.5 | 513±223 | | 0.3148 | 756±413 | 485±36.1 | | 0.7012 | 0.2222 | 0.7104 |
| | *CYP12A1* | 200±48.2 | 305±21.1 | 1.5 | 0.0003 | 66.4±30.6 | 224±53.9 | 3.4 | 0.0001 | 0.0006 | 0.0022 |
| | *CYP12A2* | 26.3±6.0 | 33.2±3.0 | 1.3 | 0.0091 | 18.0±12.0 | 16.7±4.2 | | 0.9452 | 0.2561 | 0.0003 |
| | *GAPDH* | 6,610±162 | 5,810±1238 | | 0.1871 | 6,750±1089 | 5,670±721 | | 0.1812 | 0.3984 | 0.7206 |

Mean mRNA transcript copy number×1000 is per 20 ng of total RNA. P-values between comparisons of male and female flies and treatment were obtained by a F-test followed by Student's Significant difference t-Test. Significant P-values (<0.05) are underlined.

[a]P-values obtained by comparisons of constitutive and spinosad induced gene expression within strain.

[b]P-values obtained by comparisons of gene expression between sexes.

**Table 2.** P450 gene expression of insecticide-susceptible and -resistant housefly strains upon sugar (constitutive) and spinosad exposure measured by quantitative real-time PCR.

| Gene | Strain | Male | | | | Female | | | |
|---|---|---|---|---|---|---|---|---|---|
| | | Constitutive | Spinosad | Fold- change | P-value[a] | Constitutive | Spinosad | Fold- change | P-value[a] |
| *CYP6A1* | WHO-SRS | 8.9±4.3 | 25.9±3.1 | 2.9 | 0.0006 | 1.9±1.2 | 3.2±0.3 | 1.7 | 0.0411 |
| | 766b | 15.6±8.9 | 33.5±11.5 | 2.1 | 0.0148 | 8.7±0.6 | 8.7±1.1 | | 0.8357 |
| | 791a | 0.8±0.4 | 0.6±0.3 | | 0.3939 | 0.6±0.3 | 0.7±0.2 | | 0.5622 |
| | 845b | 55.4±23.6 | 49.6±19.9 | | 0.8075 | 25.9±10.4 | 10.9±5.0 | 0.4 | 0.0021 |
| *CYP6A36* | WHO-SRS | 6.2±1.1 | 7.2±1.8 | | 0.3829 | 3.9±1.8 | 8.4±5.0 | | 0.0593 |
| | 766b | 15.1±5.8 | 21.2±5.3 | | 0.0721 | 11.5±5.3 | 11.4±3.0 | | 0.6126 |
| | 791a | 148±92.2 | 188±38.3 | | 0.4452 | 101±58.2 | 48.9±8.9 | | 0.2949 |
| | 845b | 34.4±22.3 | 48.7±15.8 | | 0.5362 | 46.3±22.9 | 53.4±16.2 | | 0.4385 |
| *CYP6A37* | WHO-SRS | 118±44.9 | 310±22.9 | 2.6 | 0.0034 | 53.3±5.2 | 141±18.1 | 2.6 | 0.0007 |
| | 766b | 199±110 | 283±70.8 | | 0.3374 | 95.1±27.1 | 114±46.4 | | 0.9015 |
| | 791a | 101±72.7 | 130±18.5 | | 0.9452 | 45.8±29.2 | 35.6±14.9 | | 0.5287 |
| | 845b | 202±48.5 | 254±68.9 | | 0.1042 | 156±36.5 | 218±54.3 | | 0.1321 |
| *CYP6D1* | WHO-SRS | 1,090±331 | 1,730±1,010 | | 0.9999 | 284±58.3 | 1,320±174 | 4.7 | 0.0003 |
| | 766b | 3,480±660 | 5,810±1,110 | 1.7 | 0.0022 | 1670±250 | 1,990±519 | | 0.3439 |
| | 791a | 4,010±2,652 | 3,910±570 | | 0.9999 | 2,040±1,320 | 1,190±607 | | 0.2380 |
| | 845b | 2,360±1,029 | 1,910±289 | | 0.6503 | 1,030±312 | 1,480±849 | | 0.5787 |
| *CYP6D3* | WHO-SRS | 69.4±11.0 | 206±45.9 | 3.0 | 0.0003 | 13.5±4.7 | 57.9±19.7 | 4.3 | 0.0006 |
| | 766b | 1,220±441 | 2,020±477 | | 0.0649 | 495±79.8 | 645±38.2 | 1.3 | 0.0043 |
| | 791a | 214±154 | 249±61.1 | | 0.9999 | 122±94.6 | 30.4±12.8 | | 0.6691 |
| | 845b | 846±612 | 448±143 | | 0.6718 | 426±107 | 436±65.2 | | 0.7345 |
| *CYP6G4* | WHO-SRS | 154±35.5 | 394±128 | 2.6 | 0.0012 | 50.8±22.9 | 348±133 | 6.9 | 0.0043 |
| | 766b | 1,060±246 | 2,220±293 | 2.1 | 0.0034 | 783±120 | 378±141 | | 0.5887 |
| | 791a | 157±130 | 339±161 | | 0.1014 | 177±160 | 58.3±41.3 | | 0.5338 |
| | 845b | 1,070±762 | 685±416 | | 0.3845 | 519±323 | 893±488 | | 0.2561 |
| *CYP12A1* | WHO-SRS | 21.6±10.2 | 12.7±9.8 | | 0.1807 | 7.0±0.8 | 13.2±1.8 | 1.9 | 0.0003 |
| | 766b | 63.2±19.5 | 129.1±33.6 | 2.0 | 0.0041 | 58.9±11.1 | 59.9±9.8 | | 0.9551 |
| | 791a | 64.8±7.7 | 73.0±23.2 | | 0.7789 | 45.8±15.5 | 29.6±3.9 | | 0.1003 |
| | 845b | 69.7±30.9 | 38.9±16.0 | | 0.1014 | 34.9±14.3 | 62.8±24.0 | 1.8 | 0.0020 |
| *CYP12A2* | WHO-SRS | 36.4±7.5 | 41.5±5.5 | | 0.2786 | 10.9±3.6 | 22.5±5.0 | 2.1 | 0.0006 |
| | 766b | 124±43.9 | 344±119 | 2.8 | 0.0006 | 65.5±8.3 | 313±96.2 | 4.8 | 0.0003 |
| | 791a | 83.7±10.8 | 65.8±7.9 | 0.8 | 0.0093 | 34.3±13.3 | 29.2±8.9 | | 0.8286 |
| | 845b | 110±56.7 | 99.3±20.5 | | 0.9551 | 50.0±22.3 | 87.2±44.5 | | 0.1893 |
| *CYP4G2* | WHO-SRS | 947±517 | 1,810±611 | | 0.1014 | 2,540±1,860 | 16,680±508 | 6.6 | 0.0050 |
| | 766b | 10,550±1,444 | 17,190±4,200 | 1.6 | 0.0087 | 15,040±1,830 | 21,450±2,390 | 1.4 | 0.0065 |

**Table 2.** Cont.

| Gene | Strain | Male | | | | Female | | | |
|---|---|---|---|---|---|---|---|---|---|
| | | Constitutive | Spinosad | Fold- change | P-value[a] | Constitutive | Spinosad | Fold- change | P-value[a] |
| | 791a | 12,770±671 | 5,520±725 | 0.4 | 0.0021 | 18,960±9,490 | 6,440±1,630 | 0.3 | 0.0006 |
| | 845b | 3,660±2,659 | 4,640±1,500 | | 0.4598 | 7,900±4,000 | 9,530±1,910 | | 0.1589 |
| *GAPDH* | WHO-SRS | 4,860±680 | 8,570±1,510 | 1.8 | 0.0006 | 4,550±319 | 8,700±676 | 1.9 | 0.0003 |
| | 766b | 9,550±662 | 10,930±3,680 | | 0.9999 | 7,460±546 | 7,430±1,060 | | 0.6126 |
| | 791a | 11,430±4,981 | 10,820±1,120 | | 0.7012 | 10,140±4,210 | 7,060±1,320 | | 0.9999 |
| | 845b | 3,780±918 | 4,670±1,220 | | 0.1673 | 4,560±1,060 | 5,090±534 | | 0.1501 |

Mean mRNA transcript copy number ×1000 is per 20 ng of total RNA. P-values between comparisons of treatments were obtained by a F-test for determination of variance differences followed by Welch Significant difference Test. Significant P-values (<0.05) are underlined.

[a]P-values obtained by comparisons of constitutive and spinosad induced gene expression within strain.

The 791spin strain has a sex-dependent resistance profile with regard to spinosad. Females were resistant, while males were considered within the range of susceptibility. This could indicate that the responsible gene(s) would be highly expressed in females compared to males in this strain due to the sex-specific resistance pattern. This was only the case for one single gene, *CYP4G2* (P< 0.0014). Sex determination in 791spin is due to a factor on autosome III [9], which led to a hypothesis of the spinosad resistance being due to a factor on the corresponding copy of autosome III. *CYP4G2* is located on autosome III, and is induced by permethrin in the permethrin-resistant ALHF houseflies [18]. Multiple carbon-oxygen bonds, including aldehydes, are present in spinosad [1]. Spinosad or initial degradation products of spinosad could be substrates for the *CYP4G2* enzyme, which is a decarbonylase enzyme catalyzing cleavage of long-chain aldehydes to hydrocarbons with release of $CO_2$ [12]. The increased constitutive expression in females compared to males fits with *CYP4G2* as a contributor to spinosad resistance in 791spin. Crossover in male houseflies is rare and it could be suggested that the male factor is connected to a certain low-expressed allele of the *CYP4G2* gene, located on the same chromosome. Then females might have another version of the allele, which is expressed at a higher level. Then females would have two highly-expressed copies of *CYP4G2*, while males have one highly-expressed copy and one low-expressed copy (linked with the M factor). This could explain the sex-specific difference in spinosad resistance.

*CYP6G4* is a possible ortholog of the *CYP6G1* gene in *D. melanogaster* and constitutive overexpression of *CYP6G1* is causing DDT and neonicotinoid resistance in the fruit fly [21,22]. A similar role for *CYP6G4* in houseflies could be suggested. Recently, *CYP6G4* has shown to be over-expressed in a pyrethroid resistant housefly strain from China, but no causal link was established [30]. In this study, *CYP6G4* expression was higher in the 791spin strain compared to the susceptible strain, 2-fold in males and almost 15-fold in females. However, *CYP6G4* gene expression in 791spin was lower than that of the spinosad susceptible strains. *CYP6G4* gene expression was not significantly different between sexes, regardless of treatment (Table 1). Spinosad treatment caused a gene expression increase in males of the 791spin strain; whereas spinosad treatment decreased *CYP6G4* expression in females, but neither of the effects were significant (P<0.70). This is contradicted by the other spinosad susceptible strains having a higher expression level of *CYP6G4* than 791spin, so a possible role of *CYP6G4* seems to be minor. Further investigations, especially description of the *CYP6G4* alleles, are needed to elucidate the role of *CYP6G4* in this strain as well as its potential role in xenobiotic metabolism and its importance as a housefly insecticide resistance gene.

Constitutive *CYP12A1* expression was 3-fold higher in male houseflies compared to females for both WHO-SRS and 791spin. Furthermore, 791spin flies had an expression level of approximately 10-fold that of WHO-SRS. The sex-dependent difference in *CYP12A1* gene expression remained after spinosad exposure (P<0.0022) in 791spin, but was eliminated in WHO-SRS (Table 1). Spinosad increased *CYP12A1* expression in 791spin males (P<0.0003) and females (P<0.0001) 1.5- and 3.4-fold, respectively (Table 1). The high constitutive expression of *CYP12A1* in 791spin flies could indicate importance in spinosad resistance, but the fact that male expression is 3-fold higher than female constitutive expression indicates the opposite.

In 791spin males, the constitutive *CYP12A2* expression was increased 1.3-fold by spinosad exposure (P<0.0091), while in the resistant females, no effect of short-term spinosad exposure was

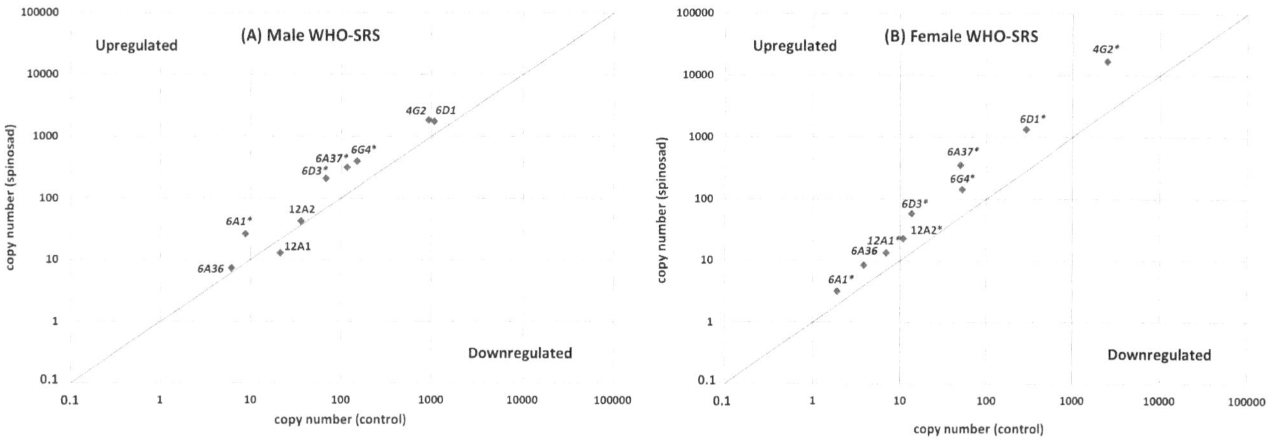


**Figure 1. WHO-SRS male (A) and female (B) spinosad-influenced gene expression as a function of the constitutive gene expression.** The line represents no effect of spinosad exposure. Genes in right-lower and left-upper corner are down-regulated and up-regulated by spinosad, respectively. Points marked with * are significantly different from the separation line by $\alpha = 0.05$.

observed (P>0.95). Expression in 791spin is similar to WHO-SRS expression, and lower than the other strains (Table 1 and 2), indicating no specific role for *CYP12A2* in spinosad resistance in 791spin.

Increased expression of *CYP6A36* has like *CYP6D1* been associated with pyrethroid resistance in the USA, whereas *CYP6A37* did not differ between ALHF and a susceptible strain [10,18]. No significant effect of spinosad treatment on *CYP6A36* gene expression was observed for any of the five strains (P>0.072), with the exception of 791spin females. Here spinosad caused a 4-fold increase in *CYP6A36* gene expression. Furthermore, 791spin had a low *CYP6A36* expression compared to the other strains of this study irrespective of sex (Table 2). The very low expression of CYP6A36 in 791spin disqualifies *CYP6A36* as responsible for spinosad resistance, while the specific female induction points in the other direction. Spinosad exposure caused a significant decrease (P<0.0019) in *CYP6A37* gene expression for male 791spin flies (Figure 2). Independent of treatment, *CYP6A37* expression was significantly lower in both WHO-SRS (P<0.0034) and 791spin (P<0.0007) females than in males (Table 1). This suggests no role for *CYP6A37* in sex-specific spinosad resistance in 791spin.

The epoxidation activity of the cytochrome P450 CYP6A1 enzyme is possibly linked to organochlorine and organophosphate resistance, and also has juvenile hormone I and III as substrate [32]. The data for *CYP6A*1 expression show a higher expression level in 791spin flies compared to WHO-SRS flies. Spinosad exposure increases expression levels in WHO-SRS 2.9- and 1.7.fold in males and females, respectively (P<0.041). A similar effect is observed for 791spin males, where expression was induced 1.7-fold (P<0.020). However, female expression remained at a similar level (P>0.54). The induction of *CYP6A1* gene expression by spinosad in susceptible WHO-SRS flies, combined with the lack of effect in 791spin females, indicate a possible role for *CYP6A1*. However, in 791spin male constitutive expression is higher than female expression, which indicates no role for *CYP6A1* in 791spin.

Prior results of feeding bioassays and gene expression studies in 791spin and its parental 791a strain suggest an involvement of cytochrome P450 enzymes in housefly spinosad resistance, which seems to be specifically female-linked [8,17]. These synergist and

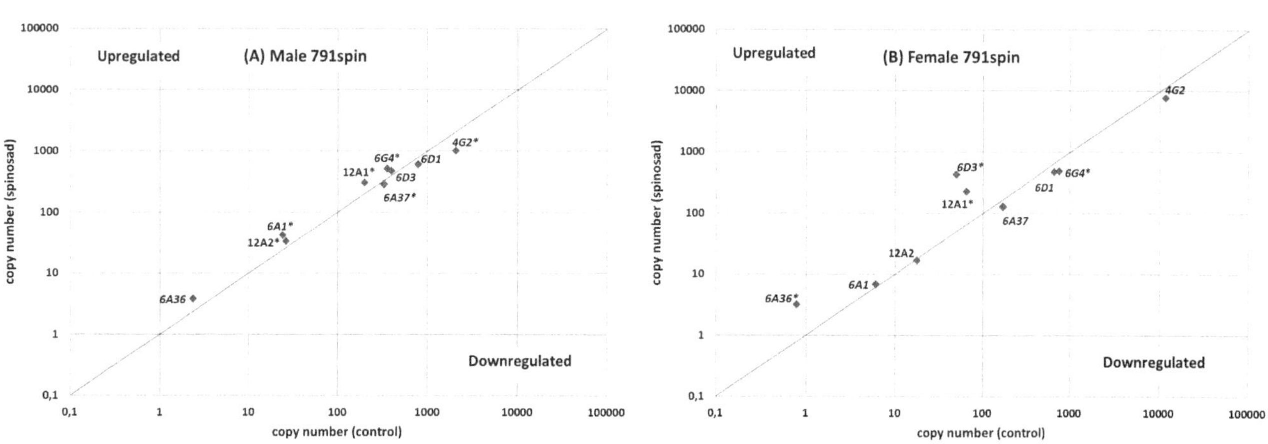


**Figure 2. 791spin male (A) and female (B) spinosad-influenced gene expression as a function of the constitutive gene expression.** The line represents no effect of spinosad exposure. Genes in right-lower and left-upper corner are down-regulated and up-regulated by spinosad, respectively. Points marked with * are significantly different from the separation line by $\alpha = 0.05$.

gene expression studies indicated a minor role of cytochrome P450 genes *CYP6D1* and *CYP6D3* in spinosad resistance in male and female houseflies, respectively. *CYP6D1* is over-expressed in flies resistant to pyrethroids and neonicotinoids [17,19,20]. A wide range of substrates is known for the CYP6D1 enzyme including the organophosphate chlorpyrifos and the synthetic pyrethroids deltamethrin and cypermethrin [33–35]. *CYP6D3* has been used as a reference to *CYP6D1* and is also located on chromosome I [36]. In this study, male constitutive expression of *CYP6A1* and *CYP6D3* was higher in 791spin than in WHO-SRS flies. Likewise, female constitutive expression of *CYP6A1* was also higher in resistant flies compared to susceptible. This correlates with the pervious study of *CYP6A1*, *CYP6D1* and *CYP6D3* [8]. However, here male *CYP6D1* and female *CYP6D3* expression in 791spin did not differ from WHO-SRS. Markussen et al. (2012) suggested a role for *CYP6D1* and *CYP6D3* in male and female resistance, respectively. However, here the opposite is indicated by a high *CYP6D1* and *CYP6D3* expression in 791spin females and males, respectively. This further complicates the determination of the importance of these two genes in spinosad resistance.

Increased expression has been linked to elevated resistance by increased enzymatic degradation of insecticide e.g. degradation of deltamethrin by CYP6D1 [19,20]. But, what does down-regulation signify? It could be hypothesized that down-regulation is due to reduced energy costs. Perhaps down-regulation is a part of a coupled system, so that groups of P450 genes are up-regulated and others are down-regulated. An example of this is a transcription factor pathway in Drosophila where 20% of differential expressed genes (including multiple resistance associated P450s) are genes targeted by the transcription factor CncC [37].

Alternatively the action is more direct and a given P450 enzyme activates the insecticide, making it more toxic. This is e.g. shown by bioassay experiments with neonicotinoids. The addition of the synergist PBO to feeding test with imidacloprid increased toxicity 9-fold, whereas toxicity of thiamethoxam decreased in male houseflies [17]. Thiamethoxam has also been shown as a pro-insecticide in plants [38]. Down-regulation will thus have a direct survival effect.

## Effect of spinosad in a new field strain

The field population 845b was included in this study to examine a neonicotinoid- and spinosad-susceptible field population. In general, high levels of P450 gene expression was found in this newly acquired field population for eight out of nine P450 genes in both males and females. Female expression of *CYP6D3* was more than 30-fold higher in 845b than in WHO-SRS (Table 2). This was despite 845b proving susceptible to pyrethrin synergized by PBO, imidacloprid and spinosad by test with discriminating doses [29]. Short-term spinosad exposure had no effect on gene expression of P450 genes with a few exceptions. In females, *CYP6A1* expression was decreased 2.5-fold ($P<0.0021$) and *CYP12A1* expression was increased 1.8-fold ($P<0.0020$). None of the other seven P450 genes were affected by spinosad in females ($P>0.13$). Furthermore, spinosad caused no effect in 845b males for any of the nine P450 genes tested here ($P>0.10$).

The high gene expression levels observed in this newly-collected strain could be due to the energy requirements involved in maintaining an alert detoxification system. The 845b strain still represent life in the field, while the other strains used for this study have adapted to the laboratory, properly losing some of the defense mechanisms involved in being a wild fly. The pattern of decreased expression in strains adapted to the laboratory was observed for the 845b strain in a two-year study of gene expression [39].

## Effect of spinosad in an insecticide-susceptible reference strain

The highest constitutive P450 expression in the susceptible reference strain WHO-SRS was observed for *CYP4G2* in both sexes, whereas *CYP6A36* and *CYP6A1* were the lowest expressed genes (Table 1).

In general, spinosad exposure caused an increase in gene expression of the nine P450 genes described here for both males and females (Figure 1). *GAPDH* was doubled by short-term spinosad exposure (Table 1). When taking that into consideration and normalizing to *GAPDH* levels, only a few P450 genes were affected by spinosad. In males, the induction of gene expression was no higher than 3-fold with the non-normalized data, while female induction ranged from 1.7-fold to almost 7-fold. In no case did spinosad exposure decrease the expression level significantly.

*CYP4G2* expression was increased 6.6-fold by spinosad in females ($P<0.0050$). A similar increase ($P<0.0043$) was observed in expression of *CYP6G4*, while *CYP12A1* and *CYP12A2* were increased approximately 2-fold in females ($P<0.0006$).

Furthermore, spinosad caused an increase in expression of *CYP6A1* ($P<0.041$), *CYP6A37* ($P<0.0034$) and *CYP6D3* ($P<0.0006$) for both males and females. However, no effect of spinosad was observed for male *CYP6D1*, *CYP12A1* and *CYP12A2* expression as well as for *CYP6A36* expression.

The most remarkable effect by spinosad was observed for *CYP6G4*, where WHO-SRS female expression was increased almost 7-fold ($P<0.0043$) and male expression 2.6-fold ($P<0.0012$) as a result of short-term spinosad exposure. This could indicate a role for *CYP6G4* in spinosad resistance or at least a role in xenobiotic response. In most cases, gene expression of P450 genes was induced in the susceptible strain, suggesting spinosad affecting several genes, not all responsible for the resistance mechanism.

## Effect of spinosad in a neonicotinoid-resistant laboratory strain

In the neonicotinoid-resistant laboratory population 766b the highest constitutive gene expression was observed for *CYP4G2* and *CYP6D1*, for both males and females (Table 2). Noteworthy is the 7-fold and 15-fold higher constitutive expression of *CYP6G4* in males and females, respectively, compared to WHO-SRS. The 766b female *CYP4G2*, *CYP6D3* and *CYP12A2* expression increased (1.4- to 4.8-fold), upon spinosad treatment ($P<0.0043$). Male *CYP6D3* gene expression was not affected by spinosad ($P>0.065$), while *CYP4G2* and *CYP12A2* expression increased in females 1.6-fold ($P<0.0087$) and 2.8-fold ($P<0.0006$), respectively. Furthermore, *CYP6A1*, *CYP6D1* and *CYP6G4* were induced in males ($P<0.015$).

Furthermore, spinosad treatment caused gene expression in males to increase 2-fold for *CYP12A1* ($P<0.0041$), whereas females didn't change *CYP12A1* expression ($P>0.96$).

*CYP6D1* appears to be involved in male neonicotinoid resistance while female resistance seems to be linked to over-expression of the *CYP6D3* gene [17]. Both *CYP6D1* and *CYP6D3* expression proved higher in 766b than in WHO-SRS. *CYP6D1* in males was induced by spinosad, while *CYP6D3* was induced in females, indicating possible roles for these genes in resistance.

The high *CYP6G4* gene expression observed in 766b in comparison with WHO-SRS could indicate a role of *CYP6G4* in 766b neonicotinoid resistance, but not in spinosad resistance, since 766b is susceptible to spinosad. Additionally, previous studies has shown that neonicotinoid resistance in strain 766b responded to PBO, indicating metabolism associated resistance as well as reduced expression of the nicotinic acetylcholine-receptor subunit Mdα2 [17,40].

## Effect of spinosad in a multi-resistant laboratory strain

In the multi-resistant laboratory strain 791a the *CYP6A1* gene was expressed at the lowest level of the genes described in this strain, whereas *CYP4G2* expression was expressed at the highest level, regardless of treatment (Table 2). Spinosad decreased *CYP4G2* gene expression ($P<0.0021$). Likewise, the *CYP12A2* gene was decreased by spinosad treatment in males, while the other seven genes were unaffected by spinosad ($P>0.10$).

## Conclusion

It can be concluded that there is very clear difference between the expression of P450 genes in male compared to female houseflies. Most of the genes tested where spinosad caused an effect were up-regulated by spinosad exposure. In the present investigation a marked difference between a true field strain ($F_1$) and the laboratory reared strains was observed.

Four requirements were purposed to help identify possible resistance genes in this study; 1) higher constitutive expression in 791spin compared to WHO-SRS, 2) spinosad induction of P450 expression in WHO-SRS or other spinosad-susceptible strains, while 3) no significant effect was observed in 791spin, 4) female constitutive expression in 791spin is higher than male expression. *CYP4G2*, *CYP6A1*, *CYP6D3* and *CYP6G4* have an expression pattern somewhat fitting the requirements and therefore the resistance profile of 791spin, but only *CYP4G2* fit all four requirements significantly, making it the most likely candidate. The overall high expression level of *CYP4G2* throughout the strains also indicates importance of this gene. However, the data on 791spin are not conclusive concerning spinosad resistance and e.g. pointing to a single spinosad resistance gene. Small contributions from multiple P450s with different enzymatic capabilities could be speculated to do the job in 791spin.

Analyzing the expression of metabolic detoxification genes rarely gives a clear and unambiguous answer to which enzymes are involved in resistance [41]. Furthermore, the P450s ability to attack the highly complicated spinosad molecule is still unresolved. There is an indication of *CYP6G4* as an insecticide-resistance gene, where involvement in spinosad resistance cannot be rejected. The high expression levels of P450 genes in flies from a field population compared to established laboratory strains presented in this study questions the use of laboratory strains beyond resistance gene identification and warrants studies of P450 expression in field populations, since only the heterogeneous nature of field populations would allow for the selection of the rare variants corresponding to resistance alleles likely to trigger control failure [42]. Laboratory strains can only be used for assessing field control issues if the starting genetic variation is present in the field strain adapted to the laboratory. Differential expression is both up- and down-regulated. Increased expression has been functionally linked to elevated resistance, but it is still puzzling what down-regulation signifies. Is it a part of an interconnected regulatory network? Is it linked to the overall energy budget? Or is it linked to toxicology dynamics of the P450 system?

The differences in expression of minor and major insecticide resistance genes are some of the important tools in pesticide resistance management aiming to limit or prevent development of resistance by controlling factors, which may lead to resistance. This study serves as a stepping stone for the dissection of P450 expression in houseflies in relation to xenobiotics.

## Acknowledgments

The authors would like to thank Folmer Jensen, Lars Damberg and Claus Dahl for providing helpful technical assistance during housefly sampling and breeding.

## Author Contributions

Conceived and designed the experiments: DHH KMVJ MK. Performed the experiments: DHH. Analyzed the data: DHH KMVJ MK. Wrote the paper: DHH KMVJ MK.

## References


1. Sparks TC, Crouse GD, Durst G (2001) Natural products as insecticides: the biology, biochemistry, and quantitative structure-activity relationships of spinosyns and spinosoids. Pest Management Science 57: 896–905.
2. Watson GB (2001) Actions of insecticidal spinosyns on gamma-aminobutyric acid responses from small-diameter cockroach neurons. Pesticide Biochemistry and Physiology 71: 20–28.
3. Sparks TC, Dripps JE, Watson GB, Paroonagian D (2012) Resistance and cross-resistance to the spinosyns - A review and analysis. Pesticide Biochemistry and Physiology 102: 1–10.
4. Scott JG (2008) Unraveling the mystery of spinosad resistance in insects. Journal of Pesticide Science 33: 221–227.
5. Kristensen M, Jespersen JB (2004) Susceptibility of spinosad in Musca domestica (Diptera: Muscidae) field populations. Journal of Economic Entomology 97: 1042–1048.
6. Shono T, Scott JG (2003) Spinosad resistance in the housefly, Musca domestica, is due to a recessive factor on autosome 1. Pesticide Biochemistry and Physiology 75: 1–7.
7. Shi J, Zhang L, Gao XW (2011) Characterisation of spinosad resistance in the housefly Musca domestica (Diptera: Muscidae). Pest Management Science 67: 335–340.
8. Markussen MDK, Kristensen M (2012) Spinosad resistance in female Musca domestica L. from a field-derived population. Pest Management Science 68: 75–82.
9. Højland DH, Scott JG, Jensen KM, Kristensen M (2013) Autosomal male determination in a spinosad-resistant housefly strain from Denmark. Pest Manag Sci.
10. Zhu F, Li T, Zhang L, Liu N (2008) Co-up-regulation of three P450 genes in response to permethrin exposure in permethrin resistant house flies, Musca domestica. BMC Physiology 8: 18.
11. Feyereisen R (2012) Insect CYP Genes and P450 Enzymes. In: Gilbert LI, editor. Insect Molecular Biology and Biochemistry: Elsevier, Amsterdam. 236–316.
12. Qiu Y, Tittiger C, Wicker-Thomas C, Le Goff G, Young S, et al. (2012) An insect-specific P450 oxidative decarbonylase for cuticular hydrocarbon biosynthesis. Proceedings of the National Academy of Sciences of the United States of America 109: 14858–14863.
13. Andersen JF, Utermohlen JG, Feyereisen R (1994) Expression of House-Fly Cyp6a1 and Nadph-Cytochrome P450 Reductase in Escherichia-Coli and Reconstitution of an Insecticide-Metabolizing P450 System. Biochemistry 33: 2171–2177.
14. Feyereisen R, Andersen JF, Carino FA, Cohen MB, Koener JF (1995) Cytochrome-P450 in the House-Fly - Structure, Catalytic Activity and Regulation of Expression of Cyp6a1 in an Insecticide-Resistant Strain. Pesticide Science 43: 233–239.
15. Carino FA, Koener JF, Plapp FW, Feyereisen R (1994) Constitutive Overexpression of the Cytochrome-P450 Gene Cyp6a1 in a House-Fly Strain with Metabolic Resistance to Insecticides. Insect Biochemistry and Molecular Biology 24: 411–418.
16. Sabourault C, Guzov VM, Koener JF, Claudianos C, Plapp FW, et al. (2001) Overproduction of a P450 that metabolizes diazinon is linked to a loss-of-function in the chromosome 2 ali-esterase (Md alpha E7) gene in resistant houseflies. Insect Molecular Biology 10: 609–618.
17. Markussen MDK, Kristensen M (2010) Cytochrome P450 monooxygenase-mediated neonicotinoid resistance in the house fly Musca domestica L. Pesticide Biochemistry and Physiology 98: 50–58.
18. Zhu F, Feng JN, Zhang L, Liu N (2008) Characterization of two novel cytochrome P450 genes in insecticide-resistant house-flies. Insect Molecular Biology 17: 27–37.

19. Liu N, Scott JG (1998) Increased transcription of CYP6D1 causes cytochrome P450 mediated insecticide resistance in house fly. Insect Biochemistry and Molecular Biology 28: 531–535.
20. Kasai S, Scott JG (2000) Overexpression of cytochrome P450CYP6D1 is associated with monooxygenase-mediated pyrethroid resistance in house flies from Georgia. Pesticide Biochemistry and Physiology 68: 34–41.
21. Daborn P, Boundy S, Yen J, Pittendrigh B, Ffrench-Constant R (2001) DDT resistance in Drosophila correlates with Cyp6g1 over-expression and confers cross-resistance to the neonicotinoid imidacloprid. Molecular Genetics and Genomics 266: 556–563.
22. Daborn PJ, Yen JL, Bogwitz MR, Le Goff G, Feil E, et al. (2002) A single P450 allele associated with insecticide resistance in Drosophila. Science 297: 2253–2256.
23. Guzov VM, Unnithan GC, Chernogolov AA, Feyereisen R (1998) CYP12A1, a mitochondrial cytochrome P450 from the house fly. Archives of Biochemistry and Biophysics 359: 231–240.
24. Kristensen M, Huang J, Qiao CL, Jespersen JB (2006) Variation of Musca domestica L. acetylicholinesterase in Danish housefly populations. Pest Management Science 62: 738–745.
25. Kristensen M, Jespersen JB (2008) Susceptibility to thiamethoxam of Musca domestica from Danish livestock farms. Pest Management Science 64: 126–132.
26. Kristensen M, Jespersen JB (2003) Larvicide resistance in Musca domestica (Diptera: Muscidae) populations in Denmark and establishment of resistant laboratory strains. Journal of Economic Entomology 96: 1300–1306.
27. Kristensen M, Jespersen JB, Knorr M (2004) Cross-resistance potential of fipronil in Musca domestica. Pest Management Science 60: 894–900.
28. Kristensen M, Spencer AG, Jespersen JB (2001) The status and development of insecticide resistance in Danish populations of the housefly Musca domestica L. Pest Management Science 57: 82–89.
29. Højland DH, Jensen KM, Kristensen M (2013) A comparative study of P450 gene expression in field and laboratory Musca domestica L. strains. Pest Manag Sci.
30. Gao Q, Li M, Sheng CF, Scott JG, Qiu XH (2012) Multiple cytochrome P450s overexpressed in pyrethroid resistant house flies (Musca domestica). Pesticide Biochemistry and Physiology 104: 252–260.
31. R Development Core Team (2012) R: A Language and Environment for Statistical Computing. In: Computing RFfS, editor.
32. Scott JG, Wen ZM (2001) Cytochromes P450 of insects: the tip of the iceberg. Pest Management Science 57: 958–967.
33. Wheelock GD, Scott JG (1992) The Role of Cytochrome-P450ipr in Deltamethrin Metabolism by Pyrethroid-Resistant and Susceptible Strains of House-Flies. Pesticide Biochemistry and Physiology 43: 67–77.
34. Wheelock GD, Scott JG (1992) Anti-P450lpr Antiserum Inhibits Specific Monooxygenase Activities in Lpr House-Fly Microsomes. Journal of Experimental Zoology 264: 153–158.
35. Zhang ML, Scott JG (1996) Cytochrome b(5) is essential for cytochrome P450 6D1-mediated cypermethrin resistance in LPR house flies. Pesticide Biochemistry and Physiology 55: 150–156.
36. Kasai S, Scott JG (2001) Expression and regulation of CYP6D3 in the house fly, Musca domestica (L.). Insect Biochemistry and Molecular Biology 32: 1–8.
37. Misra JR, Lam G, Thummel CS (2013) Constitutive activation of the Nrf2/Keap1 pathway in insecticide-resistant strains of Drosophila. Insect Biochem Mol Biol 43: 1116–1124.
38. Nauen R, Ebbinghaus-Kintscher U, Salgado VL, Kaussmann M (2003) Thiamethoxam is a neonicotinoid precursor converted to clothianidin in insects and plants. Pesticide Biochemistry and Physiology 76: 55–69.
39. Højland DH, Jensen KMV, Kristensen M (2014) Adaptation of Musca domestica L. field population to laboratory breeding causes transcriptional alterations Plos One.
40. Markussen MDK, Kristensen M (2010) Low expression of nicotinic acetylcholine receptor subunit Md alpha 2 in neonicotinoid-resistant strains of Musca domestica L. Pest Management Science 66: 1257–1262.
41. Saavedra-Rodriguez K, Suarez AF, Salas IF, Strode C, Ranson H, et al. (2012) Transcription of detoxification genes after permethrin selection in the mosquito Aedes aegypti. Insect Molecular Biology 21: 61–77.
42. Ffrench-Constant RH (2013) The Molecular Genetics of Insecticide Resistance. Genetics 194: 807–815.

# 9

# Quick and Label-Free Detection for Coumaphos by Using Surface Plasmon Resonance Biochip

**Ying Li[1,2], Xiao Ma[3], Minglu Zhao[3], Pan Qi[4], Jingang Zhong[1,3*]**

**1** Key Laboratory of Optoelectronic Information and Sensing Technologies of Guangdong Higher Education Institutes, Jinan University, Guangzhou, Guangdong, China, **2** Pre-university Department, Jinan University, Guangzhou, Guangdong, China, **3** Department of Optoelectronic Engineering, Jinan University, Guangzhou, Guangdong, China, **4** Department of Electronics Engineering, Guangdong Communication Polytechnic, Guangzhou, Guangdong, China

## Abstract

Coumaphos is a common organophosphorus pesticide used in agricultural products. It is harmful to human health and has a strictly stipulated maximum residue limit (MRL) on fruits and vegetables. Currently existing methods for detection are complex in execution, require expensive tools and are time consuming and labor intensive. The surface plasmon resonance method has been widely used in biomedicine and many other fields. This study discusses a detection method based on surface plasmon resonance in organophosphorus pesticide residues. As an alternative solution, this study proposes a method to detect Coumaphos. The method, which is based on surface plasmon resonance (SPR) and immune reaction, belongs to the suppression method. A group of samples of Coumaphos was detected by this method. The concentrations of Coumaphos in the samples were 0 μg/L, 50 μg/L, 100 μg/L, 300 μg/L, 500 μg/L, 1000 μg/L, 3000 μg/L and 5000 μg/L, respectively. Through detecting a group of samples, the process of kinetic reactions was analyzed and the corresponding standard curve was obtained. The sensibility is less than 25 μg/L, conforming to the standard of the MRL of Coumaphos stipulated by China. This method is label-free, using an unpurified single antibody only and can continuously test at least 80 groups of samples continuously. It has high sensitivity and specificity. The required equipments are simple, environmental friendly and easy to control. So this method is promised for a large number of samples quick detection on spot and for application prospects.

**Editor:** Sabato D'Auria, CNR, Italy

**Funding:** This work was supported by National Natural Science Foundation of China (NSFC) under the Grant No. 41206081, Key Laboratory of Integrated Marine Monitoring, Guangdong Natural Science Foundation under the Grant No. 8451063201000076, and Applied Technologies for Harmful Algal Blooms under the Grant No. MATHAB20120208, and breakthroughs in key areas of Guangdong and Hong Kong tender No. 2005A20501001. The funders had no role in study design, data collection and analysis, decision to publish, or preparation of the manuscript.

**Competing Interests:** The authors have declared that no competing interests exist.

* Email: tzjg@jnu.edu.cn

## Introduction

Organophosphorus pesticide (OPPs) is a type of phosphate with different substituent groups that cause the inhibition of acetylcholinesterase to produce an insecticidal effect [1–3]. China is one of the world's leading users of pesticide, nearly 70% of which is organophosphorus. Organophosphorus pesticide often has high levels of neurotoxicity in human and animals through gradual accumulation and regular intake. Organophosphorus pesticide may lead to symptoms such as neurological disorders, tremors, language disorders and even death [4–15]. Due to its high levels of toxicity, many organophosphorus pesticides have been banned or highly limited in use by most countries. Efforts are being made to revamp restrictions and regulations of organophosphorus pesticide residues in imported fruits and vegetables across America, Europe and Asia, which makes improved detection methods urgent and necessary.

Coumaphos is one of the organophosphorus pesticides. Coumaphos is harmful to human health and has a strictly stipulated maximum residue limit (MRL) on fruits and vegetables. China stipulates that the residues of Coumaphos in vegetables and fruits must be lower than 0.05 mg/kg [16]. Currently, it can be detected through methods of chromatography such as low pressure gas chromatography-mass spectroscopy (LP GC-MS) [17–20], high performance liquid chromatography (HPLC) [21–23] and thin layer chromatography (TLC) [24–25]. Additionally the wave spectrum method [26–28] and enzyme-linked immunosorbent assay (ELISA) [29–30], are also used.

Although these methods are highly sensitive and accurate, the sample pretreatment procedures are time-consuming and labor intensive, complex in execution and require expensive equipment. Furthermore, they do not enable real-time detection and rapid screening of large amount of samples. In contrast, the surface plasmon resonance biochip method is quick, highly sensitive, and label-free, and has been widely used in pharmaceutical analysis, food analysis, environment monitoring and many other fields [31–33]. This study discusses a detection technique based on surface plasmon resonance in organophosphorus pesticide residues, for example Coumaphos. This research adopts the self-developed portable surface plasmon resonance biochip detector by using of the specificity of immunoreaction to study the detection of higher toxicity organophosphorus pesticide-Coumaphos, to analyze the process of kinetic reactions and to establish its standard curve. The technique, which is based on surface plasmon resonance (SPR) and immune reaction, belongs to the suppression method. Comparing with ELISA methods and other methods, this method is label-free and has high specificity, and the sample pretreatment procedure is simple. The utilized equipment is cheap, simple, environmental friendly and easy to control. It achieves the continuousdetection and the quick screening of a large number of samples. This method can be applied in places where real-time quick detection and quality control is needed such as supermarkets, bazaars and factories.

## Materials and Methods

### Instrument and equipment, materials and reagents

The self-constructed portable SPR biochip detector is a sensitive and accurate angle scanning device. The use of this device is simple and easy, with a precision of up to 0.002°. It can also be controlled easily by LabView and has a user friendly interface [34].

Coumaphos standard (0.02 g/mL, the molecular weight of 362.78) is obtained from Dr. Ehrenstorfer GmbH (Augsburg, Germany). $H_{11}$-OVA (Coumaphos-ovalbumin, 83 mg/L) and unpurified monoclonal antibody ascites of Coumaphos (5 mg/mL) are produced by the College of Food Science, South China Agricultural University, as described in the previous work [35–36]. $HS(CH_2)_{10}COOH$ (mercapto-undecanoic acid) and $HS(CH_2)_6OH$ (mercapto-hexanoate), N-hydroxysuccinimide (NHS), N-ethyl-N'-(dimethylaminopropyl) carbodiimide (EDC), Ethanolamine (Eth), and Sodium dodecyl sulfate (SDS) are purchased from Sigma, USA. Other reagents are purchased from Beijing Chemical Reagent Company. The PBS buffer (2 mmol/L $NaH_2PO_4$, 2 mmol/L $Na_2HPO_4$, 150 mmol/L NaCl, pH 7.4) is used as immune reaction buffer. Coumaphos antigen and monoclonal antibody ascites of Coumaphos are diluted to the appropriate concentration solution with PBS buffer.

### Preparation of Biochip

A gold film (thickness of 50 nm) is deposited on the surface of a piece of glass (diameter of 20 mm and thickness of 1 mm). The biochip with gold film is attached to the instrument. The flow cell is installed, and the PBS buffer is injected into it. The biochip will be self-assembledwhen the baseline is stable for a few minutes. The ethanol solution of mercaptoundecanoic acid (0.1 mmol/L $HS(CH_2)_{10}COOH$ and 0.9 mmol/L $HS(CH_2)_6OH$ (Mercapto-hexyl hexanoate)) are injected into the flow cell. The gold film is chemically modified with self-assembled monolayers (SAMs) by the prepared solution for 2 hours. The sulfhydryl terminal of the modified liquid can interact with gold and they bond to each other (S-Au bond), and then adsorb on the gold film stably and orderly. C-terminus can be activated to active ester by EDC/NHS as active groups. Active ester can react with proteins (antibody or antigen) to form amide linkage which will fix protein. After the biochip is modified, the PBS buffer is injected to wash the biochip for forming the baseline of the whole response. The 0.1 mol/L

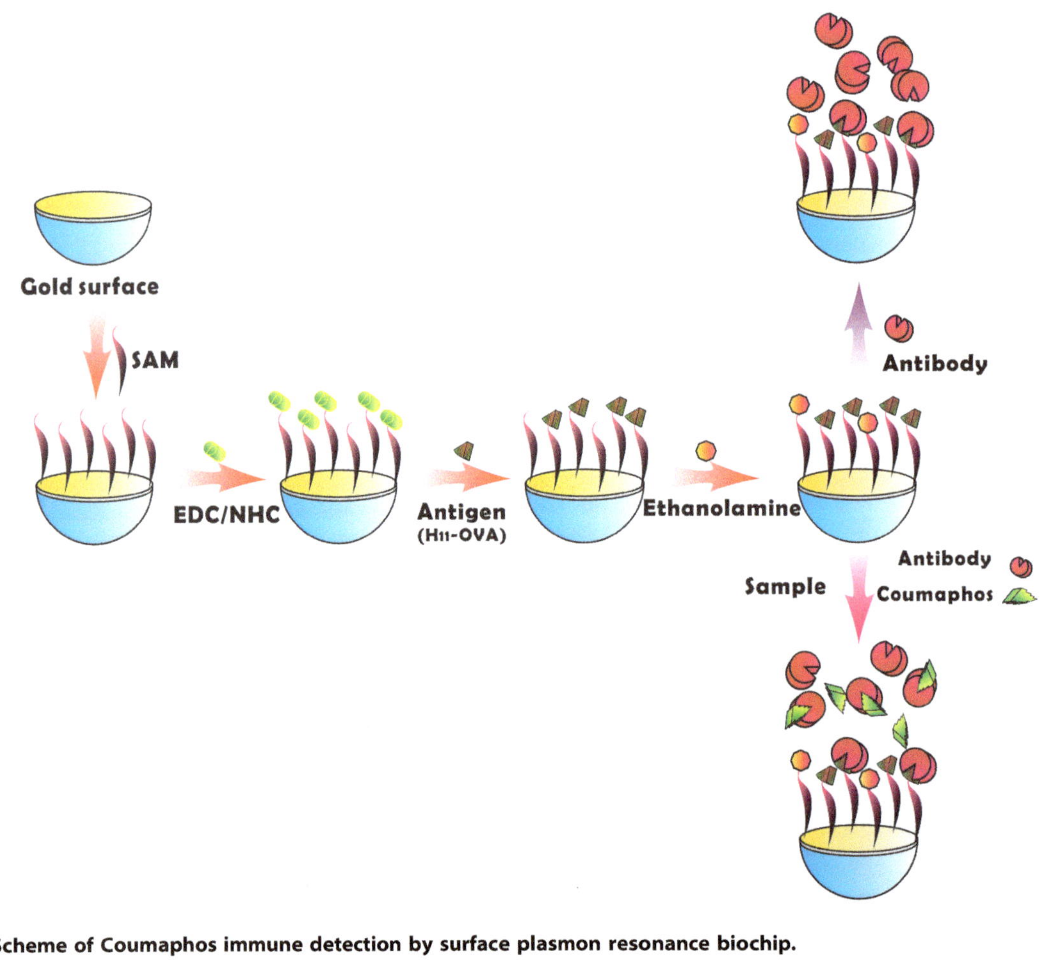


**Figure 1. Scheme of Coumaphos immune detection by surface plasmon resonance biochip.**

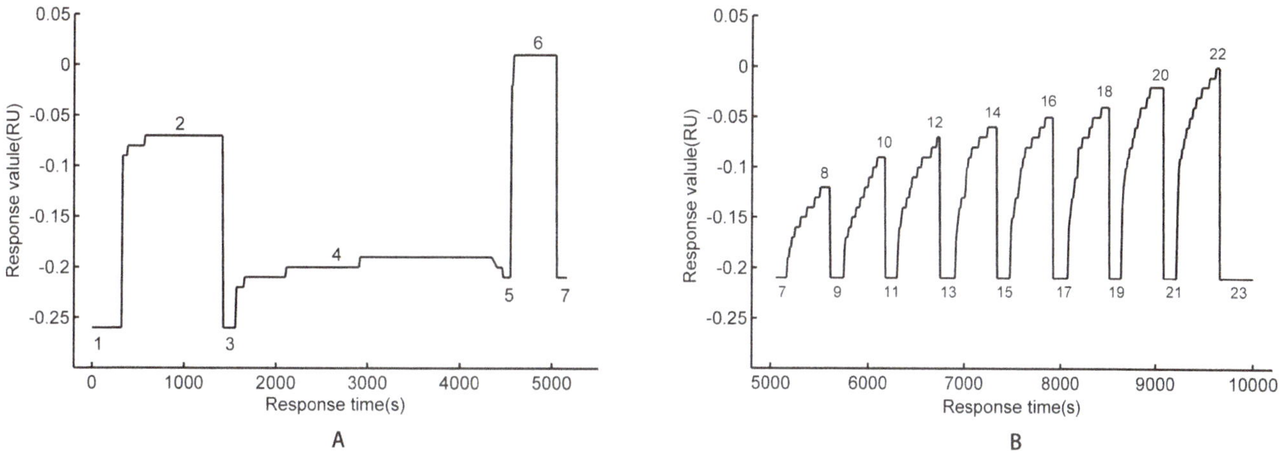


**Figure 2. The process of a group of Coumaphos molecules detected by suppression method.** A) The preparation of the biochip. 1: PBS; 2: active ester; 3: PBS; 4: The fixation of the bioprobe $H_{11}$-OVA; 5: PBS; 6: inactivated; 7: PBS. B) Detecting a group of samples by suppression method. 8, 10, 12, 14, 16, 18, 20, 22 plot samples 5000 μg/L, 3000 μg/L, 1000 μg/L, 500 μg/L, 300 μg/L, 100 μg/L, 50 μg/L and 0 μg/L, respectively. 7, 9, 11, 13, 15, 17, 19, 21, 23 are PBS.

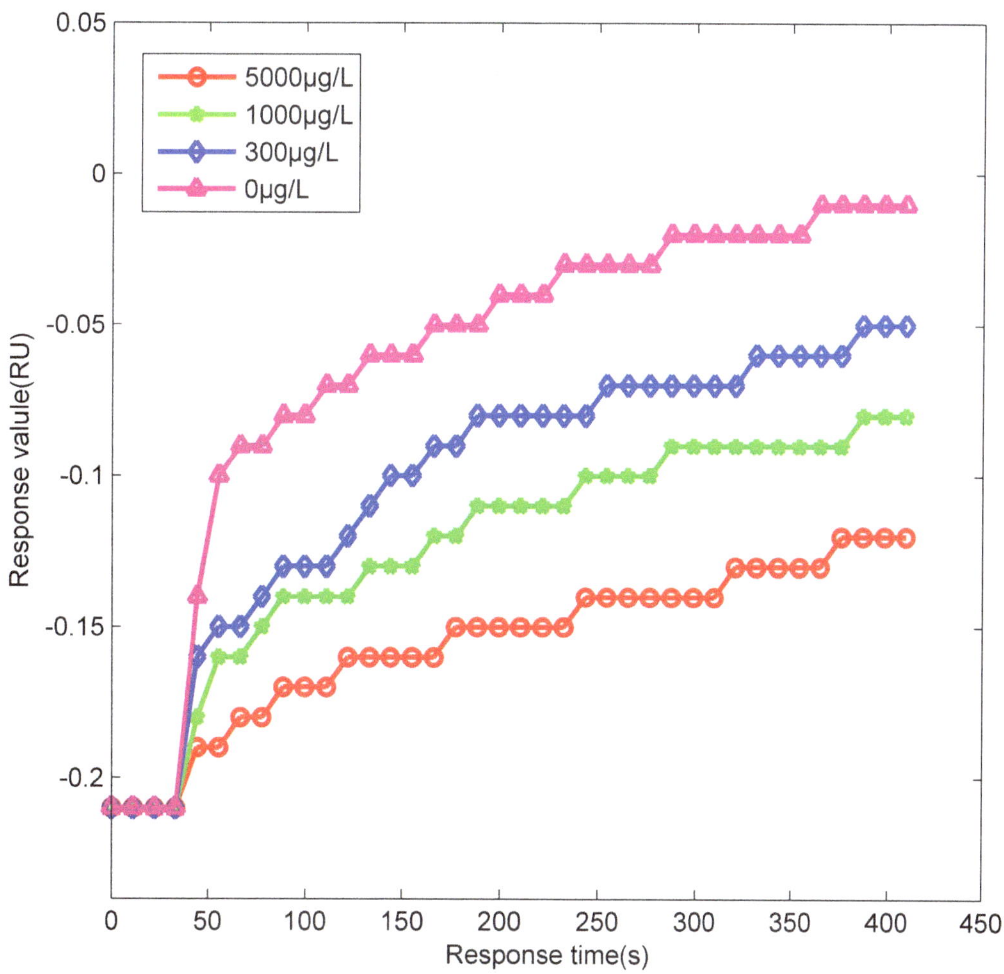


**Figure 3. The dynamic curve of Coumaphos molecules detected by suppression method.** The concentrations of samples are 0 μg/L, 300 μg/L, 1000 μg/L and 5000 μg/L.

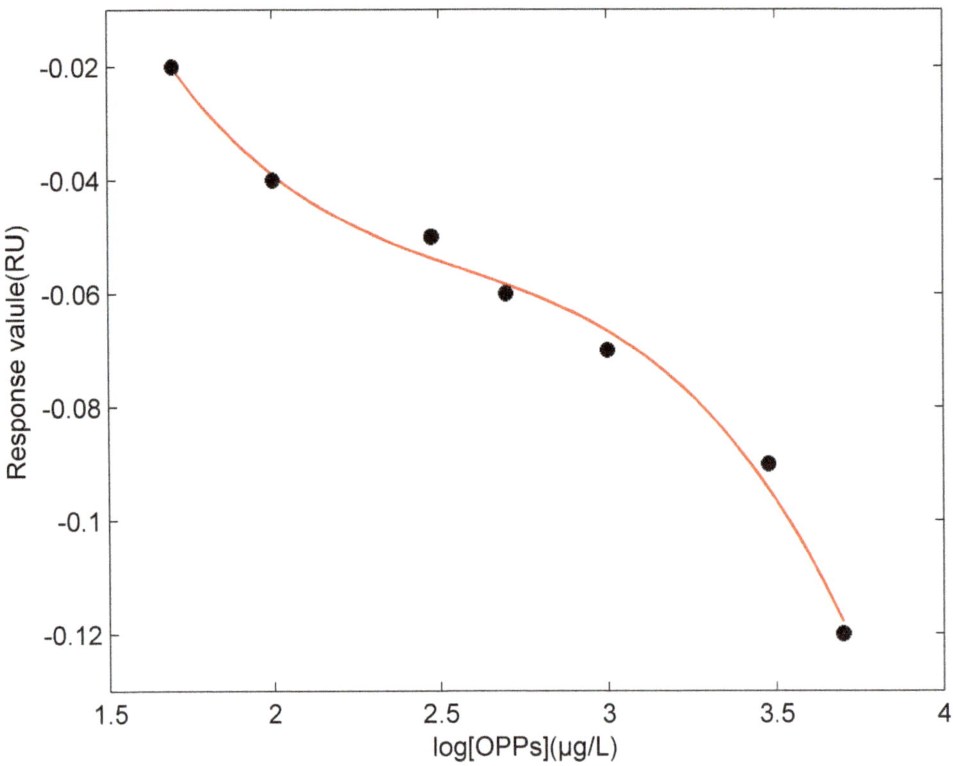


**Figure 4. Standard curve of detecting Coumaphos molecules by suppression method.** The y-coordinate represents the response value, and the x-coordinate is the concentration of Coumaphos. The curve appears approximately to be an inverted "S". The value of $IC_{50}$ is 1000 μg/L.

NHS and 0.1 mol/L EDC (1:1, V/V) are added on to the surface of the biochip after the baseline being stable to activate the biochip for 15 minutes. The bioprobe $H_{11}$-OVA is fixed after washed by PBS for 2 minutes. The $H_{11}$-OVA which has been diluted 15 times as bioprobe is fixed on the surface of the biochip. The response value of the SPR increases significantly. Thirty minutes later, the biochip is washed by PBS for two minutes. The response value of the SPR after fixing is higher than it was before fixing, meaning that the effect of the fixed bioprobe is better. The ethanolamine (pH 8.5, 1 mol/L) is added for the inactivation of the remaining ester bond. The PBS is used to wash the gold film 5–7 minutes after for inactivation. The preparation of the biochip is complete and it can be used for the following detection. The process of Coumaphos immune detection by surface plasmon resonance biochip is shown in Fig. 1.

### Suppression Detection

If the probe of biochip is the antibody of Coumaphos, Coumaphos, with a molecular weight of 362.78, can be combined with the antibody that is fixed on the surface of the biochip. This is the direct detection of Coumaphos, but the effect of the refractive index changing around the biochip surface is small. The effect of the direct detection will be unsatisfied because the variation of resonance peak is small. For the trace detection of substances with small molecular weight like Coumaphos, the suppression method was proposed. The bioprobe of the biochip is $H_{11}$-OVA. 5 minutes after the antibodies were mixed with different concentrations of Coumaphos, the resultant mixtures are added on the surface of the biochip. The SPR effect is detected and the kinetic process can be analyzed. If the concentration of Coumaphos molecules is low, more antibodies of Coumaphos will combine with the probes $H_{11}$-OVA on the surface of the biochip, and then the response value of SPR will become larger. The concentration of Coumaphos molecules is inversely proportional to the response value of SPR. That is the meaning of the method of suppression detection.

## Results and Discussion

The process of a group of Coumaphos molecules detected by suppression method is shown in Fig. 2. It plots the baseline of PBS in Marking 1 in Figure 2A, the activation of the surface of the biochip, and c-terminus activates into active ester in Marking 2, and the washing by PBS after activation in Marking 3. At that time the response value of the SPR falls back to baseline. The fixation of the bioprobe $H_{11}$-OVA is shown in Marking 4, and the response value increases drastically. The fixed probe is monitored in real time. The response value no longer increases after about 45 minutes, meaning the bioprobe fixed on the biochip has been saturated. The fixation of the bioprobe is complete and is washed by PBS as shown in the Marking 5. At this time the response value of the SPR decreases slightly but is higher than the baseline in Marking 3, which suggests that the fixation effect is good. In Marking 6, the rest of the ester linkages are inactivated and confined with ethanolamine in order to avoid the formation of nonspecific adsorption. Marking 7 is the cleaning process by PBS to get rid of the substances that are not fixed. With all the operations above, the preparation of the biochip is complete and can be used for the detection of samples.

The antibodies against Coumaphos are diluted with PBS, standing for 5 minutes to ensure sufficient reaction after the antibody was mixed with samples of Coumaphos molecules. The concentration of the antibodies in the mixture is 10 mg/L, the concentrations of Coumaphos molecules are 0 μg/L, 50 μg/L, 100 μg/L, 300 μg/L, 500 μg/L, 1000 μg/L, 3000 μg/L and 5000 μg/L, the mixtures are added on the surface of the biochip. Then, the dynamic variation of the SPR is recorded. As Markings 8, 10, 12, 14, 16, 18, 20, 22 plot the dynamic curve of samples in Figure 2B, the concentrations of Coumaphos molecules are 5000 μg/L, 3000 μg/L, 1000 μg/L, 500 μg/L, 300 μg/L, 100 μg/L, 50 μg/L and 0 μg/L, respectively. Thus it can be seen when the concentrations of the samples decrease, the response value of SPR is increasing.

There is a competitive relationship between the Coumaphos molecules in the mixture and the probe $H_{11}$-OVA on the surface of the biochip, as both of them can be combined with the antibody in the mixture. The Coumaphos molecules suppress the combination between the antibody and the probe $H_{11}$-OVA on the surface of the biochip. The antibody of Coumaphos is macromolecule, and Coumaphos is micromolecule. The molecular weight of the antibody is larger than Coumaphos, and the weight of $H_{11}$-OVA-antibody is larger than Coumaphos-antibody. After the combination between the antibody and the probe $H_{11}$-OVA on the surface of the biochip, the antigen-antibody conjugates on the surface of the biochip increase while the SPR response value also increases. The concentration of Coumaphos molecules is inversely proportional to the response value of SPR. If the concentration of Coumaphos molecules is low more antibodies of Coumaphos will combine with the probes on the surface of the biochip, and then the response value of SPR, namely the resonance angle, will become larger, as is shown in Fig. 2 and Fig. 3. Six minutes after the immune reaction, the SDS-HCl solution is used to elute the combination of the antigen and antibody. The PBS is used to wash the SDS-HCl after two minutes. The response value of SPR can be returned to the baseline, as is shown in Markings 9, 11, 13, 15, 17, 19, 21, 23 of Figure 2B.

Figure. 3 shows the dynamic curve of the partial samples, the concentrations of samples are 0 μg/L, 300 μg/L, 1000 μg/L and 5000 μg/L, respectively. With the concentration of Coumaphos molecules increasing, the response value of SPR declines.

In this case, the biochip with the fixation to the bioprobe can be used in the continuous detection of 80 groups of samples. The response value of SPR declines significantly and the sensitivity decreases after the detection of 80 groups of samples, signaling the bioprobe fixed on the surface of the biochip is damaged. The biochip regeneration can be realized after 0.1 mol/L HCl is added on the surface of the biochip. The sensor biochip can still be used after self-assembling and the fixation of bioprobe.

Figure 4 shows the standard curve of immunoreaction of Coumaphos molecules tested by suppression method with the SPR biochip. The y-coordinate represents the response value of the immune reaction at 5 minutes, and the x-coordinate is the concentration of Coumaphos. The curve appears approximately to be an inverted "S". The standard deviation is less than 10% after the repeated detections of each concentration (3 times). The value of $IC_{50}$ is 1000 μg/L. The detection limit is less than 25 μg/L. After detections of the SPR response value, the concentration of Coumaphos in the sample could be obtained by checking the standard curve. This method could contribute greatly to the quality control of fruits and vegetables and in-field real time detection.

To prepare different concentrations of Coumaphos standard solution, the concentrations of samples are 50 μg/L, 100 μg/L, 500 μg/L, 1000 μg/L, respectively, as the true value of samples. The content of Coumaphos molecules in Coumaphos standard solution is detected by using the SPR biochip in suppression method. And the estimated value of the concentrations of

**Table 1.** The analysis of the detection of Coumaphos molecules in Coumaphos standard solution by using the SPR biochip in suppression method.

| Concentration (μg/L) | concentration measured SPR detector (μg/L) | | | | standard deviation | absolute deviation | relative deviation |
|---|---|---|---|---|---|---|---|
| | estimated value | | | average value | | | |
| 50 | 50.2 | 49.7 | 49.7 | 49.867 | 0.289 | −0.133 | 0.58% |
| 100 | 100.9 | 100.5 | 100.4 | 100.600 | 0.265 | 0.600 | 0.26% |
| 500 | 501.2 | 502.1 | 504.4 | 502.567 | 1.650 | 2.567 | 0.33% |
| 1000 | 998.5 | 1002.4 | 1004.5 | 1001.800 | 3.045 | 1.800 | 0.30% |

The absolute deviation of the real value compares and the estimated value is smaller, and so is the relative deviation.

Coumaphos molecules can be calculated by the combination of the data detected by SPR detector and the standard curve shown in Fig. 4. The concentration of the Coumaphos standard is regarded as the real value and the data calculated by the standard curve is regarded as the estimated value. The real value compares with the estimated value, as is shown in Table 1. The absolute deviation of the real value compares and the estimated value is smaller, and so is the relative deviation. The SPR biochip method is effective, and the experimental method is feasible, and may successfully be used in food quality control and real-time detection.

## Conclusions

Organophosphorus, specifically Coumaphos, has become one of the main pesticides for the prevention and control of plant diseases and insect pests. Excessive pesticide residues in fruits and vegetables, however, threaten the health of human beings and animals directly. As the inspection of food imports and exports becomes stricter, it is urgent and necessary to build a simple, effective and cheap detection method of pesticide residue for avoiding pesticide poisoning and protect the health of humans [37].

The detection of Coumaphos by using the biochip of SPR in suppression method mentioned in this paper, has a lot of advantages including inexpensive equipment, label-free detection, high accuracy, specificity and low cost. It also does not require a second antibody, chemical substances, such as fluorescence dye, and use unpurified antibody ascites. It can be used to detect pesticide residues in fruits and vegetables and quickly provide quantitative results. The equipment utilized is inexpensive, portable, environmentally friendly and easy to control. It ensures the achievement of real-time detection and rapid screening of large amount of samples making it ideal for usage in environments such as supermarkets, bazaars and factories.

In this paper, the adopted working concentration of antibody is 10 mg/L, if the working concentration of antibody is reduced, the detection limit of the sample of small Coumaphos molecules can also be reduced.

## Acknowledgments

We thank Zong Dai and Yun Chen for the technical assistance.

## Author Contributions

Conceived and designed the experiments: YL JZ. Performed the experiments: YL XM. Analyzed the data: YL MZ PQ. Contributed reagents/materials/analysis tools: YL. Contributed to the writing of the manuscript: YL MZ. Designed the software used in analysis: PQ XM.

## References


1. Eversole JW, Lilly JH, Shaw FR (1965) Toxicity of droppings from Coumaphos-Fed hens to little house fly larvae. Journal of Economic Entomology 58: 709–710.
2. Pardío VT, Ibarra NDJ, Waliszewski KN, López KM (2007) Effect of coumaphos on cholinesterase activity, hematology, and biochemical blood parameters of bovines in tropical regions of Mexico. Journal of Environmental Science and Health, Part B: Pesticides, Food Contaminants, and Agricultural Wastes 42: 359–366. doi:10.1080/03601230701310500. PubMed: 17474014.
3. Johnson RM, Pollock HS, Berenbaum MR (2009) Synergistic interactions between In-Hive miticides in apis mellifera. Journal of Economic Entomology 102: 474–479. doi:http://dx.doi.org/10.1603/029.102.0202. PubMed: 19449624.
4. Balayiannis G, Balayiannis P (2008) Bee Honey as an environmental bioindicator of pesticides' occurrence in six agricultural areas of Greece. Arch Environ Contam Toxicol 55: 462–470. doi:10.1007/s00244-007-9126-x. PubMed: 18231699.
5. Yavuz H, Guler GO, Aktumsek A, Cakmak YS, Ozparlak H (2010) Determination of some organochlorine pesticide residues in honeys from Konya, Turkey. Environ Monit Assess 168: 277–283. doi:10.1007/s10661-009-1111-6. PubMed: 19685151.
6. Das YK, Kaya S (2009) Organophosphorus Insecticide Residues in Honey Produced in Turkey. Bull Environ Contam Toxicol 83: 378–383. doi:10.1007/s00128-009-9778-5. PubMed: 19452111.
7. Choudhary A, Sharma DC (2008) Pesticide residues in honey samples from Himachal Pradesh (India). Bull Environ Contam Toxicol 80: 417–422. doi:10.1007/s00128-008-9426-5. PubMed: 18506381.
8. Škerl MIS, Bolta ŠV, Česnik HB, Gregorc A (2009) Residues of pesticides in Honeybee (Apis mellifera carnica) bee bread and in pollen loads from treated apple orchards. Bull Environ Contam Toxicol 83: 374–377. doi:10.1007/s00128-009-9762-0. PubMed: 19434347.
9. Weick J, Thorn RS (2002) Effects of acute sublethal exposure to Coumaphos or Diazinon on acquisition and discrimination of odor stimuli in the honey bee (Hymenoptera: Apidae). Journal of Economic Entomology 95: 227–236. doi:http://dx.doi.org/10.1603/0022-0493-95.2.227. PubMed: 12019994.
10. Huang Z (2001) Mite zapper a new and effective method for Varroa Mite control. American Bee Journal: 730–732.
11. Savage EP, Keefe TJ, Mounce LM, Heaton RK, Lewis JA, et al. (1998) Chronic neurological sequelae of acute Organophosphate Pesticide poisoning. Archives of Environmental Health 43: 38–45. doi:10.1080/00039896.1988.9934372. PubMed: 3355242.
12. Popendorf WJ, Leffingwell JT (1982) Regulating OP pesticide residues for farmworker protection. Residue Reviews 82: 125–201. doi:10.1007/978-1-4612-5709-7_3. PubMed: 7051208.
13. Marrs TC (1993) Organophosphate poisoning. Pharmacology & Therapeutics 58: 51–66. doi:0163-7258/9. PubMed: 8415873.
14. Hill EF, Fleming WJ (1982) Anticholinesterase poisoning of birds: Field monitoring and diagnosis of acute poisoning. Environmental Toxicology and Chemistry 1: 27–38. doi:10.1002/etc.5620010105.
15. Eddleston M, Szinicz L, Eyer P, Buckley N (2002) Oximes in acute organophosphorus pesticide poisoning: a systematic review of clinical trials. Oxford Journals 95: 275–283. doi:10.1093/qjmed/95.5.275. PubMed: 11978898.
16. Ministry of Health of the People's Republic of China and the Ministry of Agriculture of the People's Republic of China (2012) National food safety standard Maximum residue limits for pesticides in food. GB 2763–2012.
17. Albero B, Sánchez-Brunete C, Tadeo JL (2004) Analysis of pesticides in honey by Solid-Phase Extraction and Gas Chromatography-Mass Spectrometry. Journal of Agricultural and Food Chemistry 52: 5828–5835. doi:10.1021/jf049470t. PubMed: 15366828.
18. Rissato SR, Galhiane MS, de Almeida MV, Gerenutti M, Apon BM (2007) Multiresidue determination of pesticides in honey samples by gas chromatography–mass spectrometry and application in environmental contamination. Food Chemistry 101: 1719–1726. doi:10.1016/j.foodchem.2005.10.034.
19. Jimenez JJ, Bernal JL, del Nozal MJ, Martın MT, Mayorga AL (1998) Solid-phase microextraction applied to the analysis of pesticide residues in honey using gas chromatography with electron-capture detection. Journal of Chromatography A 829: 269–277. PII: S0021-9673(98)00826-7. PubMed: 9923084.
20. Pang GF, Fan CL, Liu YM, Cao YZ, Zhang JJ, et al. (2006) Multi-residue method for the determination of 450 pesticide residues in honey, fruit juice and wine by double-cartridge solid-phase extraction/gas chromatography-mass spectrometry and liquid chromatography-tandem mass spectrometry. Food Additives and Contaminants 23: 777–810. doi:10.1080/02652030600657997. PubMed: 16807205.
21. Grimalt S, Sancho JV, PozoÓJ, Garca-Baudin JM, Fernndez-Cruz ML, et al. (2006) Analytical study of Trichlorfon residues in Kaki Fruit and Cauliflower aamples by Liquid Chromatography-Electrospray Tandem Mass Spectrometry. Journal of Agricultural and Food Chemistry 54: 1188–1195. doi:10.1021/jf052737j. PubMed: 16478235.
22. Famiglini G, Palma P, Termopoli V, Trufelli H, Cappiello A (2009) Single-step LC/MS method for the simultaneous determination of GC-Amenable Organochlorine and LC-Amenable Phenoxy Acidic Pesticides. Analytical Chemistry 81: 7373–7378. doi:10.1021/ac9008995. PubMed: 19663448.
23. Sannino A, Bolzoni L, Bandini M (2004) Application of liquid chromatography with electrospray tandem mass spectrometry to the determination of a new generation of pesticides in processed fruits and vegetables. Journal of Chromatography A 1036: 161–169. doi:10.1016/j.chroma.2004.02.078. PubMed: 15146917.
24. Lawrence JF (1976) Fluorogenic labeling of organophosphate pesticides with dansyl chloride: Application to residue analysis by high-pressure liquid chromatography and thin-layer chromatography. Journal of Chromatography A 121: 343–351. PubMed: 932151.
25. Butz S, Stan HJ (1995) Screening of 265 pesticides in water by Thin-Layer Chromatography with automated multiple development. Analytical Chemistry 67: 620–630. doi:10.1021/ac00099a021.
26. Janotta M, Karlowatz M, Vogt F, Mizaikoff B (2003) Sol–gel based mid-infrared evanescent wave sensors for detection of organophosphate pesticides in aqueous

solution. Analytica Chimica Acta 496: 339–348. doi:10.1016/S0003-2670(03)01011-0.

27. Tanner PA, Leung KH (1996) Spectral interpretation and qualitative analysis of organophosphorus pesticides using FT-Raman and FT-Infrared Spectroscopy. Applied Spectroscopy 50: 565–571. doi:0003-7028/96/5005-0565.
28. Viveros L, Paliwal S, McCrae D, Wild J, Simonian A (2006) A fluorescence-based biosensor for the detection of organophosphate pesticides and chemical warfare agents. Sensors and Actuators B: Chemical 115: 150–157. doi:10.1016/j.snb.2005.08.032. PubMed: 17616234.
29. Kim YJ, Cho Y, Lee HS, Lee YT (2003) Investigation of the effect of hapten heterology on immunoassay sensitivity and development of an enzyme-linked immunosorbent assay for the organophosphorus insecticide fenthion. Analytica Chimica Acta 494: 29–40. doi:10.1016/j.aca.2003.07.003.
30. Brun EM, Garcés-García M, Puchades R, Maquieira Á (2004) Enzyme-linked immunosorbent assay for the organophosphorus insecticide fenthion. Influence of hapten structure. Journal of Immunological Methods 295: 21–35. doi:10.1016/j.jim.2004.08.014. PubMed: 15627608.
31. Moeller N, Mueller-Seitz E, Scholz O, Hillen W, Petz M, et al. (2007) A new strategy for the analysis of tetracycline residues in foodstuffs by a surface plasmon resonance biosensor. European Food Research and Technology 224: 285–292. doi:10.1007/s00217-006-0392-z.
32. Nogues C, Leh H, Langendorf CG, Law RHP, Buckle AM, et al. (2010) Characterisation of peptide microarrays for studying antibody-antigen binding using surface plasmon resonance imagery. PLoS ONE 5: e12152. doi:10.1371/journal.pone.0012152. PubMed: 20730101.
33. Kuo YC, Ho JH, Yen TJ, Chen HF, Lee OKS (2011) Development of a surface plasmon resonance biosensor for real-time detection of Osteogenic differentiation in live mesenchymal stem cells. PLoS ONE 6: e22382. doi:10.1371/journal.pone.0022382. PubMed: 21818317.
34. Li Y, Qi P, Ma X, Zhong JG (2014) Quick detection technique for clenbuterol hydrochloride by using surface plasmon resonance biosensor. European Food Research and Technology 239: 195–201. doi:10.1007/s00217-014-2201-4.
35. Dai Z, Liu H, Shen YD, Su XP, Xu ZL, et al. (2012) Attomolar determination of Coumaphos by electrochemical displacement immunoassay coupled with oligonucleotide sensing. Analytical Chemistry 84: 8157–8163. doi:10.1021/ac301217s. PubMed: 22934793.
36. Xu ZL, Wang Q, Lei HT, Eremin SA, Shen YD, et al. (2011) A simple, rapid and high-throughput fluorescence polarization immunoassay for simultaneous detection of organophosphorus pesticides in vegetable and environmental water samples. Analytica Chimica Acta 708: 123–129. doi:10.1016/j.aca.2011.09.040.
37. Yin GH, Sun ZN, Liu N, Zhang L, Song YZ, et al. (2009) Production of double-stranded RNA for interference with TMV infection utilizing a bacterial prokaryotic expression system. Applied Microbiology and Biotechnology 84: 323–333. Doi:10.1007/s00253-009-1967-y. PubMed: 19330324.

# 10

# Perinatal Exposure of Mice to the Pesticide DDT Impairs Energy Expenditure and Metabolism in Adult Female Offspring

**Michele La Merrill[1,2,3]*, Emma Karey[1], Erin Moshier[2], Claudia Lindtner[3,4], Michael R. La Frano[5,6], John W. Newman[5,6,7], Christoph Buettner[3,4]**

**1** Department of Environmental Toxicology, University of California Davis, Davis, California, United States of America, **2** Department of Preventive Medicine, Mount Sinai School of Medicine, New York, New York, United States of America, **3** Metabolism Institute, Mount Sinai School of Medicine, New York, New York, United States of America, **4** Division of Endocrinology, Diabetes and Bone Disease, Mount Sinai School of Medicine, New York, New York, United States of America, **5** West Coast Metabolomic Center, University of California Davis, Davis, California, United States of America, **6** Department of Nutrition, University of California Davis, Davis, California, United States of America, **7** Obesity and Metabolism Research Unit, USDA-ARS-Western Human Nutrition Research Center, Davis, California, United States of America

## Abstract

Dichlorodiphenyltrichloroethane (DDT) has been used extensively to control malaria, typhus, body lice and bubonic plague worldwide, until countries began restricting its use in the 1970s. Its use in malaria control continues in some countries according to recommendation by the World Health Organization. Individuals exposed to elevated levels of DDT and its metabolite dichlorodiphenyldichloroethylene (DDE) have an increased prevalence of diabetes and insulin resistance. Here we hypothesize that perinatal exposure to DDT disrupts metabolic programming leading to impaired metabolism in adult offspring. To test this, we administered DDT to C57BL/6J mice from gestational day 11.5 to postnatal day 5 and studied their metabolic phenotype at several ages up to nine months. Perinatal DDT exposure reduced core body temperature, impaired cold tolerance, decreased energy expenditure, and produced a transient early-life increase in body fat in female offspring. When challenged with a high fat diet for 12 weeks in adulthood, female offspring perinatally exposed to DDT developed glucose intolerance, hyperinsulinemia, dyslipidemia, and altered bile acid metabolism. Perinatal DDT exposure combined with high fat feeding in adulthood further impaired thermogenesis as evidenced by reductions in core temperature and in the expression of numerous RNA that promote thermogenesis and substrate utilization in the brown adipose tissue of adult female mice. These observations suggest that perinatal DDT exposure impairs thermogenesis and the metabolism of carbohydrates and lipids which may increase susceptibility to the metabolic syndrome in adult female offspring.

**Editor:** Barbara T. Alexander, Univ Mississippi Medical Center, United States of America

**Funding:** This research was supported through funding by the National Institute of Health (R00 ES019919, R03 DK082724, U24 DK092993, U24 DK097154, T32 ES007059, and P60 DK020541), the American Diabetes Association, and United States Department of Agriculture - Agricultural Research Service Intramural Project 5306-51530-019-00D. The funders had no role in study design, data collection and analysis, decision to publish, or preparation of the manuscript.

**Competing Interests:** The authors have declared that no competing interests exist.

* Email: mlamerrill@ucdavis.edu

## Introduction

The prevalence of type 2 diabetes (T2D) is rising worldwide [1], creating a tremendous burden for health care systems in developed and developing nations. The estimated cost associated with T2D was $245 billion last year in the United States alone [2]. This rise in the incidence of T2D is largely attributed to increasing obesity rates [3]. Obesity and T2D are characterized by insulin resistance and are often associated with perturbations in energy expenditure [4,5].

Numerous epidemiological studies have suggested that exposure to persistent organic pollutants (POPs) contribute to the rising incidence of T2D [6]. For example, exposure to the organochlorine pesticide DDT has been found to be associated with increased odds of T2D [7]. Further, exposure to DDE, the primary metabolite of DDT, has also been associated with increased odds of excess body weight, insulin resistance, and dyslipidemia in people [8]. Because the World Health Organization still recommends the use of DDT for malaria control, DDT's potential contribution to the pathogenesis of T2D continues to be relevant to public health. Furthermore, obesity and T2D have increased at alarming rates in countries where DDT is still in use. For instance, the prevalence of diabetes among rural Indian adults increased 3.8 fold from 1989 to 2006 [9] and although a confluence of factors contribute to this rise, the contribution of endocrine disrupting chemicals cannot be ruled out without further study. Since epidemiological associations cannot establish causality, we tested whether exposure to DDT and DDE predisposes to T2D in mice at doses comparable to human exposures [10–13].

Both obesity and T2D are characterized by impaired energy expenditure and reduced thermogenesis, which are often the result of reduced brown adipose tissue (BAT) activity [5]. Mouse studies also suggest that impaired thermogenesis may precede insulin resistance [14]. Therefore, increasing energy expenditure by activating BAT thermogenesis may hold promise as a therapeutic strategy for managing both obesity and T2D [15–18].

Perturbations in the perinatal environment can impair metabolic programming that can consequently increase susceptibility to obesity and T2D in adulthood. A classic example of the importance of the perinatal environment on the future risk of T2D has been historically provided by the Dutch famine. Fetuses carried to term that were exposed to fetal malnutrition suffered from increased rates of the metabolic syndrome and glucose intolerance as adult offspring [19,20]. The consequences of impaired fetal development on metabolic programming can be modeled in rodents, where fetal under-nutrition results in the development of the metabolic syndrome and glucose intolerance in adult offspring [21,22]. Here we confirm the hypothesis that perinatal exposure to DDT impairs energy expenditure and metabolism leading to insulin resistance and the metabolic syndrome in adult offspring. We demonstrate for the first time that developmental exposure to DDT impairs BAT thermogenesis.

## Methods

### DDT exposure

To mimic the commercial formulation of DDT used as a pesticide in the US prior to its ban, 77.2% p,p′-DDT (98.5% purity neat, AccuStandard, New Haven, CT) and 22.8% o,p′-DDT (100% purity neat, AccuStandard) were dissolved at a final concentration of 0.17 g DDT mixture/L organic olive oil [23], hereafter referred to as DDT. We selected the dosing period to span liver and white- and brown- adipose tissue ontogenesis and programming [24–28]. From 11.5 days post coitus (determined by vaginal plug) to postnatal day (PND) 5, we administered 1.7 mg DDT/kg body weight (per os, n = 15 dams, 10 µL solution/g mouse), or equivalent volume of olive oil vehicle (hereafter referred to as vehicle or VEH, per os, n = 14 dams) daily to primigravid C57BL/6J mice (Jackson Laboratories, Bar Harbor, ME, Fig. S1). To determine the internal dose after transplacental and lactational transfer to pups [29], 4 additional dams were treated identically with DDT, and 24 hours after the final dose their sera was analyzed for o,p′-DDT, p,p′-DDT, and p,p′-DDE by gas chromatography using electron capture detection with detection limits of 0.36, 0.27 and 0.12 pg/µL serum respectively [30].

### Study Design

We culled litters to 6 pups after the final DDT dose on PND 5, to minimize infanticide common among C57BL/6J dams while normalizing lactational transfer of DDT and maternal behavior effects (Fig. S1). We evaluated timing of puberty daily by monitoring vaginal opening (VO) and preputial separation (PPS), in all female and male mice respectively, beginning at PND 18. On PND 21 we weaned pups (1 cage/sex/litter/treatment). We assessed body composition, energy homeostasis, metabolic parameters, and glucose tolerance up to 6 months of age when we randomized mice into 2 study arms to assess: 1) energy balance or 2) the metabolic consequences of high fat feeding (Fig. S1).

For the energy balance study, 10 females from 10 litters and 8 metabolic chambers were available. Energy balance was assessed by indirect calorimetry in 1 female mouse/litter (4 mice/perinatal treatment) for 7 days after 2 days of acclimation. Cold tolerance was assessed 3 days after indirect calorimetry in 5 female mice/perinatal treatment, as described below.

For the high fat feeding study, littermates were randomized to high fat diet (HFD, 4.73 kcal/g, 20% protein, 35% carbohydrate, and 45% fat per kcal; D12451, Research Diets, New Brunswick, NJ) or a low fat diet (LFD, 3.85 kcal/g, 20% protein, 70% carbohydrate, and 10% fat per kcal; D12450B, Research Diets) feeding following their glucose tolerance test (GTT) at 6 months of age for a total of 12 weeks (n = 8 mice/sex/litter per perinatal treatment and diet combination).

All mice had access to food and water *ad libitum* (other than during stated fasting periods) in sterile ventilated cages in an American Association for the Accreditation of Laboratory Animal Care-approved facility. At the end of all the studies, mice were sacrificed by exsanguination under isoflurane anesthesia. All animal experiments were conducted under the guidelines on humane use and care of laboratory animals for biomedical research published by the US National Institutes of Health. The procedures of this study were approved by the Mount Sinai School of Medicine Institutional Animal Care and Use Committee protocol #LA11-00116.

### Body composition

We measured body weight and percent fat mass by MRI (3in1-EchoMRI, Echo Medical Systems, Houston, TX) of each offspring monthly in 1–6 month old mice (n = 15 DDT- and n = 14 vehicle-exposed litters). The body mass of all mice was also measured at PND 5. During HF feeding, mice were weighed weekly and we used MRI to determine body composition at 8 months of age (n = 8 mice/sex/litter per perinatal treatment and diet combination).

### Energy homeostasis

We measured food intake in mice on a monthly basis from 2–6 months (n = 15 DDT- and n = 14 vehicle- exposed litters). In 2 month old mice, each mouse/sex/litter/treatment was singly housed and food intake was assessed daily over a 5 day period. In group- housed mice aged 3–6 months, we measured food intake per cage over a 2 day period. In the HF feeding study, we measured weekly food intake of singly housed mice fed LFD or HFD. Food intake (g or kcal) was calculated as the difference in food weight divided by the number of mice that had access to it and the number of days they had access, and transformed to kcal when comparing HFD and LFD (SAS 9.2, SAS Institute, Cary, NC). We measured core temperature monthly in all mice aged 2–5 and 8 months old by inserting a thermocouple into the rectum (BAT-10, Physitemp, Clifton, NJ).

For the energy balance studies, we housed 4 females/perinatal treatment (1 mouse/litter) individually in metabolic chambers, which continuously measured oxygen consumption, carbon dioxide production, food intake, and 3 plane-locomotor activity through open-circuit, indirect calorimetry (CLAMS, Columbus Instruments, Columbus, OH). During this 9 day period, we provided mouse powdered diet (NIH 07) and water *ad libitum* while they were maintained at 21–22°C for a 12-hour light/dark cycle. The respiratory exchange ratio (RER) was calculated as the quotient of $vCO_2/vO_2$.

Three days after the indirect calorimetry was completed, we individually housed 5 female mice/perinatal treatment (1 mouse/litter) at 4°C with bedding enrichment. Core body temperature was measured immediately upon exposure to 4°C, and every 30 minutes thereafter until 90 minutes, when 50% of the mice reached core temperatures of <30°C after being kept for 90-minutes at 4°C. Two days later, we housed each of the 10 female mice for 1 hour at 4°C prior to sacrifice.

## GTT

Glucose tolerance was assessed in 1 mouse/sex/litter/perinatal treatment aged 2, 3 and 6 months (randomized selection from n = 15 DDT- and n = 14 vehicle- exposed litters) and 1 mouse/sex/litter for each perinatal treatment and adult diet combination at eight months (randomized selection from n = 8 litters/treatment). Two-month-old mice received 2.0 mg D-glucose/g mouse body weight (i.p.), since at young age mice (as well as humans) are more glucose tolerant, and all other mice received 1.0 mg D-glucose/g mouse body weight (i.p.). For all GTT, we fasted mice for 5 hours, and collected sub-mandibular blood to determine fasting blood glucose (AlphaTrak, Abbott Laboratories, Abbott Park, IL) and fasting plasma metabolic parameters. We measured blood glucose 15, 30, 45, 60, 90, and 120 minutes after a D-glucose bolus for GTT. We determined total AUC using the trapezoidal approach (PROC SQL, SAS 9.4, SAS Institute, Cary, NC).

## Metabolic parameters

Trunk blood glucose was measured in all mice culled on PND 5. Sub-mandibular blood was collected from mice fasted for 5 hours (1 mouse/sex/litter/treatment) at ages 1–6 months old (randomized selection from n = 15 DDT- and n = 14 vehicle- exposed litters) and 8 months old (randomized selection from n = 8 litters/treatment) to determine blood glucose (AlphaTrak) and plasma insulin by ELISA (Mercodia, Uppsala, Sweden). Fasting plasma triglycerides, total cholesterol, and non-esterified fatty acids (NEFA) were evaluated enzymatically (Wako Chemicals, Richmond, VA). Homeostasis model assessment- insulin resistance (HOMA-IR) was calculated as the product of fasting glucose and insulin divided by 22.5 [31].

## Pathology

Livers were harvested from 1 female fed HFD per perinatal treatment group (randomized selection from n = 1 female/litter in 7 VEH and 8 DDT litters) into OCT medium, frozen 5 μM thick sections were stained with oil red O, and fields with representative staining of the sample and the least clear space (vessels, ducts, etc.) were selected. Histopathological evaluations were assessed by a veterinary pathologist who was blinded to the treatments.

## Hepatic lipids

Hepatic triglycerides and cholesterol esters were measured after a modified Folch extraction (randomized selection from n = 1 female/litter in 7 VEH and 8 DDT litters fed HFD) [32]. Total cholesterol [33] and triglycerides [34] were measured enzymatically from these livers (Fisher Diagnostics, Middletown, VA).

## Hepatic bile acids

A 15 mg liver sample (randomized selection from n = 1 female/litter in 7 VEH and 8 DDT litters fed HFD) was pulverized on dry ice, enriched with deuterated bile acid analogs, butylated hydroxytoluene and ethylinediaminetetraacetic acid, and extracted twice with 500 μL methanol. The combined extract was dried, reconstituted in 100 μL methanol:acetonitrile (50:50, v/v) with internal standards 1-phenyl-3-hexanoic acid urea and 1-cyclohexyl-3-dodecanoic acid urea (Sigma-Aldrich, St. Louis, MO) and filtered through 0.1 μm filters. Tissue bile acids were isolated and quantified by LC-MS/MS using modifications of a previously published method [35]. Separation of bile acids was achieved using reversed phase UPLC with a 1.7 μm, 2.1×100 mm Acquity BEH C18 column (Waters, Milford, MA) using a 16 min gradient (Solvent A = 0.1% formic acid; Solvent B = 0.1% formic acid in acetonitrile) on a Waters Acquity UPLC (Waters) and detected on an API 4000 QTrap (AB Sciex, Framingham, MA, USA) by multiple reaction monitoring (MRM) after negative mode electrospray ionization. Residues were quantified against 7 point curves of authentic standards purchased from Steraloid Inc. (Newport, RI), Sigma-Aldrich (St. Louis, MO), and Medical Isotopes, Inc. (Pelham, NH) using internal standard methodologies.

## Western blot analysis

Protein expression was assessed as previously described [36]. Briefly, liver was homogenized and 10–20 μg of protein was subjected to Western blot analysis with primary antibodies against fatty acid synthase (FAS, BD Biosciences, San Jose, CA), phospho-acetyl-CoA carboxylase (ACC), ACC, phospho- adenosine triphosphate citrate lyase (ATPCL), ATPCL, adipose triglyceride lipase (ATGL), AKT $Thr^{308}$ and $Ser^{473}$, AKT, phospho- glycogen synthase kinase 3 (GSK3), GSK3, phospho-ERK, and ERK (all antibodies from Cell Signaling Technology, Danvers, MA unless stated otherwise) and insulin receptor (IR) beta (Santa Cruz Biotechnology, Dallas, TX). The secondary antibody was DyLight conjugated goat anti-rabbit or anti-mouse IgG (ThermoScientific). Signal intensity of proteins of interest was divided by the intensity of the housekeeping protein HSC 70 (Santa Cruz Biotechnology) to determine fold change (Odyssey, LI-COR, Lincoln, NE). We have confirmed the specificity of ATGL in adipose tissue specific ATGL knockout vs. wild type mice, and all other antibodies have been characterized and validated in prior peer review publications [(Table S1), [37–41]].

## Semi-quantitative PCR

mRNA was extracted with the RNeasy kit (Qiagen, Hilden, Germany) to synthesize cDNA using reverse transcription PCR (Applied Biosystems, Foster City, CA). Semi-quantitative PCR was performed using proprietary TaqMan primers for transcripts: peroxisome proliferative activated receptor gamma coactivator 1 alpha (*Ppargc1a*), iodothyronine type II deiodinase (*Dio2*), twist-related protein 1 (*Twist1*) and uncoupling protein 1 (*Ucp1*) in BAT using TATA box binding protein (*Tbp*) as an endogenous control; 3-hydroxy-3-methylglutaryl-Coenzyme A reductase (*Hmgcr*), low density lipoprotein receptor (*Ldlr*), and cytochrome P450 7a1 (*Cyp7a1*) in livers with glucuronidase beta (*Gusb*) as an endogenous control (Applied Biosystems). Additional semi-quantitative PCR was performed in BAT using SYBR Green with published primers: glucose transporter type 4 (*Glut4*, F: CATCCCACAAGGCACCCCTC, R: CATGCCACCCACAGAGAAGA [42]), lipoprotein lipase (*Lpl*, F:TGGAGAAGCCATCCGTGTG, R: TCATGCGAGCACTTCACCAG [42]), adipose triglyceride lipase (*Pnpla*, F: TGTTTCAGACGGAGAGAACGTC, R: TGAGAATGGGGACACTGTGATG [43]), and carnitine palmitoyltransferase 2 (*Cpt2*, F: CCAGCGGATAAACCACAACA, R: AAGCCACCGAGGCTCACTG [42]) using *Tbp* (F: GAAGCTGCGGTACAATTCCAG, R: CCCCTTGTACCCTTCACCAAT [44]) as endogenous control. Endogenous controls did not vary significantly across treatment groups. The $2^{-ddCT}$ method was used to approximate relative transcript fold change between treatment groups (coefficient of replicate variation <10%) [45].

## Statistical analyses

Phenotypes were assessed at specific ages by modeling the fixed effect of perinatal DDT and the random effect of litter while stratifying by gender (PROC MIXED, SAS). Longitudinal analyses of body mass, fat mass, and temperature were conducted using the age-specific model as described above, with the addition of age and an age*treatment term as fixed effects, and individual mouse identifiers as random effects. To assess circadian effects during indirect calorimetry, photoperiods and a photoperiod*perinatal treatment term were included as fixed effects in the age-specific model. Body weight and fat mass were considered as potential covariates in indirect calorimetry analyses [46] but because these body composition variables were not significantly different between perinatal treatments and caused no reduction in data significance or difference in interpretation, these covariates were not included in the final indirect calorimetry models. When examining phenotypes from the HF feeding study, diet and a diet*perinatal treatment term were included as fixed effects in the age-specific and longitudinal models described above. GTTs, fasting metabolic parameters, protein expression, mRNA, lipids, and bile acids were evaluated without random effects because only one mouse per litter was included (PROC GLM, SAS). To evaluate the potential covariate behavior of hepatic bile acids and other metabolic factors which may influence or be influenced by cholesterol metabolism, imDEV v 1.4.2 [47] was employed to normalize data, transform variables to mean centers with unit variance, and conduct partial least squares discriminate analysis (PLS DA; oscores pls). The dependent variable was the perinatal treatment group while the independent variables were hepatic bile acids, *Cyp7a1* fold-change, hepatic cholesterol and triglycerides, plasma cholesterol, % body fat, and age. Measures with variable importance in projection (VIP) scores >1 were considered important discriminators of treatment [48]. Timing of puberty was evaluated using survival analysis with treatment as a fixed effect and litter as a repeated event (PROC PHREG, SAS). In all models, we used a significance threshold of $p<0.05$ for main effects and $p<0.1$ for interaction effects.

## Results

### Perinatal DDT exposure increases adiposity in young adults

To examine whether perinatal DDT exposure causes the metabolic syndrome in adult mouse offspring, dams were exposed to DDT from mid-gestation to PND 5. The resulting internal doses were 2.2 (±0.1) ng p,p'-DDE/mL, 51.1 (±10.2) ng p,p-DDT/mL, while o,p'-DDT was undetected [mean (± SEM)]. These exposures were sufficiently low as to have no impact on length of gestation, weight gain during gestation, or litter size (Fig. S2) and fall within the range of contemporary human exposures [10,49]. Perinatal DDT exposure decreased body weight and blood glucose in male mice (Fig. S3A–B) and decreased glucose in female mice (Fig. 1B) on PND 5. Although DDT and DDE are ligands of the estrogen and androgen receptors [50], the timing of puberty in either sex was not significantly affected by the perinatal DDT exposure (Fig. 1C, Fig S3C). Perinatal DDT exposure caused a transient increase in body weight and adiposity of young adult female mice (Fig. 2A–C) but no differences in female lean mass (Fig. 2D), or male body composition were observed (Fig. S3D–E). Further, glucose tolerance, fasting glucose, insulin, and lipid levels were not altered by perinatal DDT exposure in either sex during their first 6 months of post-weaning life (Fig. S4, Table S2). These data suggest that mice exposed to perinatal DDT manifested only subtle features of the metabolic syndrome until adulthood. Since perinatal DDT exposure had no effect on food intake of offspring (Fig. 2E, Fig. S3F), the increased adiposity transiently observed among young female mice exposed to DDT could have resulted from impairments in thermogenesis and energy expenditure.

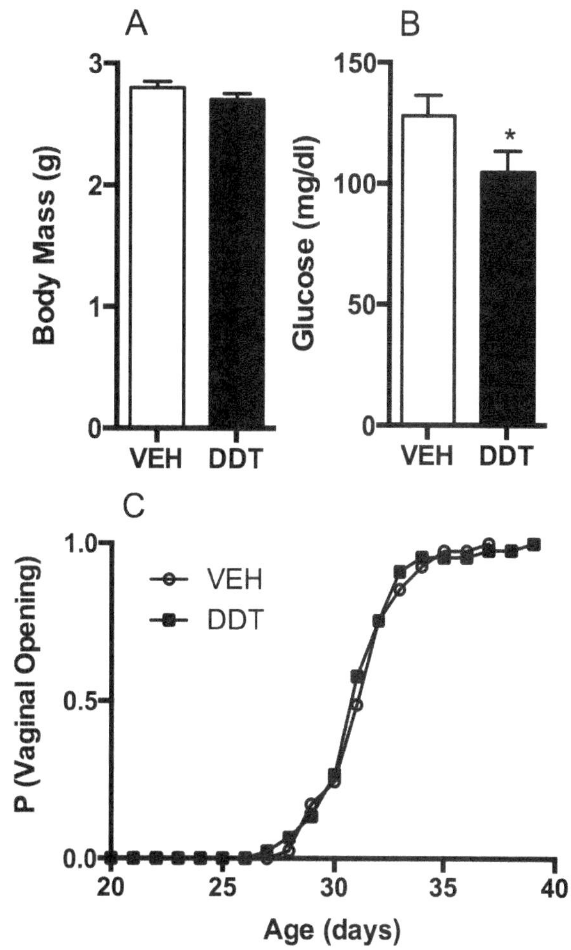


**Figure 1. Effect of perinatal DDT on the development of female mice.** (A) Female body mass at PND5 (n = 15 DDT- and n = 14 VEH-exposed litters). (B) Trunk blood glucose in PND5 females culled from litters >6 pups (n = 7 DDT- and n = 7 VEH- exposed litters). (C) Proportion of mice with vaginal opening as a function of age (n = 15 DDT- and n = 14 VEH- exposed litters). *p<0.05 DDT vs. vehicle controls. Data are represented as least squares (LS) means + SEM.

### Perinatal DDT exposure reduces energy expenditure

To assess thermogenesis, core body temperature was measured longitudinally. Perinatal DDT exposure did not affect the core body temperature of male mice (Fig. S3G), while it decreased the core temperature of female mice, an effect that became more pronounced with age (Fig. 2F). At 6 months, female mice perinatally exposed to DDT had decreased oxygen consumption (Fig. 3A) and a more pronounced RER reduction in the light period than in the dark period, indicative of greater fasting lipid utilization (Fig. 3B), although food intake appeared equivocal across treatment groups (Fig. 3C). Female mice perinatally exposed to DDT also had decreased energy expenditure, while physical activity was not altered (Fig. 3D–E), suggesting that decreased energy expenditure was the result of decreased

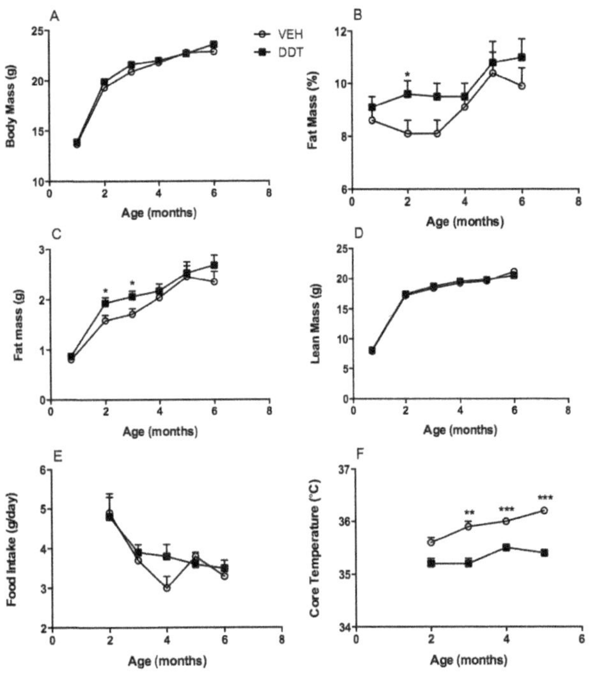


**Figure 2. Effect of perinatal DDT on body composition and energy balance in female mice.** Monthly (A) body mass (p = 0.05 at 2 months, p = 0.08 at 3 months), (B) percent adiposity (p = 0.05 at 3 months), (C) fat mass, (D) lean mass, and (E) food intake (A–E: n = 15 DDT- and n = 14 VEH- exposed litters). (F) Monthly core temperature (1 female/litter at 2 month time point, all females/litter thereafter, p = 0.07 at 2 months, age*DDT pi = 0.07). *p<0.05, **p<0.001, ***p<0.0001 DDT vs. vehicle controls. Data are represented as LS means + SEM.

thermogenesis. To further establish that female mice perinatally exposed to DDT had compromised thermogenesis, we evaluated cold tolerance in female mice exposed to 4°C. Female mice with perinatal DDT exposure exhibited marked cold intolerance, necessitating the termination of the cold tolerance test after 90 minutes (Fig. 3F).

To determine if high fat feeding, a commonly used metabolic stressor to model human obesity, could unveil a metabolic phenotype caused by perinatal DDT exposure, mice were fed either LFD or HFD for 12 weeks beginning at 6 months of age. HF feeding increased caloric intake, adiposity, and body mass equally in all groups (Fig. 4A–C, Fig. S5), indicating that any metabolic alterations observed in HF fed mice that were perinatally exposed to DDT occurred independently of caloric intake and/or body composition.

Perinatal DDT exposure reduced core temperature in 9 month old female mice (Fig. 4D). HFD feeding led to higher body core temperatures in non-exposed mice, however this thermogenic adaptation to HFD was absent in mice perinatally exposed to DDT (Fig. 4D). Perinatal DDT exposure paradoxically reduced body core temperature by more than 1°C in mice fed HFD ($p < 0.0001$). Thus, perinatal DDT impairs thermogenesis which is further worsened by HFD, indicating that developmental DDT exposure increases susceptibility to HFD induced- impairment in energy balance.

We next examined if perinatal DDT exposure altered molecular mechanisms controlling thermogenesis in BAT of 9 month old female mice (Fig. 5). Compared to LFD-fed control mice, adult HF feeding and perinatal DDT exposure caused a reduction in mRNA expression of *Ppargc1a* (Fig. 5A), a master regulator of thermogenesis. Indeed HFD and DDT exposures reduced the expression of *Dio2* (Fig. 5B), which encodes an enzyme that locally activates thyroid hormone by deiodinating T4 to T3 to stimulate thermogenesis in BAT. However there was no significant change in *Ucp1* RNA expression (Fig. 5C), a protein that uncouples respiration to increase thermogenesis [51,52]. Consistent with the regulatory role of PGC1A in substrate utilization [14,53–56], perinatal DDT exposure and adult HF feeding decreased the expression of glucose transporter *Glut4*, lipoprotein transporter *Lpl* (no diet effect), lipolysis rate-limiting *Pnpla*, and the fatty acid oxidation shuttle *Cpt2* (Fig. 5D–G). Because transcription of *Ppargc1a* and its downstream targets was suppressed, we investigated negative regulation of *Ppargc1a* across DDT and HFD treatments. Indeed, expression of *Twist1*, a negative regulator of *Ppargc1a* [57], was increased in BAT of 9 month old mice exposed perinatally to DDT and fed a HFD during adulthood (Fig. 5H).

## Perinatal exposure to DDT increases susceptibility to HFD- induced insulin resistance

Female mice maintained on a LFD and exposed perinatally to DDT had equivocal glucose tolerance compared to vehicle treated controls (Fig. 6A). However, after HF feeding for 12 weeks, female mice exposed perinatally to DDT developed glucose intolerance and insulin resistance as evidenced by higher AUC during GTT, fasting hyperinsulinemia, and increased HOMA-IR (Fig. 6A–C). Of note, this constellation is often observed in prediabetic patients that commonly have the metabolic syndrome. Further, the expression and activation states of AKT and ERK, proteins involved in insulin signaling, were moderately decreased in female mice perinatally exposed to DDT and fed a HFD (Fig. 6D–E), although it remains unclear how much these moderate decreases in hepatic insulin signaling contribute to the observed glucose intolerance and insulin resistance. However, male mice perinatally exposed to DDT developed glucose intolerance when fed LFD, but not HFD (Fig. S8A–B). Thus, perinatal exposure to DDT increases the susceptibility to HFD- induced glucose intolerance in female but not male mice (Fig. 6, Fig. S6). The sex differences will require further studies in the future to test if this is estrogen dependent, but of note, female mice (and humans) are somewhat protected from HF feeding- induced glucose intolerance [58].

## Perinatal DDT exposure and high fat feeding cause mild dyslipidemia in female mice

To evaluate whether perinatal DDT exposure impaired lipid metabolism in adult HFD fed- mice, we measured fasting triglycerides, cholesterol, and NEFAs (Fig. 7, Fig. S6). Perinatal DDT exposure significantly increased fasting cholesterol in female mice fed HFD, although it had a negligible effect on cholesterol in female mice fed LFD (Fig. 7A). A similar, but non-significant trend was observed in fasting triglycerides levels in female mice (Fig. 7B), whereas NEFAs were not altered by perinatal DDT or HFD in female mice (Fig. 7C). Thus perinatal DDT exposure combined with HFD caused mild dyslipidemia in female (Fig. 7A) but not male mice (Fig. S6E-G). We also examined whether perinatal DDT exposure altered the expression of mRNA and proteins associated with the regulation of lipid homeostasis in livers from female mice fed HFD. mRNA expression of *Ldlr*, a cholesterol transporter, was significantly decreased, whereas that of *Cyp7a1*, the rate limiting enzyme in the synthesis of bile acids from cholesterol, was significantly increased by perinatal DDT exposure (Fig. 7D). Changes in hepatic bile acid metabolism among females fed HFD were consistent with these changes in *Cyp7a1* expression

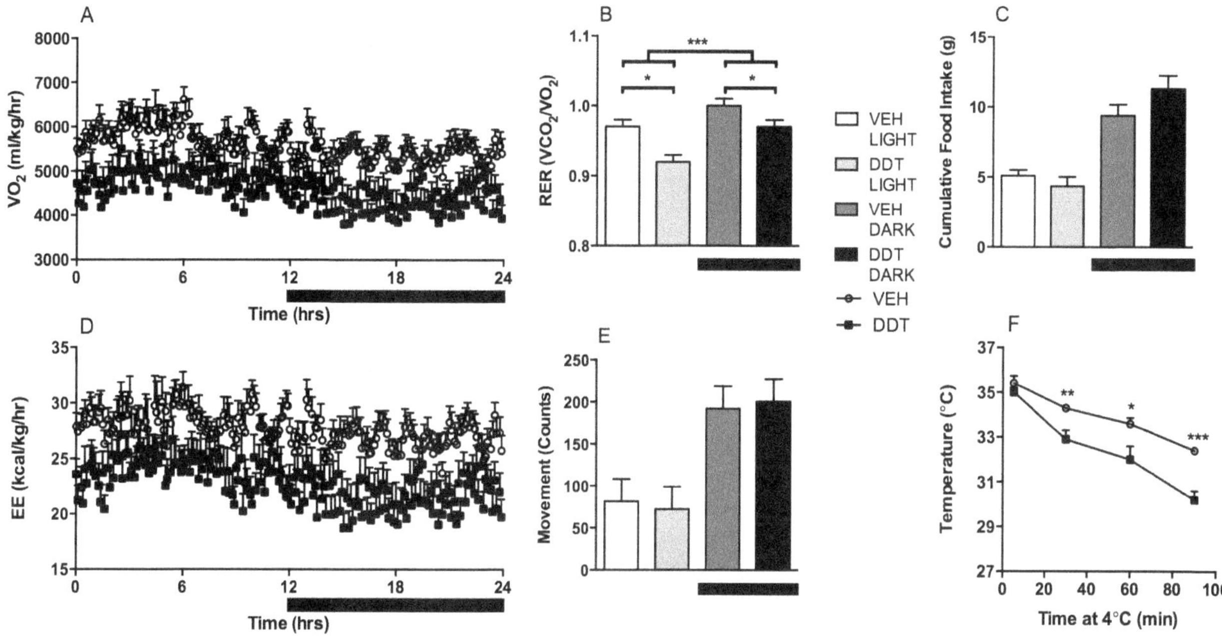


**Figure 3. Effect of perinatal DDT on energy expenditure and cold tolerance in six month old female mice.** (A) Oxygen consumption, (B) RER (period*DDT $p_{interaction}<0.0001$), (C) cumulative food intake, (D) energy expenditure, and (E) movement (over 5 day period and 4 female/treatment, 1 mouse/litter). (F) Core temperature (5 females/treatment, 1 female/litter). *p<0.05, **p<0.01, ***p<0.001 DDT vs. vehicle controls. Data are represented as LS means + SEM.

due to perinatal DDT exposure. Specifically, conjugated bile acids, including GLCA, TLCA, and TDCA, and the C27 precursor TriHCA, increased with DDT treatment and these conjugated bile acid changes were closely correlated with hepatic *Cyp7a1* expression and plasma cholesterol (VIP>1, Fig. 7E, Fig. S7, Table S3). Conversely, unconjugated bile acids, including HCA, HDCA, and β-MCA, were relatively lower in the DDT exposed versus control groups and were closely associated with hepatic lipids (VIP>1, Fig. 7E, Fig. S7, Table S3). Hepatic protein expression of two lipogenic enzymes ACC and ATPCL was also increased, consistent with increased hepatic lipogenesis as a consequence of perinatal DDT exposure (Fig. 7F). This did not appear to contribute to steatosis, as hepatic lipid content was not increased due to perinatal DDT exposure among female mice fed HFD (Fig. S8).

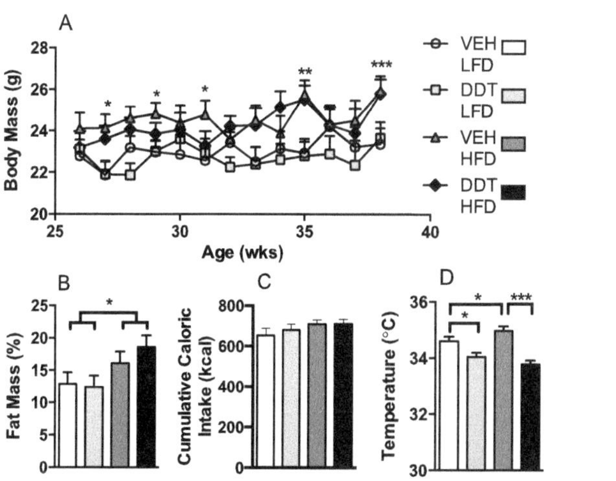


**Figure 4. Effects of perinatal DDT and adult HFD feeding on body composition and energy balance in female mice (n = 1 female/litter in 7 VEH and 8 DDT litters).** (A) Weekly body mass (HFD*age pi<0.0001). (B) Percent adiposity in 8 month old mice. (C) Cumulative caloric intake over 12 weeks of HFD or LFD feeding. (D) Core temperature in 8 month old mice (DDT*HFD pi = 0.01). *p<0.05, **p<0.01, ***p<0.0001 A–B) LFD vs. HFD only; D) significance between groups indicated by bars. Data are represented as LS means + SEM.

## Discussion

Here we investigated the hypothesis that perinatal DDT exposure impairs energy expenditure and metabolism in adult mice. Our results suggest that perinatal DDT exposure decreases energy expenditure, and impairs the regulation of thermogenesis, glucose-, and lipid- homeostasis, which contribute to the development of insulin resistance and the metabolic syndrome in adult female mice.

### Obesity and thermogenesis

Perinatal DDT exposure increased adiposity in young adult female mice. This transient increase in adiposity may have resulted from the profoundly impaired thermogenesis we observed after DDT exposure, which was greater than what was observed in UCP1 knock-out mice [59]. Impaired thermogenesis may have been a consequence of reduced energy supply and utilization in BAT as perinatal DDT exposure significantly decreased expression of lipolytic *Pnpla* and substrate transporters (*Glut4*, *Lpl*, *Cpt2*) [14,53–56]. Because food intake and physical activity were equal among groups, the reduced energy expenditure would be predicted to increase fat mass more profoundly than what we observed. Whether increased caloric content in feces accounts for this discrepancy will require further study [60].

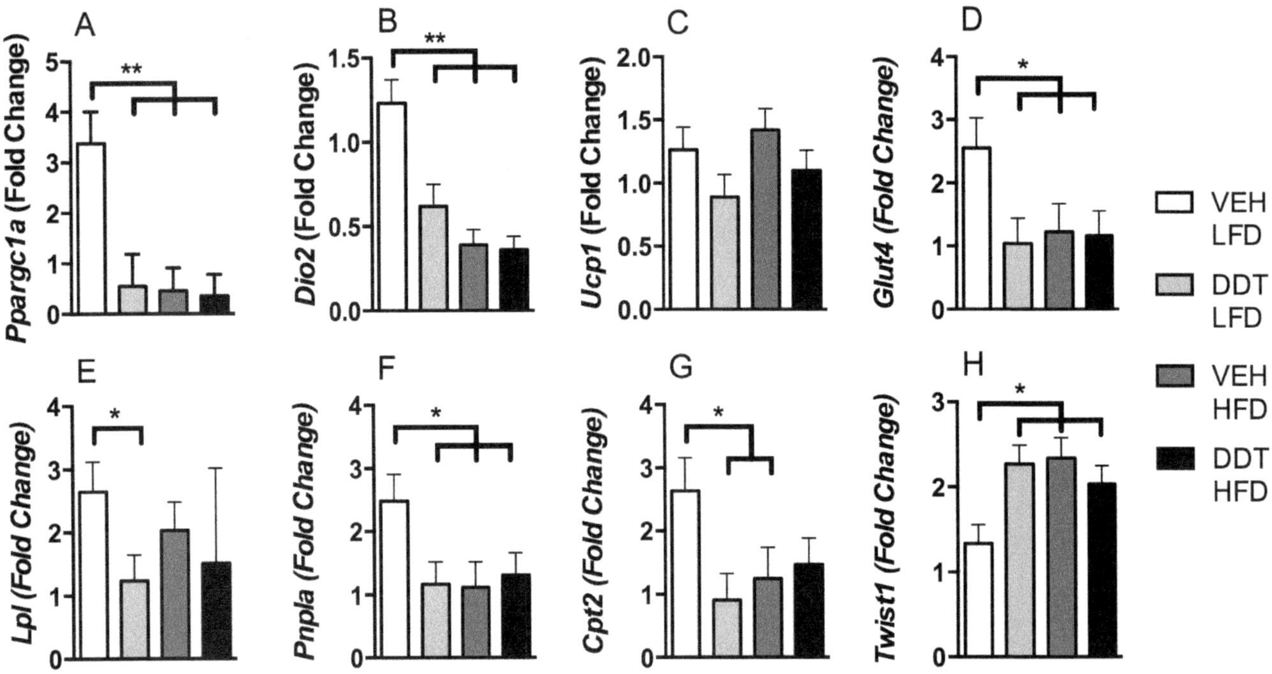


**Figure 5. Effects of perinatal DDT and adult HFD feeding on RNA expression in BAT from 9 month old female mice (n = 1 female/litter in 8 VEH + LFD, 8 DDT + LFD, 7 VEH + HFD and 8 DDT + HFD litters).** (A) *Ppargc1a*- (DDT*HFD pi<0.01), (B) *Dio2*- (DDT*HFD pi<0.01), (C) *Ucp1*-, (D) *Glut4*- (DDT*HFD pi<0.05), (E) *Lpl*- (DDT*HFD pi<0.05), (F) *Pnpla*- (DDT*HFD pi<0.05), (G) *Cpt2*- (DDT*HFD pi<0.05), and (H) *Twist1*- (DDT*HFD pi<0.1) fold change (relative to LFD + VEH and adjusted by *Tbp*). *p<0.05, **p<0.01 significance between groups indicated by bars. Data are represented as LS means + SEM.

## Insulin resistance and thermogenesis

Epidemiological studies have demonstrated associations between adult DDT and DDE exposures and diabetes in humans [7,8,61,62]. Similarly, adult rodents exposed to high doses of DDT had impaired insulin secretion, glucose intolerance, and elevated gluconeogenesis [63,64]. Here we find that female mice perinatally exposed to low doses of DDT and subsequently fed a HFD in adulthood developed impaired glucose tolerance, increased fasting insulin, and HOMA-IR, indicative of insulin resistance and increased risk for T2D. We have observed that *Ppargc1a* expression in BAT of adult female offspring was reduced after perinatal DDT exposure. *Ppargc1a* plays an important role linking thermogenesis to risk of T2D as illustrated by genetic ablation studies of *Ppargc1a* in mice, after which thermogenesis was impaired and susceptibility to HFD-induced insulin resistance, but not obesity, was increased [14]. Whether the reduced *Ppargc1a* expression in BAT of DDT-exposed mice accounts for their reduced thermogenesis and insulin resistance needs further study. It is plausible that the effects of perinatal DDT on thermogenesis and insulin resistance in mice may be relevant to humans who suffered elevated DDT exposure perinatally.

## Lipid homeostasis and adaptive energy metabolism

We also observed that perinatal DDT exposure caused a modest increase in circulating lipids after HFD feeding, which may in part be due to increased hepatic de novo lipogenesis as suggested by the increased protein expression of key lipogenic enzymes in the liver. Previous studies suggest that acute DDT administration increases hepatic cholesterol and triglyceride synthesis, resulting in their increased circulation in both rats and monkeys [65,66]. It is plausible that DDT and HFD exposures caused higher hepatic de novo lipogenesis, which was compensated for by increased lipid utilization, as evidenced by reduced RER, and increased hepatic bile acid conjugation, to prevent the development of hepatic steatosis in the female mice here [67,68].

## Developmental origins of metabolic disruption

Several lines of evidence implicate developmental origins in DDT-induced metabolic syndrome. First, when the developmental stage of newborn mice was equivalent to that of a human fetus at term, pups exposed to DDT exhibited reduced body weights and glucose levels. This may contribute to their increased susceptibility to the metabolic syndrome since human studies have demonstrated that low birth weight and low neonatal glucose can be associated with the metabolic syndrome in adulthood [20,69,70]. Second, perinatal DDT exposure was associated with early-adult obesity in female offspring here and across generations in other rodent studies [71,72]. While sex differences are common in mouse models of metabolic disease [73], and both DDT and DDE have estrogenic and anti-androgenic properties [50] the basis for the sex differences observed in our study will have to be delineated in future studies. Further, these data corroborate a prospective study of women where higher cord blood levels of DDE were associated with significantly increased BMI of the adult daughters in a dose dependent manner [74].

## Conclusions

Our study is the first to establish a link between developmental exposure to DDT and a moderately increased risk for the metabolic syndrome in adult offspring. Further, we know of no other studies implicating impaired thermogenesis in toxicant-induced insulin resistance and the metabolic syndrome. Given that circulating DDT and DDE levels in our mice fell within the range of DDT and DDE found in serum of contemporary people living

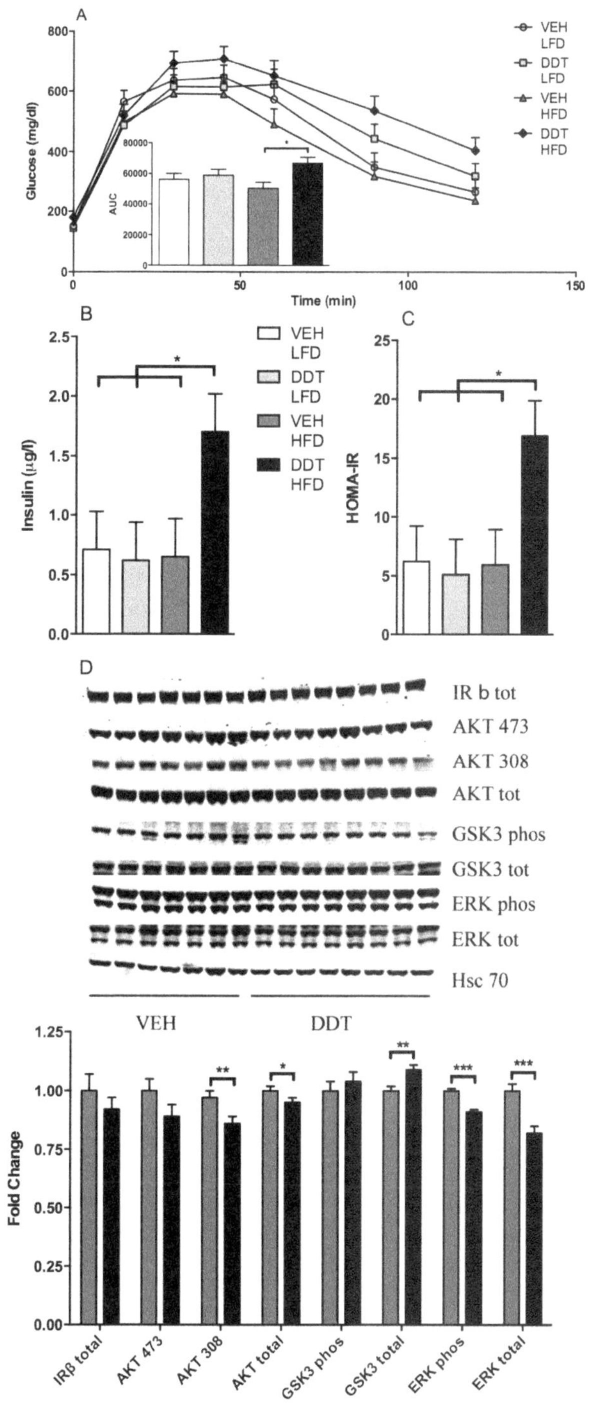


**Figure 6. Effects of perinatal DDT and adult HFD feeding on glucose homeostasis in adult female mice (n = 1 female/litter in 8 VEH + LFD, 8 DDT + LFD, 7 VEH + HFD and 8 DDT + HFD litters).** (A) Glucose tolerance test and resulting AUC (DDT*HFD pi = 0.08), (B) fasting insulin (DDT*HFD pi = 0.08) and (C) HOMA-IR (DDT*HFD pi = 0.07) in 8 month old mice. (D) Hepatic IRβ total, AKT 473 phosphorylation, AKT 308 phosphorylation, AKT total, GSK 3 phosphorylation, GSK 3 total, ERK phosphorylation, and ERK total fold change (relative to LFD + VEH and adjusted by HSC 70) as assessed by Western blot analysis (n = 1 female/litter in 7 VEH and 8 DDT litters) in 9 month old mice fed HFD. *p<0.05, **p<0.01, ***p<0.001 significance between groups indicated by bars. Data are represented as LS means + SEM.

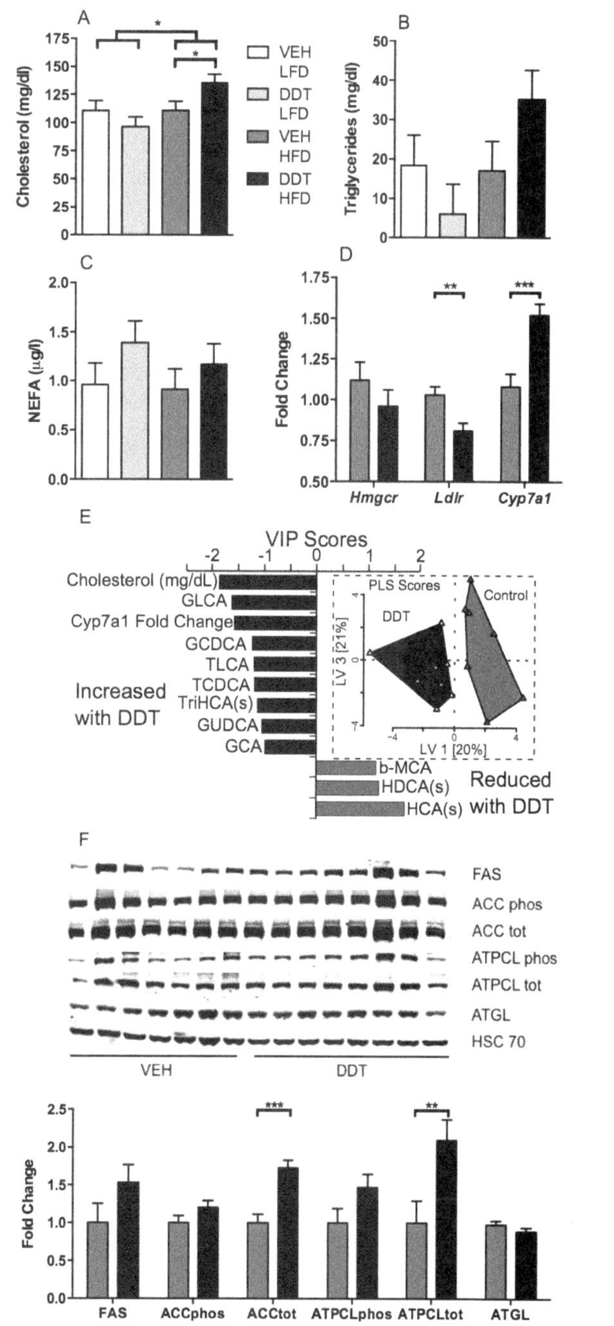


**Figure 7. Effects of perinatal DDT and adult HFD feeding on lipid homeostasis in adult female mice (n = 1 female/litter in 8 VEH + LFD, 8 DDT + LFD, 7 VEH + HFD and 8 DDT + HFD litters).** Fasting- (A) total cholesterol (HFD*DDT pi = 0.02), (B) triglycerides (HFD*DDT pi = 0.16), and (C) NEFA in 8 month old mice. (D) Hepatic *Hmgcr*-, *Ldlr*-, and *Cyp7a1*- fold change (relative to LFD + VEH and adjusted by *Gusb*) in 9 month old female mice fed HFD. (E) Bile acid metabolites discriminated by DDT or VEH exposure (VIP>1). Inset: The PLS-DA Scores Plot demonstrates clear discriminatory separation of mice with or without perinatal DDT exposure along latent variable 1 (LV1). Symbols represent individual animals. Metabolite abbreviations followed by "(s)" indicate "screening" data without the benefit of authentic calibration standards. F) Hepatic FAS- (p = 0.14), ACC phosphorylation - (p = 0.13), ACC total-, ATPCL phosphorylation - (p = 0.07), ATPCL total-, and ATGL- (p = 0.22) fold change (relative to LFD + VEH and adjusted by HSC 70) in 9 month old female mice fed HFD. *p<0.05, **p<0.01, ***p<0.001 significance between groups indicated by bars. Data are represented as LS means + SEM.

in northern Europe and in malaria-infested regions, and of the pregnant mothers of US adults who are currently in their mid-fifties [7,49,75,76] this is an important observation. Thus, it is likely that perinatal DDT exposure represents a risk factor for reduced energy expenditure in adult humans, even decades after DDT usage had ceased. The exact mechanism of this metabolic toxicity is unknown and how much it contributes to insulin resistance and dyslipidemia in the current obesity and diabetes epidemic will require further studies. But even small dysmetabolic effects of DDT and DDE are likely to be important from a public health perspective due to the ubiquity of these exposures among aging US and European populations as well as current populations of some developing and circumpolar countries. Given that the ban of DDT has reduced DDT exposure among the US population by about 100 fold [77], replacing DDT in malaria control should decrease DDT exposures over time and may reduce the susceptibility to the metabolic syndrome in future generations.

## Supporting Information

**Figure S1 Schematic of study design.** 1.7 mg DDT/kg body weight was administered to dams daily on gestational day 11 through postnatal day 5 and litters were culled to 6 mice at postnatal day 6. Mice were metabolically screened from 2–6 months when they were randomized into 2 independent study arms: energy balance and HFD or LFD feeding. Indirect calorimetry and cold tolerance were assessed in 6-month-old mice. Metabolic parameters (e.g. body composition, fasting lipids, glucose, and insulin, GTT) were assessed during 12 weeks of HFD or LFD feeding until mice were sacrifice at 9 months of age.

**Figure S2 Effect of prenatal DDT on gestation.** (A) Gestation length, (B) gestation weight gain, and (C) litter size (n = 15 DDT litters and n = 14 VEH litters). Data are represented as LS means ± SEM.

**Figure S3 Effect of perinatal DDT on body composition and energy balance in male mice.** (A) Body mass at PND5 (n = 15 DDT- and n = 14 vehicle- exposed litters). (B) Trunk glucose in PND5 males culled from litters >6 pups (n = 7 DDT and n = 7 VEH litters). (C) Proportion of males with preputial separation as a function of age (n = 15 DDT and n = 14 VEH litters). Monthly (D) body mass and (E) percent adiposity, (F) food intake, and (G) rectal temperature ((n = 15 DDT and n = 14 VEH litters, 1 mouse/litter at 2 month time point, all males/litter thereafter). Data are represented as LS means ± SEM. *$p<0.05$ DDT vs. vehicle controls.

**Figure S4 Effects of perinatal DDT on glucose tolerance.** GTT in (A) female and (B) male 2 month old mice (2.0 mg glucose/kg body mass, 1.5 mg glucose/kg body mass, n = 15 DDT-exposed litters and n = 14 oil-exposed litters, 1 mouse/litter), in (C) female and (D) male 3 month old mice (1.0 mg glucose/kg body mass, 1.5 mg glucose/kg body mass, n = 12 litters/treatment, 1 mouse/litter), and in (E) female and (F) male 6 month old mice (1.0 mg glucose/kg body mass, n = 15 DDT-exposed litters and n = 14 oil-exposed litters, 1 mouse/litter). Data are represented as LS means ± SEM.

**Figure S5 Joint effects of perinatal DDT and HFD on body composition and energy balance in male mice.** (A) Weekly body mass (n = 8 litters/treatment, 1 mouse/litter, HFD*age pi<0.0001). (B) Percent adiposity in 8 month old mice (n = 8 litters/treatment, 1 mouse/litter). (C) Cumulative food intake over 12 week HFD and LFD feeding period (n = 8 litters/treatment, 1 mouse/litter). (D) Rectal temperature in 8 month old mice (n = 8 litters/treatment, 1 mouse/litter). *$p<0.05$, **$p<0.01$, ***$p<0.0001$ LFD vs. HFD. Data are represented as LS means ± SEM.

**Figure S6 Joint effects of perinatal DDT and HFD on glucose and lipid homeostasis in male mice.** (A) Glucose tolerance test at 1 mg glucose/kg body mass and (B) corresponding AUC, in 8 month old mice (n = 8 litters/treatment, 1 mouse/litter). (C) Insulin, (D) HOMA-IR, (E) total cholesterol, (F) triglycerides, and (G) NEFA (μg/l plasma) in 8 month old mice fasted for 5 hrs (n = 8 litters/treatment, 1 mouse/litter). LFD vs. HFD. *$p<0.05$ significance between groups indicated by bars. Data are represented as LS means ± SEM.

**Figure S7 Partial Least Squares Discrimination Analysis (PLS-DA) of hepatic bile acids, triglycerides, and other metabolic factors which may influence or be influenced by cholesterol metabolism segregate treatment groups.** (B) Relative weight of independent variable to this segregation are shown in the PLS Loadings Plot. Hierarchical cluster analyses were used to segregate variables and 4 unique clusters are displayed with unique colors. Variables with the variable importance in projection VIP scores >0.8 or <−0.8 are labeled, >1 or <−1 shown as triangles. (C) All LV1 VIP scores. Metabolite abbreviations followed by "(s)" indicate "screening" data relative areas generated from the expected precursor > product ion mass transition in the LC-MS/MS system, without the benefit of authentic calibration standards.

**Figure S8 Lipid content of adult female mice fed HFD.** (A) Oil red O stained liver from 9 month old female mice fed HFD for 12 weeks with exposure to perinatal vehicle (top) or DDT (bottom, bar = 50 microns). (B) Hepatic triglycerides and cholesterol in 9 month old female mice fed HFD. Data are represented as LS means + SEM.

**Table S1 Table of Information about Western Blot Antibodies.**

**Table S2 Effects of perinatal DDT on metabolic parameters in mice fed regular chow.** LS Means (± SEM) shown.

**Table S3 Liver Bile Acid concentrations (nmol/g).** Values are reported as mean ± SEM).

## Acknowledgments

We wish to thank Leslie Roteta for help with parts of the animals studies and Drs. Gary Schwartz and Derek LeRoith for sharing their BAT-10 and Echo-MRI respectively; Drs. Myrto Petreas and June-Soo Park for quantitating sera DDTs; Drs. James Godbold, Philip Landrigan, and Michael Gallo for their comments. The USDA is an equal opportunity provider and employer.

## Author Contributions

Conceived and designed the experiments: MLM. Performed the experiments: MLM EK CL MLF. Analyzed the data: MLM EK EM CL MLF JN. Contributed reagents/materials/analysis tools: MLM JN CB. Contributed to the writing of the manuscript: MLM EK EM CL MLF JN CB.

## References


1. Shaw JE, Sicree RA, Zimmet PZ (2010) Global estimates of the prevalence of diabetes for 2010 and 2030. Diabetes Res Clin Pract 87: 4–14.
2. American Diabetes A (2013) Economic costs of diabetes in the U.S. in 2012. Diabetes Care 36: 1033–1046.
3. InterAct C, Langenberg C, Sharp SJ, Schulze MB, Rolandsson O, et al. (2012) Long-term risk of incident type 2 diabetes and measures of overall and regional obesity: the EPIC-InterAct case-cohort study. PLoS Med 9: e1001230.
4. Gallagher EJ, Leroith D, Karnieli E (2010) Insulin resistance in obesity as the underlying cause for the metabolic syndrome. Mount Sinai Journal of Medicine 77: 511–523.
5. Ouellet V, Routhier-Labadie A, Bellemare W, Lakhal-Chaieb L, Turcotte E, et al. (2011) Outdoor temperature, age, sex, body mass index, and diabetic status determine the prevalence, mass, and glucose-uptake activity of 18F-FDG-detected BAT in humans. Journal of Clinical Endocrinology and Metabolism 96: 192–199.
6. Taylor KW, Novak RF, Anderson HA, Birnbaum LS, Blystone C, et al. (2013) Evaluation of the association between persistent organic pollutants (POPs) and diabetes in epidemiological studies: a national toxicology program workshop review. Environ Health Perspect 121: 774–783.
7. Cox S, Niskar AS, Narayan KM, Marcus M (2007) Prevalence of self-reported diabetes and exposure to organochlorine pesticides among Mexican Americans: Hispanic health and nutrition examination survey, 1982–1984. Environ Health Perspect 115: 1747–1752.
8. Lee DH, Steffes MW, Sjodin A, Jones RS, Needham LL, et al. (2011) Low Dose Organochlorine Pesticides and Polychlorinated Biphenyls Predict Obesity, Dyslipidemia, and Insulin Resistance among People Free of Diabetes. PLos One 6: e15977.
9. Ramachandran A, Snehalatha C, Shetty AS, Nanditha A (2012) Trends in prevalence of diabetes in Asian countries. World J Diabetes 3: 110–117.
10. Herrera-Portugal C, Ochoa H, Franco-Sanchez G, Yanez L, Diaz-Barriga F (2005) Environmental pathways of exposure to DDT for children living in a malarious area of Chiapas, Mexico. Environmental Research 99: 158–163.
11. Bouwman H, Becker PJ, Cooppan RM, Reinecke AJ (1992) Transfer of DDT used in malaria control to infants via breast milk. Bull World Health Organ 70: 241–250.
12. Kreiss K, Zack MM, Kimbrough RD, Needham LL, Smrek AL, et al. (1981) Cross-sectional study of a community with exceptional exposure to DDT. JAMA 245: 1926–1930.
13. Farhang L, Weintraub JM, Petreas M, Eskenazi B, Bhatia R (2005) Association of DDT and DDE with birth weight and length of gestation in the Child Health and Development Studies, 1959–1967. American Journal of Epidemiology 162: 717–725.
14. Kleiner S, Mepani RJ, Laznik D, Ye L, Jurczak MJ, et al. (2012) Development of insulin resistance in mice lacking PGC-1alpha in adipose tissues. Proc Natl Acad Sci U S A 109: 9635–9640.
15. Fisher FM, Kleiner S, Douris N, Fox EC, Mepani RJ, et al. (2012) FGF21 regulates PGC-1alpha and browning of white adipose tissues in adaptive thermogenesis. Genes Dev 26: 271–281.
16. Hondares E, Gallego-Escuredo JM, Flachs P, Frontini A, Cereijo R, et al. (2013) Fibroblast growth factor-21 is expressed in neonatal and pheochromocytoma-induced adult human brown adipose tissue. Metabolism.
17. Zhao Y, Dunbar JD, Kharitonenkov A (2012) FGF21 as a therapeutic reagent. Adv Exp Med Biol 728: 214–228.
18. Hinoi E, Nakamura Y, Takada S, Fujita H, Iezaki T, et al. (2013) Growth differentiation factor-5 promotes brown adipogenesis in systemic energy expenditure. Diabetes.
19. Ravelli AC, van der Meulen JH, Michels RP, Osmond C, Barker DJ, et al. (1998) Glucose tolerance in adults after prenatal exposure to famine. Lancet 351: 173–177.
20. Hult M, Tornhammar P, Ueda P, Chima C, Bonamy AK, et al. (2010) Hypertension, diabetes and overweight: looming legacies of the Biafran famine. PLoS One 5: e13582.
21. Dahri S, Snoeck A, Reusens-Billen B, Remacle C, Hoet JJ (1991) Islet function in offspring of mothers on low-protein diet during gestation. Diabetes 40 Suppl 2: 115–120.
22. Fernandez-Twinn DS, Wayman A, Ekizoglou S, Martin MS, Hales CN, et al. (2005) Maternal protein restriction leads to hyperinsulinemia and reduced insulin-signaling protein expression in 21-mo-old female rat offspring. Am J - Physiol Regul Integr Comp Physiol 288: R368–373.
23. Ecobichon DJ, Saschenbrecker PW (1968) Pharmacodynamic study of DDT in cockerels. Canadian Journal of Physiology and Pharmacology 46: 785–794.
24. Nnodim JO, Lever JD (1985) The pre- and postnatal development and ageing of interscapular brown adipose tissue in the rat. Anat Embryol (Berl) 173: 215–223.
25. Asahina K, Tsai SY, Li P, Ishii M, Maxson RE, Jr., et al. (2009) Mesenchymal origin of hepatic stellate cells, submesothelial cells, and perivascular mesenchymal cells during mouse liver development. Hepatology 49: 998–1011.
26. Han J, Lee JE, Jin J, Lim JS, Oh N, et al. (2011) The spatiotemporal development of adipose tissue. Development 138: 5027–5037.
27. Lee JS, Ward WO, Knapp G, Ren H, Vallanat B, et al. (2012) Transcriptional ontogeny of the developing liver. BMC Genomics 13: 33.
28. Wang QA, Tao C, Gupta RK, Scherer PE (2013) Tracking adipogenesis during white adipose tissue development, expansion and regeneration. Nat Med 19: 1338–1344.
29. You L, Gazi E, Archibeque-Engle S, Casanova M, Conolly RB, et al. (1999) Transplacental and lactational transfer of p,p′-DDE in Sprague-Dawley rats. Toxicology and Applied Pharmacology 157: 134–144.
30. Sholtz RI, McLaughlin KR, Cirillo PM, Petreas M, Park JS, et al. (2011) Assaying organochlorines in archived serum for a large, long-term cohort: implications of combining assay results from multiple laboratories over time. Environ Int 37: 709–714.
31. Matthews DR, Hosker JP, Rudenski AS, Naylor BA, Treacher DF, et al. (1985) Homeostasis model assessment: insulin resistance and beta-cell function from fasting plasma glucose and insulin concentrations in man. Diabetologia 28: 412–419.
32. Folch J, Lees M, Sloane Stanley GH (1957) A simple method for the isolation and purification of total lipides from animal tissues. Journal of Biological Chemistry 226: 497–509.
33. Allain CC, Poon LS, Chan CS, Richmond W, Fu PC (1974) Enzymatic determination of total serum cholesterol. Clinical Chemistry 20: 470–475.
34. Trinder P (1969) Determination of blood glucose using an oxidase-peroxidase system with a non-carcinogenic chromogen. Journal of Clinical Pathology 22: 158–161.
35. Garcia-Canaveras JC, Donato MT, Castell JV, Lahoz A (2012) Targeted profiling of circulating and hepatic bile acids in human, mouse, and rat using a UPLC-MRM-MS-validated method. J Lipid Res 53: 2231–2241.
36. Lindtner C, Scherer T, Zielinski E, Filatova N, Fasshauer M, et al. (2013) Binge drinking induces whole-body insulin resistance by impairing hypothalamic insulin action. Sci Transl Med 5: 170ra114.
37. Hou X, Xu S, Maitland-Toolan KA, Sato K, Jiang B, et al. (2008) SIRT1 regulates hepatocyte lipid metabolism through activating AMP-activated protein kinase. J Biol Chem 283: 20015–20026.
38. Scherer T, O'Hare J, Diggs-Andrews K, Schweiger M, Cheng B, et al. (2011) Brain insulin controls adipose tissue lipolysis and lipogenesis. Cell Metab 13: 183–194.
39. Novo CL, Polese C, Matheus N, Decottignies A, Londono-Vallejo A, et al. (2013) A new role for histone deacetylase 5 in the maintenance of long telomeres. FASEB J 27: 3632–3642.
40. Green HF, Nolan YM (2012) GSK-3 mediates the release of IL-1beta, TNF-alpha and IL-10 from cortical glia. Neurochem Int 61: 666–671.
41. Zaytseva YY, Rychahou PG, Gulhati P, Elliott VA, Mustain WC, et al. (2012) Inhibition of fatty acid synthase attenuates CD44-associated signaling and reduces metastasis in colorectal cancer. Cancer Res 72: 1504–1517.
42. Yu XX, Lewin DA, Forrest W, Adams SH (2002) Cold elicits the simultaneous induction of fatty acid synthesis and beta-oxidation in murine brown adipose tissue: prediction from differential gene expression and confirmation in vivo. FASEB J 16: 155–168.
43. Qiao A, Liang J, Ke Y, Li C, Cui Y, et al. (2011) Mouse patatin-like phospholipase domain-containing 3 influences systemic lipid and glucose homeostasis. Hepatology 54: 509–521.
44. Bostrom P, Wu J, Jedrychowski MP, Korde A, Ye L, et al. (2012) A PGC1-alpha-dependent myokine that drives brown-fat-like development of white fat and thermogenesis. Nature 481: 463–468.
45. Livak KJ, Schmittgen TD (2001) Analysis of relative gene expression data using real-time quantitative PCR and the 2(-Delta Delta C(T)) Method. Methods 25: 402–408.
46. Butler AA, Kozak LP (2010) A recurring problem with the analysis of energy expenditure in genetic models expressing lean and obese phenotypes. Diabetes 59: 323–329.
47. Grapov D, Newman JW (2012) imDEV: a graphical user interface to R multivariate analysis tools in Microsoft Excel. Bioinformatics 28: 2288–2290.
48. Mevik BH, Wehrens R (2007) The pls package: Principal component and partial least squares regression in R. Journal of Statistical Software. 18: 1–23.
49. Rylander C, Sandanger TM, Brustad M (2009) Associations between marine food consumption and plasma concentrations of POPs in a Norwegian coastal population. J Environ Monit 11: 370–376.
50. Sohoni P, Sumpter JP (1998) Several environmental oestrogens are also anti-androgens. Journal of Endocrinology 158: 327–339.
51. Uldry M, Yang W, St-Pierre J, Lin J, Seale P, et al. (2006) Complementary action of the PGC-1 coactivators in mitochondrial biogenesis and brown fat differentiation. Cell Metab 3: 333–341.
52. Hall JA, Ribich S, Christoffolete MA, Simovic G, Correa-Medina M, et al. (2010) Absence of thyroid hormone activation during development underlies a permanent defect in adaptive thermogenesis. Endocrinology 151: 4573–4582.
53. CR B, JG N, J L, XX H, GP H, et al. (2007) Modest PGC-1alpha overexpression in muscle in vivo is sufficient to increase insulin sensitivity and palmitate oxidation in subsarcolemmal, not intermyofibrillar, mitochondria. J Biol Chem 283: 4228–4240. Epub 2007 Dec 4212.
54. Koves TR, Sparks LM, Kovalik JP, Mosedale M, Arumugam R, et al. (2013) PPARgamma coactivator-1alpha contributes to exercise-induced regulation of intramuscular lipid droplet programming in mice and humans. J Lipid Res 54: 522–534.

55. Bastin J, Lopes-Costa A, Djouadi F (2011) Exposure to resveratrol triggers pharmacological correction of fatty acid utilization in human fatty acid oxidation-deficient fibroblasts. Hum Mol Genet 20: 2048–2057.
56. Rohas LM, St-Pierre J, Uldry M, Jager S, Handschin C, et al. (2007) A fundamental system of cellular energy homeostasis regulated by PGC-1alpha. Proc Natl Acad Sci U S A 104: 7933–7938.
57. Jun HJ, Gettys TW, Chang JS (2012) Transcriptional Activity of PGC-1alpha and NT-PGC-1alpha Is Differentially Regulated by Twist-1 in Brown Fat Metabolism. PPAR Res 2012: 320454.
58. Clegg DJ (2012) Minireview: the year in review of estrogen regulation of metabolism. Mol Endocrinol 26: 1957–1960.
59. Stefl B, Janovska A, Hodny Z, Rossmeisl M, Horakova M, et al. (1998) Brown fat is essential for cold-induced thermogenesis but not for obesity resistance in aP2-Ucp mice. Am J Physiol 274: E527–533.
60. Kim JY, Moon KD, Seo KI, Park KW, Choi MS, et al. (2009) Supplementation of SK1 from Platycodi radix ameliorates obesity and glucose intolerance in mice fed a high-fat diet. J Med Food 12: 629–636.
61. Rylander L, Rignell-Hydbom A, Hagmar L (2005) A cross-sectional study of the association between persistent organochlorine pollutants and diabetes. Environ Health 4: 28.
62. Patel CJ, Bhattacharya J, Butte AJ (2010) An environment-wide association study (EWAS) on Type 2 Diabetes Mellitus. PLos One 5: e10746.
63. Yau DT, Mennear JH (1977) The inhibitory effect of DDT on insulin secretion in mice. Toxicology and Applied Pharmacology 39: 81–88.
64. Kacew S, Singhal RL (1974) Role of cyclic adenosine 3′:5′-monophosphate in the action of 1,1,1-trichloro-2,2-bis-(p-chlorophenyl)ethane (DDT)on hepatic and renal metabolism. Biochemical Journal 142: 145–152.
65. Nagaoka S, Masaki H, Aoyama Y, Yoshida A (1986) Effects of excess dietary tyrosine or certain xenobiotics on the cholesterogenesis in rats. Journal of Nutrition 116: 726–732.
66. Sanyal S, Agarwal N, Dudeja PK, Mahmood A, Subrahmanyam D (1982) Effect of a single oral dose of DDT on lipid metabolism in protein-calorie malnourished monkeys. Indian Journal of Biochemistry and Biophysics 19: 111–114.
67. Rolo AP, Palmeira CM, Wallace KB (2002) Interactions of combined bile acids on hepatocyte viability: cytoprotection or synergism. Toxicol Lett 126: 197–203.
68. Attili AF, Angelico M, Cantafora A, Alvaro D, Capocaccia L (1986) Bile acid-induced liver toxicity: relation to the hydrophobic-hydrophilic balance of bile acids. Med Hypotheses 19: 57–69.
69. Bazaes RA, Salazar TE, Pittaluga E, Pena V, Alegria A, et al. (2003) Glucose and lipid metabolism in small for gestational age infants at 48 hours of age. Pediatrics 111: 804–809.
70. Barker DJ, Hales CN, Fall CH, Osmond C, Phipps K, et al. (1993) Type 2 (non-insulin-dependent) diabetes mellitus, hypertension and hyperlipidaemia (syndrome X): relation to reduced fetal growth. Diabetologia 36: 62–67.
71. Tomatis L, Turusov V, Day N, Charles RT (1972) The effect of long-term exposure to DDT on CF-1 MICE. International Journal of Cancer 10: 489–506.
72. Skinner MK, Manikkam M, Tracey R, Guerrero-Bosagna C, Haque M, et al. (2013) Ancestral dichlorodiphenyltrichloroethane (DDT) exposure promotes epigenetic transgenerational inheritance of obesity. BMC Medicine 11: 228–244.
73. Guarner-Lans V, Rubio-Ruiz ME, Perez-Torres I, Banos de MacCarthy G (2011) Relation of aging and sex hormones to metabolic syndrome and cardiovascular disease. Exp Gerontol 46: 517–523.
74. Karmaus W, Osuch JR, Eneli I, Mudd LM, Zhang J, et al. (2009) Maternal levels of dichlorodiphenyl-dichloroethylene (DDE) may increase weight and body mass index in adult female offspring. Occupational and Environmental Medicine 66: 143–149.
75. Bouwman H, Cooppan RM, Becker PJ, Ngxongo S (1991) Malaria control and levels of DDT in serum of two populations in Kwazulu. Journal of Toxicology and Environmental Health 33: 141–155.
76. La Merrill M, Cirillo PM, Terry MB, Krigbaum NY, Flom JD, et al. (2013) Prenatal Exposure to the Pesticide DDT and Hypertension Diagnosed in Women before Age 50: A Longitudinal Birth Cohort Study. Environ Health Perspect 121: 594–599.
77. Eskenazi B, Chevrier J, Rosas LG, Anderson HA, Bornman MS, et al. (2009) The Pine River statement: human health consequences of DDT use. Environ Health Perspect 117: 1359–1367.

11

# Determination of 17 Organophosphate Pesticide Residues in Mango by Modified QuEChERS Extraction Method Using GC-NPD/GC-MS and Hazard Index Estimation

**Ashutosh K. Srivastava[1], Satyajeet Rai[1], M. K. Srivastava[1], M. Lohani[2], M. K. R. Mudiam[1], L. P. Srivastava[1]***

**1** Pesticide Toxicology Laboratory, CSIR-Indian Institute of Toxicology Research, (Council of Scientific and Industrial Research, Govt. of India), Mahatma Gandhi Marg, Lucknow, India, **2** Department of Biotechnology, Integral University, Lucknow, India

## Abstract

A total of 162 samples of different varieties of mango: Deshehari, Langra, Safeda in three growing stages (Pre-mature, Unripe and Ripe) were collected from Lucknow, India, and analyzed for the presence of seventeen organophosphate pesticide residues. The QuEChERS (Quick, Easy, Cheap, Effective, Rugged and Safe) method of extraction coupled with gas chromatography was validated for pesticides and qualitatively confirmed by gas chromatography- mass spectrometry. The method was validated with different concentrations of mixture of seventeen organophosphate pesticides (0.05, 0.10, 0.50 mg $kg^{-1}$) in mango. The average recovery varied from 70.20% to 95.25% with less than 10% relative standard deviation. The limit of quantification of different pesticides ranged from 0.007 to 0.033 mg $kg^{-1}$. Out of seventeen organophosphate pesticides only malathion and chlorpyriphos were detected. Approximately 20% of the mango samples have shown the presence of these two pesticides. The malathion residues ranged from ND-1.407 mg $kg^{-1}$ and chlorpyriphos ND-0.313 mg $kg^{-1}$ which is well below the maximum residues limit (PFA-1954). In three varieties of mango at different stages from unpeeled to peeled sample reduction of malathion and chlorpyriphos ranged from 35.48%–100% and 46.66%–100% respectively. The estimated daily intake of malathion ranged from 0.032 to 0.121 μg $kg^{-1}$ and chlorpyriphos ranged from zero to 0.022 μg $kg^{-1}$ body weight from three different stages of mango. The hazard indices ranged from 0.0015 to 0.0060 for malathion and zero to 0.0022 for chlorpyriphos. It is therefore indicated that seasonal consumption of these three varieties of mango may not pose any health hazards for the population of Lucknow, city, India because the hazard indices for malathion and chlorpyriphos residues were below to one.

**Editor:** Aditya Bhushan Pant, Indian Institute of Toxicology Research, India

**Funding:** The financial assistance for the publication of manuscript is provided by GAP-168 MPRNL (Monitoring of Pesticide Residues at National Level) Government of India, as well as the CSIR-SRF Contingency. The funders had no role in study design, data collection and analysis, decision to publish, or preparation of the manuscript.

**Competing Interests:** The authors have declared that no competing interests exist.

* E-mail: laxmanprasad13@gmail.com

## Introduction

Mango (*Mangifera indica*) is one of the most common and highly consumable tropical fruits of India. It is rich in carotenoid, minerals, carbohydrates and vitamins. India ranked first in mango production in the world during 2010-11 [1]. Lucknow, the capital of Uttar Pradesh, India, is the largest producer of mangoes producing around 3469.5 metric tons, the productivity being about 12.8 tones/hectare [2]. There are many insect pest pressures for mangos grown in this region of India requiring the use of pesticides to increase the productivity [3]. Therefore, for obtaining good quality and high productivity of mango fruits, the commercial cultivation of mango receive frequent application of various contact and systemic pesticides throughout the cropping season [4]. Most pesticide residues find their way into the human body through fruits, vegetables, cereals, water and other food commodities. Thus, analysis of pesticide residues in food commodities and other environmental samples have become an essential requirement for consumers, producers and food quality control authorities [5]. Due to increased use of pesticide in the orchard, pesticide residues may remain in the raw fruits and their products such as juices, nectar, jellies and ice cream 'pose to be poisonous hazards to human health owing to their toxicity' [6–7]. To increase foreign trade under WTO regime, it is imperative to produce pesticide free mango [8]. Among the various pesticides, organophosphates (OPs), are the most extensively used insecticides in many crops including mango. Due to low persistence and high bio-efficiency of organophosphates, many farmers regularly use this group of pesticides for various vegetables and fruit crops. The continuous use of pesticides has caused deleterious effects the ecosystem [9]. Because of wide spread use of pesticides, the presence of their toxic residues have been reported in various environmental component/commodities [5,10–11]. Public awareness of health hazards posed by pesticide residues in fruits and vegetables has led to the development of many analytical methods. [5,7,8,12–16]. Method validation is an important requirement in chemical analysis. The analyst must generate information to show that a method intended for this purpose is capable of providing adequate specificity, accuracy and precision at relevant analyte concentrations in appropriate matrices. In the present study, an attempt has been made to validate modified QuEChERS method using ethyl acetate (EtOAc) for the extraction. In the method QuEChERS is reported that the acetonitrile is not compatible with system of gas chromatography due to high volume spray, which significantly increases the internal pressure of the chromatographic system. So we adapt the QuEChERS method employing the ethyl acetate solvent.

Seventeen organophosphate pesticides like dichlorvos, phorate, phorate-sulfone, phorate-sulfoxid, dimethoate, diazinon, methyl-parathion, chlorpyrifos-methyl, fenitrothion, malathion, chlorpyrifos, chlorfenvinfos, profenofos, ethion, edifenophos, anilophos and phosalone in mango fruits were analysed by Gas chromatography using Nitrogen Phosphorus Detector (GC-NPD). Majority of these pesticides are being used in mango orchards during spraying [17–18]. The validated method has been applied to determined 17 OPs residues in three delicious varieties of mango like Deshehari, Langra and Safeda of Malihabad, Lucknow, Uttar Pradesh, India as these varieties are prone to insect pests and consumed largely (Fig. 1), determining estimated daily intake (EDI) and hazard index of detected OPs residues for the consumption of mangoes by local population of Lucknow, India.

## Materials and Methods

### Ethics Statement

No specific permissions were required for these locations/activities. There is no requirement for ethical permission for this study.

### Chemicals and Pesticide Standards

All solvents like n-hexane, acetone and ethyl acetate (HPLC grade) were purchased from Sigma-Aldrich. Co. USA, Spectrochem Pvt. Ltd. India and were glass distilled before use. Acetone was refluxed over potassium permanganate for 4 hr and then distilled. Sodium chloride (NaCl), anhydrous sodium sulphate ($Na_2SO_4$) and magnesium sulphate ($MgSO_4$) were procured from Himedia Pvt. Ltd. India. Before use sodium sulphate ($Na_2SO_4$) and magnesium sulphate ($MgSO_4$) were purified with acetone and baked for 4 hr @ 600°C in muffle Furness to remove possible phthalate impurities. Primary secondary amine (PSA) bondasil 40 μm part 12213024 of Varian was used.

Pesticide standards (Dichlorvos 98.9%, phorate 96.0%, phorate-sulfone 96.8%, phorate-sulfoxid 94.8%, Dimethoate 99.6%, diazinon, methyl-parathion 99.7%, chlorpyrifos-methyl 99.9%, fenitrothion 95.2%, Malathion 97.2%, chlorpyriphos 99.9%, chlorofenvinfos 98.7%, profenofos 95.0%, Ethion 97.8%, Edifenphos 99.9%, anilophos 97.5% and phosalone 95.2%) were purchased from Supelco Sigma Aldrich USA, Fluka Sigma-Aldrich Schweis and Rankem Pvt. Ltd. New Delhi, India.

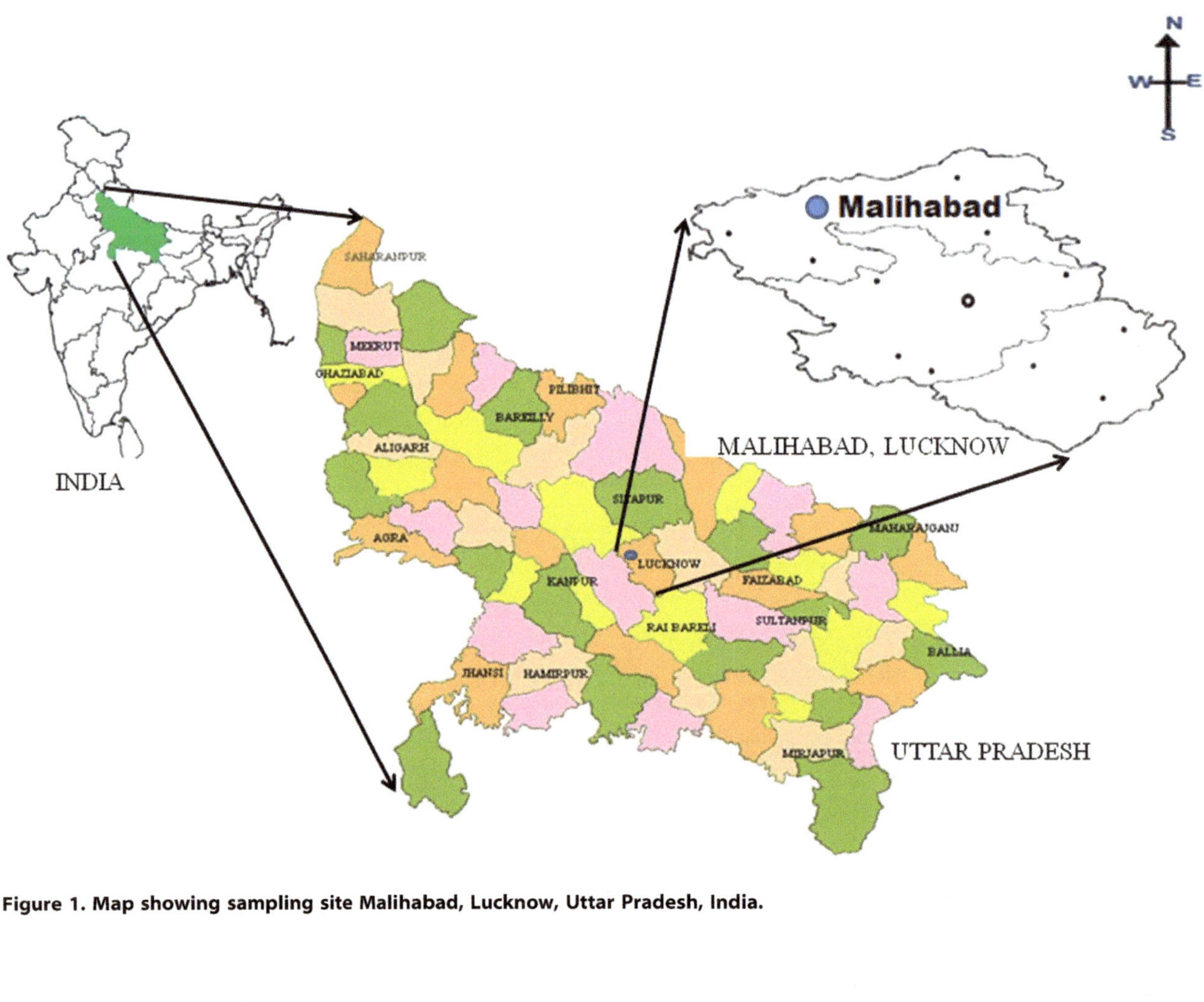


**Figure 1. Map showing sampling site Malihabad, Lucknow, Uttar Pradesh, India.**

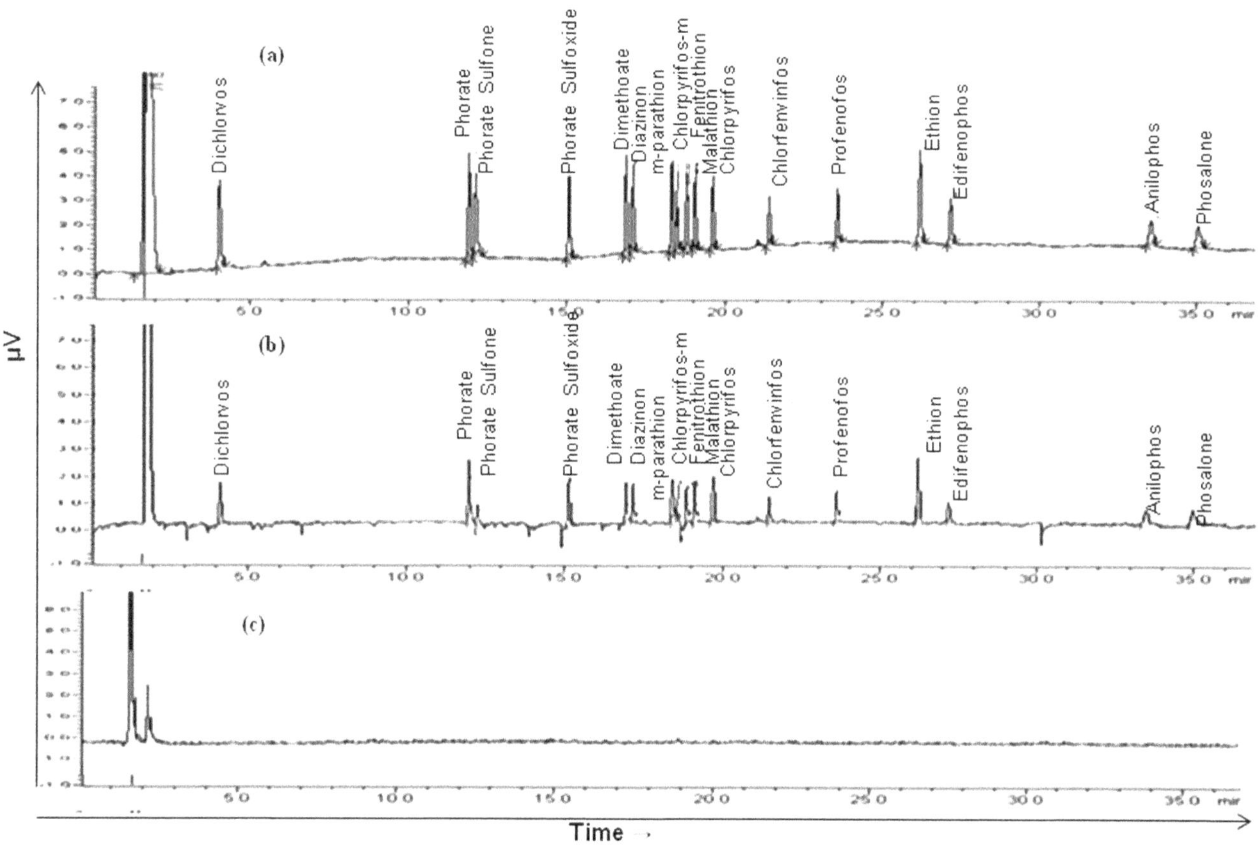


**Figure 2. GC-NPD Chromatogram of (a) standard solution of 17 pesticides (0.5 mg $l^{-1}$). (b) spiked mango sample (0.10 mg $kg^{-1}$) and (c) blank mango sample.**

## Sample collection

A total of 162 mango samples like Deshehari, Langra and Safeda of pre mature, unripe and ripe, from three different orchards of Malihabad, Lucknow, Uttar Pradesh, India were collected. (Fig. 1). Three batches of pre-mature, unripe and ripe mangoes (6 samples per batch) of individual varieties were taken for analysis. Samples of Mangoes (6 samples x 3 conditions/ batches x 3 varieties x 3 orchards = 162 samples) were brought to the laboratory and analyzed as soon as possible or stored in refrigerator at 4±2°C until analysis.

## QuEChERS Sample Preparations

The unpeeled and peeled mango samples (50 g) of each variety were chopped and grinded in Warring blander. Macerated sample of each mango in triplicates were extracted and cleaned by QuEChERS method as follows: 10±0.1 g macerated sample was mixed with 10 ml ethyl acetate, 4 g activated anhydrous $MgSO_4$, 1.0 g activated NaCl in centrifuge tube and shaken for 10 min. at 50 rpm on rotospin. The extract was centrifuged for 10 min at 8,000 rpm. 1 ml supernatant of extract was cleaned with the mixture of 50 mg PSA, 150 mg anhydrous $MgSO_4$ and 10 mg activated charcoal (application of activated charcoal in the stage of clean up to remove pigments from matrix). The extract was again shaken for 10 min at 50 rpm on rotospin and centrifuged for 10 min at 8,000 rpm. The supernatant was collected in GC vial for the analysis. Steps of extraction and clean up using a longer agitation related to the use of ethyl acetate solvent, which is less efficient than the acetonitrile with regard to power extraction of pesticides in the partitioning process is documented [12]. The use of larger time in the agitation may have been an attempt to maximize the extraction of pesticides, but promoted a significant increase in analysis time compared with the original method. 1 µl clean extract was injected in gas chromatography equipped with-nitrogen phosphorus detector (NPD) for the analysis OP pesticide residues. (Fig. 2)

## GC-NPD Analysis

Residues were analyzed on Shimadzu GC-2010 equipped with fused silica capillary column, DB-1 (30 mt.×0.25 mm. id) coated with 100% dimethylpolysiloxane (0.25 µm film thickness) using NPD. General operating conditions were as follows; Injector port temperature: 250°C; detector temperature 280°C; carrier gas Nitrogen ($N_2$); flow 1.46 ml $min^{-1}$; Hydrogen ($H_2$) makeup is 30 ml $min^{-1}$ and zero air 60 ml $min^{-1}$, (zero air has less than 0.1 ppm hydrocarbons, to decreases the background noise level and gives the baseline much better stability, considerably increasing detector sensitivity and ensuring precise analytical results) column temperature program: initially 130°C hold 2 min, increase at 5°C/min to 170°C hold 3 min, then increase 220°C $min^{-1}$ at 5°C $min^{-1}$, hold for 14 min; injection volume: 1 µl split ratio 1:5. The total run time was 37 min and Shimadzu, GC Solution software was used for instrument control and data analysis. Quantification of the pesticides was done by peak area using the standard method.

**Table 1.** Fortification experiment of organophosphate pesticides residues from spiked mango at different levels (recovery and repeatability), limit of detection (LOD), limit of quantification (LOQ), maximum residues limits (MRL) and acceptable daily intake (ADI).

| P Pesticides | Fortification level (mg $kg^{-1}$) | Recovered | (%) Recovery (n = 3) | LOD (mg $kg^{-1}$) | LOQ (mg $kg^{-1}$) | %RSD[a] | MRL[b] (mg $kg^{-1}$) | ADI[c] (mg $kg^{-1}day^{-1}$) |
|---|---|---|---|---|---|---|---|---|
| Dichlorvos | 0.05 | 0.039±3.22 | 78.90 | | | | | |
| | 0.10 | 0.080±3.22 | 80.23 | 0.003 | 0.010 | 4.42 | 0.1 | 0.004 |
| | 0.50 | 0.423±3.22 | 84.57 | | | | | |
| Phorate | 0.05 | 0.036±3.22 | 72.33 | | | | | |
| | 0.10 | 0.075±3.22 | 75.45 | 0.002 | 0.007 | 3.17 | 0.05 | 0.0005 |
| | 0.50 | 0.391±3.22 | 78.20 | | | | | |
| Phorate Sulfone | 0.05 | 0.035±3.22 | 70.22 | | | | | |
| | 0.10 | 0.071±3.22 | 71.45 | 0.010 | 0.033 | 3.84 | 0.05 | 0.0005 |
| | 0.50 | 0.373±3.22 | 74.60 | | | | | |
| Phorate Sulfoxide | 0.05 | 0.040±3.22 | 80.35 | | | | | |
| | 0.10 | 0.080±3.22 | 80.00 | 0.003 | 0.010 | 5.35 | 0.05 | 0.0005 |
| | 0.50 | 0.426±3.22 | 85.25 | | | | | |
| Dimethoate | 0.05 | 0.039±3.22 | 78.23 | | | | | |
| | 0.10 | 0.080±3.22 | 80.45 | 0.003 | 0.010 | 6.95 | 2.0 | 0.002 |
| | 0.50 | 0.425±3.22 | 85.10 | | | | | |
| Diazinon | 0.05 | 0.043±3.22 | 85.28 | | | | | |
| | 0.10 | 0.086±3.22 | 86.30 | 0.001 | 0.004 | 9.50 | NA | 0.02 |
| | 0.50 | 0.450±3.22 | 90.00 | | | | | |
| m-Parathion | 0.05 | 0.045±3.22 | 91.11 | | | | | |
| | 0.10 | 0.088±3.22 | 88.35 | 0.001 | 0.004 | 3.23 | 0.2 | 0.003 |
| | 0.50 | 0.467±3.22 | 93.45 | | | | | |
| Chlorpyrifos-m | 0.05 | 0.043±3.22 | 85.25 | | | | | |
| | 0.10 | 0.084±3.22 | 83.70 | 0.003 | 0.010 | 4.91 | NA | 0.01 |
| | 0.50 | 0.444±3.22 | 88.90 | | | | | |
| Fenitrothion | 0.05 | 0.036±3.22 | 72.90 | | | | | |
| | 0.10 | 0.070±3.22 | 70.20 | 0.010 | 0.033 | 5.37 | 0.5 | 0.005 |
| | 0.50 | 0.372±3.22 | 74.50 | | | | | |
| Malathion | 0.05 | 0.041±3.22 | 81.25 | | | | | |
| | 0.10 | 0.082±3.22 | 82.50 | 0.002 | 0.008 | 8.49 | 4.0 | 0.02 |
| | 0.50 | 0.428±3.22 | 85.55 | | | | | |
| Chlorpyrifos | 0.05 | 0.045±3.22 | 90.57 | | | | | |
| | 0.10 | 0.092±3.22 | 92.40 | 0.001 | 0.004 | 6.70 | 0.5 | 0.01 |
| | 0.50 | 0.476±3.22 | 95.25 | | | | | |
| Chlorfenvinfos | 0.05 | 0.044±3.22 | 88.10 | | | | | |

**Table 1.** Cont.

| P Pesticides | Fortification level (mg kg$^{-1}$) | Recovered | (%) Recovery (n = 3) | LOD (mg kg$^{-1}$) | LOQ (mg kg$^{-1}$) | %RSD[a] | MRL[b] (mg kg$^{-1}$) | ADI[c] (mg kg$^{-1}$day$^{-1}$) |
|---|---|---|---|---|---|---|---|---|
| | 0.10 | 0.091±3.22 | 90.75 | 0.003 | 0.010 | 6.48 | NA | NA |
| | 0.50 | 0.466±3.22 | 93.25 | | | | | |
| Profenofos | 0.05 | 0.042±3.22 | 83.25 | | | | | |
| | 0.10 | 0.080±3.22 | 80.25 | 0.004 | 0.013 | 8.68 | NA | NA |
| | 0.50 | 0.417±3.22 | 83.45 | | | | | |
| Ethion | 0.05 | 0.040±3.22 | 80.25 | | | | | |
| | 0.10 | 0.080±3.22 | 80.40 | 0.005 | 0.017 | 8.23 | 2.0 | NA |
| | 0.50 | 0.428±3.22 | 85.70 | | | | | |
| Edifenophos | 0.05 | 0.046±3.22 | 92.34 | | | | | |
| | 0.10 | 0.090±3.22 | 90.40 | 0.001 | 0.005 | 7.23 | NA | NA |
| | 0.50 | 0.467±3.22 | 93.50 | | | | | |
| Anilophos | 0.05 | 0.039±3.22 | 77.72 | | | | | |
| | 0.10 | 0.079±3.22 | 79.45 | 0.002 | 0.007 | 8.18 | NA | NA |
| | 0.50 | 0.416±3.22 | 83.20 | | | | | |
| Phosalone | 0.05 | 0.042±3.22 | 84.22 | | | | | |
| | 0.10 | 0.080±3.22 | 80.50 | 0.003 | 0.010 | 9.61 | 5.0 | 0.02 |
| | 0.50 | 0.431±3.22 | 86.25 | | | | | |

[a]RSD-Relative Standard Deviation.
[b]PFA - Prevention of Food Adulteration Act Govt. of India, 1954 [31].
[c][32].
NA- Not Available.

**Table 2.** Level of malathion and chlorpyrifos residues (mg $kg^{-1}$) in different species of mango fruits and their percent reduction after peeling.

| **Mango** | **Pre mature n=6** | | **Unripe n=6** | | **Ripe n=6** | | **Samples [A]** |
|---|---|---|---|---|---|---|---|
| | **Malathion Mean ± SD** | **Chlorpyrifos Mean ± SD** | **Malathion Mean ± SD** | **Chlorpyrifos Mean ± SD** | **Malathion Mean ± SD** | **Chlorpyrifos Mean ± SD** | **Detected/Analysed** |
| **Deshehari** | | | | | | | **14/54** |
| Unpeeled | 0.969±0.050{3} | 0.090±0.035{4} | 0.062±0.020{3}*** | 0.028±0.005{2}** | 0.021±0.010{2}*** | ND | |
| | (ND-1.185) | (ND-0.223) | (ND-0.197) | (ND-0.113) | (ND-0.304) | | |
| Peeled | 0.011±0.005{3}$^{###}$ | 0.048±0.050{4} | 0.040±0.025{3}* | 0.014±0.025{2} | ND | ND | |
| | (ND-0.046) | (ND-0.142) | (ND-0.128) | (ND-0.057) | | | |
| % reduction | 98.86 | 46.66 | 35.48 | 50.00 | 100 | ND | |
| **Langra** | | | | | | | **11/54** |
| Unpeeled | 0.049±0.030{2} | 0.160±0.070{2} | 0.679±0.125{5}*** | ND | 0.213±0.050{2}*** | ND | |
| | (ND-0.294) | (ND-0.313) | (ND-1.407) | | (ND-0.853) | | |
| Peeled | ND | ND | 0.272±0.015{5}$^{###}$ | ND | ND | ND | |
| | | | (ND-0.932) | | | | |
| % reduction | 100 | 100 | 59.94 | ND | 100 | ND | |
| **Safeda** | | | | | | | **7/54** |
| Unpeeled | ND | ND | 0.480±0.015{5} | ND | 0.147±0.080{2} | ND | |
| | | | (ND-0.908) | | (ND-0.591) | | |
| Peeled | ND | ND | 0.139±0.025{5}$^{###}$ | ND | 0.035±0.650{2} | ND | |
| | | | (ND-0.239) | | (0.105) | | |
| % reduction | ND | ND | 71.04 | ND | 76.19 | ND | |
| **Total Samples Detected/ Analysed** | | | | | | | **32/162** |

[A]Total samples of each variety of mango.
Values within small parentheses ( ) indicates range of residues and middle {} indicates number of samples detected with pesticides.
* $p<0.05$, ** $p<0.01$, *** $p<0.001$, $^{###}p<0.001$.

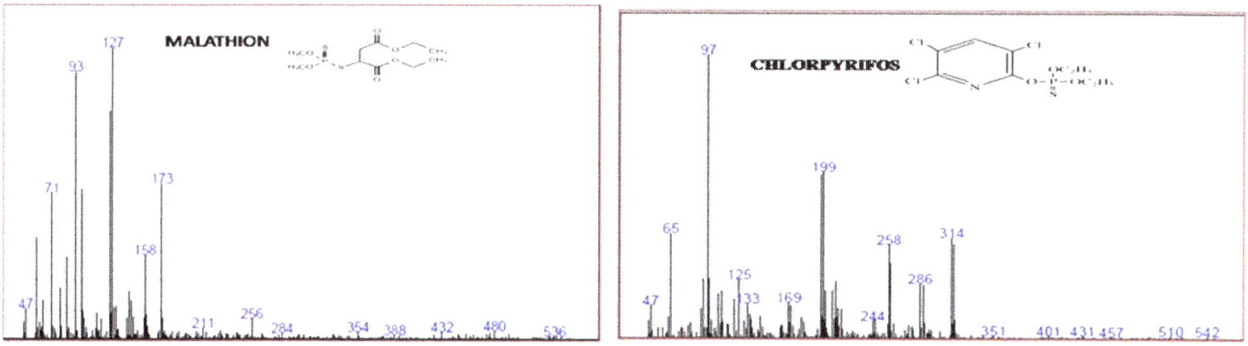


**Figure 3. GC-MS chromatogram showing confirmation of malathion and chlorpyrifos.**

## GC-MS Confirmation

A Perkin Elmer GC-MS consisting of Auto system XL Gas Chromatograph with a Turbo Mass Spectrometer was used for analysis. The column used in this study is Elite-5MS fused-silica capillary column (30 m×0.32 mm I.D., 0.25 mm film thickness). Carrier gas used was helium (purity 99.999%) with a flow rate of 1.6 ml min$^{-1}$. A 1 µl aliquot of the extract was injected using the splitless mode. The oven temperature program is 100°C for 1 min and then @ 20°C min$^{-1}$ to 210°C and hold for 1 min; then @45°C min$^{-1}$ to 300°C and hold for 1 min. The total runtime of the GC is 9.5 min. Base peaks 173, 158, 127 m/z for malathion as target ion as well as 314, 286, 258 m/z for chlorpyriphos were noticed as qualifiers in selective ion mode (SIM) for analysis. The injector temperature was set at 300°C. The transfer line and source temperature was set at 280°C and 230°C respectively. Solvent delay for MS is 5 min.

## Method validation

The validation of the analytical method was performed by the accuracy, precision, linearity and limit of detection (LOD), quantification (LOQ). All the analysis was carried out using the same blank samples of mango fruit.

Accuracy and precision data were obtained with recovery studies by spiking samples with pesticide standards at levels of 0.05, 0.1, and 0.5 mg kg$^{-1}$. The spiked and control samples were analyzed in five replicates. Precision of the method was evaluated through the relative standard deviations (%RSD) associated with pesticides measurements during recovery.

Linearity was determined by plotting calibration curve with standard solutions in n-hexane containing five different concentrations (0.025, 0.05, 0.1, 0.5 and 1.0 mg L$^{-1}$). Five injections were made at each of the five concentration levels.

The limits of detection (LOD) and Quantification (LOQ) were calculated according standard guidelines [19–20]. Five independent analysis of mango samples spiked with mixture of 17 OPs at the level of 0.05, 0.1, and 0.5 mg kg$^{-1}$ were performed for the percentage recovery of each pesticides. The LOD and LOQ were calculated at the spiking level of 0.1 mg kg$^{-1}$ from the standard deviation of this determination. Table 1

In order to maintain analytical quality control for each sample batch a spiked sample (similarly used in the recovery study) was analyzed simultaneously. Batch results were considered unsatisfactory when the sample used as quality control had low recovery.

## Hazard index estimation

Health risk estimations were performed on the basis of pesticide analysis data obtained in present study and annual fruit intake per person. The estimated daily intake (EDI) of pesticides is the toxicological criteria for their exposure. It is calculated as per international guidelines [21–23] using the equation: EDI = C x F/ D x W where C is the mean of individual pesticide concentration (mg kg$^{-1}$); F is mean annual intake of fruit per person (kg); D is days in a year (365 days) and W is the mean human body weight (60 kg). The annual intake as the total fruits per person is 9.5 kg per year according to Indian survey perform in years 2005–2006 [24–26]. The EDI (mg kg$^{-1}$ day$^{-1}$) as obtained were used to estimate the hazards index by dividing them to their corresponding value of known acceptable daily intake (Hazard index = EDI/ ADI).

## Statistical Analysis

The data were statically analyzed by using one way ANOVA. Criterion of significance was taken as $P<0.05$, $P<0.01$ and $P<0.001$. All statistical calculations have been done using IBM SPSS statistics version 20.

# Results and Discussion

## Validation of QuEChERS method

Fig. 2 (a.b,c) showing the representative chromatogram for standard of OP mixture, spiked and blank mango fruit samples. Adequate separation of the 17 OPs was achieved. No interference peaks were obtained in the blank sample chromatogram at the same retention time of the target compounds.

All OPs showed linearity ranged (0.025 to 1.0 mg kg$^{-1}$) with the coefficient co-relation more than 0.997 (r). The relative standard deviation (RSD) of the three replicate injections ranged from 3.17 to 9.50% showing good repeatability.

Table 1, shows that recovery data and repeatability of 17 OPs analyzed at three different spiking levels. The recovery ranged from 70.20 to 93.50%. The overall recovery was more than 70% (except phorate sulfone and fenitrothion) at lower spiking level 0.05 mg kg$^{-1}$ with RSD below 10% which represent satisfactory repeatability of the method for all pesticides (Barakat et al. 2007). It is observed that LOD and LOQ for 17 pesticides ranged from 0.001 to 0.010 mg kg$^{-1}$ and 0.007 to 0.033 mg kg$^{-1}$ respectively. It is interesting to note that, LOQ analyzed of all 17 OPs was lower than their respective MRL's established by PFA 1954, Govt. of India (Table 1). Few studies have been reported for the presence of organophosphate residues in mango using various extraction technique used in gas chromatographic analysis [13,20,27–28].

In the present study, the QuEChERS sample preparation and GC analysis using ethyl acetate solvent has been validated. The results indicate that the QuEChERS sample preparation method

**Table 3.** Estimated daily intake and hazard index of malathion and chlorpyriphos residues in mango.

| Mango | Average Pesticide concentration in three varieties of Mangos (ΣC) (μg kg$^{-1}$) | | Mean annual intake of fruit [a,b] per-person/day (kg) (F) | No. of Days in year (D) | Average weight of person (W) | ADI (μg kg$^{-1}$ bw daily) | | EDI (μg kg$^{-1}$ bw daily) | | Hazard index | | Hazard Risk | |
|---|---|---|---|---|---|---|---|---|---|---|---|---|---|
| | Mala | CPF | | | | Mala | CPF | Mala | CPF | Mala | CPF | Mala | CPF |
| Pre-Mature | 172 | 50 | 9.5 | 365 | 60 | 20 | 10 | 0.075 | 0.022 | 0.0037 | 0.0022 | No | No |
| Unripe | 279 | 7 | 9.5 | 365 | 60 | 20 | 10 | 0.121 | 0.003 | 0.0060 | 0.0003 | No | No |
| Ripe | 69 | 0 | 9.5 | 365 | 60 | 20 | 10 | 0.030 | 0.00 | 0.0015 | 0.00 | No | No |

Mala = Malathion; CPF = Chlorpyriphos.
[a]From National Nutrition Monitoring Board (2008).
[b]National Sample Survey Organization (2000).

coupled with GC-NPD analysis is suitable for the determination of 17 OPs viz; dichlorvos, phorate, phorate-sulfone, phorate-sulfoxide, dimethoate, diazinon, methyl-parathion, chlorpyrifos-methyl, fenitrothion, malathion, chlorpyrifos, chlorofenvinfos, profenofos, ethion, edifenophos, anilophos and phosalone. Successful use of QuEChERS method has been reported for the analysis of pesticide residues in high sugar content matrices like honey and sugarcane juice [27,29]. Similarly the present study revealed good result by applying QuEChERS method for the analysis of various organophosphate in ripe mango fruit which are enriched in high sugar content.

## Pesticide residue in various stages of the mango

Pesticides are commonly sprayed to mango trees at pre-mature, unripe and one month before the harvest. Therefore, mangoes of all three stages were studied. Analysis was done in samples of three different varieties of mangos (deshehari, langra and safeda) in triplicates for the presence of pesticides residues and values are given in Table 2. Out of seventeen analyzed pesticide only two pesticides malathion and chlorpyrifos were detected in three varieties of mangoes.

### Deshehari

**Comparison between peeled and unpeeled samples.** The results of Table 2 showed the presence of malathion and chlorpyrifos residues in different stages of mango. The mean concentration of malathion and chlorpyrifos was 0.969 (ND-1.185) and 0.090 (ND-0.223) mg kg$^{-1}$ respectively in pre-mature unpeeled samples. However, malathion and chlorpyrifos residues were 0.011(ND-0.046) and 0.048 (ND-0.142) mg kg$^{-1}$ in the pre-mature peeled mango. It is observed that there was a reduction of 98.86% malathion and 46.66% chlorpyrifos after the peeling of samples. The level of malathion residue in unpeeled to peeled mango samples was significantly ($P<0.001$) reduced and no significant variation was observed for chlorpyrifos. However in unripe unpeeled mangoes the mean concentration of malathion was 0.062 (ND-0.197) mg kg$^{-1}$ and chlorpyrifos was 0.028 (ND-0.113) mg kg$^{-1}$. The residues of malathion and chlorpyrifos were 0.040(ND-0.128) mg kg$^{-1}$ and 0.014 (ND-0.057) mg kg$^{-1}$ in unripe peeled mangoes. The percent reduction of malathion was 35.48 and chlorpyrifos was 50.00 through peeling in terms of mean values. The variation in level of malathion and chlorpyrifos was not significant. In ripe unpeeled mangoes, the residue of malathion 0.021 mg kg$^{-1}$ (ND-0.304) was observed while chlorpyrifos was below to detectable level. None of these two pesticides were detected after the peeling of ripe mangoes. It is interesting to note that in ripe mangoes, the OP residues were least as compared to unripe or pre-mature mangoes. It may be because of judicious use of pesticides in the latent harvesting period of mangoes.

**Comparison between pre-mature, unripe and ripe mango samples.** The variation of malathion residues in unpeeled unripe and ripe samples was significant ($P<0.001$) in comparison to pre-maturate samples. Similarly, residues of chlorpyrifos was also significant ($P<0.05$) in unripe Deshehari mango. Whereas only malathion was significant ($P<0.01$) in unripe peeled mango samples.

### Langra

**Comparison between peeled and unpeeled samples.** The mean concentration of malathion and chlorpyrifos was 0.049 (ND-0.294) mg kg$^{-1}$ and 0.160 (ND-0.313) mg kg$^{-1}$, respectively in pre-mature unpeeled langra mango samples. The cent percent malathion and chlorpyrifos residues were reduced from unpeeled to peeled samples. However in unripe unpeeled mango the mean concentration of malathion was 0.679 (ND-1.407) mg $kg^{-1}$ and 0.272 (ND-0.932) mg $kg^{-1}$ in peeled mango samples. The percent reduction of malathion was 59.94 through peeling with respect to their mean values. The reduction in the level of malathion was significant (P<0.001) in unpeeled to peeled mango samples. Chlorpyrifos was not detected in unripe, unpeeled and peeled mango samples. In ripe mango malathion was detected 0.213 (ND-0.853) mg $kg^{-1}$ in unpeeled sample with 100% reduction from unpeeled to peeled samples. However, chlorpyrifos was again not detected in ripe unpeeled and peeled samples.

**Comparison between pre-mature, unripe and ripe mango samples.** The variation of malathion residues in unpeeled samples of pre-maturate, unripe and ripe mangoes was significant (P<0.001), if compared with premature Langra mango samples.

## Safeda

**Comparison between peeled and unpeeled samples.** Malathion and chlorpyrifos residues were detected in unripe and ripe stages of mangoes. In unripe unpeeled mango malathion residue was 0.480(ND-0.908 mg $kg^{-1}$) and in peeled samples the residue was 0.139(ND-0.239 mg $kg^{-1}$) with 71.04% reduction of their mean values after peeling of the samples. The reduction of malathion was significant (P<0.001) in unpeeled to peeled mango samples. In ripe mangoes, malathion was 0.147(ND-0.591 mg $kg^{-1}$) in unpeeled samples and 0.035(ND-0.105 mg $kg^{-1}$) in peeled samples with 76.19% reduction is seen in their mean values after peeling. Malathion and chlorpyrifos residues were not detected in unpeeled and peeled pre-mature mangoes. However, chlorpyrifos was below detection limit in all three stages of safeda mango.

**Comparison between pre-mature, unripe and ripe mango samples.** The malathion and chlorpyrifos residues in unpeeled and peeled pre-maturate, unripe and ripe mangoes was not significant in safeda mango samples.

The texture, sugar and water content of mango may vary on the maturity, varieties and cultivars of the fruits. The texture and water content play an important role in trapping of pesticides and recovery efficiency.

The results of the method validation indicated that the QuEChERS sample preparation coupled with the GC-NPD analysis is suitable for the determination of 17 OPs in mango samples. In the present study none of pesticide residues accept malathion and chlorpyriphos were detected in 162 mango samples. Therefore, malathion and chlorpyrifos were further confirmed on GC-MS (Fig. 3).

Table 3 shows average estimated daily intake and hazard index for malathion and chlorpyrifos in pre-mature, unripe and ripe mango fruits. Hazard indices were calculated 0.0037, 0.0060 and 0.0015 for malathion and 0.0022, 0.0003 and zero for chlorpyrifos respectively in pre-mature, unripe and ripe mangoes. It is therefore indicated that the consumption of mango may not pose health hazards for the population of Lucknow city, India as hazard indices for malathion and chlorpyrifos in all three stages of mangoes were below one [30]. The total concentration of malathion and chlorpyrifos residues were 0.520 and 0.057 mg $kg^{-1}$ which are below to maximum residues limits, 4.0 mg $kg^{-1}$ and 0.5 mg $kg^{-1}$ [31]. It is further stated that there is no health risk associated with malathion and chlorpyrifos residues after the consumption of mango fruits.

## Conclusion

The validated QuEChERS method applied in the present study fulfils the established criteria for sensitivity and confident identification of organophosphorus pesticides at low level in matrix with high sugar content like mango. The results revealed that none of OPs pesticides except traces of malathion and chlorpyrifos residues were present in mango fruits. It is also observe that peeling has great influence in reduction of pesticide residues in the pulp. To avoid adverse effects on public health it is necessary to set up control measures so as to make sure that each pesticide residues should be below MRL in the fruits to be marketed. Therefore the study has explored the significant information regarding the analysis of 17 OPs residues in different varieties of mango fruits of Malihabad, Lucknow region of India. It is further observed that mango fruits appear to be safe from OP pesticides residues as revealed by EDI and hazard index. In this manner one can assume that there is no apparent risk to the health of consumers of mango, as two detected pesticides have hazard index below to one. It is therefore, suggested that regular evaluation of pesticide residues should be carried out on mango fruits at national level for the planning of future policies about the use of pesticides in mango orchards and enables pesticide free fruits.

## Acknowledgments

The authors are grateful to the Dr. K.C. Gupta Director, CSIR-Indian Institute of Toxicology Research (IITR), Lucknow for his keen interest and providing research facilities. The financial assistance to Ashutosh Kumar Srivastava funded by Council of Scientific and Industrial Research, New Delhi as SRF is also acknowledged. IITR communication no. is 3050.

## Author Contributions

Conceived and designed the experiments: AKS SR LPS. Performed the experiments: AKS SR MKS. Analyzed the data: AKS SR MKS ML LPS. Contributed reagents/materials/analysis tools: LPS MKRM. Wrote the paper: AKS SR MKS LPS. Performed the GC-MS confirmation of pesticides: MKRM.

## References

1. Food and Agricultural Organization of United Nations: Economic and Social Department: The Statistical Division, 2010.http://en.wikipedia.org/wiki/Mango.
2. Biswas BC, Kumar L (2011) Revolution in mango production: Success Stories of some Farmers. New Delhi, *Fertilizer Marketing News* 1–3.
3. Mohapatra S, Ahuja AK, Deepa M, Sharma D (2010) Residues of acephate and its metabolite methamidophos in/on mango fruit (Mangiferae indica L.). Bull Environ Contam Toxicol 86: 101–104.
4. Banerjee K, Oulkar DP, Patil SB, Jadhav MR, Dasgupta S, et al. (2009) Multiresidue determination and uncertainty analysis of 87 pesticides in mango by liquid chromatography-tandem mass spectrometry. J Agri Food Chem 57: 4068–4078.
5. Srivastava AK, Trivedi P, Srivastava MK, Lohani M, Srivastava LP (2011) Monitoring of pesticide residues in market basket samples of vegetable from Lucknow City, India: QuEChERS method. Environ Monit Assess 176: 465–472.
6. Filho AM, Santos FND, Pereira PADP (2011) Multi-residue analysis of pesticide residues in mangoes using solid-phase microextraction coupled to liquid chromatography and UV–Vis detection. J Sepr Sci DOI 10.1002/jssc.201100341.
7. Perret D, Gentili A, Sergi SM, Dascenzo G (2002) Validation of a method for the determination of multiclass pesticide residues in fruit juices by liquid chromatography/tandem mass spectrometry after extraction by matrix solid-phase dispersion. J AOAC Int 85: 724–730.
8. Mukherjee I, Singh S, Sharma PK, Jaya M, Gopal M, et al. (2007) Extraction of multi-class pesticide residues in mango fruits (mangiferae indica L.): application of pesticide residues in monitoring of mangoes. Bull Environ Contam Toxicol 78:380–383.

9. Sharma D, Nagpal A, Pakade YB, Katnoria JK (2010) Analytical methods for estimation of organophosphorus pesticide residues in fruits and vegetables. A review Talanta 82: 1077–1089.
10. Bhanti M, Taneja A (2007) Contamination of vegetables of different seasons with organophosphorous pesticides and related health risk assessment in northern India. Chemosphere 69: 63–68.
11. Wang L, Liang Y, Jiang X (2008) Analysis of eight organophosphorus pesticide residues in fresh vegetables retailed in agricultural product market of Nanjing, China. Bull of Environ Contam Toxicol 81: 377–382.
12. Anastassiades M, Lehotay SJ, Stajnbaher D, Schenck FJ (2003) Fast and easy multiresidue method employing acetonitrile extraction/partitioning and "dispersive solid phase extraction" for the determination of pesticides residues in produce. J AOAC Int 86: 412–431.
13. Aysal P, Ambrus AR, Lehotay SJ, Andrew C (2007) Validation of an efficient method for the determination of pesticide residues in fruits and vegetables using ethyl acetate for extraction. J Environ Sci and Health Part B42: 481–490.
14. Lehotay SJ (2007) Determination of pesticide residues in foods by acetonitrile extraction and partitioning with magnesium sulphate: Collaborative study. J AOAC Int 90: 485–520.
15. Payá P, Anastassiades M, Mack D, Sigalova I, Tasdelen B, et al. (2007) Analysis of pesticide residues using the Quick Easy Cheap Effective Rugged and Safe (QuEChERS) pesticide multiresidue method in combination with gas and liquid chromatography and tandem massspectrometric detection. Anal and Bio anal Chem 389: 1697–1714.
16. Furlani RPZ, Marcilio KM, Leme FM, Tfouni SAV (2010) Analysis of pesticide residues in sugarcane juice using QuEChERS sample preparation and gas chromatography with electron capture detection. Food Chem 126: 1283–1287.
17. Pathak MK, Fareed M, Bihari V, Reddy MMK, Patel DK, et al. (2011) Nerve conduction studies in sprayers occupationally exposed to mixture of pesticides in a mango plantation at Lucknow, North India. Toxicological & Environmental Chemistry 93: 1, 188–196
18. Pathak MK, Fareed M, Srivastava AK, Pangtey BS, Bihari V, et al. (2013) Seasonal variations in cholinesterase activity, nerve conduction velocity and lung function among sprayers exposed to mixture of pesticides. Environmental Science and Pollution Research DOI 10.1007/s11356-013-1743-5.
19. Taylor JK (1987). In Quality assurance of chemical measurements. Chelsea, USA: Lewis Publishers, Inc. 31. PFA (1954) Prevention of Food Adulteration Act 1954. Act No. 37 with Prevention of Food Adulteration Rules 1955 and Notification and Commodity Index 16th edition. Eastern Book Company Publishing Pvt. Ltd. Lucknow, India.
20. Saadati N, Pauzi Abdullah MD, Zakaria Z, Belin S, Sany T, et al. (2013) Limit of detection and limit of quantification development procedures for organochlorine pesticides analysis in water and sediment matrices. Chemistry Central Journal 7: 63–73.
21. World Health Organization: WHO (1997) Guideline for predicting dietary intake of pesticide residues (revised). Global Environment Monitoring System-Food Contamination and Assessment Programme (GEMS/Food) in collaboration with codex committee on pesticide residues.
22. Food and Agriculture Organization: FAO (2002) Submission and evaluation of pesticide residues data for the estimation of maximum residues levels in food and feed (Ist edition). Rome: Food and Agriculture Organization. <http://www.fao.org/ag/AGP/AGPP/Pesticide/p.htm>.
23. Osman KA, Al-Humaid AI, Al-Rehiayani A, Al-Redhaiman KN (2011) Estimated daily intake of pesticide residues exposure by vegetables grown in greenhouse in Al-Qassim region Saudi Arabia. Food Control 22: 947–953.
24. National Sample Survey Organization (NSSO). 1975–2000; http://mospi.nic.in/mospi_nsso_rcpt_pubn.htm; last accessed on 24/0907, Ministry of Statistics and Programme Implementation, Govt. of India.
25. European Commission (2007) Method validation and quality control procedures for pesticide residues analysis in food and feed, Document No. SANCO/2007/3131. Availablefrom:<www.ec.europa.eu/plant/prtection/resources/qualcontrol_en.pdf>
26. National Nutrition Monitoring Bureau (NNMB). 1979–2008. *NNMB Report (7.7)*: National Institute of Nutrition Hyderabad India.
27. Barakat AA, Badaway HMA, Salama E, Attallah E, Maatook G (2007) Simple and rapid method of analysis for determination of pesticide residues in honey using dispersive solid phase extraction and GC determination. J Food Agri & Environ 5: 97–100.
28. Raju MB, Rao CN, Kumar GVR, Rao BT, Krishna PM, et al. (2011) Determination of multiclass pesticide residues in mangoes by Liquid chromatography–tandem mass spectrometry. J App Bio and Pharm Tech2: 279–289.
29. Prati P, Camargo GA (2008) Characteristics of sugarcane juice and your influence in the beverage stability. Bio Eng2: 37–44.
30. Darko G, Akoto O (2008) Dietary intake of organophosphorous pesticide through vegetables from Kumasi, Ghana. Food and Chem Toxicol 46: 3703–3706.
31. PFA (1954) Prevention of Food Adulteration Act 1954. Act No. 37 with Prevention of Food Adulteration Rules 1955 and Notif ication and Commodity Index (16th ed.). Lucknow: Eastern Book.
32. Sharma KK (2007) Pesticide residues Analysis Manual. Directorate of Information and Publications of Agriculture, New Delhi, 2007.

# 12

# Biological Differences between Brackish and Fresh Water-Derived *Aedes aegypti* from Two Locations in the Jaffna Peninsula of Sri Lanka and the Implications for Arboviral Disease Transmission

**Ranjan Ramasamy*, Pavilupillai J. Jude, Thabothiny Veluppillai, Thampoe Eswaramohan, Sinnathamby N. Surendran***

Department of Zoology, Faculty of Science, University of Jaffna, Jaffna, Sri Lanka

## Abstract

The mainly fresh water arboviral vector *Aedes aegypti* L. (Diptera: Culicidae) can also undergo pre-imaginal development in brackish water of up to 15 ppt (parts per thousand) salt in coastal areas. We investigated differences in salinity tolerance, egg laying preference, egg hatching and larval development times and resistance to common insecticides in *Ae. aegypti* collected from brackish and fresh water habitats in Jaffna, Sri Lanka. Brackish water-derived *Ae. aegypti* were more tolerant of salinity than fresh water-derived *Ae. aegypti* and this difference was only partly reduced after their transfer to fresh water for up to five generations. Brackish water-derived *Ae. aegypti* did not significantly discriminate between 10 ppt salt brackish water and fresh water for oviposition, while fresh water-derived *Ae. aegypti* preferred fresh water. The hatching of eggs from both brackish and fresh water-derived *Ae. aegypti* was less efficient and the time taken for larvae to develop into pupae was prolonged in 10 ppt salt brackish water. *Ae. aegypti* isolated from coastal brackish water were less resistant to the organophosphate insecticide malathion than inland fresh water *Ae. aegypti*. Brackish and fresh water-derived *Ae. aegypti* however were able to mate and produce viable offspring in the laboratory. The results suggest that development in brackish water is characterised by pertinent biological changes, and that there is restricted genetic exchange between coastal brackish and inland fresh water *Ae. aegypti* isolates from sites 5 km apart. The findings highlight the need for monitoring *Ae. aegypti* developing in coastal brackish waters and extending vector control measures to their habitats.

**Editor:** Immo A. Hansen, New Mexico State University, United States of America

**Funding:** The authors have no support or funding to report.

**Competing Interests:** The authors have declared that no competing interests exist.

* Email: ranjanramasamy@yahoo.co.uk (RR); noble@jfn.ac.lk (SNS)

## Introduction

*Aedes aegypti* L. (Diptera: Culicidae) is the principal mosquito vector of yellow fever and, together with the closely related *Aedes albopictus* Skuse, a primary vector of dengue and chikungunya [1–6]. Dengue is the most common arboviral disease of humans, with 50 million annual cases in more than 100 countries, an increasing incidence and global spread, and a 2.5% fatality rate in severe dengue [4,5]. There is presently no licensed vaccine or specific anti-viral drug for dengue [5]. Yellow fever has a zoonotic reservoir, is endemic to Africa and South America with the potential to spread to Asia and, although an effective live-attenuated vaccine is available, responsible for 200,000 cases and 30,000 deaths in the world every year [6]. Chikungunya, a debilitating but not often fatal arboviral disease, occurs in Asia, Africa and the Americas, and caused a recent epidemic in temperate Europe [3]. A vaccine against chikungunya is not yet available. Dengue and chikungunya are endemic in Sri Lanka with a high incidence in the northern Jaffna peninsula [7,8].

The control of dengue and chikungunya therefore relies heavily on surveillance for *Ae. aegypti* and *Ae. albopictus* and larviciding and eliminating pre-imaginal development habitats [3,5]. The two vectors have been long regarded to lay eggs and undergo pre-imaginal development only in fresh water collections (e.g. blocked drains, roof gutters, flower pot bases, leaf axils) near human settlements that are the principal targets for control measures worldwide [2,5,9]. Recent data however show that both vectors can also develop in brackish water collections in coastal areas of Sri Lanka and Brunei Darussalam [10–14]. Fresh, brackish and saline waters are respectively defined as containing <0.5, 0.5 to 30, and >30 ppt (parts per thousand) salt [10]. Pre-imaginal stages of *Ae. aegypti* and *Ae. albopictus* develop in brackish water in discarded containers and domestic wells of up to 15 ppt and 14 ppt salinity respectively in northern and eastern Sri Lanka [10–12]. Salinity tolerance in the two *Aedes* vectors therefore has implications for the effective control of arboviral diseases particularly in a future context of rising sea levels increasing the extent of coastal brackish water habitats [14–16].

*Aedes aegypti* larvae from a long-established laboratory colony tolerate a short-term increase in salinity of up to 10 ppt through a reversible osmoconformation mechanism involving the accumulation of amino acids and ions in the haemolymph [17]. *Aedes aegypti* are able to oviposit in brackish water of up to 18 ppt in field conditions [10]. *Aedes aegypti* collected from coastal locations are able to undergo preimaginal development in brackish water in the laboratory [10,18]. Some other mosquito species have evolved to tolerate salinity through genetic changes. Salinity-tolerant mosquito larvae possess cuticles that are less permeable to water than freshwater forms, and their pupae have thickened and sclerotized cuticles that are impermeable to water and ions [19]. Salinity-tolerant species have also evolved various physiological mechanisms to cope with salinity in the larval environment. *Aedes taeniorhynchus* larvae ingest the surrounding fluid and excrete $Na^+$ and $Cl^-$ from the posterior rectum to produce a hyperosmotic urine [19]. *Culex tarsalis* larvae accumulate amino acids and trehalose in the hemolymph to maintain iso-osmolarity in brackish waters in an osmoconformation process [20]. Larvae of the malaria vector *Anopheles albimanus* are able to differentially localize sodium-potassium ATPase in rectal cells in fresh or saline water for likely osmoregulation through ion excretion [21]. There is evidence to suggest that a similar adaptation accompanied speciation of salinity-tolerant *Anopheles merus* within the predominantly fresh water *Anopheles gambiae* complex of malaria vectors in Africa [22].

There is presently no information on possible genetic and physiological changes associated with brackish water development in field *Ae. aegypti* populations. This information is relevant for developing more effective measures for controlling dengue, chikungunya and other arboviral diseases. It has been proposed that differences in salinity tolerance between *Ae. aegypti* isolates from northern and eastern Sri Lanka may be due to genetic variation [10].

We hypothesised that *Ae. aegypti* developing in coastal brackish water habitats differ from inland fresh water *Ae. aegypti* in the Jaffna peninsula in salinity tolerance, oviposition preference, egg hatching and larval development times and insecticide resistance. We tested this by measuring these characteristics in the laboratory in *Ae. aegypti* collected from brackish and fresh water habitats at two locations in the Jaffna peninsula. We also determined whether differences in salinity tolerance between brackish and fresh water-derived *Ae. aegypti* were due to genetic changes by investigating the reversibility of salinity tolerance on transferring brackish water isolates to fresh water and *vice versa* and maintaining such reversal colonies for up to five generations.

## Materials and Methods

### Ethical statement

The owners were informed of the nature of the study and their verbal informed consent obtained when larval collections were carried out in private property. Permission was not required for larval collections in public land as this did not involve endangered species or protected areas. The care and use of mice were according to WHO guidelines (WHO/LAB/88.1) and the protocol for using anaesthetised Balb/c mice for feeding mosquitoes was approved by the Animal Ethics Review Committee of the University of Jaffna (AERC/2014/02).

### Mosquito isolates

*Ae. aegypti* larvae were collected from brackish water of 2–8 ppt salinity in domestic wells and water tanks and from ovitraps containing brackish water of 10 ppt salinity in the Kurunagar coast of Jaffna city (9°39′N: 80°1′E) in northern Sri Lanka (Figure 1) as previously described [10,12]. The larvae were maintained at 10 ppt salinity in the insectary of the Department of Zoology at ambient temperature (28±2°C) with fish meal powder provided twice a day as larval food. Emergent adults were used to establish a self-mating colony of *Ae. aegypti* that was maintained in 10 ppt salinity. At the same time, fresh water ovitraps (0 ppt salinity) were used to collect *Ae. aegypti* larvae in Thirunelvely (9°41′N: 80°1′E) in the centre of the Jaffna peninsula (Figure 1), the larvae subsequently maintained in fresh water (0 ppt salinity) and the emergent adults used to establish a self-mating colony of fresh water *Ae. aegypti*. The mosquitoes were fed every three days on Balb/c mouse blood and 10% glucose pledgets were provided at other times. Two separate collections of *Ae. aegypti* were made at Kurunagar and Thirunelvely in October 2012 (onset of the rainy season) and February 2013 (dry season) for establishing brackish and fresh water colonies for the respective duplicate experiments 1 and 2 described below.

For experiments 1 and 2 that investigated the reversibility of salinity tolerance, reversal colonies were independently generated from the previously established brackish and fresh water colonies of *Ae. aegypti*. From the 10 ppt salinity brackish water colony, a reversal colony was established by transferring approximately 200 larvae into fresh water and maintaining subsequent generations in fresh water. Similarly, a brackish water reversal colony was established from the fresh water colony by transferring approximately 200 larvae to brackish water containing 10 ppt salt. After having successfully maintained the reversal colonies for two and five generations, the larval progenies from all four colonies maintained in parallel for each experiment, *viz.* original brackish water colony in 10 ppt, brackish water colony transferred to fresh water (reversal colony), original fresh water colony in 0 ppt and fresh water colony transferred to brackish water at 10 ppt (reversal colony) were used to determine their comparative salinity tolerance.

### Comparative salinity tolerance of brackish and fresh water *Ae. aegypti* and its reversibility

In each of two independent experiments performed in October–November 2012 and February–March 2013, first instar larvae of *Ae. aegypti* from the four colonies were exposed to salinities of 0, 2, 4, 6, 8, 10, 12, 14, 18 and 20 ppt following methods previously described [10]. The required salinities were obtained by adding tap water to seawater. Salinity was measured using a refractor salinometer (Atago, Japan). Twenty larvae in 150 ml capacity plastic containers containing 100 ml of water of specific salinity were maintained at room temperature (28±2°C) until their emergence as adults. Three replicates using 20 larvae each were performed in parallel at each salinity level for the four colonies in both experiments. Plastic lids were used to cover the containers to minimize evaporation. Test media were changed on alternate days. Larvae were fed twice daily with locally available powered fish meal. Mortality of larvae and the numbers of emerging adults were determined.

### Oviposition preference of brackish and fresh water *Ae. aegypti*

Fifty blood-fed three day-old females from the original brackish and fresh water colonies after eight generations in the laboratory were released into separate mosquito cages (one for brackish water and the other for fresh water-derived females). Each cage was provided with three fresh water and three 10 ppt salinity brackish water egg laying surfaces. The numbers of eggs laid on each type

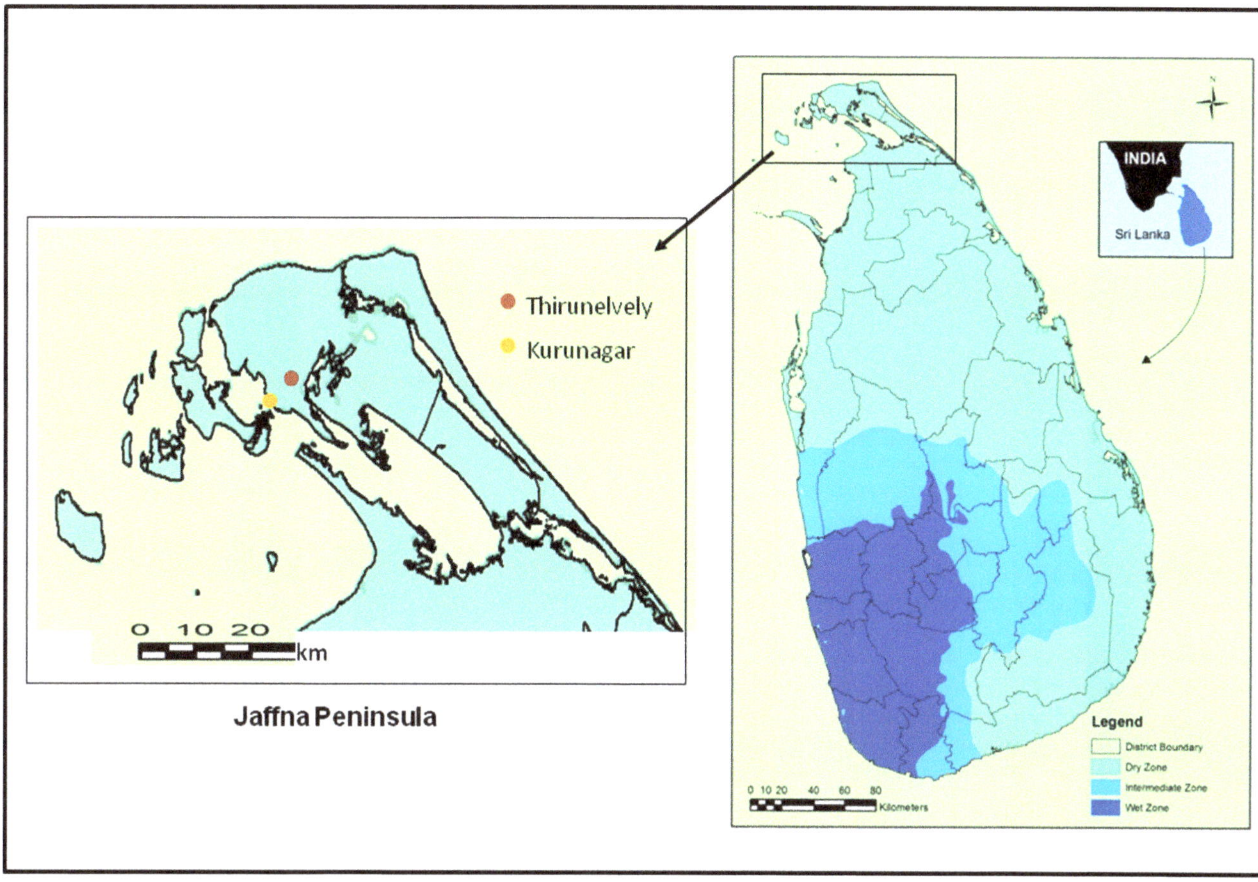


**Figure 1. Map of Sri Lanka showing the respective location of brackish and fresh water larval collection sites at Kurunagar and Thirunelvely.** The two sites are located in the northern Jaffna peninsula. The map also shows the boundaries of the different administrative districts and the dry, intermediate and wet rainfall zones in the country.

of egg-laying surface were determined after three days. The experiment was repeated three times.

## Hatching of *Ae. aegypti* eggs and larval development times in brackish and fresh water

One hundred and fifty eggs from each of the four colonies *viz.* the two original colonies and the two reversal colonies described above (fifth generation of reversal colonies for experiment 1 and second and fifth generations of reversal colonies for experiment 2) were collected and stored at ambient temperature (28±2°C) for 1–2 weeks. They were then flooded as appropriate with 10 ppt brackish water or fresh water to initiate hatching. The numbers of eggs hatching into larvae were determined after 48 hours. Developed larvae were fed twice daily with locally available powered fish meal. The duration of larval development were then determined by counting the numbers of pupae produced at different times. Three replicates were run in parallel for each determination.

## Ability of brackish and fresh water-derived *Ae. aegypti* to interbreed

After eight generations in the laboratory, 50 three and four day-old females from the brackish water-derived *Ae. aegypti* and 50 males from fresh water-derived *Ae. aegypti* colonies were allowed to mate naturally in a mosquito cage under standard insectary conditions (12 hour dark and light; 28±2°C). After 2 days and over-night starvation, they were fed on mouse blood and a fresh water egg-laying surface supplied. Fifty females from the fresh water colony were similarly mated with 50 males from the brackish water colony. The laid eggs were counted and allowed to hatch in fresh water and the emerging F1 male adult progeny from each cross were back-crossed with respective females from the original parent colony to test the viability of F1 males.

## Susceptibility of brackish and fresh water isolates of *Ae. aegypti* to insecticides

Separate collections of larvae from brackish water containers (2–4 ppt salinity) in coastal Kurunagar and fresh water ovitraps (0 ppt salinity) in inland Thirunelvely respectively were made between November 2011 and March 2012 as previously described [10] for testing their susceptibility to insecticides. The first generation of adults emerging from the field-collected larvae were exposed to three insecticides commonly used in the Jaffna peninsula, *viz.* permethrin (0.25%), propoxur (0.1%) and malathion (4%). Susceptibility bioassays were conducted using WHO standard bioassay kits as described previously [23]. Based on availability, 10 to 20 females, aged two to three days, were exposed to insecticide impregnated papers for one hour. Three replicate determinations were made for each insecticide. Papers impregnated with the solvent alone were used as controls. Dead mosquitoes were counted after a recovery period of 24 hours.

## Statistical analysis

The concentration of salt causing 50% mortality in the first instar larvae to adult transition ($LC_{50}$) was determined with 95% confidence limits using the Minitab 14 statistical software (Minitab Inc, PA, USA) as previously described [10]. $LC_{50}$ ratio tests were done to further determine the significance of $LC_{50}$ variations between test populations as described by Wheeler *et al.* [24]. Two-tailed Student's t tests were performed to determine the significance of differences in mean numbers of eggs laid and eggs hatching into larvae, and the mean percentage susceptibility to insecticides using the Minitab statistical software.

# Results

## Salinity tolerance of fresh and brackish water *Ae. aegypti*

The results from the two independent experiments performed at different times on the effect of salinity on the first instar larvae to adult transition in the different colonies are presented in Figure 2. They show that brackish water-derived first instar *Ae. aegypti* larvae maintained at 10 ppt salinity demonstrated 100% survival to adulthood at salinities up to 12 ppt (Figure 2 B&D) while 100% survival in fresh water-derived *Ae. aegypti* maintained in fresh water was only observed up to 8 ppt salinity (Figure 2 A–D).

These findings were also reflected in the maximal salinity tolerance defined as the highest salinity that permitted any larvae to survive to become adults. For example, some larvae from the brackish water colony maintained in 10 ppt salinity survived 20 ppt salt (Figure 2 D) while larvae from the fresh water colony maintained in fresh water could maximally tolerate only 16 ppt salt (Figure 2 C&D).

Furthermore, brackish water-derived *Ae. aegypti* even after five generations of maintenance in fresh water had higher maximal salinity tolerance (18 ppt, Figure 2 C&D) and show 100% survival at higher salinity (10 ppt, Figure 2 D) than fresh water-derived *Ae. aegypti* maintained in fresh water. Similarly, fresh water-derived *Ae. aegypti* after five generations of maintenance in 10 ppt brackish water showed 100% survival at lower salinity (8 ppt Figure 2 C&D) and lower maximal salinity tolerance (16 ppt Figure 2 C&D) than the brackish water colony maintained at 10 ppt salinity throughout.

We also determined the $LC_{50}$ for salinity tolerance as the ppt salt that causes 50% mortality in the transition from first instar larvae to adults for statistical analysis (Table 1). The $LC_{50}$ of the original brackish water colony that continued to be maintained at 10 ppt salinity remained significantly greater than the $LC_{50}$ for the fresh water colony maintained in fresh water for up to five generations, based on non-overlapping 95% confidence intervals in both experiments. This was confirmed by $LC_{50}$ ratio tests [24] at the $p<0.001$ level of significance (Table S1). Furthermore, the fresh water colony that was transferred to 10 ppt salinity always had lower $LC_{50}$ than the original brackish water colony maintained at 10 ppt salinity at the $p<0.005$ level of significance

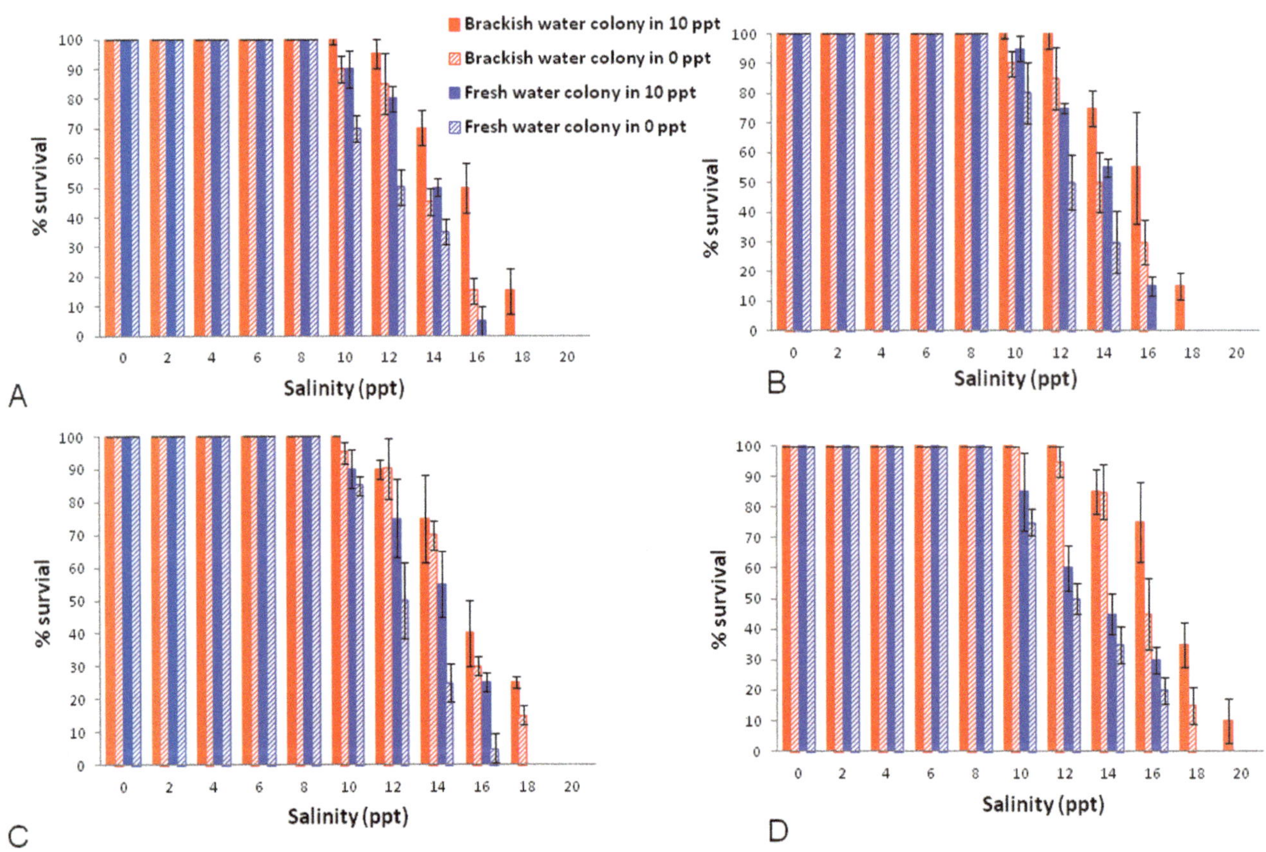


**Figure 2. Salinity tolerance of brackish and fresh water-derived *Ae. aegypti* from Kurunagar and Thirunelvely respectively.** The mean percent survival of first instar larvae to adulthood at each salinity level together with standard errors of the means of triplicate determinations are shown for the different colonies. Larvae were derived from brackish and fresh water colonies maintained at the original salinity and after reversal of salinity for two and five generations. A. Experiment 1 - second generation; B. Experiment 2 - second generation; C. Experiment 1 - fifth generation; D. Experiment 2 - fifth generation.

(Table S1). These findings show that even after five generations of adaptation to brackish water, the fresh water-derived *Ae. aegypti* never become as salinity tolerant as the brackish water-derived *Ae. aegypti*. Conversely, the brackish water-derived *Ae. aegypti* that were transferred to fresh water remained significantly more salinity tolerant in terms of $LC_{50}$ at the $p<0.005$ level than the original fresh water colony maintained in fresh water (Table S1). These findings show that even after five generations of adaptation to fresh water, the brackish water-derived *Ae. aegypti* remain more salinity tolerant than fresh water-derived *Ae. aegypti*. The $LC_{50}$ values for brackish water-derived *Ae. aegypti* that were transferred to fresh water were always higher than that of the fresh water colony transferred to 10 ppt brackish water in every comparison but the differences were not statistically significant (Table S1).

The transfer of brackish water-derived *Ae. aegypti* to fresh water always resulted in lower salinity tolerance than the original brackish water colony that was reflected in a reduced $LC_{50}$ that was significant at the $p<0.05$ level in three out of the four comparisons (Table S1). Significant lowering of $LC_{50}$ was seen even after two generations in fresh water in experiment 1 (Tables 1 and S1). Similarly the transfer of fresh water-derived *Ae. aegypti* to brackish water resulted in the development of greater salinity tolerance that was reflected in higher $LC_{50}$ values than the original fresh water colony that was significant at the $p< 0.01$ level after two and five generations of reversal (Table S1).

The $LC_{50}$ for salinity tolerance in the respective original colonies were not significantly different between experiments 1 and 2 after two generations but diverged for the original brackish water colonies after five generations in the laboratory (Tables 1 and S2). The $LC_{50}$ values between the corresponding reversal colonies in the two experiments also increasingly diverged with increasing number of generations (Tables 1 and S2).

In summary, the results in the two individual experiments show that brackish water-derived *Ae. aegypti* are more salinity tolerant than fresh water-derived *Ae. aegypti* and that this difference was partly reduced when the respective salinities at which they were being maintained were reversed for up to five generations.

## Preferences of brackish and fresh water-derived *Ae. aegypti* to oviposit on brackish and fresh water surfaces

The preferences of blood-fed female *Ae. aegypti* from the two original colonies, maintained in 10 ppt brackish water and 0 ppt fresh water for eight generations after field collection, to oviposit on 10 ppt brackish or fresh water surfaces are shown in Table 2. The results show that female fresh water-isolated *Ae. aegypti* significantly preferred fresh water over brackish water for oviposition. In contrast, females from brackish water-derived *Ae. aegypti* tended to lay more eggs in brackish than fresh water surfaces but the difference was not statistically significant.

## Hatching of *Ae. aegypti* eggs in brackish and fresh water

The numbers of eggs hatching from an initial number of 150 eggs after 48 h from the original brackish water colony, the original fresh water colony, the reversal brackish water colony transferred to fresh water and the reversal fresh water colony transferred to brackish water in their respective salinities are shown in Table 3. Egg hatching was quantitatively determined for the 5$^{th}$ generation of the colonies in experiment 1 and for both 2$^{nd}$ and 5$^{th}$ generations in experiment 2. The numbers of eggs hatching in fresh water were significantly higher than in 10 ppt brackish water regardless of the colony from which the eggs were derived. The original brackish water-derived *Ae. aegypti* colony tended to have more eggs hatching in 10 ppt brackish water than the reversal colony of fresh water-derived *Ae. aegypti* transferred to 10 ppt salt brackish water, but the differences were not statistically significant.

## Larval development times of *Ae. aegypti* in brackish and fresh water

Figure 3 shows the time taken for larvae to develop into pupae in brackish and fresh water in the experiments. The emergence of pupae at different times after hatching was quantitatively determined for the 5$^{th}$ generation of the colonies in experiment 1 and for both 2$^{nd}$ and 5$^{th}$ generations in experiment 2. The results show that larval development occurred more slowly in brackish water of 10 ppt salinity than fresh water, regardless of the brackish or fresh water origin of the colonies from which the eggs were derived. Larval development in fresh water showed a sharper peak of pupation at days seven to nine than in brackish water where there was a more prolonged emergence of pupae with broader peaks at days ten to twelve. Delayed larval development in 10 ppt salinity brackish water was also reflected in the median time to pupation. For the 5$^{th}$ generation in experiment 1, the median times to pupation were 9.5, 10.5, 9.5 and 11.0 days after hatching for the original fresh water colony maintained in fresh water, the reversal fresh water colony transferred to 10 ppt salinity, the reversal brackish water colony transferred to fresh water and the original brackish water colony maintained in 10 ppt salinity respectively. The corresponding median times to pupation for the four colonies were 9.5, 11.5, 9.0, 12.0 and 10.0, 11.5, 9.5 and 11.5 days for the 2$^{nd}$ and 5$^{th}$ generations respectively in experiment 2.

**Table 1.** $LC_{50}$ for salinity tolerance of *Aedes aegypti* colonies.

| Experiment number and the generation of larvae tested | Colonies | | | |
|---|---|---|---|---|
| | **Brackish water colony maintained at 10 ppt salinity** | **Reversal brackish water colony transferred to 0 ppt salinity** | **Reversal fresh water colony transferred to 10 ppt salinity** | **Fresh water colony maintained at 0 ppt salinity** |
| Experiment 1, 2$^{nd}$ generation | 15.6 (14.9 - 16.3) | 13.7 (13.0–14.4) | 13.4 (12.8–14.1) | 12.1 (11.4–12.8) |
| Experiment 2, 2$^{nd}$ generation | 15.6 (14.9–16.4) | 15.0 (14.2–15.8) | 13.9 (13.1–14.6) | 12.3 (11.6–13.0) |
| Experiment 1, 5th generation | 15.9 (15.2–16.6) | 14.1 (13.3–14.8) | 13.8 (13.1–14.5) | 12.2 (11.5–12.9) |
| Experiment 2, 5th generation | 17.1 (16.4–17.8) | 15.8 (15.1–16.5) | 15.4 (14.6–16.2) | 13.0 (12.2–13.8) |

$LC_{50}$ is the salt concentration in parts per thousand (ppt) that results in 50% mortality in the transition from first instar larvae to adults.The 95% confidence intervals of the $LC_{50}$ values are shown in parentheses. The original brackish and fresh water colonies were derived from *Ae. aegypti* collected in Kurunagar and Thirunelvely respectively in the Jaffna peninsula.

**Table 2.** Preference of blood-fed females from brackish and fresh water- derived *Ae. aegypti* for oviposition on brackish and fresh water surfaces.

| Experiment | Numbers of eggs laid | | | |
|---|---|---|---|---|
| | Brackish water colony (10 ppt) | | Fresh water colony (0 ppt) | |
| | Fresh water surfaces (0 ppt salinity) | Brackish water surfaces (10 ppt salinity) | Fresh water surfaces (0 ppt salinity) | Brackish water surfaces (10 ppt salinity) |
| Experiment 1 | 371 | 466 | 696 | 171 |
| Experiment 2 | 266 | 406 | 498 | 101 |
| Experiment 3 | 486 | 485 | 627 | 118 |
| Mean ± sd | 374±110 | 452±41 | 607±101 | 130±37 |
| Student's t test value T and probability P | T=1.15, P=0.31 | | T=7.68, P=0.001 | |

The results are the mean number of eggs laid in the two types of surfaces ± standard deviation in the three separate experiments. A two-tailed unpaired Student's t test was performed to determine the probability of significant differences in the mean number of eggs laid by females of each colony on brackish and fresh water surfaces. The brackish and fresh water colonies were derived from *Ae. aegypti* collected in Kurunagar and Thirunelvely respectively in the Jaffna peninsula.

## Interbreeding between brackish and fresh water-derived *Ae. aegypti*

Because of the differences between the brackish and fresh water-derived *Ae. aegypti*, we investigated the possible existence of a reproductive barrier between the two populations. In the mating experiment between brackish water-derived females and fresh water-derived males, 657 eggs were collected and of these 73.5% hatched into larvae. In the back cross between these F1 males and parental brackish water-derived females, 544 egg were produced and these showed 74.6% hatchability. The mating of brackish water-derived males and fresh water-derived females produced 749 eggs with 75.8% hatchability, while 923 eggs were produced in the backcross between these F1 males and parental fresh water-derived females with 77.3% hatchability. The results show that there are no reproductive barriers between the coastal brackish water and inland fresh water populations of *Ae. aegypti* derived from Kurunagar and Thirunelvely respectively.

## Differential susceptibility to insecticides in brackish and fresh water *Ae. aegypti* isolates

The insecticide bioassays show that *Ae. aegypti* collected from fresh water ovitraps in inland Thirunelvely were significantly more resistant to malathion than *Ae. aegypti* collected from brackish water habitats in coastal Kurunagar (Table 4). The differences between the brackish and fresh water-derived *Ae. aegypti* were not statistically significant for propoxur and permethrin. Both brackish and fresh water-derived *Ae. aegypti* were relatively resistant to the carbamate insecticide propoxur.

## Discussion

Wallis showed in 1954 that long-established laboratory colonies of *Ae. aegypti* prefer to oviposit in fresh water or 2.5 ppt salinity and that oviposition is inhibited at higher salinities [25]. Wallis also found that the chemosensors for sodium chloride relevant to oviposition were located in all the tarsomeres in *Ae. aegypti* legs [25]. Previous studies in Sri Lanka showed that *Ae. aegypti* can oviposit in water of up to 18 ppt salt in containers and wells in the peri-urban environment as well as ovitraps placed in the field [10–12]. However egg laying in field ovitraps decreased with increasing salinity [10]. The present findings show that *Ae. aegypti* derived from fresh water ovitraps in Thirunelvely and maintained in fresh water for eight generations clearly prefer fresh water to 10 ppt brackish water for oviposition. In contrast, *Ae. aegypti* collected from brackish water at Kurunagar and maintained at 10 ppt salinity for eight generations did not significantly differentiate between 10 ppt brackish and fresh water surfaces for oviposition. The results suggest that there has been an adaptation to enable greater oviposition in brackish water in brackish water-derived *Ae.*

**Table 3.** Hatching of *Ae. aegypti* eggs in brackish and fresh water.

| Experiment | Number of eggs out of 150 from different colonies hatching into larvae at 48 h | | | |
|---|---|---|---|---|
| | Original fresh water colony at 0 ppt salinity | Reversal fresh water colony at 10 ppt salinity | Reversal brackish water colony at 0 ppt salinity | Original brackish water colony at 10 ppt salinity |
| 5th generation Experiment 1 | 127±8 | 63±10* | 125±9 | 86±13* |
| 2nd generation Experiment 2 | 128±4 | 61±17* | 123±11 | 74±23*e |
| 5th generation Experiment 2 | 123±7 | 73±9* | 121±8 | 74±12* |

Results are the mean numbers of eggs from three replicate experiments ± standard deviation that hatched into larvae at 48 h from 150 original eggs. Unpaired Student's t tests were performed to determine the significance of differences in the mean numbers of eggs hatching in brackish and fresh water for the corresponding original and reversal colonies (columns 1 *vs* 2 and 3 *vs* 4 respectively) in the three different experiments. The asterisks indicate that in every such comparison significantly more eggs hatched in fresh water compared to brackish water at $p<0.05$. Corresponding comparisons of the pairs of means in columns 1 *vs* 3 and 2 *vs* 4 were not statistically significant with $p>0.05$. The original brackish and fresh water colonies were derived from *Ae. aegypti* collected in Kurunagar and Thirunelvely respectively in the Jaffna peninsula.

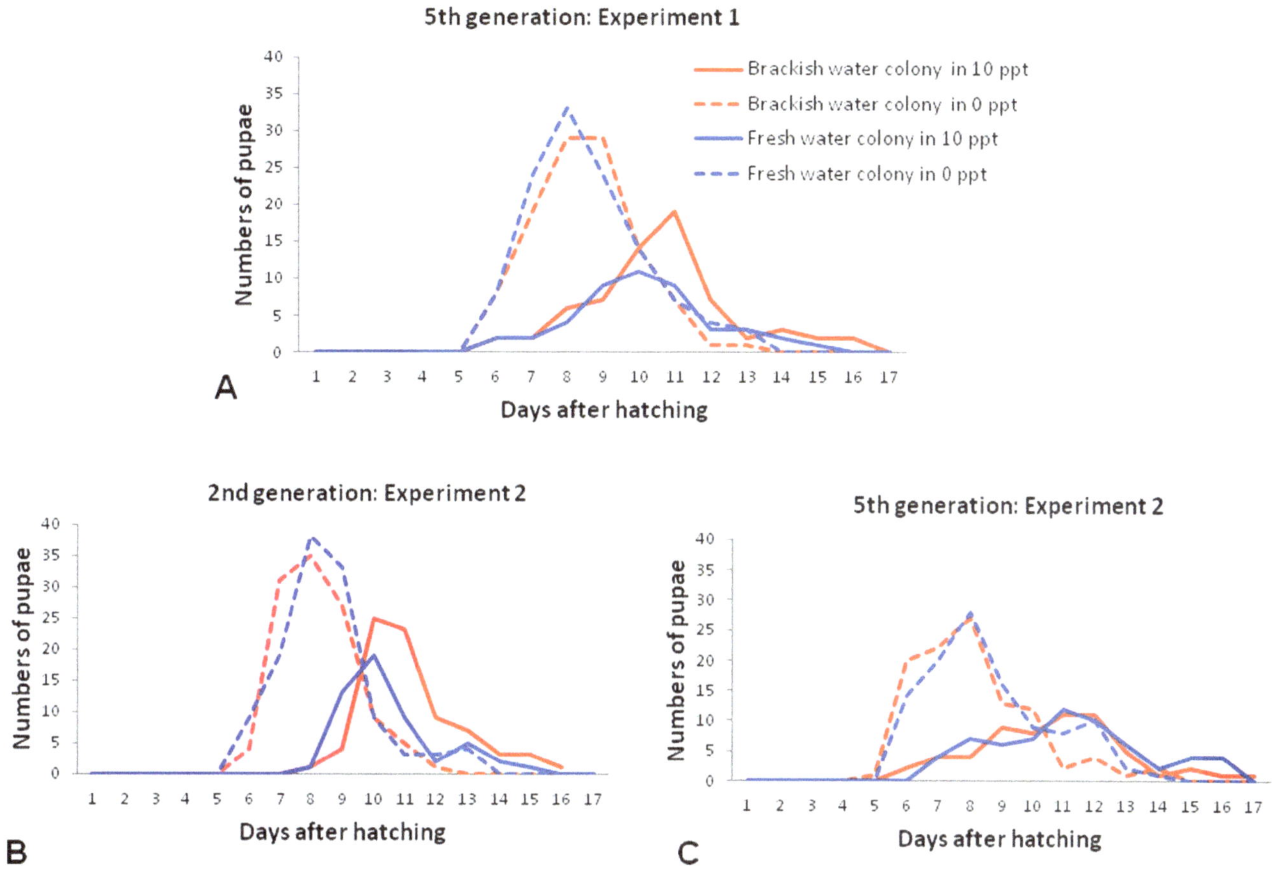


**Figure 3. Larval development times of *Ae. aegypti* in fresh and 10 ppt salt brackish water.** The results show the mean numbers of *Ae. aegypti* pupae emerging on different days after hatching of eggs derived from the brackish and fresh water colonies maintained at the original salinity and after reversal of salinity for up to five generations. The results are the means of three replicate determinations. A. Experiment 1 with fifth generation larvae, B. Experiment 2 with second generation larvae, C. Experiment 2 with fifth generation larvae. The original brackish and fresh water colonies were derived from *Ae. aegypti* collected in Kurunagar and Thirunelvely respectively in the Jaffna peninsula.

*aegypti* that could involve alterations in the physiological mechanisms for sensing and responding to sodium chloride. It is possible that a greater adaptation to salinity in oviposition may be demonstrable at salinities <10 ppt in brackish water-derived *Ae. aegypti*. This merits further investigation as there are many brackish water habitats with <10 ppt salinity where *Ae. aegypti* larvae are found in Kurunagar [10,12]

The present findings also show that the hatching of eggs from the brackish or fresh water-derived *Ae. aegypti* is inhibited by 10 ppt salt demonstrating that there has been no significant adaptation to 10 ppt salinity in brackish water-derived *Ae. aegypti* in this respect even after five generations of maintenance at 10 ppt salt in the laboratory. It is possible, as discussed above for oviposition, that significant adaptation to salinity in egg

**Table 4.** Resistance of field-collected brackish and fresh water populations of *Ae. aegypti* from Kurunagar and Thirunelvely respectively to common insecticides.

| Insecticide | % Resistant | | Student's t test value T and Probability P |
|---|---|---|---|
| | Coastal brackish water population from Kurunagar | Inland fresh water population from Thirunelvely | |
| Permethrin (0.25%) | 19±6 | 30±9 | T = 1.31, P = 0.32 |
| Propoxur (0.1%) | 74±13 | 78±17 | T = 0.24, P = 0.83 |
| Malathion (4%) | 34±8 | 79±14 | T = 6.24, P = 0.02 |

The results are the means ± standard deviation of three replicate determinations. A two-tailed Student's t test for matched samples was performed to determine the probability of significant differences in the mean resistance of brackish and fresh water populations to the three insecticides.

hatching may be demonstrable at salinities <10 ppt. Our observations are compatible with recent findings in Florida, USA which showed that high salinities inhibited egg hatching in *Ae. aegypti* [18].

Studies in the USA have shown that larval development times are prolonged and pupal mass reduced at high salinities in *Ae. aegypti* [18,26]. *Ae. aegypti* larvae from long-established laboratory colonies were able to survive at salinity up to 17.5 ppt [26] while larvae from short – term colonies showed positive growth at 10 ppt salinity [18]. Our findings show that larval development times are prolonged at 10 ppt salinity in both brackish and fresh water-derived *Ae. aegypti*. This shows that brackish water-derived *Ae. aegypti* from Kurunagar have not detectably adapted to salinity in this respect, and this remains so even after five generation of maintenance at 10 ppt salinity in the laboratory. As discussed above for oviposition and egg hatching, it is possible that evidence of adaptation in larval development times may be demonstrable at <10 ppt salinity, and this merits further investigation.

Our previous study suggested that *Ae. aegypti* isolated in Thirunelvely in the Jaffna peninsula differed in salinity tolerance from *Ae. aegypti* isolated in Batticaloa in the eastern province of mainland Sri Lanka [10]. The difference was attributed to the genetic variation caused by adaptation to the generally higher ground water salinity in the Jaffna peninsula [10]. We therefore looked for differences in salinity tolerance between *Ae. aegypti* collected from coastal brackish water and inland fresh water habitats in Kurunagar and Thirunelvely respectively within the Jaffna peninsula. Our findings show that coastal brackish water-derived *Ae. aegypti* from Kurunagar are more tolerant of salinity than inland fresh water-derived *Ae. aegypti* from Thirunelvely in the Jaffna peninsula. Our findings also suggest that the difference in salinity tolerance between coastal and inland *Ae. aegypti* isolates from Kurunagar and Thirunelvely respectively are only partially reduced after five generations of reversal of salinity in the laboratory. Biological changes associated with adaptation to brackish water in *Ae. aegypti* that are not readily reversible are likely to be caused by underlying genetic and/or epigenetic changes that can only be characterised through appropriate genetic and molecular biological studies. However, the present data also show that brackish water-derived *Ae. aegypti* from Kurunagar have not yet undergone biological changes that overcome less efficient hatching of eggs and slower larval development in 10 ppt brackish water.

It is possible that one or more of the physiological and structural changes associated with salinity tolerance previously demonstrated in *Ae. aegypti* [17] and other mosquito species [19–22] contribute to the greater salinity tolerance of brackish water-derived *Ae. aegypti* from coastal Kurunagar. The underlying genetic and physiological mechanisms need to be elucidated in more detailed investigations.

Collection of larvae for experiments 1 and 2 and the respective experiments were performed four months apart. While there were no significant differences in the $LC_{50}$ after two generations between the respective original colonies in the two experiments, the $LC_{50}$ between corresponding colonies in the two experiments tended to diverge with increasing number of generations and reversal of salinity. Differential responses in the complex biological mechanisms underlying responses to salinity changes and laboratory colonisation may be responsible for these differences between the two experiments undertaken four months apart. However, it is clear that within each experiment, brackish water-derived *Ae. aegypti* always showed significantly higher salinity tolerance than fresh water-derived *Ae. aegypti* and that this difference was only partly reduced by reversing the corresponding salinities for five generations.

Our findings show that brackish water-derived *Ae. aegypti* from Kurunagar remain relatively susceptible to malathion (an organophosphate insecticide) and permethrin (a pyrethroid). These and related insecticides of the two classes may therefore be useful for controlling brackish water-derived coastal populations of *Ae. aegypti* in Kurunagar. The differential susceptibility to malathion also suggests that there are genetic differences between brackish water-derived *Ae. aegypti* from coastal Kurunagar and fresh water-derived *Ae. aegypti* from inland Thirunelvely in the Jaffna peninsula. Malathion was widely used for indoor residual spraying of houses to control malaria vectors in the Jaffna peninsula from the mid 1970s until early 2000s, when it was replaced with pyrethroids. The organophosphate Temephos is presently used as a larvicide for *Aedes* dengue vectors solely in fresh water habitats. Other organophosphate insecticides continue to be used in inland areas of the peninsula for controlling agricultural pests. Coastal isolates of *Anopheles subpictus* in Sri Lanka were recently shown to be less resistant to malathion and pyrethroids than inland isolates, and this was attributed to the widespread use of insecticides in inland areas [23]. It is therefore possible that the use of organophosphate insecticides in predominantly inland locations in the Jaffna peninsula has led to inland fresh water *Ae. aegypti* populations developing greater resistance to the malathion, which can be mediated by mutations in the target acetylcholinesterase and changes in glutathione S-transferase and carboxylesterases that metabolise malathion [23].

However, further sampling studies are needed to determine how representative the phenotypic differences observed between brackish water-derived *Ae. aegypti* from Kurunagar and fresh water-derived *Ae. aegypti* from Thirunelvely in the Jaffna peninsula are of brackish and fresh water-derived *Ae. aegypti* populations elsewhere in Sri Lanka or other countries. There is also a need to determine the relative vectorial capacities of brackish and fresh water *Ae. aegypti* populations at the two locations in the Jaffna peninsula and elsewhere.

Genetic differences between *Ae. aegypti* populations in Venezuela [27] and North Queensland [28] have been documented using polymorphic DNA markers. Similarly microsatellite analysis has shown habitat-based population structuring in the closely related *Ae. albopictus* over short distances in Reunion island in the South Indian Ocean [29]. Coastal Kurunagar (the origin of brackish water *Ae. aegypti* used in the study) and inland Thirunelvely (the origin of fresh water *Ae. aegypti*) in the Jaffna peninsula are 5 km apart. *Ae. aegypti* are short distance migrants that normally lay eggs within 1 km from the site of a blood meal [30,31]. The present findings are therefore compatible with the possibility that genetic differences have developed, despite the relatively short distance involved, due to restricted gene flow between coastal brackish water populations of *Ae. aegypti* from Kurunagar and inland fresh water populations from Thirunelvely. An analogous situation has been observed in South-West Australia where larvae of coastal marsh populations of *Aedes camptorhynchus* (a vector of Ross River virus) tolerate greater salinity (52 ppt, i.e. hypersalinity) than inland populations (30 ppt, i.e. approaching the average salinity of sea water), probably due to genetic changes in osmoregulatory mechanisms [32]. However, the present data show that likely genetic differences between the coastal and inland isolates of *Ae. aegypti* from Kurunagar and Thirunelvely respectively in the Jaffna peninsula have not yet become a barrier to reproduction. There is a possibility that brackish water *Ae. aegypti* can develop into a separate species given sufficient time and physical isolation from the original fresh water species. The evolution of salinity-tolerant *Anopheles melas* and *Anopheles merus* in Africa within the *Anopheles gambiae* complex provides a precedent for such a process [33].

It is not presently clear to what extent the development of *Ae. aegypti* in brackish water in Kurunagar has been driven by selective pressure exerted by dengue control measures applied to fresh water habitats or the opportunistic exploitation of a recent proliferation of anthropogenic ecological niches near human dwellings in coastal areas. It is likely that both are contributing factors. However the evidence suggests that the utilisation of brackish water habitats by *Ae. aegypti* may be associated at the present time with less efficient hatching of eggs and slower larval development at higher salinities that are presumably balanced by the other advantages.

Analysis of the temporal relationship between dengue incidence and rainfall suggests that monsoonal rains are important drivers of dengue transmission in the Jaffna district [14]. Coastal Kurunagar has a high incidence of dengue [10]. It is therefore possible that coastal brackish water *Aedes* vectors in the Jaffna peninsula constitute an unappreciated, perennial source of dengue transmission that promotes increased transmission following the monsoon. Similar considerations may apply to coastal locations elsewhere in Sri Lanka and other dengue-endemic countries.

The present findings further support assertions [10–16,18] that existing guidelines on dengue control [5] need to be extended to target brackish water habitats of *Ae. aegypti* and *Ae. albopictus* in the urban and peri-urban environment. They also highlight a global need for more research into the genetic and physiological basis for salinity adaptation in vectors and the role of salinity tolerant vectors in disease transmission, and the formulation of appropriate mitigating measures in a future context of rising sea levels [14–16].

## Supporting Information

**Table S1 Statistical comparison of $LC_{50}$ values for salinity tolerance between the different original and reversal colonies of *Aedes aegypti* after two and five generations in the laboratory in Experiments 1 and 2.**

**Table S2 Statistical comparison of $LC_{50}$ values for salinity tolerance between corresponding colonies of *Aedes aegypti* from the two different collections used in Experiments 1 and 2.**

## Author Contributions

Conceived and designed the experiments: RR SNS. Performed the experiments: PJJ TV TE SNS. Analyzed the data: SNS RR. Wrote the paper: RR SNS.

## References


1. Weaver SC, Reisen WK (2010) Present and future arboviral threats. Antivir Res 85: 328.
2. Walter Reed Biosystematics Unit (2013) Keys to medically important mosquito species. Silver Spring, MA, USA. Smithsonian Institution. http://wrbu.org/command_aors_MQ.html. Accessed 26 January 2014.
3. World Health Organization (2014) Fact Sheet No. 327 – Chikungunya. http://www.who.int/mediacentre/factsheets/fs327/en/. Accessed 5 April 2014.
4. World Health Organization (2013) Fact Sheet No. 117 – Dengue and severe dengue. http://www.who.int/mediacentre/factsheets/fs117/en/. Accessed 26 January 2014.
5. World Health Organization (2009). Dengue guidelines for diagnosis, treatment, prevention and control. http://www.whqlibdoc.who.int/publications/2009/9789241547871_eng.pdf. Accessed 26 January 2014.
6. World Health Organisation (2013). Fact sheet No 100. Yellow fever. http://www.who.int/mediacentre/factsheets/fs100/en/. Accessed 26 January 2014.
7. Ministry of Healthcare and Nutrition of Sri Lanka (2013) Monthly distribution of suspected dengue cases from 2004 to 2010 by district. http://www.epid.gov.lk/Dengue_updates.htm. Accessed 13 June 2013.
8. Surendran SN, Kannathasan S, Kajatheepan A, Jude PJ (2007) Chikungunya type fever outbreak: some aspects related to this new epidemic in Jaffna district, northern Sri Lanka. Trop Med Hlth 35: 249–252.
9. Barraud PJ (1934) Diptera Vol V Family *Culicidae*. Tribes Megarhinini and Culicini. In: Sewell RBS, Edwards PW editors, The fauna of British India, including Ceylon and Burma. pp 28–426.
10. Ramasamy R, Surendran SN, Jude PJ, Dharshini S, Vinobaba M (2011) Larval development of *Aedes aegypti* and *Aedes albopictus* in peri-urban brackish water and its implications for transmission of arboviral diseases. PLoS Negl Trop Dis 5(11): e1369. doi:10.1371/journal.pntd.0001369.
11. Jude PJ, Thamasegaram T, Sivasubramanyam G, Senthilnathan M, Kannathasan S, et al. (2012) Salinity-tolerant larvae of mosquito vectors in the tropical coast of Jaffna, Sri Lanka and the effect of salinity on the toxicity of *Bacillus thuringiensis* to *Aedes aegypti* larvae. Parasit Vect 5:269. doi: 10.1186/1756-3305-5-269.
12. Surendran SN, Jude PJ, Thabothiny V, Raveendran S, Ramasamy R (2012) Pre-imaginal development of *Aedes aegypti* in brackish and fresh water urban domestic wells in Sri Lanka. J Vect Ecol 37(2): 471–473.
13. Idris FHj, Usman A, Surendran SN, Ramasamy R (2013) Detection of *Aedes albopictus* pre-imaginal stages in brackish water habitats in Brunei Darussalam. J Vect Ecol 38: 197–199.
14. Ramasamy R, Surendran SN (2013) Global environment changes and salinity adaptation in mosquito vectors. ISBN 978-3-8484-2290-6 Saarbrucken: Lambert Academic Publishing. 100 p.
15. Ramasamy R, Surendran SN (2011) Possible impact of rising sea levels on vector-borne infectious diseases. BMC Infect Dis 11: 18.
16. Ramasamy R, Surendran SN (2012) Global climate change and its potential impact on disease transmission by salinity-tolerant mosquito vectors in coastal zones. Front Physiol 3: 198. doi: 10.3389/fphys.2012.00198.
17. Edwards HA (1982) Free amino acids as regulators of osmotic pressure in aquatic insect larvae. J Exp Biol 101: 153–160.
18. Yee DA, Himel E, Reiskind MH, Vamosi SM (2014) Implications of saline concentrations for the performance and competitive interactions of the mosquitoes *Aedes aegypti* (*Stegomyia aegypti*) and *Aedes albopictus* (*Stegomyia albopictus*). Med Vet Entomol 28: 60–69.
19. Bradley TJ (1987) Physiology of osmoregulation in mosquitoes. Annu Revs Entomol 32: 439–462.
20. Garrett MA, Bradley TJ (1987) Extracellular accumulation of proline, serine and trehalose in the haemolymph of osmoconforming brackish-water mosquitoes. J Exp Biol 129: 231–238.
21. Smith KE, Van Ekeris LA, Okech BA, Harvey WR, Linser PJ (2008) Larval anopheline mosquito recta exhibit a dramatic change in localization patterns of ion transport proteins in response to shifting salinity: a comparison between anopheline and culicine larvae. J Exp Biol 211(19): 3067–3076.
22. White BJ, Kundert PN, Turissini DA, Van Ekeris LA, Linzer PJ, et al. (2013) Dose and developmental response of *Anopheles merus* larvae to salinity. J Exp Biol 216(18): 3433–3441.
23. Surendran SN, Jude PJ, Weerarathne TC, Karunaratne SHPP, Ramasamy R (2012) Variations in susceptibility to common insecticides and resistance mechanisms among morphologically identified sibling species of the malaria vector *Anopheles subpictus* in Sri Lanka. Parasit Vect 5: 34. doi:10.1186/1756-3305-5-34.
24. Wheeler MW, Park RM, Bailer AJ (2006) Comparing median lethal concentration values using confidence interval overlap or ratio tests. Environ Toxicol Chem 25: 1441–1444.
25. Wallis RC (1954) A study of ovipostion activity of mosquitoes. Am J Hyg 60(2): 135–168.
26. Clark TM, Flis BJ, Remold SK (2004) Difference in the effects of salinity on larval growth and developmental programs of a freshwater and a euryhaline mosquito species (Insecta: Diptera, Culicidae). J Exp Biol 207: 2289–2295.
27. Herrera F, Urdaneta U, Rivero J, Zoghbi K, Ruiz J, et al. (2006) Population genetic structure of the dengue mosquito *Aedes aegypti* in Venezuela. Mem Inst Oswaldo Cruz 101(6): 625–633.
28. Endersby NM, Hoffman AA, White VA, Ritchie SA, Johnson PH, et al. (2001) Changes in the genetic structure of *Aedes aegypti* (Diptera: Culicidae) populations in Queensland, Australia, across two seasons: implications for potential mosquito releases. J Med Entomol 48(5): 999–1007.
29. Delatte H, Toty C, Boyer S, Bouetard A, Bastien F, et al. (2013). Evidence of habitat structuring *Aedes albopictus* populations in Reunion Island. PLoS Negl Trop Dis 7(3): e2111. doi:10.1371/journal.pntd.0002111.
30. Honorio NH, Silva WdaC, Leite PJ, Goncalves JM, Lounibos LP, et al. (2003) Dispersal of *Aedes aegypti* and *Aedes albopictus* (Diptera: Culicidae) in an urban

endemic dengue area in the state of Rio de Janeiro, Brazil. Mem Inst Oswaldo Cruz 98(2): 191–198.

31. Harrington LC, Scott TW, Lerdthusnee K, Coleman RC, Costero A, et al. (2005) Dispersal of the dengue vector *Aedes aegypti* within and between rural communities. Am J Trop Med Hyg 72(2): 209–220.
32. Van Schie C, Spafford H, Carver S, Weinstein P (2009). Salinity tolerance of *Aedes camptorhynchus* (Diptera: Culicidae) from two regions in southwestern Australia.
33. Coluzzi M, Sabatini A (1969). Cytogenetic observations on the salt water species, *Anopheles merus* and *Anopheles melas* of the Gambiae complex. Parassitologia 11: 177–187.

# 13

# Development of a Freeze-Dried Fungal Wettable Powder Preparation Able to Biodegrade Chlorpyrifos on Vegetables

**Jie Liu[1], Yue He[2], Shaohua Chen[3], Ying Xiao[1], Meiying Hu[1], Guohua Zhong[1]***

**1** Laboratory of Insect Toxicology, and Key Laboratory of Pesticide and Chemical Biology, Ministry of Education, South China Agricultural University, Guangzhou, P.R. China, **2** Guangdong Zhuhai Supervision Testing Institute of Quality and Metrology, Zhuhai, P.R. China, **3** Guangdong Province Key Laboratory of Microbial Signals and Disease Control, South China Agricultural University, Guangzhou, P.R. China

## Abstract

Continuous use of the pesticide chlorpyrifos has resulted in harmful contaminations in environment and species. Based on a chlorpyrifos-degrading fungus *Cladosporium cladosporioides* strain Hu-01 (collection number: CCTCC M 20711), a fungal wettable powder preparation was developed aiming to efficiently remove chlorpyrifos residues from vegetables. The formula was determined to be 11.0% of carboxymethyl cellulose-Na, 9.0% of polyethylene glycol 6000, 5.0% of primary alcohol ethoxylate, 2.5% of glycine, 5.0% of fucose, 27.5% of kaolin and 40% of freeze dried fungi by response surface methodology (RSM). The results of quality inspection indicated that the fungal preparation could reach manufacturing standards. Finally, the degradation of chlorpyrifos by this fungal preparation was determined on pre-harvest cabbage. Compared to the controls without fungal preparation, the degradation of chlorpyrifos on cabbages, which was sprayed with the fungal preparation, was up to 91% after 7 d. These results suggested this freeze-dried fungal wettable powder may possess potential for biodegradation of chlorpyrifos residues on vegetables and provide a potential strategy for food and environment safety against pesticide residues.

**Editor:** Raul Narciso Carvalho Guedes, Federal University of Viçosa, Brazil

**Funding:** The authors are grateful to the National Natural Science Foundation (No.30871660 and No. 31371960) of China for the financial support. The funders had no role in study design, data collection and analysis, decision to publish, or preparation of the manuscript.

**Competing Interests:** The authors have declared that no competing interests exist.

* Email: guohuazhong@scau.edu.cn

These authors contributed equally to this work.

## Introduction

Chlorpyrifos is a broad-spectrum, moderately toxic organophosphorus pesticide [1]. Since been first introduced into market in 1960s, chlorpyrifos has been globally used in pest control in agriculture and home for its acute neurotoxic effects through acetylcholinesterase inhibition and consequent cholinergic hyper stimulation [2,3]. However, increasing evidences indicate that chronic exposure to chlorpyrifos can cause persisting neurobehavioural dysfunction [4], even with low doses, which can elicit chronic toxicity [5]. There is ample evidences that chlorpyrifos for controlling insects agriculturally or residentially can adversely affect non-target organisms like fish [6], bees [7], silkworms and rats [8]. These potential threats lead to a great concern of environment and food safety and human health for its potential toxic [9,10]. Although the strict limitation of chlorpyrifos residue launched in many countries, the use of the pesticide is still extensive. Thus, it is urgent to develop an efficient and convenient strategy able to eliminate the chlorpyrifos residues on vegetables and soils so that the food and environment safety can be guaranteed.

Biodegradation, using living microorganisms or active enzymes to detoxify and degrade pollutants, has received attention as a safe, efficient and cost-effective approach to clean up contaminated environment [11–15]. The degradation of chlorpyrifos by microorganisms in pure cultures has been well investigated [16,17] and all the published studies concluded that using active microorganisms could be a promising approach for chlorpyrifos degradation on farm products [16,18]. However, field conditions adversely influence microbial ingredients and impair the degradation capacity, so little is known on chlorpyrifos degradation by microorganisms in agricultural application. How to apply these pesticide-degrading microorganisms in field conditions has always been the greatest obstacle. In fact, with the reported methods [19], like using cell-free extracts [18], it was still difficult to reach the expected performance and hard to put into field application.

The practical application of microorganisms with pesticide degrading capacity could refer to the microbial pesticides like *Bacillus thuringiensis* (BT), which have already been developed many commercial formulations including oil solutions, wettable powder and effervescent tablets [20]. Development of microbial preparations capable of degrading pesticides mostly depends on the persistence of active compounds, which could be improved by the formulation design and selection of assistant agents. However, the extraction of active proteins increases the manufacture cost and the producing procedure may impact the enzyme ability [21], thus a cost-efficient, effective and safe method of microbial preparation should be established.

**Table 1.** Effects of protectants on chlorpyrifos degradation by freeze-dried fungi.

| Protectants | Amount (%) | Degradation rate (%) |
|---|---|---|
| Fucose | 2.5 | 44.9±5.3[i] |
| | 5.0 | 50.1±6.6[g] |
| | 7.5 | 60.2±5.2[d] |
| Glycine | 2.5 | 48.7±1.6[h] |
| | 5.0 | 52.5±1.8[f] |
| | 7.5 | 56.5±6.3[e] |
| Glycerol | 2.5 | 68.1±1.7[c] |
| | 5.0 | 72.0±1.7[b] |
| | 7.5 | 78.8±1.5[a] |
| CK | N/A | 34.6±1.0[j] |

Note: The data presented are means of three replicates with standard deviation, which is within 5% of the mean. Different letters indicate significant differences ($p<0.05$, LSD test). CK represents the control group without protectants.

In this study, using a chlorpyrifos degrading fungus *Cladosporium cladosporioides* strain Hu-01 (collection number: CCTCC M 20711), a fungal preparation able to eliminate chlorpyrifos residues was developed through a process of vacuum freeze dehydration with appropriate agents. In addition, the multiple factors could affect the preparation quality was evaluated and checked to ensure its being qualified for industrial manufacture and market promotion. The objectives of this study were to test and determine the capacity of the preparation to degrade chlorpyrifos residue on vegetables, and to explore a potential approach for chlorpyrifos degradation by microorganisms in field conditions.

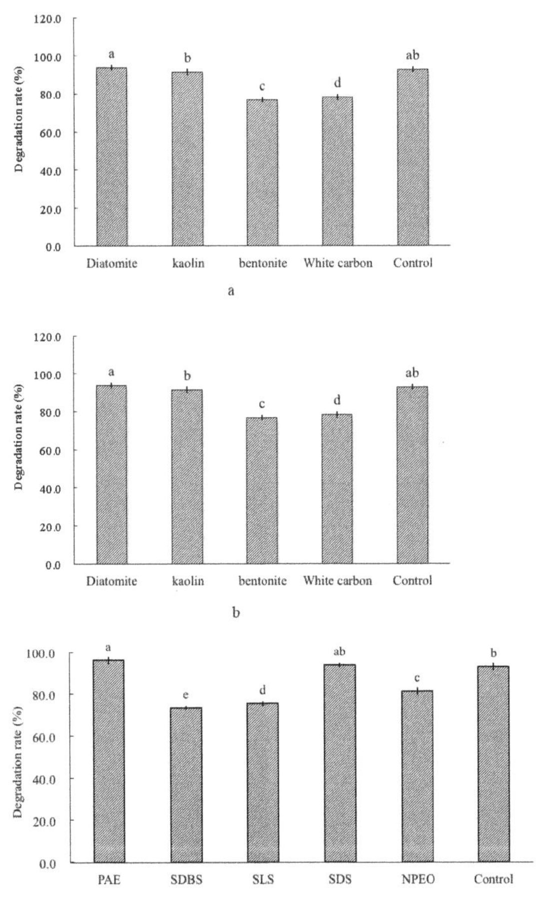


**Figure 1. Effects of carriers, dispersants or surfactants on chlorpyrifos degradation added in freeze dried fungi.** Note: a: effects of carriers; b: effects of dispersants; c: effects of surfactants.

## Materials and Methods

### Chemicals and fungus

Technical grade chlorpyrifos (96% purity) was obtained from Dow AgroSciences, USA. All other chemicals and solvents used were analytical grade.

Activated sludge samples were collected as inoculum from a pesticide-manufacturing wastewater treatment system and chlorpyrifos-degrading fungus *Cladosporium cladosporioides* Hu-01 was screened and isolated [16]. This strain was deposited in China Center for Type Culture Collection under the collection No. CCTCC M 20711. The constant-temperature culture method was referred to Gao et al. [22] and the conditions were inoculum amount 0.1% (wet fungi), 28°C and 150 rpm on a rotary shaker for 5 days. Previously, the degradation of chlorpyrifos by strain Hu-01 was determined. 92.7% of chlorpyrifos at the initial concentration of 25 mg·L$^{-1}$ was degraded in half hour, which revealed that strain Hu-01 can significantly reduce chlorpyrifos residue.

### Chemical analysis

The analysis method of chlorpyrifos was referred to Gao et al. [22]. Chlorpyrifos was analyzed on an Agilent 1100 High Performance Liquid Chromatography (HPLC) (Agilent, USA) equipped with a Hypersil ODS2 $C_{18}$ reversed phase column (4.6 nm × 250 mm, 5 mm). A mixture of methanol and water (90:10, *v/v*) was used as the mobile phase at a flow rate of 1.0 mL·min$^{-1}$. The injection volume was 10 μL.

**Table 2.** The dispersity of CMC-Na and PEG 6000 in different contents.

| Dispersants | Content (%) | Dried fungi (%) | Dispersity(%) |
|---|---|---|---|
| CMC-Na | 7 | 48 | 63.2±1.1[b] |
| | 9 | 46 | 69.0±1.2[a] |
| | 11 | 44 | 68.9±0.9[a] |
| PEG 6000 | 7 | 48 | 53.3±0.4[c] |
| | 9 | 46 | 69.2±0.7[a] |
| | 11 | 44 | 68.9±1.1[a] |

Note: The data presented are means of three replicates with standard deviation, which is within 5% of the mean. Different letters indicate significant differences ($p<0.05$, LSD test).

Recoveries of 2.5, 5, 10, 25, and 50 mg·L$^{-1}$ of chlorpyrifos were determined from 86.3% to 106.2% and relative standard deviation (RSD) ranged from 0.6% to 4.1%.

## Procedure of freeze-dried fungi powder

After culturing for 7 d, hyphae were collected by filtering. Being pre-freezed at −20°C for 24 h, the fungi samples were immediately vacuumed and freeze-dried for another 24 h. Then, those samples were kept in normal temperature and the preliminary preparation was done. The degradation of chlorpyrifos by dried powder was conducted by adding 0.1 g of fungi powder to 5 mL of 25 mg·L$^{-1}$ chlorpyrifos in phosphate buffered solution and the residual concentration was determined in half hour by HPLC.

## Determination of protectants and carriers

To maintain the microbial activity, protectants were added before dehydration and each fungi-protectant sample was moved to freeze dehydration machine for 24 h until the humidity fell to 2% or 3%. The degradation of 25 mg·L$^{-1}$ chlorpyrifos was then investigated by HPLC. Control groups without protectants were set and the HPLC results leaded to the available protective agents and the suitable dosages.

The biocompatibility of carriers including diatomite, kaolin, bentonite and white carbon were investigated by mixing 50 mL of 10% fungi solution with 500 mg·L$^{-1}$ carriers, adjusting chlorpyrifos at 25 mg·L$^{-1}$ and degradation rate of chlorpyrifos was determined after 24 h. Carriers were individually mixed with freeze-dried fungi ($v$: $v=2{:}3$) and then the status of blocking, fluidity and cost of each carrier-fungi powder sample was investigated by investigation and surveys. Control groups without carriers were set and the observation results leaded to the available carriers and the suitable dosages.

## Determination and composition optimization of dispersants and surfactants

The biocompatibility of fungi powder with assistant agents was tested including dispersants: gelatin, polyethylene glycol 6000 (PEG 6000), polyvinylpyrrolidone (PVP), carboxymethyl cellulose-Na (CMC-Na), CMC and Polyvinyl alcohol (PVA); and surfactants: primary alcohol ethoxylate (PAE), sodium dodecyl benzene sulfonate (SDBS), sodium dodecyl sulfate (SDS) and nonylphenol ethoxylate (NPEO). 50 mL of 10% fungi solution was mixed with 50 mg·L$^{-1}$ dispersant or surfactant, adjusting chlorpyrifos at 25 mg·L$^{-1}$ and degradation rate of chlorpyrifos was determined.

For the dispersants, the selection was carried out by pouring 5.000 g of each dispersant into 250 mL of standard hard water at 30°C ±2°C and letting it sit for 30 min after oscillation. Then move 9/10 of upper liquid out and hydrate the left 25 mL solution. Repeat 3 times. The weight of each sample was record and the dispersion rate was calculated by **Eq. (1)**.

$$W_1 = m_1 - m_2 \times 10 \times 100 \quad (1)$$

where $W_1$ is dispersity (%), $m_1$ is the average mass of sample and $m_2$ is the average mass of hydrated sample of 25 mL left in bottom.

For the surfactants, wetting time of each one was determined. Pour 5.0 g of each surfactant-fungi powder sample into 100 mL of standard hard water. The duration it took to be fully wet was recorded and repeated 5 times. Then the average time was calculated and the available surfactants were determined.

Response surface methodology (RSM) based on the Central Composite Rotatable Design (CCRD) was applied to optimize the each key component as a variable [16]. Theoretically, the interactions among variables could significantly affect the dispersion of fungal powder preparation in solution and the degradation of chlorpyrifos can be influenced accordingly [23,24]. In this experiment, three assistant agents for optimizing were PAE (3.0% to 7.0%), CMC-Na (5.0% to 13.0%) and PEG 6000 (5.0% to 13.0%). The range and center point values of three independent variables were based on the results of preceding experiments. The dependent variable was the dispersion rate determined by adding each 5.000 g agent via **Eq. (1)**. The data were analyzed using RSM of the statistic analysis system (SAS) software (Version 9.0) to fit the following quadratic polynomial equation (**Eq. (2)**).

$$Y_i = b_0 + \sum b_i X_i + \sum b_{ii} X_i^2 + \sum b_{ij} X_i X_j \quad (2)$$

where $Y_i$ is the predicted response (dispersion (%) ), $X_i$ and $X_j$ are variables, $b_0$ is the constant, $b_i$ is the linear coefficient, $b_{ij}$ is the interaction coefficient, and $b_{ii}$ is the quadratic coefficient.

## Preparation of wettable freeze-dried fungi powder

The procedure of fungi preparation was carried out as following. Applied protectants were individually mixed with hyphae filtrated from fermented solution before freeze dehydration, and then the freeze-dried powder sample was mingled with other assistant agents. After altering pH to 6~7 and smashing till 95% of it sieved to 63 μm, the final preparation was done and ready for degradation performance inspection.

## Preparation quality inspection

Protein concentration was measured by the method of Coomassie blue staining method with bovine serum albumin as a standard using a spectrophotometer (Shimadzu, Japan).

**Table 3.** Central composite rotatable design (CCRD) matrix and the response of dependent variable for dispersity of preparation (*Y*).

| Run | Contents of variables (%) | | | Dispersity (%) |
|---|---|---|---|---|
| | $X_1$ | $X_2$ | $X_3$ | $Y$ |
| 1 | 11.0 | 11.0 | 6.0 | 61.42±0.84 |
| 2 | 11.0 | 7.0 | 4.0 | 61.13±0.66 |
| 3 | 7.0 | 11.0 | 4.0 | 64.09±1.25 |
| 4 | 7.0 | 7.0 | 6.0 | 65.09±0.58 |
| 5 | 11.0 | 11.0 | 4.0 | 60.57±1.84 |
| 6 | 11.0 | 7.0 | 6.0 | 62.49±0.48 |
| 7 | 7.0 | 11.0 | 6.0 | 65.62±1.10 |
| 8 | 7.0 | 7.0 | 4.0 | 64.37±0.55 |
| 9 | 11.0 | 11.0 | 4.0 | 56.87±1.62 |
| 10 | 11.0 | 7.0 | 6.0 | 63.11±1.14 |
| 11 | 7.0 | 11.0 | 6.0 | 65.04±1.04 |
| 12 | 7.0 | 7.0 | 4.0 | 65.82±1.68 |
| 13 | 11.0 | 11.0 | 6.0 | 61.87±1.22 |
| 14 | 11.0 | 7.0 | 4.0 | 61.23±1.73 |
| 15 | 7.0 | 11.0 | 4.0 | 64.20±0.98 |
| 16 | 7.0 | 7.0 | 6.0 | 68.67±1.05 |
| 17 | 9.0 | 9.0 | 5.0 | 65.92±2.97 |
| 18 | 9.0 | 9.0 | 5.0 | 69.83±0.91 |
| 19 | 9.0 | 9.0 | 5.0 | 68.58±0.79 |
| 20 | 9.0 | 9.0 | 5.0 | 71.04±0.85 |
| 21 | 5.0 | 9.0 | 5.0 | 59.98±0.69 |
| 22 | 13.0 | 9.0 | 5.0 | 72.07±2.02 |
| 23 | 9.0 | 5.0 | 5.0 | 59.05±1.93 |
| 24 | 9.0 | 13.0 | 5.0 | 70.41±0.54 |
| 25 | 9.0 | 9.0 | 3.0 | 60.68±17.99 |
| 26 | 9.0 | 9.0 | 7.0 | 72.66±2.09 |
| 27 | 9.0 | 9.0 | 5.0 | 72.04±0.27 |
| 25 | 9.0 | 9.0 | 3.0 | 60.68±17.99 |
| 26 | 9.0 | 9.0 | 7.0 | 72.66±2.09 |
| 27 | 9.0 | 9.0 | 5.0 | 72.04±0.27 |
| 28 | 9.0 | 9.0 | 5.0 | 71.83±0.40 |
| 29 | 9.0 | 9.0 | 5.0 | 72.60±0.49 |
| 30 | 9.0 | 9.0 | 5.0 | 72.51±0.63 |
| 31 | 9.0 | 9.0 | 5.0 | 72.46±0.67 |
| 32 | 9.0 | 9.0 | 5.0 | 71.60±0.85 |

Note: $X_1$: CMC-Na; $X_2$: PEG 6000; $X_3$: PAE; $Y$: dispersity. The data presented are means of three replicates with standard deviation, which is within 5% of the mean.

Wetting ability was determined by the time that 5.0 g of fungal powder took to be fully wet. Dispersion ability was inspected by the dispersion rate of fungal powder in solution. The size of powder granules was tested through the standard mesh of 325 (≥ 95%). Humidity was detected by weighing 0.5 g of fungal preparation and culture vessel together. After 2 h in dryer at 100°C, calculate the average humidity content using the weight of cooled preparation and culture vessel and repeated 3 times.

Storage stability was checked by storing for 15 d and 1 to 5 months at 4°C and 25°C, respectively, then reacting with chlorpyrifos at 25 mg·L$^{-1}$ for 30 min and the degradation performance was detected to illustrate the stability of the preparation. The stability in low temperature was tested at 0°C in 1 h and 7 d, respectively.

## Degradation performance of freeze-dried fungal powder

Each test area of 4 m$^2$ was divided from a cabbage filed without any pesticides applied before and 45% chlorpyrifos emulsifiable concentrate (EC) was sprayed at 225, 350, 450, 700 and 900 g·L$^{-1}$, respectively, with the spray amount of 50 L·667 m$^{-2}$. After 3 d, the fungal preparation was sprayed at the amount of 200 g·667 m$^{-2}$ against the controls sprayed only water. The samples investigated were taken randomly in 1, 3 and 7 d. 25.0 g of each vegetable sample soaking in 50.0 mL of

**Table 4.** Analysis of variance (ANOVA) for the fitted quadratic polynomial model for dispersity of preparation.

| Source | DF | SS | MS | *F* value | *Pr>F** |
|---|---|---|---|---|---|
| Model | 9 | 718.08 | 35.90 | 6.805 | 0.001 |
| Linear | 3 | 334.91 | 66.98 | 12.695 | 0.000 |
| Square | 3 | 342.93 | 68.59 | 12.999 | 0.000 |
| Interaction | 3 | 40.24 | 4.02 | 0.763 | 0.661 |
| Error | 13 | 58.04 | 5.28 | | |
| Total | 22 | 776.12 | | | |

Note: DF refers to degrees of freedom; SS refers to sum of sequences; MS refers to mean square.
**P* Level less than 0.05 indicate the model terms are significant.

acetonitrile were decolored by 0.2 to 0.8 g of active carbon. After filtrating, partitioning with NaCl, clean-up, and concentration, residues dissolved in *n*-hexane was analyzed by Gas Chromatography (GC) HP-6890 with ECD and a capillary column of BD-1701, 30 m×0.32 mm×1.0μm. The temperatures of injection port, column and detector were 100°C, 220°C and 250°C, respectively. The flow of gas was as follows: nitrogen, 60 $mL \cdot min^{-1}$; air, 60 m $L \cdot min^{-1}$; hydrogen, 30 m $L \cdot min^{-1}$.

### Data analysis

All of the experiments were carried out in triplicate, and the results were the means of three replicates. Standard deviations were also determined using Statistic Analysis System (SAS). The significance ($p<0.05$) of differences was treated statistically by one-, two-, or three-way analysis of variance (ANOVA) and evaluated by post hoc comparison of means using lowest significant differences (LSD).

### Ethics Statement

No specific permits were required for the described field studies. No specific permissions were required for these locations. We confirm that the location is not privately-owned or protected in any way. We confirm that the field studies did not involve endangered or protected species.

## Results

### Chlorpyrifos degradation by freeze-dried fungi

After freeze dehydration, the degradation of 10% freeze dried fungi powder was determined and simply 40.9±0.3% of chlorpyrifos at the initial concentration of 25 $mg \cdot L^{-1}$ was degraded, which was significantly lower than the degradation rate of 92.5±0.1% before freeze dehydration.

### Protectants and carriers

To preserve the biodegradation ability of strain Hu-01, the available protective agents were selected. From the results in **Table 1**, it was shown that three protectants, fucose, glycin and glycerol, largely protected the freeze fungi from inactivation. With the increase of protectant content, the degradation of chlorpyrifos was gradually enhanced, which was significantly differed from the controls. 7.5% of glycerol revealed the best protective ability and nearly 80% of chlorpyrifos was degraded. However, the sample with glycerol was difficult to smash and dehydrate, so glycerol was not the available option. Fucose and glycine also presented as good protectants and more than half of chlorpyrifos was degraded when they reached the maximum testing amount. Concerning the factors above, 5.0% of fucose and 2.5% of glycine were selected to be the protective compounds in this fungal preparation for their protective performance, effective cost and a reasonable distribution of assisted agents.

The biocompatibility of tested carriers in **Fig. 1(a)** indicated that they were compatible with active fungi and no obvious

**Table 5.** Effect estimates for the fitted quadratic polynomial model for dispersity of preparation.

| Term | Estimate | Standard Error | *t* value | *Pr>/t/** |
|---|---|---|---|---|
| $X_1$ | 2.64 | 0.47 | 5.633 | 0.000 |
| $X_2$ | 1.66 | 0.47 | 3.550 | 0.005 |
| $X_3$ | 1.83 | 0.47 | 3.909 | 0.002 |
| $X_1*X_1$ | −1.94 | 0.42 | −4.566 | 0.001 |
| $X_1*X_2$ | 0.17 | 0.57 | 0.301 | 0.769 |
| $X_1*X_3$ | 0.12 | 0.57 | 0.201 | 0.844 |
| $X_2*X_2$ | −2.26 | 0.42 | −5.330 | 0.000 |
| $X_2*X_3$ | 0.22 | 0.57 | 0.391 | 0.703 |
| $X_3*X_3$ | −1.78 | 0.42 | −4.186 | 0.002 |

Note: $X_1$: CMC-Na; $X_2$: PEG 6000; $X_3$: PAE.
**P* Level less than 0.05 indicate the model terms are significant.

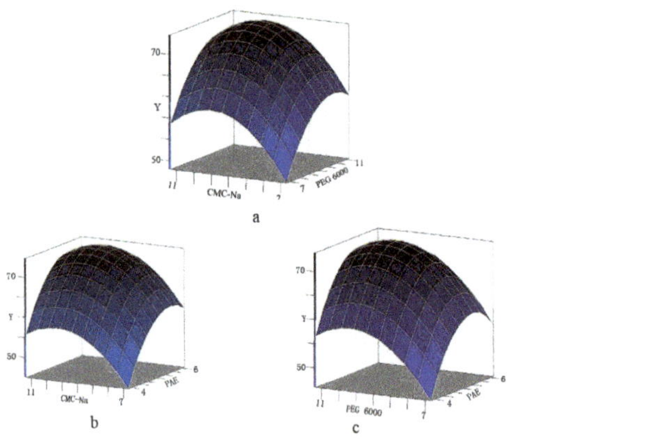


**Figure 2. Response surface plots showing the effects of three variables on dispersity of the fungal preparation.** Note: a: the effects of CMC-Na ($X_1$) and PEG 6000 ($X_2$) on dispersity ($Y$) of the preparation; b: the effects of CMC-Na ($X_1$) and PAE ($X_3$) on dispersity ($Y$) of the preparation; c: the effects of PEG 6000 ($X_2$) and PAE ($X_3$) on dispersity ($Y$) of the preparation.

disadvantages observed. From the observed results, the samples pre-treated by 27.5% of kaolin were easy to smash into powder with good fluidity while other three carriers were lack of fluidity or easily, even seriously blocking. As a result, kaolin was chosen to be the carrier in this process.

## Selection and composite optimization of dispersants and surfactants

The biocompatibility of dispersants and surfactants was investigated at first. The results in **Fig. 1(b and c)** illustrated that two categories of assistant agents were compatible with active fungi and no obvious disadvantages observed, which revealed that these agents in the present study contributed more or less to remain the degradation capacity of the strain and were suitable for further screening.

Then, four surfactants, SDS, SDBS, PAE and NPEO were evaluated. The time that each surfactant took to be fully wet was 63.7±2.8 s, 49.3±3.4 s, 10.6±0.7 s and 14.1±2.0 s, respectively, which indicated that PAE's wetting period was shorter than the other three surfactants, so PAE was selected to be the surfactant in the fungal preparation.

Dispersants in the microbial preparation remarkably improve the thermodynamic stability. The results indicated that the samples with 5% of CMC-Na or PEG 6000 were the best in dispersion and ones with CMC, PAE and NPEO followed with obvious sediments or floccules. Hence, CMC-Na and PEG 6000 were recommended to be the available dispersants. The results from **Table 2** showed that 9% or 11% of CMC-Na or PEG 6000 revealed more excellent performance in dispersion with no significant differences.

Based on CCRD, RSM was employed to investigate the interactive effects of significant variables including CMC-Na ($X_1$), PEG 6000 ($X_2$) and PAE ($X_3$) on dispersion performance. The design matrix and experimental responses for ingredient formulation are shown in **Table 3**. Subsequently, the data from **Table 3** were assessed by response surface regression procedure of SAS software package, and also the results of the quadratic polynomial model fitting in the term of ANOVA were shown in **Table 4**. By applying the multiple regression analysis on the experimental data, the following quadratic polynomial model equation (**Eq. (3)**) was delivered to explain the dispersion ($Y$) of

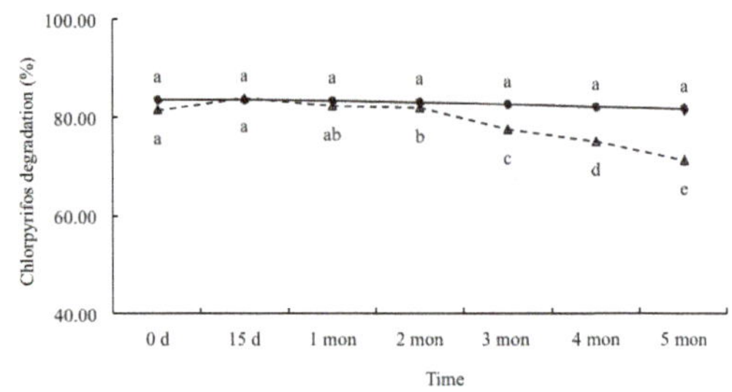


**Figure 3. Chlorpyrifos degradation trends reflecting the stability of fungal preparation in normal and low temperature for different periods.** Note: The data presented are means of three replicates with standard deviation, which is within 5% of the mean. Different letters indicate significant differences ($p<0.01$, LSD test). ●, chlorpyrifos degradation by fungal preparation stored at 4°C; △, chlorpyrifos degradation by fungal preparation stored in normal temperature.

dried fungal preparation ($p>0.05$):

$$Y = -64.04 + 8.50 \times X_1 + 9.20 \times X_2 + 16.16 \times X_3 - 0.43 \times X_1^2 + 0.043 \times X_1 \times X_2 + 0.057 \times X_1 \times X_3 - 0.51 \times X_2^2 + 0.112187 \times X_2 \times X_3 - 1.58 \times X_3^2 \quad (3)$$

where $Y$ is the predicted dispersity (%) of the fungal preparation; $X_1$, $X_2$ and $X_3$ are the coded values for the CMC-Na, PEG 6000 and PAE, respectively.

The statistical significance of **Eq. (3)** was also evaluated by performing $F$-test and $t$-test (**Table 4 and 5**). The statistical analysis indicated that the model linear term coefficient of $X_1$, $X_2$ and $X_3$ and the quadric term coefficient of $X_1$, $X_2$ and $X_3$ showed significant effects ($p<0.05$) on dispersion. Thus, the quadratic polynomial equation (**Eq. (4)**) was modified to be:

$$Y = -64.04 + 8.50X_1 + 9.20X_2 + 16.16X_3 - 0.43X_1^2 - 0.51X_2^2 - 1.58X_3^2 \quad (4)$$

The adequacy of two models was examined by the determination coefficients ($R^2 = 0.9252$), which suggested that the predicted values of the models were well correlated with the experimental values. The three-dimensional (3D) response surfaces in **Fig. 2** (**a**, **b** and **c**) show the respective effects of $X_1$ and $X_2$, $X_1$ and $X_3$, $X_2$ and $X_3$ on dispersion of fungal preparation while keeping the third variable at 0. The model predicted a maximum dispersion of 70.3% at the stationary point. Accordingly, the optimized formulation was determined by the predict model and previous experiment results to be 11.0% of CMC-Na, 9.0% of PEG 6000, 5.0% of PAE, 2.5% of glycine, 5.0% of fucose, 27.5% of kaolin and 40% of freeze dried fungi.

## Preparation quality inspection

To evaluate the quality of fungal wettable powder, the degradation of chlorpyrifos by optimized preparation was checked first and the results showed that it eliminated 76.1±0.4% of chlorpyrifos at the initial concentration of 25 mg/L in 30 minutes, which can meet the practical requirements. As the amount and concentration of total protein indirectly reflected the content of active protein, the equation (**Eq. (5)**) of the standard curve was obtained ($R^2 = 0.9966$) by Coomassie blue staining method:

**Table 6.** The size, suspensibility and chlorpyrifos degradation of fungal preparation in normal and low temperature storage.

| | Size(%) | Dispersity (%) | chlorpyrifos Degradation (%) |
|---|---|---|---|
| 0 h | $95.4\pm0.6^a$ | $77.8$ n$\pm1.6^b$ | $83.4\pm2.1^c$ |
| 0°C, 1 h | $95.3\pm1.0^a$ | $74.7\pm1.4^b$ | $85.7\pm2.2^c$ |
| 0°C, 7 d | $94.9\pm1.7^a$ | $74.6\pm1.3^b$ | $85.6\pm2.4^c$ |

Note: The data presented are means of three replicates with standard deviation, which is within 5% of the mean. Different letters indicate significant differences ($p<$ 0.05, LSD test). Size: the amount of preparation granules through standard sieve in 325 meshes.

$$y=1964.6x-26.689 \quad (5)$$

where $y$ is corresponding concentration of protein ($mg\cdot L^{-1}$), and $x$ is the value of $OD_{595}$. The value of $R^2$ suggested this model was stable and credible in the range of testing concentrations and the protein amount of freeze-dried fungi powder was 10.6 $mg\cdot g^{-1}$.

The results in **Table 6** indicated that low temperature (0°C) storage rarely affected the dispersing and degrading performances, which was non-significantly different ($p>0.05$) from ones stored in normal temperature. Meanwhile, stored at 4°C in periods of 15 d, 1, 2, 3, 4 and 5 months, the degrading performance was neither degenerated nor significantly different from ones before low temperature storage. Although after 5 months, the degradation rate dropped a little, the value remained over 70%, which could be expected to meet the practical needs (**Fig. 3**). Furthermore, other quality features of the chlorpyrifos-degrading preparation were investigated individually including wettability (<1 min), dispersion (77.8±1.6%), humidity (≤1.8±0.8%), fineness (≥95% of granules sieved through 63 μm) and pH value (6 to 7). All the results indicated that the fungal preparation was physically stable and available for practical use.

## Chlorpyrifos degradation on cabbage by freeze-dried fungal powder preparation

The recoveries of chlorpyrifos on cabbage were obtained by adding chlorpyrifos at concentration of 0.01, 0.05 and 0.10 $mg\cdot kg^{-1}$ by GC and the results were 77.0%, 88.9% and 84.7%, respectively. As RSD ranged from 2.0% to 7.9%, the pre-treatment and detection methods of chlorpyrifos residue on cabbage were reliable and up the pesticide residue detection standards.

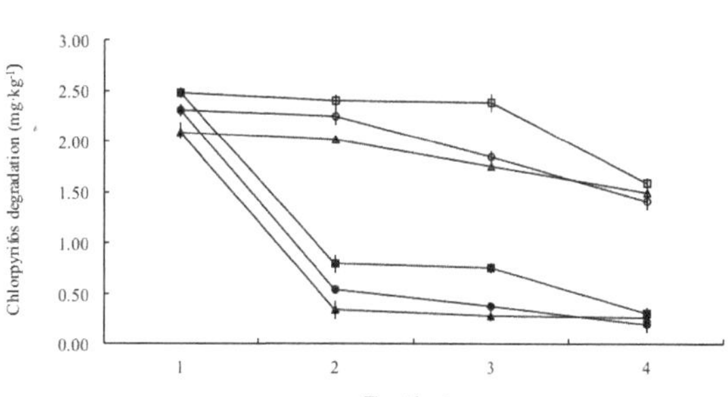


**Figure 4. The trends of degradation of chlorpyrifos by fungal wettable powder on cabbage at different initial concentrations.** The data presented are means of three replicates with standard deviation, which is within 5% of the mean: ■, degradation of chlorpyrifos at initial concentration of 225 $g\cdot ha^{-1}$; ●, degradation of chlorpyrifos at initial concentration of 350 $g\cdot ha^{-1}$; ▲, degradation of chlorpyrifos at initial concentration of 450 $g\cdot ha^{-1}$. Solid legends are treatments with fungal wettable powder while hollow ones are control groups.

From **Fig. 4**, the residual concentrations started at 2.12, 2.31 and 2.45 $mg\cdot kg^{-1}$, respectively in areas treated with 48% chlorpyrifos EC at the concentrations of 225, 350 and 450 $g\cdot ha^{-1}$. After applied with the microbial preparation in 1 d, the residual concentrations remarkably dropped to 0.31, 0.53 and 0.81 $mg\cdot kg^{-1}$, respectively while the residual concentrations of control groups were still over 2.0 $mg\cdot kg^{-1}$. In 3 d, residual concentrations decreased to 0.28, 0.37 and 0.76 $mg\cdot kg^{-1}$ while the control groups were 1.75, 1.85 and 2.38 $mg\cdot kg^{-1}$, respectively. In addition, the degradation rates grew up to nearly 87% within three days. In a period of 7 d, the chlorpyrifos residual concentrations constantly declined to 0.26, 0.21 and 0.29 $mg\cdot kg^{-1}$ while the control groups of 1.50, 1.41 and 1.59 $mg\cdot kg^{-1}$, respectively. Moreover, the degradation rates of each area increased to 87.3%, 91.0% and 88.2%, respectively.

## Discussion

Given the fact that biodegradation is an enzymatic reaction, the microbial activity of pesticide-degrading microorganisms is easily affected by external factors. Consequently, the technical difficulties of preparation design become the main obstacles of practical application and promotion. On the purpose of enhancing degradation activity and availability, the assistant agents, referred to the design of bio-pesticides, were introduced in this study. Assistant agents are all ingredients except the active compounds in microbial preparation and they help remain the extracted enzymes active and available [25,26].

In this study, four categories of assistant agents including protectants, carriers, dispersants and surfactants were compared within group and the best options were determined from the results of four independent tests. Firstly, protectants preserved the fungi cells from cell damage caused by the process of freeze dehydration. The selection of available protectants was conducted with fucose, glycine and glycerol. From the test result, glycine and fucose were chosen out among applied agents. Glycine is a well-known protective compound for preventing tissue injury and enhancing anti-oxidative ability [26,27] while fucose is safe in wound healing and protective against cytotoxicity [28,29]. Secondly, carriers are the basic composites which are required to be high absorptive, fluid and cost-efficient [30]. By the test of physical features, kaolin was the best option for the carrier of this microbial preparation. Then, concerning the actual function of a final preparation, the wettability is vital [31]. This preparation was supposed to be sprayed on the surface of vegetable leaves and it was hard to soak the leaf efficiently because most plant blades are covered with wax which is the main reason of low surface energy on leaf surface. As a result, the surfactants were introduced. Common surfactants are anionic surfactants like ABS and PAE while some are nonionic surfactants like tween. Some natural products could be used as fine surfactants as well, for instance, lignosulfonates and saponin [32]. Through tests, PAE was selected for its shortest wetting time. Lastly, dispersants are regular assistant agents, which significantly decrease the gathering of granules in dispersion system. Generally, smashed preparations are not thermodynamically stable in water, which results in the automatical gathering of dispersed granules, called flocculation, to form a stable thermodynamic system [33]. However, the granules can be efficiently stopped from flocculating by dispersants which keep the active granules suspending in solution and remarkably improve the dispersibility and penetrability. Usually, dispersants are anionic surfactants with polycyclic structure, like lignosulfonates and alkylbenzene sulfonates (ABS). In the present study, CMC-Na and PEG 6000 were chosen out from tested dispersants because of their finest dispersity and compatibility in aqueous solution.

The strategy of response surface methodology (RSM) is commonly used in formula improvement and optimization [16,34–36]. Since the function of fungal preparation is mostly relied on its property of dispersity, this strategy was applied for a better design of chosen surfactant (PAE) and dispersants (CMC-Na and PEG 6000). The mathematical model (**Eq. (4)**) was obtained from the statistics analysis, which could be effectively used to predict and optimize the dispersity of fungal wettable powder preparation within chosen factors. Consequently, the optimum preparation formula was determined to be 11.0% of CMC-Na, 9.0% of PEG 6000, 5.0% of PAE, 2.5% of glycine, 5.0% of fucose, 27.5% of kaolin and 40% of freeze dried fungi. Under this formula, the dispersion rate was above 70.0% and approximately 76.1% of chlorpyrifos was degraded within half an hour in lab condition.

Assessment of preparation quality is necessary. Apparently, the impact of its physical and biological properties on the degradation performance of fungal preparation cannot be underestimated or ignored. Therefore, several features of product quality were checked separately and the results showed that the fungal preparation would meet the practical demands. However, the results still revealed an inevitable fact that microbial preparations are more possibly affected by external factors than the chemosynthetic ones. Moreover, this shortage was also evident in the field tests. Because of the vulnerability of active ingredients, a less efficient degradation performance of this fungal preparation was detected after one day, which can be considered as an indirect consequence of environmental factors on the active enzymes. As a result, these limitations will definitely enlarge the difficulty of its practical application and market promotion.

Undeniably, the results from the field tests showed a degradation of chlorpyrifos on cabbage more than 85% within one day, which indicated that the removal was efficient by this fungal preparation. In addition, it revealed the potential of degradation within a broad range of initial chlorpyrifos concentrations, which can be necessary because of the inappropriate and excessive use of chlorpyrifos in some farming areas in China. During the field experiments, there were no adverse effects observed on cabbage growth, which suggested that the fungal preparation was safe to the plants and can be an option for large-scale use in crops treated with chlorpyrifos. Lastly, as chlorpyrifos residues were constantly reduced, it was suggested that even though the efficiency was weakened after one day, this fungal preparation showed an important feature of persistently degrading the target compound.

In conclusion, the results obtained in the present study indicated that freeze-dried fungal wettable powder preparation of *C. cladosporioides* Hu-01 efficiently reduced the residues of chlorpyrifos on cabbage, which can be used as an effective and promising approach for chlorpyrifos bioremediation on vegetables. Besides, this study also provided a safe and cost-effective procedure for microbial preparation, which can be regarded as a potential strategy for better utilization of active microorganisms in bioremediation.

## Author Contributions

Conceived and designed the experiments: JL YH MH GZ. Performed the experiments: JL YH SC YX. Analyzed the data: JL YH. Contributed reagents/materials/analysis tools: JL YH MH GZ. Wrote the paper: JL SC GZ.

## References


1. Racke KD (1994) Environmental fate of chlorpyrifos. Rev Environ Contam Toxicol 131: 1–150.
2. Amitai G, Moorad D, Adani R, Doctor BP (1998) Inhibition of acetylcholinesterase and butyrylcholinesterase by chlorpyrifos-oxon. Biochem Pharmacol 56: 293–299.
3. Sandahl JF, Baldwin DH, Jenkins JJ, Scholz NL (2005) Comparative thresholds for acetylcholinesterase inhibition and behavioral impairment in coho salmon exposed to chlorpyrifos. Environ Toxicol Chem 24: 136–145.
4. Eaton DL, Daroff RB, Autrup H, Bridges J, Buffler P, etal. (2008) Review of the toxicology of chlorpyrifos with an emphasis on human exposure and neurodevelopment. Crit Rev Toxicol 382: 1–125.
5. Haviland JA, Butz DE, Porter WP (2010) Long-term sex selective hormonal and behavior alterations in mice exposed to low doses of chlorpyrifos in utero. Reprod Toxicol 29: 74–79. dio: 10.1016/j.reprotox.2009.10.008.
6. Kavitha P, Rao JV (2008) Toxic effects of chlorpyrifos on antioxidant enzymes and target enzyme acetylcholinesterase interaction in mosquito fish, *Gambusia affinis*. Environ Toxicol Pharmacol 26: 192–198.
7. Shafiq-Ur-Rehman RS, Waliullah MIS (2012) Chlorpyrifos-induced neuro-oxidative damage in bee. Toxicol Environ Health Sci 4: 30–36.
8. Betancourt AM, Burgess SC, Carr RL (2006) Effect of developmental exposure to chlorpyrifos on the expression of neurotrophin growth factors and cell-specific markers in neonatal rat brain. Toxicol Sci 92: 500–506.
9. Bicker W, Lammerhofer M, Genser D, Kiss H, Lindner W (2005) A case study of acute human chlorpyrifos poisoning: novel aspects on metabolism and toxicokinetics derived from liquid chromatography-tandem mass spectrometry analysis of urine samples. Toxicol Lett 159: 235–251.
10. Sandhu MA, Saeed AA, Khilji MS, Ahmed A, Latif MSZ, etal. (2013) Genotoxicity evaluation of chlorpyrifos: a gender related approach in regular toxicity testing. J Toxicol Sci 38: 237–244.
11. Chen S, Hu M, Liu J, Zhong G, Yang L, etal. (2011) Biodegradation of beta-cypermethrin and 3-phenoxybenzoic acid by a novel *Ochrobactrum lupini* DG-S-01. J Hazard Mater 187: 433–440.
12. Chen S, Chang C, Deng Y, An S, Dong YH, etal. (2014) Fenpropathrin biodegradation pathway in *Bacillus* sp. DG-02 and its potentials for bioremediation of pyrethroid-contaminated soils. J Agric Food Chem 62: 2147–2157.
13. Chen S, Yang L, Hu M, Liu J (2011) Biodegradation of fenvalerate and 3-phenoxybenzoic acid by a novel *Stenotrophomonas* sp. strain ZS-S-01 and its use in bioremediation of contaminated soils. Appl Microbiol Biotechnol 90: 755–767.
14. Cycoń M, Wojcik M, Piotrowska-Seget Z (2011) Biodegradation kinetics of the benzimidazole fungicide thiophanate-methyl by bacteria isolated from loamy sand soil. Biodegradation 22: 573–583.
15. Arora PK, Sasikala C, Ramana CV (2012) Degradation of chlorinated nitroaromatic compounds. Appl Microbiol Biotechnol 93: 2265–2277.
16. Chen S, Liu C, Peng C, Liu H, Hu M, etal. (2012) Biodegradation of chlorpyrifos and its hydrolysis product 3,5,6-trichloro-2-pyridinol by a new fungal strain *Cladosporium cladosporioides* Hu-01. Plos One 7: e47205.
17. Lu P, Li Q, Liu H, Feng Z, Yan X, etal. (2013) Biodegradation of chlorpyrifos and 3,5,6-trichloro-2-pyridinol by *Cupriavidus* sp. DT-1. Bioresour Technol 127: 337–342.
18. Yu YL, Fang H, Wang X, Wu XM, Shan M, etal. (2006) Characterization of a fungal strain capable of degrading chlorpyrifos and its use in detoxification of the insecticide on vegetables. Biodegradation 17: 487–494.
19. Park J, Lee S, Lee J (2013) Sampling and selection factors that enhance the diversity of microbial collections: application to biopesticide development. Plant Pathol J 29: 144–153.

20. Carpio C, Dangles O, Dupas S, Lery X, Lopez-Ferber M, etal. (2013) Development of a viral biopesticide for the control of the Guatemala potato tuber moth *Tecia solanivora*. J Invertebr Pathol 112: 184–191.
21. Copping LG, Menn J (2000) Biopesticides: a review of their action, applications and efficacy. Pest Manag Sci 56: 651–676.
22. Gao Y, Chen S, Hu M, Hu Q, Luo J, etal. (2012) Purification and characterization of a novel chlorpyrifos hydrolase from *Cladosporium cladosporioides* Hu-01. Plos One 7: e381376.
23. Zhang C, Wang S, Yan Y (2011) Isomerization and biodegradation of beta-cypermethrin by *Pseudomonas aeruginosa* CH7 with biosurfactant production. Bioresour Technol 102: 7139–7146.
24. Chen S, Dong YH, Chang C, Deng Y, Zhang XF, etal. (2013) Characterization of a novel cyfluthrin-degrading bacterial strain *Brevibacterium aureum* and its biochemical degradation pathway. Bioresour Technol 132: 16–23.
25. Suciu NA, Ferrari T, Ferrari F, Trevisan M, Capre E (2012) Pesticide removal from waste spray-tank water by organoclay adsorption after field application: an approach for a formulation of cyprodinil containing antifoaming/defoaming agents. Environ Sci Pollut Res 19: 1229–1236.
26. Wang W, Wu Z, Dai Z, Yang Y, Wang J, etal. (2013) Glycine metabolism in animals and humans: implications for nutrition and health. Amino Acids 45: 463–477.
27. Hamburger T, Broecker-Preuss M, Hartmann M, Schade FU, de Groot H, etal. (2013) Effects of glycine, pyruvate, resveratrol, and nitrite on tissue injury and cytokine response in endotoxemic rats. J Surg Res 183: E7–E21.
28. Yamaki K, Goto M, Takano-Ishikawa Y (2009) Inhibitory effects of fucose-related sugar compounds on oxidised low-density lipoprotein uptake in macrophage cell line J774.1. Food Agric Immunol 20: 355–362.
29. Peterszeg G, Robert AM, Robert L (2003) Protection by L-fucose and fucose-rich polysaccharides against ROS-produced cell death in presence of ascorbate. Biomed Pharmacother 57: 130–133.
30. Shukla PG, Kalidhass B, Shah A, Palaskar DV (2002) Preparation and characterization of microcapsules of water-soluble pesticide monocrotophos using polyurethane as carrier material. J Microencapsul 19: 293–304.
31. Zhang G, Zhan J, Li H (2011) Selective binding of carbamate pesticides by self-assembled monolayers of calixarene lipoic acid: wettability and impedance dual-signal response. Org Lett 13: 3392–3395.
32. Matsushita Y, Yasuda S (2005) Preparation and evaluation of lignosulfonates as a dispersant for gypsum paste from acid hydrolysis lignin. Bioresour Technol 96: 465–470.
33. Dinner AR, Sali A, Smith LJ, Dobson CM, Karplus M (2000) Understanding protein folding via free-energy surfaces from theory and experiment. Trends Biochem Sci 25: 331–339.
34. Chen S, Hu W, Xiao Y, Deng Y, Jia J, etal. (2012) Degradation of 3-phenoxybenzoic acid by a *Bacillus* sp. Plos One 7: e50456.
35. Chen S, Luo J, Hu M, Lai K, Geng P, etal. (2012) Enhancement of cypermethrin degradation by a coculture of *Bacillus cereus* ZH-3 and *Streptomyces aureus* HP-S-01. Bioresour Technol 110: 97–104.
36. Chen S, Luo J, Hu M, Geng P, Zhang Y (2012) Microbial detoxification of bifenthrin by a novel yeast and its potential for contaminated soils treatment. Plos One 7: e30862.

# 14

# Actions of Agonists, Fipronil and Ivermectin on the Predominant *In Vivo* Splice and Edit Variant ($RDL_{bd}$, I/V) of the *Drosophila* GABA Receptor Expressed in *Xenopus laevis* Oocytes

**Kristin Lees[1]**, Maria Musgaard[2]**, Siros Suwanmanee[2], Steven David Buckingham[3], Philip Biggin[2], David Sattelle[3]***

1 Faculty of Life Sciences, University of Manchester, Manchester, United Kingdom, 2 Department of Biochemistry, University of Oxford, Oxford, United Kingdom, 3 Wolfson Institute for Biomedical Research, Department of Medicine, University College London, London, United Kingdom

## Abstract

Ionotropic GABA receptors are the targets for several classes of insecticides. One of the most widely-studied insect GABA receptors is RDL (resistance to dieldrin), originally isolated from *Drosophila melanogaster*. RDL undergoes alternative splicing and RNA editing, which influence the potency of GABA. Most work has focussed on minority isoforms. Here, we report the first characterisation of the predominant native splice variant and RNA edit, combining functional characterisation with molecular modelling of the agonist-binding region. The relative order of agonist potency is GABA> muscimol> TACA> β-alanine. The I/V edit does not alter the potency of GABA compared to $RDL_{bd}$. Docking calculations suggest that these agonists bind and activate $RDL_{bd}$I/V through a similar binding mode. TACA and β-alanine are predicted to bind with lower affinity than GABA, potentially explaining their lower potency, whereas the lower potency of muscimol and isoguvacine cannot be explained structurally from the docking calculations. The $A^{301}S$ (resistance to dieldrin) mutation reduced the potency of antagonists picrotoxin, fipronil and pyrafluprole but the I/V edit had no measurable effect. Ivermectin suppressed responses to GABA of $RDL_{bd}$I/V, $RDL_{bd}$ and RDLbdI/VA301S. The dieldrin resistant variant also showed reduced sensitivity to Ivermectin. This study of a highly abundant insect GABA receptor isoform will help the design of new insecticides.

**Editor:** Subhabrata Sanyal, Biogen Idec, United States of America

**Funding:** This work was supported by Merial Ltd, 3239 Satellite Blvd, Duluth, GA USA 30096 via a post-doctoral grant to KL with DBS. PCB and MM acknowledge support from the Leverhulme Trust (RPG-059). The funders had no role in study design, data collection and analysis, decision to publish, or preparation of the manuscript.

**Competing Interests:** This work was supported by Merial Ltd, via a post-doctoral grant to DBS supporting KL. There are no patents, products in development or marketed products to declare.

* E-mail: d.sattelle@ucl.ac.uk

These authors contributed equally to this work.

## Introduction

GABA-gated chloride channels ($GABA_ARs$) are of considerable interest in neuroscience, medicine and drug discovery. This is due to several factors. First, they are widely distributed throughout the nervous systems of vertebrates [1,2] and invertebrates [3–5]. Secondly, they are the targets of a large number of diverse drugs, including anxiolytic and anticonvulsant benzodiazepines [6] as well as anaesthetics such as propofol [7] and steroids [8]. Thirdly, they are targets of several classes of commercially-important insecticides, such as cyclodienes (including dieldrin) and fipronil [4], while in nematodes they are targeted by piperazine [9]. A better understanding of the molecular basis of $GABA_AR$ function will therefore impact strongly on several areas of neuroscience.

The major $GABA_AR$ of the fruit-fly, *Drosophila melanogaster*, is the RDL (resistance to dieldrin) subunit [10,11], which has played an important strategic role in our understanding of how ligands regulate $GABA_ARs$. Not only is it from a model organism with the most extensive genetic toolkit, but the receptor (unlike nearly all vertebrate $GABA_ARs$) readily expresses robust, functional homo-oligomers, assisting the interpretation of structure-function studies and molecular simulation studies aimed at understanding how ligands interact with the receptor. RDL was the first invertebrate GABA receptor to be cloned and functionally expressed, initially from the fruit fly [10,11], and subsequently from many other insects [12–14]. In most insects, cyclodiene resistance is conferred by an alanine-to-serine substitution, but an alanine-to-glycine replacement has also been observed [15]. Identical resistance alleles have been found in geographically distant regions pointing towards multiple origins of resistance [16]. This RDL $A^{301}S$ mutation results in high levels (>4000-fold) of resistance to the cyclodiene dieldrin [17] and the structurally-related plant compound picrotoxin, a mixture of the two compounds picrotoxinin and picrotin.

More importantly, *Drosophila* RDL has also been intensively studied as a model of the effects of mRNA editing and splicing – major sources of phenotypic diversity outside the genome. However, studies to date suffer from a major weakness, in that they are performed only on rare edit/splice isoforms [18]. This paper aims to address this problem.

Alternative splicing in two locations of RDL yields two variants each for exons 3 (variants a and b) and 6 (variants c and d). All four possible splice isoforms are transcribed [10]. This alternative splicing affects agonist potency [19]. In addition, *Rdl* undergoes A-to-I RNA editing [20] which leads to the substitution of four amino acid residues in functionally significant regions: R122G (extracellular N-terminal), I283V (TM1), N294D (intracellular TM1–TM2 loop) and M360V (intracellular TM3–TM4 loop). The combination of alternative splicing and RNA editing generates considerable variation in the physiology and pharmacology of RDL. For example, Es-Salah et al. [21] showed that the edit site R122G, located between ligand binding loops D and A, affects agonist potency and reduces sensitivity to fipronil. Furthermore, it is interesting to note that Fisher [22] found that the large intracellular loop between TM3 and TM4, which contains the site of the M360V edit, influences GABAR desensitisation kinetics and GABA sensitivity.

In a comprehensive analysis of the impact of RNA A-to-I editing of *Drosophila* RDL in different splice backgrounds, Jones et al. [18] showed that alternative splicing and RNA editing have a combined effect on GABA potency and the extent of RNA editing varies between splice variants and also depends on the developmental stage. The bd isoform was the predominant splice variant with transcript levels 26 times greater than the ad form (the second most abundant variant) and it was the only splice variant to be edited at all four sites. Of the four edits, I283V was the most abundant across all life stages. However, despite being the commonest variant and therefore the most abundant form of RDL exposed to insecticides, its physiology and pharmacology are not well studied. Prior to appreciation of the potential impact of editing and splicing, most studies used the $RDL_{ac}$ variant which has been shown to be the least abundantly expressed of the four splice variants. Here we provide the first detailed characterisation of heterologously expressed $RDL_{bd}$I/V, the most abundant native form of the *Drosophila* RDL GABA receptor.

## Materials and Methods

The $RDL_{bd}$ splice variant with and without the I283V edit [18] was the starting point for site-directed mutagenesis. The Quik Change Site-Directed Mutagenesis Kit (Stratagene) was used to introduce an alanine to serine mutation (*gcgctc* to *tcgctc*) into the $Rdl_{bd}$I283V ($Rdl_{bd}$I/V) variant to produce the dieldrin-resistant mutations $Rdl_{bd}$I283V$A^{301}$S ($Rdl_{bd}$I/V$A^{301}$S). To help the reader differentiate between a natural editing site and a mutation we have put the mutation in superscript. The sequences of the forward and reverse primers were:

5′ CAACGCCGGCGCGTGTG**T**CGCTCGGTGTGA-CAACC 3′ (sense)

5′ GGTTGTCACACCGAGCG**A**CA-CACGCGCCGGCGTTG 3′ (antisense)

(altered nucleotide in bold). The presence of the mutation was confirmed by sequencing. Templates were linearised with *Xba*I, and capped RNA was synthesised using the T7 mMessage Machine Kit (Ambion).

### Two-electrode Voltage-clamp Electrophysiology

Stage V-VI oocytes from *Xenopus laevis* were purchased from the European Xenopus Resource Centre at the University of Portsmouth and defolliculated manually after a 1 h incubation with collagenase type 1A (2 mg/ml) (Sigma) and injected with 50 ng (at 1 ng/nl) cRNA encoding $RDL_{bd}$, $RDL_{bd}$I/V or $RDL_{bd}$I/V$A^{301}$S. Oocytes were incubated in filter-sterilized, standard oocyte saline (SOS) consisting of: 100 mM NaCl, 2 mM KCl, 1.8 mM $CaCl_2$, 1 mM $MgCl_2$, 5 mM HEPES (pH 7.6) and supplemented with 100 U/ml penicillin, 100 μg/ml streptomycin, 50 μg/ml gentamicin and 2.5 mM sodium pyruvate at 18°C. Currents were recorded 24–48 h post-injection. Oocytes from at least 3 animals were used for each experiment.

Xenopus oocytes were voltage clamped at −100 mV under 2-electrode voltage clamp. Only oocytes which yielded stable responses to at least three control concentrations of GABA applied at 2 min intervals were used. Agonists were obtained from Sigma (UK) and dissolved in SOS. Concentration-response curves were constructed by challenging the oocyte to increasing concentrations of agonist for 5 s at a flow-rate of 4 ml/min with at least 2 min between challenges to reduce the effects of desensitisation. Responses were normalised to a maximal GABA-evoked response.

Picrotoxin (PTX) (Sigma, UK), ivermectin (IVM) (Sigma, UK), fipronil and pyrafluprole (gifts from Dr Lance Hammerland, Merial Animal Health) were first dissolved in dimethylsulphoxide (DMSO) and then diluted in SOS to the required concentrations. Responses to GABA were not affected by DMSO at its final concentration (1% v/v). To determine inhibitory activity, the antagonists were added after three successful control applications of 1 mM GABA. Oocytes were pre-treated with the antagonists for 3 min followed by co-application of antagonist and GABA for 5 s.

### Data Analysis

Data are presented as mean ± SEM of three to five independent experiments. Agonist oocyte data were normalised to the maximal current obtained from that oocyte and non-linear regression was used to fit the concentration-response data for each oocyte obtain the $pEC_{50}$ values, fitting the normalised data to a 3-parameter sigmoidal function:

$$I = \frac{I_{\max}}{\left(I_{\max} + e^{\frac{-(x-EC_{50})}{b}}\right)}$$

where $I$ is the normalized current, $I_{max}$ is the maximum current evoked by that compound, $x$ is the concentration, $EC_{50}$ is the concentration that evokes half the maximum current, and $b$ is the slope. To obtain the concentration-inhibition relationship, the response after co-application of agonist and antagonist was normalised to the control response against GABA and analysed to obtain the $IC_{50}$ value. Statistical tests were performed using a 1-way ANOVA on the $EC_{50}$s with post-hoc *t*-tests with a significance level of $P<0.05$.

### Homology Modelling

Models of $RDL_{bd}$ with I283V and additionally R122G and N294D were constructed using MODELLER 9.9 [23] based on the structure of the *Caenorhabditis elegans* GluClα glutamate-gated chloride channel [24]. The sequence alignment was generated using ClustalW [25,26]. The best models were selected based on the MODELLER objective function, the DOPE score and PROCHECK Ramachandran plot statistics. R122G and

N294D are both located more than 20 Å from the agonist binding site and will thus presumably have no influence on the binding models.

### Agonist Docking

The five agonists, GABA, TACA, β-alanine, isoguvacine and muscimol, were constructed in Maestro version 9.3 (Schrödinger, LLC, New York, NY, 2012 academic version). All agonists were docked in their zwitterionic form in which they are expected to be found at physiological pH [27–30]. Since Autodock Vina does not account for potential flexibility in ring structures, five different ring conformations for isoguvacine were generated in Maestro.

Protein and ligands were prepared for docking using AutoDock Tools [31] and docking calculations were performed with AutoDock Vina [32]. The centre of the search box was defined from the centre of mass of E204 (chain E) and R111 (chain A), and the size of the box was $17\times17\times20$ Å$^3$. A maximum of 20 generated binding models for each input structure was requested with a maximum energy difference between the best and worst binding models of 5 kcal/mol. The sidechains of E204 (chain E) and R111 (chain A), in each end of the binding pocket, were treated as flexible in the ligand docking.

### Statistical Analyses

All numerical data are presented as mean ± standard error of the mean. Statistical tests of significance are given in the text.

### Ethics

Since no experiments were conducted on living animals or on humans, ethics approval for this study is not required.

## Results

### Agonist Profile – functional and Modelling Studies

Oocytes expressing $RDL_{bd}$, $RDL_{bd}I/V$ or $RDL_{bd}I/VA^{301}S$ produced concentration-dependent inward currents in response to bath-applied GABA over the range $10^{-6}$ to $10^{-3}$ M (Fig. 1A). We could not obtain saturation of responses to muscimol, isoguvacine or TACA within the limits of solubility (see Fig. 1B). Nonetheless, fits to the curves allowed the $EC_{50}$s to be estimated and there is no statistically significant difference ($F(2,95)=1.3$, $P=0.27$) for GABA $pEC_{50}$s between the three $RDL_{bd}$ subtypes ($RDL_{bd}$: $4.24\pm0.09$, $RDL_{bd}I/V$: $4.34\pm0.07$, and $RDL_{bd}I/VA^{301}S$:$4.34\pm0.06$). Maximum current amplitudes were recorded above ~1 mM GABA.

Five GABA analogues (GABA, TACA, β-alanine, muscimol and isoguvacine) were tested for agonist activity on $RDL_{bd}I/V$ (Fig. 1). Muscimol, TACA and β-alanine induced concentration-dependent inward currents. Responses to isoguvacine did not saturate at 3 mM and produced non-specific responses at 10 mM in uninjected oocytes. Responses to TACA, muscimol and β-alanine also failed to saturate at 6 mM, 3 mM and 10 mM respectively, and were insoluble at higher concentrations. It is therefore not clear where the true saturation concentration is, so estimates of $pEC_{50}$ are therefore less trustworthy. With the above caveat regarding the failure to reach saturation, fits to the concentration/response curves suggest that the relative order of agonist potency of the $RDL_{bd}I/V$ isoforms is as follows ($pEC_{50}$s in parentheses): GABA ($4.42\pm0.05$)>muscimol ($4.02\pm0.04$) > TACA ($3.58\pm0.04$)>β-alanine ($2.51\pm0.03$). A 1-way ANOVA revealed that these differences are statistically significant ($F(4,255)=274$, $P<0.0001$) and post-hoc unpaired *t*-tests indicated that $pEC_{50}$s of muscimol ($P<0.001$), TACA ($P=0.001$), and β-

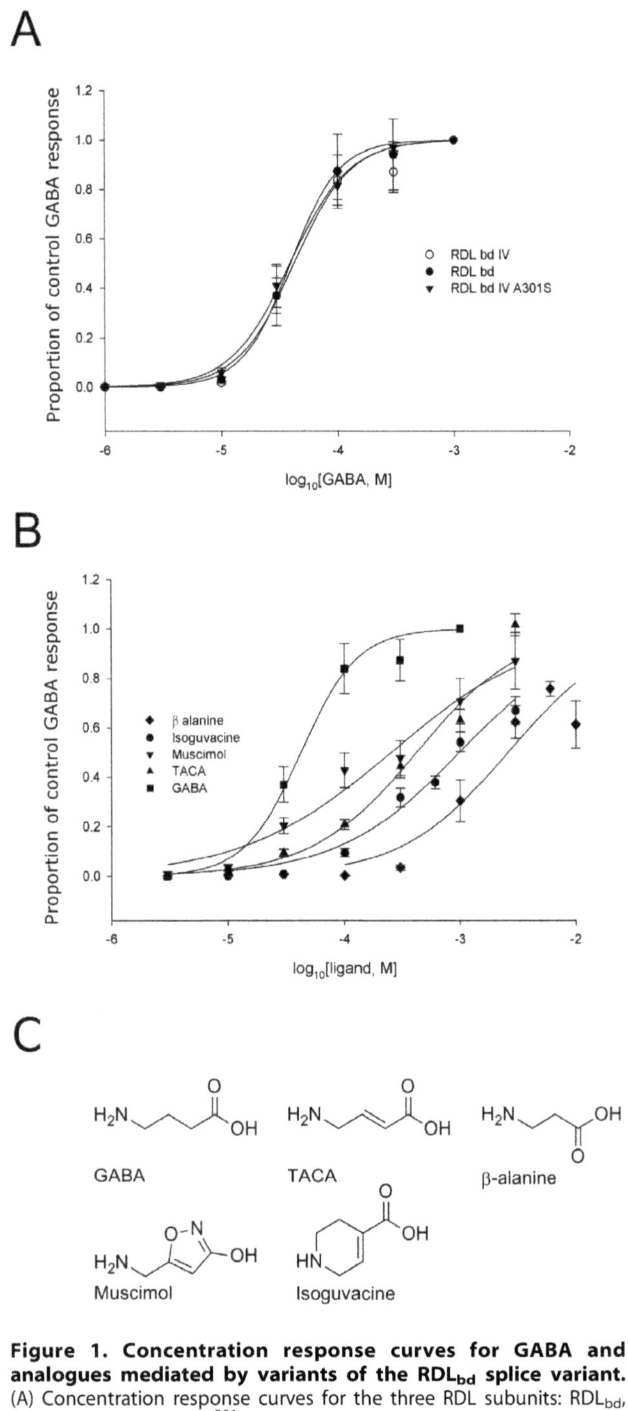


**Figure 1. Concentration response curves for GABA and analogues mediated by variants of the $RDL_{bd}$ splice variant.** (A) Concentration response curves for the three RDL subunits: $RDL_{bd}$, $RDL_{bd}IV$ and $RDL_{bd}IVA^{301}S$ to GABA (structure inserted). Data were normalised to 1 mM GABA. Each data point is the mean ± S.E.M from four to six oocytes from at least three separate frogs. Representative responses to 100 μM GABA are shown for each variant. Scale bars indicate 0.5 μA, 1 s. (B) Concentration response curves showing responses elicited by GABA and analogues at $RDL_{bd}IV$ receptors in Xenopus oocytes. GABA is the most potent agonist, and the rank order is GABA> muscimol> TACA>>β-alanine. Isoguvacine did not reach saturation. Data were normalised to the response to 1 mM GABA. Each data point represents mean±S.E.M from five to eight oocytes. (C) The chemical structures of the agonists tested in the present study: GABA (γ-amino butyric acid), TACA (trans-4-aminocrotonic acid), β-alanine, muscimol and isoguvacine.

alanine ($P = 0.0012$) were all significantly different to that of GABA.

We next set out to establish whether molecular docking simulations might explain the agonist profile of this receptor. TACA and isoguvacine share the GABA skeleton with an amine group and a carboxylic acid group connected by three carbon atoms (excluding the carbon atom in the carboxylic acid group). Muscimol contains a similar skeleton with a hydroxyl group instead of the carboxylic acid group. β-alanine has both the amine group and the carboxylic acid group but only two connecting carbon atoms. These structural similarities (See Fig. 1C) suggest that the five agonists might share a similar binding mode with overlapping positions of the amine and the acid except for β-alanine, which is possibly too short to span the distance required between the two groups.

Earlier modelling studies for the $GABA_C$ receptor predict that GABA in this receptor forms salt bridges to R104 (R111 in $RDL_{bd}$) and to E194 and E196 (E202 and E204 in $RDL_{bd}$) [33,34]. For RDL, F206 and Y254 are believed to form cation-π interactions with GABA whereas F146 does not seem to contribute to cation-π interactions [35].

To explore whether a similar binding mode for the different ligands could be identified, the five agonists were docked into the agonist binding site in the extracellular domain of a homology model of the $RDL_{bd}$ receptor. The residues R111 and E204 were used to define the binding pocket (Fig. 2). For each ligand, a variety of binding modes was generated, both inside and outside the binding site. The analysis was focused on binding modes inside the binding site, where the agonists would possibly make interactions with R111, E202 or E204, F206 and/or Y254 as suggested by previous studies for GABA.

The two best-scoring binding models of GABA (both with a predicted affinity of −5.0 kcal/mol; third-best affinity is −4.7 kcal/mol) are oriented in the same way. They illustrate a binding mode in which the carboxylate of GABA interacts with R111 through hydrogen bonds and ionic interactions, and where the amine interacts through hydrogen bonds and ionic interactions with E204 and forms hydrogen bonds to backbone oxygen atoms on S205 and F206 (Fig. 2B). In the best-scoring model, Y254 is best positioned for cation-π interactions, but also the side chains of F206 and F146 are within 4.5 Å of the positively charged amine group. This is the best scoring binding model, and additionally it fits well with earlier discussed data regarding interactions with R111, E204, Y254 and partly F206. Hence, this model is chosen as a model for GABA binding in $RDL_{bd}$.

For TACA, which is very similar to GABA except for the presence of a double bond in the carbon chain connecting the carboxylate and the amine, the three generated binding modes with the highest affinity are positioned very similar to the chosen binding model for GABA (Fig. 2C). These three binding models with the carboxylate pointing towards R111 and the amine towards E204, and thus partly sandwiched between F206 and Y254, all have predicted affinities of −4.7 kcal/mol. The fourth-best model has an affinity of −4.0 kcal/mol. The range of interactions possible for this binding mode is more or less the same as for GABA itself, except that the TACA carboxylate group additionally forms a hydrogen bond to the S176 hydroxyl group.

Despite the lack of a carbon atom in the skeleton of β-alanine relative to GABA and TACA, the three binding models for β-alanine with the highest affinities are positioned similar to the chosen binding model for GABA (Fig. 3D). The amine group superimposes with the one in GABA, and the R111 sidechain is rearranged to allow interaction with the carboxylate of the smaller β-alanine molecule. A hydrogen bond to S176 also seems likely. The predicted affinities for these three binding modes are −4.3 kcal/mol, −4.2 kcal/mol and −3.9 kcal/mol. The affinity for the fourth best model is −3.6 kcal/mol.

The binding model for muscimol with the largest predicted affinity is overall oriented in the same way as the chosen GABA binding model with the amine pointing towards E204 and the negatively charged oxygen atom pointing towards R111 (Fig. 2E). The amine forms the same interactions as the amine in GABA, while the negatively charged oxygen atom interacts with R111. The predicted affinity for this binding mode is −5.2 kcal/mol while the next best one is −5.1 kcal/mol.

Different starting conformations were used for isoguvacine as AutoDock Vina does not automatically vary the conformation of flexible rings during the docking. Thus, more poses were generated than for the other docked ligands. The best ranked binding models for each conformation can be clustered into two different binding modes, one in which the overall orientation is similar to the GABA binding model (Fig. 2F) and another, more or less rotated 90° around an axis perpendicular to the 6-membered ring, where the carboxylate group is still pointing towards R111, but the amine points directly towards backbone oxygen of F206 (Fig. 2F). Such an overall orientation is also seen for some binding modes for the other ligands, although binding modes with this orientation have a lower predicted affinity compared to modes with the amine pointing towards E204. For isoguvacine, the predicted affinity for the best model from each of the two clusters is −5.5 kcal/mol for the less GABA-like mode and −5.2 kcal/mol for the GABA-like mode. For the GABA-like binding mode, the amine interacts with E204. No other hydrogen bonds from the amine seem possible, and it is not very well positioned for forming cation-π interactions though the sidechains of Y254 and F146 are within 4.5 Å of the amine. The carboxylate interacts with R111 and S176. For the other binding mode (Fig. 2G) which is predicted to have a higher affinity, the carboxylate again interacts with R111, whereas the amine is positioned quite well for cation-π interactions with both Y254 and F206 and also forms a hydrogen bond to the backbone oxygen of F206. Judging from the scores alone, the less GABA-like binding mode (Fig. 2G) appears most physiologically relevant, whereas the GABA-like mode (Fig. 2F) seems attractive as it fits better with the proposed binding modes for the other agonists (as the comparison in Fig. 2H illustrates).

## Actions of Picrotoxin, Fipronil, Pyrafluprole and Invermectin

Given the importance of insect $GABA_ARs$ in pest control, and the paucity of studies on the overwhelmingly predominant isoform of RDL, we next considered the antagonist profile of $RDL_{bd}I/V$. The amplitude of GABA-evoked currents mediated by $RDL_{bd}$, the edited $RDL_{bd}I/V$ and the $A^{301}S$ point mutated $RDL_{bd}I/VA^{301}S$ were reduced by bath-applied PTX in a concentration-dependent manner (Fig. 3), with $pIC_{50}$ values estimated from fitting the dependency of inhibition on PTX concentration to be 6.46±0.18, 6.44±0.13 and 5.67±0.19, respectively. These values were statistically significantly different ($F(2,60) = 5.49$, $P<0.001$). While 1 μM PTX reduced GABA responses mediated by wild-type and $RDL_{bd}I/V$ receptors by more than 50%, the same concentration of PTX reduced responses of the $A^{301}S$ only by about 30% (Fig. 3).

When fipronil was applied alone (1 nM−100 μM) it induced no discernible current (data not shown). At concentrations of 10 nM to 100 μM, it reduced the response of the three receptor variants ($RDL_{bd}$, $RDL_{bd}I/V$, $RDL_{bd}I/VA^{301}S$) to GABA in a concentration-dependent manner (Fig. 4A). The $pIC_{50}$s for fipronil on the 3 isoforms were $RDL_{bd} = 6.48±0.13$; $RDL_{bd}I/V = 6.20±0.44$; $RDL_{bd}I/VA^{301}S = 5.80±0.28$. These values were not significantly

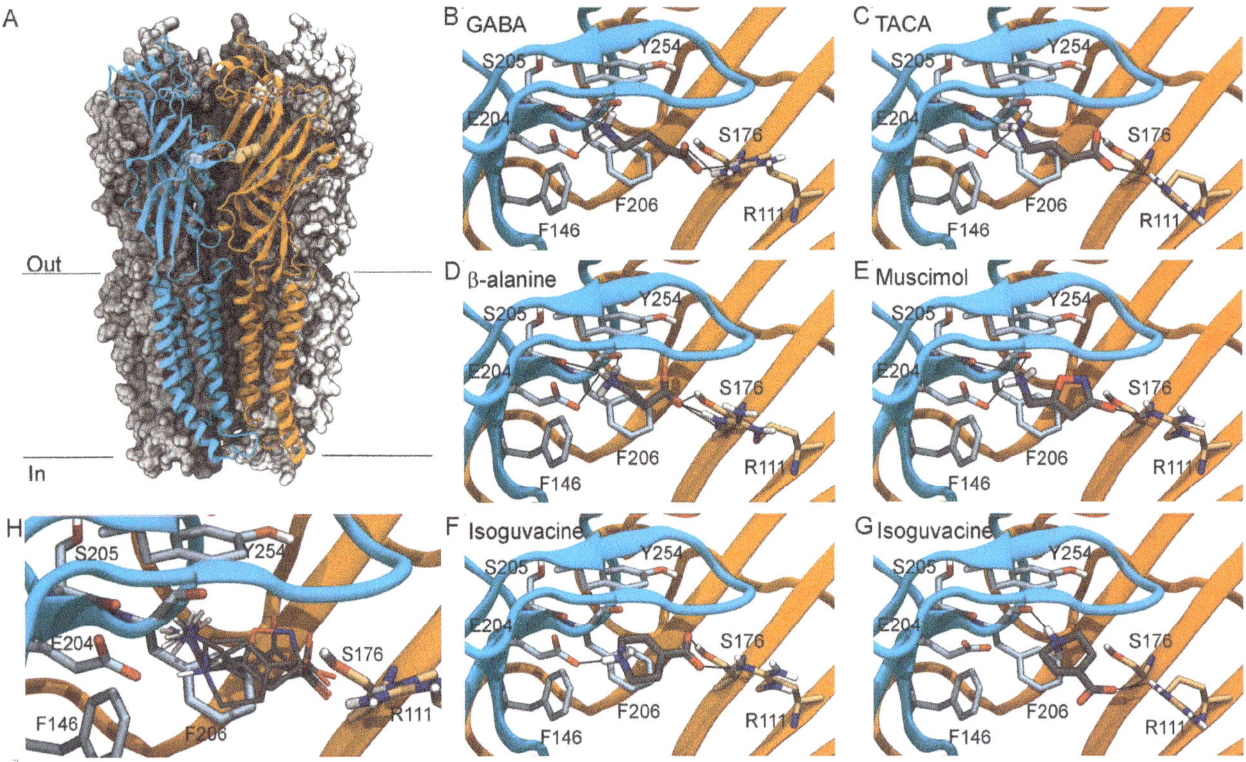


**Figure 2. Binding models for the different agonists.** (A) Overall structure of the homology model of $RDL_{bd}$. Chains A and E are shown in cartoon representation, A in orange and E in cyan. Chains B, C and D are shown in spacefill representation in three different shades of grey. The agonist-binding site is indicated by the sidechains of R111 in chain A (light orange spacefill) and D204 in chain E (light cyan spacefill). (B–G): binding models for the docked agonists. The agonists are shown with grey carbon atoms, and relevant protein sidechains from chain A and chain E are shown with light orange and light blue carbon atoms, respectively. Only polar hydrogen atoms on agonists and sidechains are included. Potential hydrogen bonds (<3 Å) between ligand and protein are marked with black lines. (H) Superposition of the binding models in (B–F). The conformations of R111 and E204 shown are from the GABA binding model.

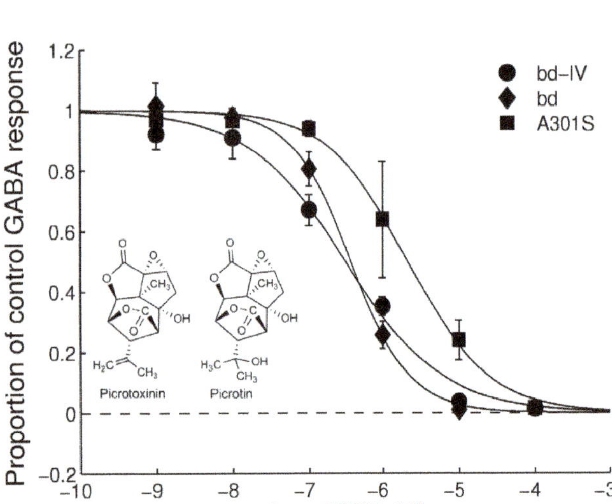


**Figure 3. Concentration-inhibition curves showing responses elicited by PTX to 1 mM GABA in RDL, $RDL_{bd}$IV and $RDL_{bd}$IV/ $A^{301}$S.** Each data point was normalised to the amplitude of the response to 1 mM GABA and represents the mean amplitude of the responses recorded in 7–13 oocytes from at least 3 separate batches. Structures of the two compounds found in picrotoxin (PTX), picrotoxinin and picrotin, are shown.

different ($F(2,55) = 1.20$, $P = 0.31$). Whereas in the wild-type and I283V receptor, 1 μM fipronil effected in a 75% reduction of the response to 50 μM GABA, this was not achieved in the $RDL_{bd}$I/ $VA^{301}$S receptor (50% reduction). Pyrafluprole also reduced the amplitude of GABA responses at concentrations of 10 nM to 100 μM (Fig. 4B). The estimated $pIC_{50}$s were $RDL_{bd} = 6.20 \pm 0.34$; $RDL_{bd}I/V = 7.20 \pm 0.36$; $RDL_{bd}I/VA^{301}S = 6.21 \pm 0.16$. These values were not significantly different one from another ($F(2,27) = 2.46$, $P = 0.10$). All 3 antagonists completely abolished GABA responses in all RDL isoforms at 100 μM.

All three RDL receptors showed no response to 10 μM IVM applied alone (data not shown). However, when co-applied with 100 μM GABA, the three RDL receptor variants showed a reduced response to GABA. When exposed to IVM, all three receptors produced a concentration-dependent suppression of amplitude to GABA (Fig. 5) with the following $pIC_{50}s = RDL_{bd} = 6.28 \pm 0.13$, $RDL_{bd}I/V = 6.33 \pm 0.12$ and $RDL_{bd}I/VA^{301}S = 5.92 \pm 0.09$. These values were significantly different one from another ($F(2,51) = 5.13$, $P < 0.001$).

## Discussion

We have characterised aspects of the pharmacology of the most abundantly expressed isoform of *Drosophila* RDL variant $RDL_{bd}$I/ V, exploring for the first time the agonist profile and the actions of other important ligands including picrotoxin, fipronil and IVM.

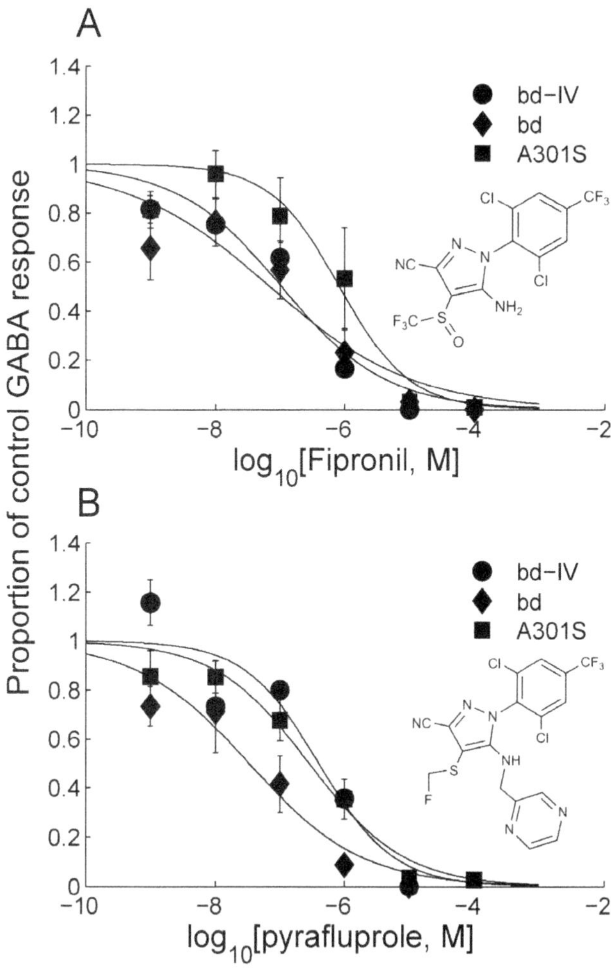


**Figure 4. Concentration-inhibition curves showing responses elicited by the phenylpyrazoles (A) fipronil and (B) pyrafluprole to 1 mM GABA in wild-type, $RDL_{bd}$IV and $RDL_{bd}IV/A^{301}S$.** Each point represents the mean amplitude of the responses recorded in 2–13 oocytes from at least 3 separate batches. Structures of the two phenylpyrazoles are inserted.

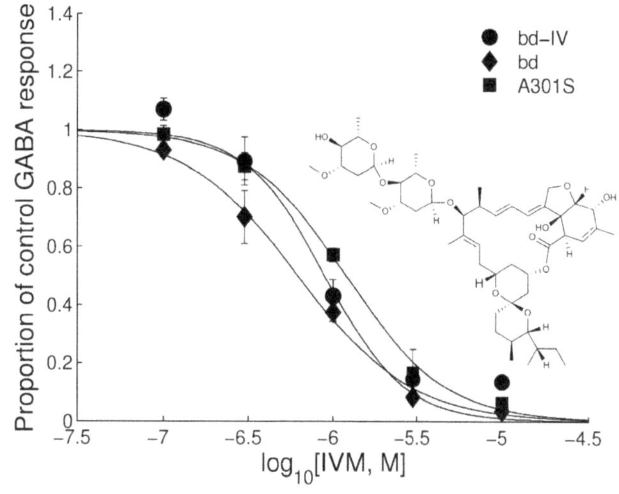


**Figure 5. Effect of ivermectin on $RDL_{bd}$.** Concentration inhibition curves showing responses elicited by IVM to 100 μM GABA in wild-type, $RDL_{bd}$IV and $RDL_{bd}IV/A^{301}S$. Each point represents the mean amplitude of the responses recorded in 2–6 oocytes from at least 3 separate batches. The structure of ivermectin is inserted.

Jones et al. [18] found the $RDL_{bd}$ splice variant to be the most abundant native variant in *D. melanogaster* and also found the single I283V edit to be the most prevalent isoform. I283V is located in the intracellular region close to the transmembrane region of TM1. That region of the subunit may contribute to the pathway of the permeating ion [36]. When compared with the $RDL_{bd}$ wild-type, the addition of this edit had no significant effect on the $EC_{50}$ of GABA. This was also observed by Jones et al. [18]. The RDL $EC_{50}$ value recorded here is very similar to previously reported $EC_{50}$ values of 56.9 μM [21] and 54 μM [18]. Here we found no statistically-significant effect of the addition of the $A^{301}S$ mutation on the GABA $EC_{50}$. This is in contrast to Le Goff et al. [15], who detected a decrease in GABA $EC_{50}$ from 70 μM to 26 μM with the addition of the $A^{301}G$ mutation in $RDL_{bd}R^{122}G$.

*Drosophila* $RDL_{ac}$ and $RDL_{bd}$I/V have a similar order of agonist potency [37], i.e. with GABA as the most potent agonist followed by muscimol, TACA, isoguvacine and lastly β-alanine. However, we did not see differences between the $pEC_{50}$ values for GABA between the 3 different isoforms of RDL that we tested. In contrast, the $EC_{50}$ values of agonists for the insect (Musca) bd type of receptors have been reported to be larger than those of $RDL_{bd}$ [38] and *Dm* $RDL_{ac}$ [39], although care is needed when comparing studies performed by different laboratories under different conditions and on different insect species. The alternative splicing of exon 6 (resulting in exons c or d) is likely to influence agonist potency as it generates diversity in loops C and F that contribute to the ligand-binding site [40].

From our docking calculations it appears likely that the five agonists studied bind to and activate $RDL_{bd}$ through the same overall binding mode as suggested by Fig. 2H. A recent study of the human $GABA_A$ receptor likewise suggests that GABA and muscimol bind with the same overall orientation [41], in a manner similar to our binding models. The lower potency of TACA and β-alanine relative to GABA might be explained by a lower binding affinity as predicted by the docking results (best affinities for GABA, TACA and β-alanine are −5.0 kcal/mol, −4.7 kcal/mol and −4.3 kcal/mol, respectively). The explanation for the lower potency of muscimol and isoguvacine relative to GABA on the other hand cannot be explained by the binding affinities predicted by the model (best affinity for muscimol is −5.2 kcal/mol and for isoguvacine either −5.5 kcal/mol for the less GABA like binding mode or −5.2 for the GABA like binding mode). The ring systems in both of these structures may interfere with some of the structural changes expected for the protein upon activation, however, the study of the dynamics is beyond the scope of this paper.

Insects showing high resistance to dieldrin also show reduced sensitivity to picrotoxin and the phenylpyrazoles, fipronil and pyrafluprole. This is also observed for $RDL_{bd}$I/V, where the $A^{301}S$ mutation causes a 6-fold resistance to picrotoxin, 3-fold resistance to fipronil and 2-fold resistance to pyrafluprole compared to wild-type. Binding of picrotoxin in the pore region close to residues corresponding to A301 has been observed in the crystal structure of the *C. elegans* GluClα glutamate-gated chloride channel [24]. The addition of the I/V edit had no significant effect on the $IC_{50}$s, which might be expected as I283V is located in TM1 and is not in contact with the pore-lining TM2 region where $A^{301}S$ is located.

IVM is known to interact with GABA receptors and several reports suggest that insect RDL and GluCl subunits co-assemble to form IVM-sensitive receptors [38,42,43]. A population of IVM-sensitive receptors is reported in *Drosophila* membranes containing *Dm*GluCl1α but lacking *Dm*RDL [42]. $RDL_{bd}$ shows sensitivity to IVM when investigated as an antagonist. *Musca domestica* $RDL_{bd}$ has also been investigated functionally [38], and the values obtained for its $pEC_{50}$ (4.0) were similar to those obtained for Drosophila $RDL_{bd}$ (4.24). However, the authors did not test for antagonist action of IVM.

It has been shown here that the I/V edit *per se* has no effect on IVM actions on $RDL_{bd}$I/V, but the addition of the $A^{301}S$ mutation resulted in 2-fold resistance. This is similar to observations by Kane et al. [44] who observed that *Drosophila* carrying the $RDLA^{301}S$ mutation were 3.3-fold resistant to IVM compared to the wild-type. Studies on invertebrate ligand-gated chloride channels have identified several point mutations that affect their IVM sensitivity (See [45] for a recent review). Recent papers on different Cys-loop receptors from a variety of species have shown that amino acid residues in the TM2–TM3 loop and in transmembrane regions contribute to high affinity interactions between the receptor and allosteric ligands such as IVM [46–48]. Lynagh and Lynch [49] identified a single mutation, A288G, in the human glycine α1 receptor, increasing IVM sensitivity almost 100-fold while *Drosophila* carrying the P299S mutation in GluCl (located in the TM2–TM3 loop) are ten-fold less sensitive to IVM [44]. Modelling by Hibbs and Gouaux [24] showed that IVM makes contact with the TM2 pore-lining helix and the TM2–TM3 loop. In *C. elegans*, GluClα Ser260 located in TM2 hydrogen bonds with IVM. It is interesting to note that for other Cys-loop receptors, including the glycine and $GABA_A$ receptors, that are directly activated by IVM [50,51], there is also a serine at the equivalent position to S250 in GluClα.

We report the first pharmacological characterization of the predominantly expressed native splice and edit variant of the *Drosophila* GABA receptor ($RDL_{bd}$I/V). The I/V edit does not impact greatly on the potency of GABA or the actions of GABA agonists. TACA and β-alanine are predicted from molecular modelling based on the *C. elegans* GluCl structure to bind with lower affinity than GABA explaining their lower potency in functional studies. The results for muscimol and isoguvacine are less readily explained structurally, and we are forced to speculate that it may relate to the efficiency with which these ligands induce the conformational change associated with the opening of the channel.

The $A^{301}S$ mutation reduced sensitivity to picrotoxin, fipronil and pyrafluprole but the I/V edit had no effect. The I/V edit did not impact on the blocking actions of IVM but the $A^{301}S$ mutation modified IVM $IC_{50}$ values. We speculate that these effects are subtle modulations of the underlying dynamics and thus will require much more detailed exploration than is possible here.

## Author Contributions

Conceived and designed the experiments: PB DS SDB. Performed the experiments: KL MM SS. Analyzed the data: PB SDB. Wrote the paper: KL SDB DS.

## References


1. Miller PS, Smart TG (2010) Binding, activation and modulation of Cys-loop receptors. Trends Pharmacol Sci 31: 161–174.
2. Smart TG (1980) Do motor-nerve terminals have gamma-aminobutyric acid receptors? Br J Pharmacol 71: 279–286.
3. Accardi MV, Beech RN, Forrester SG (2012) Nematode cys-loop GABA receptors: biological function, pharmacology and sites of action for anthelmintics. Invert Neurosci 12: 3–12.
4. Hosie AM, Aronstein K, Sattelle DB, ffrench-Constant RH (1997) Molecular biology of insect neuronal GABA receptors. Trends Neurosci 20: 578–583.
5. Sattelle DB (1990) GABA receptors of insects. Advances in insect physiology 22: 1–113.
6. Gielen MC, Lumb MJ, Smart TG (2012) Benzodiazepines modulate GABAA receptors by regulating the preactivation step after GABA binding. J Neurosci 32: 5707–5715.
7. Trapani G, Latrofa A, Franco M, Altomare C, Sanna E, et al. (1998) Propofol analogues. Synthesis, relationships between structure and affinity at GABAA receptor in rat brain, and differential electrophysiological profile at recombinant human GABAA receptors. J Med Chem 41: 1846–1854.
8. Hosie AM, Buckingham SD, Hamon A, Sattelle DB (2006) Replacement of asparagine with arginine at the extracellular end of the second transmembrane (M2) region of insect GABA receptors increases sensitivity to penicillin G. Invert Neurosci 6: 75–79.
9. Martin RJ (1982) Electrophysiological effects of piperazine and diethylcarbamazine on *Ascaris suum* somatic muscle. Br J Pharmacol 77: 255–265.
10. Ffrench-Constant RH, Rocheleau TA, Steichen JC, Chalmers AE (1993) A point mutation in a Drosophila GABA receptor confers insecticide resistance. Nature 363: 449–451.
11. Ffrench-Constant RH, Mortlock DP, Shaffer CD, MacIntyre RJ, Roush RT (1991) Molecular cloning and transformation of cyclodiene resistance in Drosophila: an invertebrate gamma-aminobutyric acid subtype A receptor locus. Proc Natl Acad Sci U S A 88: 7209–7213.
12. Anthony N, Unruh T, Ganser D, ffrench-Constant R (1998) Duplication of the Rdl GABA receptor subunit gene in an insecticide-resistant aphid, *Myzus persicae*. Mol Gen Genet 260: 165–175.
13. Bass C, Schroeder I, Turberg A, Field LM, Williamson MS (2004) Identification of the Rdl mutation in laboratory and field strains of the cat flea, *Ctenocephalides felis* (Siphonaptera: Pulicidae). Pest Manag Sci 60: 1157–1162.
14. Miyazaki M, Matsumura F, Beeman RW (1995) DNA sequence and site of mutation of the GABA receptor of cyclodiene-resistant red flour beetle, *Tribolium castaneum*. Comp Biochem Physiol B Biochem Mol Biol 111: 399–406.
15. Le Goff G, Hamon A, Berge JB, Amichot M (2005) Resistance to fipronil in *Drosophila simulans*: influence of two point mutations in the RDL GABA receptor subunit. J Neurochem 92: 1295–1305.
16. Andreev D, Kreitman M, Phillips TW, Beeman RW, ffrench-Constant RH (1999) Multiple origins of cyclodiene insecticide resistance in *Tribolium castaneum* (Coleoptera: Tenebrionidae). J Mol Evol 48: 615–624.
17. Ffrench-Constant RH, Roush RT, Mortlock D, Dively GP (1990) Isolation of dieldrin resistance from field populations of *Drosophila melanogaster* (Diptera: Drosophilidae). J Econ Entomol 83: 1733–1737.
18. Jones AK, Buckingham SD, Papadaki M, Yokota M, Sattelle BM, et al. (2009) Splice-variant- and stage-specific RNA editing of the Drosophila GABA receptor modulates agonist potency. J Neurosci 29: 4287–4292.
19. Hosie AM, Buckingham SD, Presnail JK, Sattelle DB (2001) Alternative splicing of a Drosophila GABA receptor subunit gene identifies determinants of agonist potency. Neuroscience 102: 709–714.
20. Hoopengardner B, Bhalla T, Staber C, Reenan R (2003) Nervous system targets of RNA editing identified by comparative genomics. Science 301: 832–836.
21. Es-Salah Z, Lapied B, Le Goff G, Hamon A (2008) RNA editing regulates insect gamma-aminobutyric acid receptor function and insecticide sensitivity. Neuroreport 19: 939–943.
22. Fisher JL (2004) The alpha 1 and alpha 6 subunit subtypes of the mammalian GABA(A) receptor confer distinct channel gating kinetics. J Physiol 561: 433–448.
23. Sali A, Fiser A, Yerkovich B, Eswar N, Marti-Renom M, et al. (1994) Modeller. A program for protein structure modelling by satisfaction of spatial restraints.
24. Hibbs RE, Gouaux E (2011) Principles of activation and permeation in an anion-selective Cys-loop receptor. Nature 474: 54–60.
25. Goujon M, McWilliam H, Li W, Valentin F, Squizzato S, et al. (2010) A new bioinformatics analysis tools framework at EMBL-EBI. Nucleic Acids Res 38: W695–699.
26. Larkin MA, Blackshields G, Brown NP, Chenna R, McGettigan PA, et al. (2007) Clustal W and Clustal X version 2.0. Bioinformatics 23: 2947–2948.
27. Boulanger T, Vercauteren DP, Durant F, Andre JM (1987) 3- and 5-isoxazolol zwitterions: an ab initio molecular orbital study relating to GABA agonism and antagonism. J Theor Biol 127: 479–489.
28. Brehm L, Hjeds H, Krogsgaard-Larsen P (1972) The structure of muscimol, a GABA analogue of restricted conformation. Acta Chem Scand 26: 1298–1299.
29. King MT (2003) Thermodynamics of the reduction of NADP with 2-propanol catalyzed by an NADP-dependent alcohol dehydrogenase. Arch Biochem Biophys 410: 280–286.
30. Krogsgaard-Larsen P, Falch E (1981) GABA agonists. Development and interactions with the GABA receptor complex. Mol Cell Biochem 38 Spec No: 129–146.
31. Morris GM, Huey R, Lindstrom W, Sanner MF, Belew RK, et al. (2009) AutoDock4 and AutoDockTools4: Automated docking with selective receptor flexibility. J Comput Chem 30: 2785–2791.

32. Trott O, Olson AJ (2010) AutoDock Vina: improving the speed and accuracy of docking with a new scoring function, efficient optimization, and multithreading. J Comput Chem 31: 455–461.
33. Melis C, Lummis SC, Molteni C (2008) Molecular dynamics simulations of GABA binding to the GABAC receptor: the role of Arg104. Biophys J 95: 4115–4123.
34. Osolodkin DI, Chupakhin VI, Palyulin VA, Zefirov NS (2009) Molecular modeling of ligand-receptor interactions in GABA C receptor. J Mol Graph Model 27: 813–821.
35. Lummis SC, McGonigle I, Ashby JA, Dougherty DA (2011) Two amino acid residues contribute to a cation-pi binding interaction in the binding site of an insect GABA receptor. J Neurosci 31: 12371–12376.
36. Filippova N, Wotring VE, Weiss DS (2004) Evidence that the TM1–TM2 loop contributes to the rho1 GABA receptor pore. J Biol Chem 279: 20906–20914.
37. Hosie AM, Sattelle DB (1996) Agonist pharmacology of two Drosophila GABA receptor splice variants. Br J Pharmacol 119: 1577–1585.
38. Eguchi Y, Ihara M, Ochi E, Shibata Y, Matsuda K, et al. (2006) Functional characterization of Musca glutamate- and GABA-gated chloride channels expressed independently and coexpressed in Xenopus oocytes. Insect Mol Biol 15: 773–783.
39. McGonigle I, Lummis SC (2010) Molecular characterization of agonists that bind to an insect GABA receptor. Biochemistry 49: 2897–2902.
40. Buckingham SD, Biggin PC, Sattelle BM, Brown LA, Sattelle DB (2005) Insect GABA receptors: splicing, editing, and targeting by antiparasitics and insecticides. Mol Pharmacol 68: 942–951.
41. Bergmann R, Kongsbak K, Sorensen PL, Sander T, Balle T (2013) A unified model of the GABA(A) receptor comprising agonist and benzodiazepine binding sites. PLoS One 8: e52323.
42. Ludmerer SW, Warren VA, Williams BS, Zheng Y, Hunt DC, et al. (2002) Ivermectin and nodulisporic acid receptors in *Drosophila melanogaster* contain both gamma-aminobutyric acid-gated Rdl and glutamate-gated GluCl alpha chloride channel subunits. Biochemistry 41: 6548–6560.
43. Zhao X, Salgado VL (2010) The role of GABA and glutamate receptors in susceptibility and resistance to chloride channel blocker insecticides. Pesticide biochemistry and physiology 97: 153–160.
44. Kane NS, Hirschberg B, Qian S, Hunt D, Thomas B, et al. (2000) Drug-resistant Drosophila indicate glutamate-gated chloride channels are targets for the antiparasitics nodulisporic acid and ivermectin. Proc Natl Acad Sci U S A 97: 13949–13954.
45. Wolstenholme AJ (2012) Glutamate-gated chloride channels. J Biol Chem 287: 40232–40238.
46. Krause RM, Buisson B, Bertrand S, Corringer PJ, Galzi JL, et al. (1998) Ivermectin: a positive allosteric effector of the alpha7 neuronal nicotinic acetylcholine receptor. Mol Pharmacol 53: 283–294.
47. Sattelle DB, Buckingham SD, Akamatsu M, Matsuda K, Pienaar IS, et al. (2009) Comparative pharmacology and computational modelling yield insights into allosteric modulation of human alpha7 nicotinic acetylcholine receptors. Biochem Pharmacol 78: 836–843.
48. Wolstenholme AJ (2010) Recent progress in understanding the interaction between avermectins and ligand-gated ion channels: putting the pests to sleep. Invert Neurosci 10: 5–10.
49. Lynagh T, Lynch JW (2010) An improved ivermectin-activated chloride channel receptor for inhibiting electrical activity in defined neuronal populations. J Biol Chem 285: 14890–14897.
50. Adelsberger H, Lepier A, Dudel J (2000) Activation of rat recombinant alpha(1)beta(2)gamma(2S) GABA(A) receptor by the insecticide ivermectin. Eur J Pharmacol 394: 163–170.
51. Shan Q, Haddrill JL, Lynch JW (2001) Ivermectin, an unconventional agonist of the glycine receptor chloride channel. Journal of Biological Chemistry 276: 12556–12564.

# 15

# Salivary Gland Proteome Analysis Reveals Modulation of Anopheline Unique Proteins in Insensitive Acetylcholinesterase Resistant *Anopheles gambiae* Mosquitoes

**Sylvie Cornelie[1], Marie Rossignol[1], Martial Seveno[2], Edith Demettre[2], François Mouchet[1], Innocent Djègbè[3], Philippe Marin[2], Fabrice Chandre[1], Vincent Corbel[3], Franck Remoué[3], Françoise Mathieu-Daudé[1]***

**1** Maladies Infectieuses et Vecteurs, Ecologie, Génétique, Evolution et Contrôle (MIVEGEC), UMR IRD 224-CNRS 5290-UM1-UM2, Institut de Recherche pour le Développement (IRD), Montpellier, France, **2** Institut de Génomique Fonctionnelle (IGF), UMR CNRS 5203-INSERM 661-UM1-UM2, Montpellier, France, **3** Maladies Infectieuses et Vecteurs, Ecologie, Génétique, Evolution et Contrôle (MIVEGEC), UMR IRD 224-CNRS 5290-UM1-UM2, Institut de Recherche pour le Développement (IRD), Centre de Recherche Entomologique de Cotonou (CREC), Cotonou, Bénin

## Abstract

Insensitive acetylcholinesterase resistance due to a mutation in the acetylcholinesterase (ace) encoding *ace-1* gene confers cross-resistance to organophosphate and carbamate insecticides in *Anopheles gambiae* populations from Central and West Africa. This mutation is associated with a strong genetic cost revealed through alterations of some life history traits but little is known about the physiological and behavioural changes in insects bearing the *ace-1^R^* allele. Comparative analysis of the salivary gland contents between *An. gambiae* susceptible and *ace-1^R^* resistant strains was carried out to charaterize factors that could be involved in modifications of blood meal process, trophic behaviour or pathogen interaction in the insecticide-resistant mosquitoes. Differential analysis of the salivary gland protein profiles revealed differences in abundance for several proteins, two of them showing major differences between the two strains. These two proteins identified as saglin and TRIO are salivary gland-1 related proteins, a family unique to anopheline mosquitoes, one of them playing a crucial role in salivary gland invasion by *Plasmodium falciparum* sporozoites. Differential expression of two other proteins previously identified in the *Anopheles* sialome was also observed. The differentially regulated proteins are involved in pathogen invasion, blood feeding process, and protection against oxidation, relevant steps in the outcome of malaria infection. Further functional studies and insect behaviour experiments would confirm the impact of the modification of the sialome composition on blood feeding and pathogen transmission abilities of the resistant mosquitoes. The data supports the hypothesis of alterations linked to insecticide resistance in the biology of the primary vector of human malaria in Africa.

**Editor:** Kristin Michel, Kansas State University, United States of America

**Funding:** This work was supported by the French Institut de Recherche pour le Développement (IRD). The funders had no role in study design, data collection and analysis, decision to publish, or preparation of the manuscript.

**Competing Interests:** The authors have declared that no competing interests exist.

* Email: francoise.mathieu-daude@ird.fr

## Introduction

Malaria remains one of the most important public health problems worldwide. Due to its high anthropophily *Anopheles gambiae s.l.* constitutes the main vector of malaria and the target of vector control programmes in Africa [1]. As a result, anopheles mosquito has developed resistance to the different classes of insecticides used in vector control stategies through two major mechanisms. The first mechanism is metabolic resistance, which is due to the increase of expression level or activity of detoxifying enzymes, belonging to three families: the cytochrome P450 monooxygenases, the glutathione S-transferases and the carboxyesterases [2]. The second and most studied mechanism is the target site resistance, such as the insensitive acetylcholinesterase resistance conferring a cross-resistance to carbamates and organophosphates by reducing the ability of these compounds to inhibit acetylcholinesterase in nerve synapses [3]. Two amino acid substitutions in the protein have been shown to play a role in resistance but the resistant allele encoding the G119S mutation, namely *ace-1^R^*, was reported several times in field *An. gambiae* populations [4,5,6]. In *Culex pipiens quinquefasciatus*, life history traits, reproductive capacity and physiology are strongly impacted by the *ace-1* resistance because of the interference between *ace-1^R^* allele and the general functioning of the central nervous system [7,8,9,10,11]. Little information is available so far on the impact of insecticide resistance alleles affecting other traits such as host seeking, blood feeding behaviour and pathogen transmission. A regulation of some salivary proteins was reported in *ace-1* resistant *C. p. quinquefasciatus* mosquito [12]. In this study, D7 long form protein which is known to favor the blood intake was down-regulated in resistant *Culex* mosquitoes whereas some metabolic components were up-regulated in the resistant strain [13]. For

*Anopheles* mosquitoes a large fitness cost entailed by the *ace-1*$^R$ mutation has also been hypothesized [14]. Though reports on these fitness costs are rare, they tend to confirm this hypothesis [15]. Experimental infections of *ace-1*$^R$ resistant *An. gambiae* with *Plasmodium falciparum* gametocytes revealed a slight variation in parasite prevalence suggesting an impact of this mutation on the parasite development [16].

In this work we focused on the effect of *ace-1* resistance on the salivary gland protein composition in *An. gambiae*. The main function of arthropod saliva is to help blood meal intake by releasing components with anti-coagulation, vasodilatation and platelet aggregation properties [17]. In addition, this saliva at the interface with the host could strongly influence both the human immune response and the pathogen transmission. Mosquito salivary proteins play a major role in host-vector interaction and it has been suggested that salivary content could be influenced by age, or infection status of the *Anopheles* vector [18]. Other studies have reported that salivary components could intervene in pathogen transmission [19,20]. Using a proteomic approach based on two-dimensional gel electrophoresis followed by mass spectrometry identification, we investigated whether the expression of salivary proteins would be modified in the resistant malaria vector. We performed a differential analysis of the sialome of *ace-1*$^R$ homozygous *An. gambiae* mosquito and its susceptible counterpart, having a similar genetic background. We therefore, describe and discuss the differential expression of the proteins identified as down- or up-regulated in the insecticide-resistant mosquito. We report for the first time differences in the composition of the insecticide-resistant anopheles sialome with differential expression of proteins belonging to a family exclusively expressed in anophelines. The known or putative functions of the regulated proteins in the invasion of salivary glands by the parasite or in blood meal intake by females are discussed in regards to how they could impact the vectorial capacity.

## Materials and Methods

### Mosquitoes

Two strains of *Anopheles gambiae* S molecular form were used in this study. The insecticide-susceptible strain Kisumu (Kis) originating from Kenya [21] is homozygous for the *ace-1*$^S$ allele at the *ace-1* locus, while the *ace-1*$^R$ Kisumu strain (AceRKis) is homozygous for the resistant allele *ace-1*$^R$, bearing the G119S mutation. This strain was obtained through 19 generations of back-crossing and selection, allowing introgression into the genome of the Kis strain [5]. Female mosquitoes aged 2-3 days were fed on rabbits, then maintained on a diet of glucose until salivary gland dissection between day 8 to 10 post blood meal. The protocol for mosquitoe feeding on rabbits was approved by the animal ethics committee of Languedoc Roussillon, France (Permit number CEEA-LR-13002).

### Salivary gland extract preparation

Salivary glands were pooled (15 to 20 pairs) in 30 µl of rehydratation buffer containing 7 M urea, 2 M thiourea, 4% CHAPS (3-[(3-Cholamidopropyl)dimethylammonio]-1-propanesulfonate), 0.2% tergitol, 20 mM Dithiothreitol, 1% octylβglucoside, 0.8% IPG (Immobilized pH gradient) buffer (PlusOne reagents, GE Healthcare Bio-Sciences, Uppsala, Sweden) and stored at -80°C. After several cycles of frosting-defrosting on liquid nitrogen, samples were centrifuged at 30,000 g for 15 min at 14°C. Soluble salivary proteins in the supernatant were quantified using a Bradford Protein assay (Bio-Rad Laboratories, Marnes-la-Coquette, France) compatible with the presence of urea in the samples. The protein standard curve was run in the same buffer as the SGE samples, 2 µl of rehydratation buffer being diluted in 24 µl standard assays, to compensate for interfering components from the rehydratation buffer on the protein assay. Protein concentration was estimated at an average of 5.0 ± 0.6 µg (mean ± standard deviation) per 10 gland pairs. Total protein content of the salivary glands was slightly lower in the Kis strain compared with the AceRKis strain (4.8 µg and 5.2 µg mean values, respectively). Six samples of salivary gland extracts (SGE) containing 15 µg of proteins were prepared for each strain. Four of them were used for the analytical gel replicates while the last two samples were pooled (30 µg) for loading onto the preparative gel.

### Two-dimensional gel electrophoresis (2-DE)

SGE (15 µg) were diluted to 180 µl by adding rehydratation buffer and 2% final DeStreak reagent (GE Healthcare) and loaded onto 11 cm Immobiline DryStrip gel pH 3-11 non linear (NL) (GE Healthcare). Strips were rehydrated at 20°C overnight (16 hours) and run using the following conditions: temperature 20°C; current 50 µA per strip; 60 V for 1 hour, 500 V (gradient) for 1 hour, 1000 V (gradient) for 1 hour, 6000 V (gradient) for 2 hours, and a 6000 V step up to 30000 Vhrs. Reduction and alkylation steps were performed under standard conditions [12]. The second dimension was carried out on 10-20% SDS-PAGE gels (Bio-Rad) at 30 V for 16 min followed by one hour at 200 V. After fixation, the gels were stained with LavaPurple Protein Gel & Blot Stain Kit (Fluorotechnics, Australia) according to manufacturer's instructions. Gel scanning was performed using a Typhoon 9400 imager (GE Healthcare). Preparative 2-DE gels for spot picking were carried out in the same conditions, using 30 µg of SGE. Preparative gels were stained overnight with colloidal coomassie blue (Fermentas, Saint-Remy les Chevreuse, France) and scanned using a Perfection V750 Pro scanner (EPSON). Four analytical gels and one preparative gel were run for each strain.

### 2-DE gel analysis

Progenesis SameSpots 4.1 software (Nonlinear Dynamics, Newcastle upon Tyne, United Kingdom) was used for 2-DE gel comparative analysis. Statistical tests were performed as recommended by the software provider including a Principal Component Analysis (PCA) and an ANOVA test adjusted by a False Discovery Rate (FDR) approach for all spots in both groups (Kis and AceRKis) assigning to each spot a p-value and a q-value. The statistical power of the analysis was greater than 0.8. Spots under the ANOVA test p-value of 0.01 and above the 1.5-fold minimal difference in expression levels in either direction (up and down-expression) between the two strains were selected and manually excised from the preparative gels.

### Mass spectrometry and protein identification

Tryptic digestion of in-gel proteins was performed according to the Shevchenko modified protocol [22]. Digested proteins were dehydrated in a vacuum centrifuge, solubilised in 10 µl of formic acid (2%), desalted using Zip Tips C18 (Millipore, Bedford, MA), eluted with 10 µl of 0.1% trifluoroacetic acid in 50% acetonitrile and concentrated to 3 µl volume. 0.4 µl of analyte solutions was mixed with the same volume of alpha-cyano-4-hydroxy-trans-cinnamic acid (saturated solution was prepared in 0.1% trifluoroacetic acid in 50% acetonitrile, vortexed, sonicated 30 s and microfuged 30 s, then a 1/3 dilution of the supernatant was used as the matrix). The mixture was deposited on a 384-well MALDI target using the dry-droplet procedure [23], then air dried at room temperature.

MALDI-TOF MS analyses were performed using an UltraFlex I MALDI TOF-TOF mass spectrometer (Bruker Daltonics, Bremen, Germany) in the reflectron mode with a 26 kV accelerating voltage and a 50 ns delayed extraction. Mass spectra were acquired manually using FlexControl version 3.0 software (Bruker Daltonics) (laser power ranged from 35 to 50%, 600 shots). Spectra were analyzed using FlexAnalysis version 3.0 software (Bruker Daltonics) and calibrated internally with the autoproteolysis peptides of trypsin (m/z 842.51, 1045.56, 2211.10). Peptides were selected in the mass range of 900–3000 Da. Peptide Mass Fingerprint identification of proteins was performed by searching against the Insecta entries of Swiss-Prot and TrEMBL databases (UniProt: http://www.uniprot.org/) and by using the MASCOT version 2.3 algorithm (MatrixScience: http://www.matrixscience.com/) with trypsin enzyme specificity and one trypsin missed cleavage allowed [24]. Oxidation was set as variable methionine modification for searches. A mass tolerance of 50 ppm was allowed for identification. Matching peptides with one missed cleavage were considered as pertinent when there were two consecutive basic residues or when arginine and lysine residues were in an acidic context. MASCOT scores higher than 70 were considered significant ($p<0.05$) for Swiss-Prot and TrEMBL database interrogations.

Peptide samples that could not be identified by MALDI-TOF MS analysis were subjected to nanoLC ESI MS/MS analysis. Samples were dehydrated in a vacuum centrifuge, solubilized in 2 µl of 0.1% formic acid-2% acetonitrile and analyzed online by nano-flow HPLC-nanoelectrospray ionization using a LTQ Orbitrap XL mass spectrometer (LTQ Orbitrap XL, Thermo Fisher Scientific, San Jose, CA) coupled with an Ultimate 3000 HPLC (Dionex, Amsterdam, Netherlands). Desalting and pre-concentration of samples were performed on-line on a Pepmap precolumn (0.3 mm×10 mm). A gradient consisting of 0-40% A in 30 min, 80% B in 15 min (A = 0.1% formic acid, 2% acetonitrile in water; B = 0.1% formic acid in acetonitrile) at 300 nl/min was used to elute peptides from the capillary (0.075 mm×150 mm) reverse-phase column (Pepmap, Dionex). LC-MS/MS experiments comprised cycles of 5 events; an MS1 scan with orbitrap mass analysis at 60000 resolution followed by CID of the five most abundant precursors. Fragment ions generated by CID were detected at the linear trap. Normalized collision energy of 35 eV and activation time of 30 ms were used for CID. All Spectra were recorded under positive ion mode using the Xcalibur version 2.0.7 software (Thermo Fisher Scientific). Spectra were acquired with the instrument operating in the information-dependent acquisition mode throughout the HPLC gradient. The mass scanning range was m/z 400-2000 and standard mass spectrometric conditions for all experiments were: spray voltage, 2.4 kV; no sheath and auxiliary gas flow; heated capillary temperature, 200 °C; capillary voltage, 40 V and tube lens, 120 V. For all full scan measurements with the Orbitrap detector, a lock-mass ion from ambient air (m/z 445.120024) was used as an internal calibrant as described [25].

All MS/MS spectra were searched against the Insecta entries of Swiss-Prot and TrEMBL databases (release 2010_11) by using the Proteome Discoverer version 1.2 software (Thermo Fisher Scientific) and MASCOT version 2.3 algorithm with trypsin enzyme specificity and one trypsin missed cleavage. Oxidation was set as variable methionine modification for searches. A peptide mass tolerance of 5 ppm and a fragment mass tolerance of 0.5 Da were allowed for identification. Validation of mass spectrometry data was carried out using Proteome Discoverer software ($p<0.01$ for 2 peptides or more/protein).

## Results

### Differential analysis between SGE samples from resistant and susceptible mosquitoes

Four independent biological repeats corresponding to four SGE samples of each *An. gambiae* strain, AceRKis and Kis, were used for the comparative analysis. Principal component analysis (PCA) separated eight gels according to expression variation into two distinct groups corresponding to the SGE samples of the two *An. gambiae* strains (PCA values of 56% and 12% for the two first components). A significant differential expression between the two strains was observed in 10 of the 708 spots according to ANOVA test ($p<0.01$) and a display of 1.5-fold minimal difference in expression levels. Figure 1A shows the localization of the 10 spots identified as differentially expressed by SameSpots analysis. Of these spots eight were excised from the preparative Kis and AceRKis gels, whereas the other two spots were too faint for excision (spots 572 and 658). Statistical results for the 8 spots are summarized in Table 1.

### Identification of the differentially expressed salivary proteins in the resistant AceRKis mosquitoes

The eight spot samples excised from the preparative gels were analysed by MALDI-TOF MS. A unique protein identification in the UniProtKB/Swiss-Prot or UniProtKB/TrEMBL database was assigned to four spots (Table 1), three of which were identified as isoforms of saglin [Q2TLV8_ANOGA)] and one as a long form D7 salivary protein [Q7PJ76_ANOGA]. A fifth spot was assigned to two proteins, the salivary gland 1-like 3 (SG1-like 3) [Q8WR33_ANOGA] and the TRIO protein [Q7PUJ5_ANOGA]. The three remaining spot samples were subjected to nanoLC ESI MS/MS analysis. This analysis allowed for the identification of the TRIO protein [Q7PUJ5_ANOGA] and an isoform of the Glutathione S-transferase 1 [Q93113_ANOGA]. On the other hand, the remaining spot sample (IDs 578) could not be assigned to a single protein. Results of protein identification using mass spectrometry are summarized in Table 1. The peptide sequences and the number of peptides detected by nanoLC ESI MS/MS and used for protein assignment are provided as supporting information (Table S1).

Saglin, D7 long form and Glutathione S-transferase D1 (GSTD1) proteins were identified as down-regulated in the salivary glands of the resistant AceRKis strain, while the TRIO protein was up-regulated in the resistant strain (Figure 1, B and C). Figure 2 illustrates the differential expression of the identified proteins between the two strains. The graph distinguishes saglin and TRIO proteins, which displayed high expression level differences and the lowest q-values. Saglin, TRIO and D7 long form are secreted proteins as evidenced by the presence of a signal peptide which is indicative of secretion through the classical ER-golgi pathway, as predicted by SignalP 4.0 [26].

## Discussion

The approach combining comparative 2-DE gel analysis and mass spectrometry identification allowed for the identification of five proteins regulated in the resistant AceRKis salivary gland extracts, two of them showing high differences between strains. Among these proteins, saglin presented the highest differences in protein expression with an average value of 5.7 fold for the three isoforms. Saglin was first identified as the target of monoclonal antibodies that reduce salivary gland infectivity of *Plasmodium* sporozoites, suggesting that it may represent one of the molecules involved during the invasion of salivary glands by sporozoites [27].

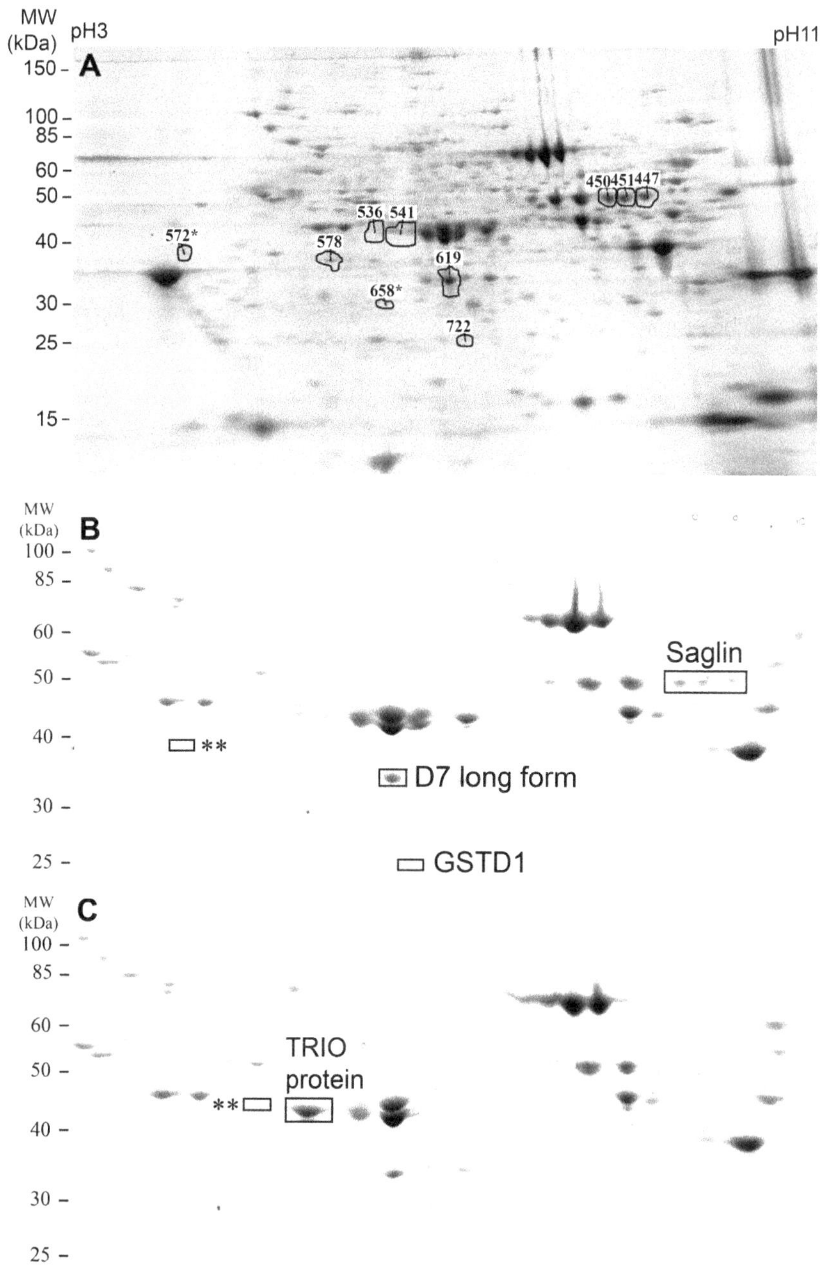


**Figure 1. 2-DE analysis of salivary gland extracts from susceptible Kis and resistant AceRKis *An. gambiae* strains.** (A) Comparative analysis and statistical tests were performed using Progenesis SameSpots software. Localization of the 10 spots under the ANOVA test p-value of 0.01 and above the 1.5-fold minimal difference in expression levels between the two groups of gels are indicated on the reference image corresponding to one of the Kis SGE extract gels. *Low intensity spots 572 and 658 could not be excised from the preparative gel. (B) Identification of the differentially expressed salivary proteins excised from the susceptible Kis and (C) resistant AceRKis gels. Protein identification was performed using MALDI-TOF MS or nanoLC ESI MS/MS analysis. The three isoforms of saglin, down-regulated in the AceRKis strain, and the TRIO protein, up-regulated in the AceRKis strain, correspond to spots 447, 450, 451 and 541, respectively. **These spots were assigned to two or three accession numbers corresponding to different proteins (see Table 1).

This 100 kDa-salivary protein was described as a disulphide bond-linked homodimer of ~50 kDa subunits, and then reported as the receptor for sporozoites in salivary gland, binding to the thrombospondin-related anonymous protein (TRAP) expressed in sporozoites and conserved in all *Plasmodium* species [28]. TRAP was shown to be essential for sporozoite gliding, cell invasion and in vivo infectivity, thus candidate target protein for malaria transmission-blocking vaccine [29,30]. Previous sialome studies showed an over-expression of saglin in *An. gambiae* females versus males [31].

Blast analysis identified the closest homologous and orthologous sequences in *An. stephensi*, *An. cracens* and *An. gambiae* (49, 47, and 32% identity, respectively), and confirmed saglin as being a member of the salivary gland-1 (SG1) protein family, described as unique to anophelines [32,33]. Sequence analysis of saglin did not reveal any functional domain, but an N-terminal signal peptide for

**Table 1.** Differentially expressed salivary gland proteins between the susceptible Kis and resistant AceRKis strains.

| Spot# | Protein ID | p-value | q-value | Fold | Accession | MW (kDa) | pI | Cover | Score* |
|---|---|---|---|---|---|---|---|---|---|
| 541 | TRIO protein SP | $2.18\times10^{-5}$ | 0.015 | 3.72 | Q7PUJ5 | 43.76 | 6.03 | 35 | - |
| | | | | | Q8WR22 | 43.72 | 6.03 | 33 | - |
| 447 | Saglin SP | $1.26\times10^{-4}$ | 0.023 | 6.38 | Q2TLV8 | 47.10 | 7.66 | 30 | 91 |
| 450 | Saglin SP | $1.59\times10^{-4}$ | 0.023 | 4.94 | Q2TLV8 | 47.10 | 7.66 | 33 | 124 |
| 451 | Saglin SP | $2.30\times10^{-4}$ | 0.031 | 5.83 | Q2TLV8 | 47.10 | 7.66 | 31 | 131 |
| 722 | GSTD1 | $7.17\times10^{-4}$ | 0.080 | 4.91 | Q93113 | 23.40 | 6.34 | 18 | - |
| 536 | SG1-like 3 SP° | $1.85\times10^{-3}$ | 0.178 | 2.68 | Q8WR33 | 30.95 | 5.94 | 39 | 81 |
| | TRIO SP | | | | Q7PUJ5 | 43.76 | 6.03 | 29 | 78 |
| 578 | Transaldolase | $2.93\times10^{-3}$ | 0.247 | 3.57 | Q7PZ95 | 36.90 | 6.29 | 12 | - |
| | Phosphatase | | | | Q7PD38 | 35.50 | 5.54 | 8 | - |
| | Translation IF | | | | Q7PP77 | 36.20 | 5.50 | 8 | - |
| 619 | D7 long form SP | $7.69\times10^{-3}$ | 0.432 | 1.90 | Q7PJ76 | 35.57 | 5.64 | 34 | 106 |

Protein identification was performed using MALDI-TOF MS or nanoLC ESI MS/MS (underlined spot#) analysis. Insecta entries of Swiss-Prot and TrEMBL databases were searched by using the MASCOT algorithm, or the Proteome Discoverer software for nanoLC ESI MS/MS spectra. All spots displayed a power >0.9. Spot# refers to the SameSpots analysis spot number (see Figure 1A); SP indicates the presence of a signal peptide for secretion as predicted by SignalP 4.0 (SP°: SP in the N-terminal homolog Q5TV62_ANOGA); Fold: fold difference in expression levels between the two strains; Accession: accession number in UniProtKB/Swiss-Prot or UniProtKB/TrEMBL databases (_ANOGA); pI: isoelectric point; Cover: indicates the amino acid coverage (%); *Mascot Score is provided for MALDI-TOF protein identification; for nanoLC ESI MS/MS based identification the peptide sequences and the number of peptides are provided as supporting information (Table S1). Translation IF: Translation Initiation Factor.

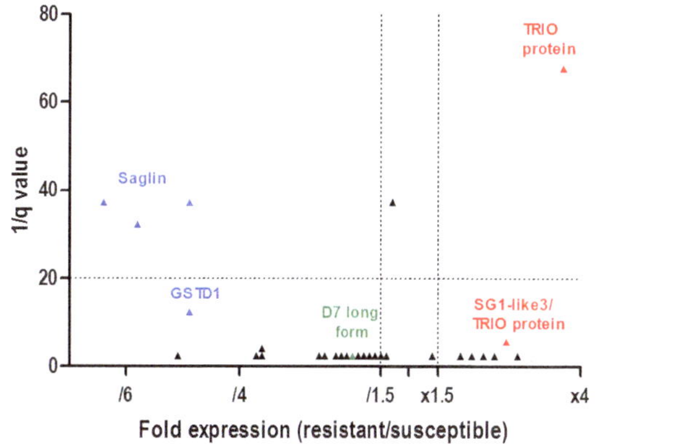


**Figure 2. Differential expression of the salivary proteins between the susceptible Kis and resistant AceRKis SGE extracts.** Differences in protein expression are represented as a function of both expression ratio (resistant/susceptible) and significance ratio (q-value). Vertical dotted lines indicate the 1.5-fold difference in expression level in either direction (x1.5 for a higher expression in the resistant strain and/1.5 for a lower expression in the resistant strain). The horizontal dotted line indicates a q-value =0.05 (or 1/q =20). Saglin and TRIO proteins display expression level differences above 3 and highly significant q-values.

ER-golgi secretion pathway and one putative N-linked glycosylation site. The three saglin isoforms identified on the 2-DE gels are thus likely unglycosylated and partially to totally glycosylated forms of the ~50 kDa subunits of the protein. A similar hypothesis of an heterogeneous mixture of unglycosylated and partially glycosylated forms of saglin was reported from 2D Western blot analysis and treatment with peptide N-glycosidase F (PNGase F) of the recombinant protein [27]. No transmembrane domain or GPI-anchoring signal indicative of cell surface protein localization could be identified. Thus, saglin appears to be a secreted extracellular protein. Whether saglin is partly present in the lumen of salivary glands and saliva remains to be determined, since experimental evidence supports the presence of saglin on the outer surface of the distal lobes of salivary gland, facing the hemocoel [34]. Sporozoite TRAP-domain A interaction with salivary gland is mediated by binding to saglin, but it is still not clear how these proteins interact and how the surface attachment could be promoted [28]. We speculate that saglin present at the basal side of the salivary gland cells could interact with other membrane bound proteins on the basal salivary-gland surface and facilitates sporozoite interaction and salivary gland invasion. Though, more than one salivary gland protein is likely to be involved during salivary gland invasion by sporozoites, a facilitation of salivary gland invasion in the strain over-expressing saglin can be envisioned. Whether this down-regulation in the resistant strain can be linked to a lesser salivary gland invasion property by the *Plasmodium* parasite should be carefully considered. With salivary gland invasion being an essential step of the *Plasmodium* life cycle in its mosquito vector, the insecticide resistant AceRKis strain might be less susceptible to *Plasmodium* infectivity. A recent study on the vectorial competence of *ace-1* resistant *An. gambiae* for *Plasmodium falciparum* reported a low increase in prevalence of infection in the resistant mosquito regarding the oocyste stage in the midgut but not the sporozoite stage in the head and thorax [16]. The role of saglin in salivary gland invasion could thus be one factor contributing to the control of sporozoite load in this compartment. Further studies targeting both the parasite quantification in the salivary glands and the characterization of saglin partners likely to participate in salivary gland invasion by the pathogen would shed some light on the relevance of the differential expression of saglin in the susceptibility to parasite infection of *An. gambiae*.

In contrast with saglin, the TRIO protein presents a significant up-regulation in the insecticide resistant AceRKis strain. cDNA encoding this protein was identified among the salivary gland transcripts of *An. gambiae* adult females [35]. Expression data showed an over-expression of *An. gambiae* TRIO in adult female versus male, in salivary gland versus whole body extract, and in non-blood-fed versus blood-fed female [31,36]. This female-enriched salivary gland gene product is thus likely to play a role in blood feeding, but not in the digestion of the blood meal. Rather it could be involved in the blood feeding process in relation either with anti-hemostatic/immunomodulation properties or with the different steps preceding the blood meal *i.e.* probing and blood uptake. Thus up-regulation observed in the resistant strain could testify for some differences in blood meal uptake between the sensitive and resistant mosquitoes. Sequence alignments and similarity searches performed in the present study indicate that this secreted protein is another member of the SG1 family sharing only 16% identity with saglin. Biochemical studies targeting secretion properties, localization, and function of TRIO could shed light on its role in the salivary glands of *Anopheles*, and on the significance of its up-regulation in the biology of the insecticide-resistant *An. gambiae* strain.

Surprinsingly a Glutathione S-transferase was found as down-regulated in the resistant AceRKis strain. Glutathione S-transferases (GSTs) are a diverse family of enzymes, playing a central role in detoxification of both endogenous and xenobiotic compounds [37]. *An. gambiae* GSTs have been studied for their implication in resistance to the organochlorine insecticide DDT. Expression profiling experiments identified members of the epsilon-class of GSTs but none of the delta-class GSTs as upregulated in DDT resistant strains, and functional studies confirmed the high DDT-metabolizing ability of one GST member of the epsilon-class [38,39]. The GST identified in the present study is a member of the delta class, named GSTD1 or agGSTd1-6 [39,40]. Structural studies revealed differences in the substrate-binding site conformation that explained the very low DDT-detoxifying activity of this GST [41,42]. Thus, the down-regulation of the protein in the resistant strain might not be linked to a decreased insecticide detoxifying activity, but the relevance of this difference in expression observed between the two strains is unclear. Besides their detoxifying function GSTs are involved in a wide range of biological processes including protection against oxidative stress [43]. Their contribution to the removal of toxic oxygen free radicals and maintainance of the redox status of the cell represents a function that deserves further attention to better understand the role of this enzyme family, particularly in relation to parasite infection and vector competence.

Although the down-regulation of the secreted D7 long form protein observed in the resistant strain is low, this protein deserves some comment. Indeed, its ortholog in *Culex pipiens quinquefasciatus,* named long form D7 cluster12 salivary protein, was previously recorded as differentially expressed between the *ace1-*$^{R}$ insecticide-resistant and the susceptible strains [12]. D7 protein family is widely conserved in blood sucking diptera, and its expression is restricted to female salivary glands suggesting a potential role in the blood meal process [44]. This family is ancestrally related to Odorant Binding Protein family, but is a separate branch as supported by the structural analysis of *Anopheles* D7 proteins members [13,45]. The *Anopheles* D7 family is composed by three long form proteins (27-30 kDa) and five short form proteins, named D7r1-D7r5. The short forms (D7r1-r4) have been demonstrated binding biogenic amines, like serotonin, norepinephrin and histamine, highlighting functions in the inhibition of itching, burning and vasoconstriction induced by the bite [13]. D7r1 possess also anticlotting activity, as its ortholog in *An. stephensi*, hamadarin. Function of the *Anopheles* species D7 long form recently became elucidated. A structure/function study on the long form D7 L1 from *An. stephensi* has shown that this protein binds cystenyl-leukotrienes and analogues of tromboxane A2. These molecules are released during early inflammatory response following injury, contributing to pain, increase of vascular permeability and platelet activation in the host [46]. The high similarity observed between the D7L1 and the *An. gambiae* ortholog (60%), especially for amino acid residues involved in ligand binding pocket (80% identity), strongly suggests that the *An. gambiae* D7 long form could have similar functions in blood feeding. The down regulation of this D7 long form protein may thus alter the efficacy of the blood feeding process, with less inhibition of platelet activation and anticlotting activity of the host. The blood feeding might be interrupted hence leading the mosquito to bite several hosts/times to achieve a complete blood meal. Biting behavioural studies comparing resistant and susceptible strains of *An. gambiae* as well as RNAi studies would help to determine the implication of this D7 long form protein in the blood feeding success.

The resistant AceRKis strain was obtained by introgression of the *ace-1* G119S allele (*ace-1*$^R$ allele) from a resistant population into the Kisumu genome. After 19 generations of introgression, the two *An. gambiae* strains should share the same genetic background for most of their genome, mainly differing in their genotype at the *ace-1* locus: the Kis strain is homozygous for the *ace-1*$^S$ allele and the AceRKis strain is homozygous for the *ace-1*$^R$ allele. Comparative studies of the two strains should thus reveal changes in characters or phenotypes that could be essentially attributed to the G119S mutation. Nevertheless, the genetic differences between the two strains might not be totally negligible: the presence of inversions, founder effects, and the genes closely linked to *ace-1* could have prevented part of the expected recombination. Regarding chromosomal inversions, differences in the parental strains could hamper recombination and limit introgession [47]. Indeed, loci inside a polymorphic inversion and in linkage disequilibrium with the resistant allele may be selected with this allele. The origin of the Kisumu strain is East Africa, where *An. gambiae* is polymorphic for the chromosomal inversions 2Rb and 2La [48]. The insecticide resistant parental strain was colonized from wild mosquitoes from Bobo-Dioulasso (Burkina Faso) where the S form of *An. gambiae* has been shown to also possess the 2Rb and 2La inversions at high frequency [49]. Thus, as the two parental populations share the same inversions, they should not prevent recombination between the two strains, except if a founder effect randomly fixed different inversions before or during the introgression process. Moreover, as the *ace-1* gene is located in the telomeric region of the chromosome 2R (region 7C), thus far from the closest inversion (2Rb, between sections 11C and 13A), these inversions are not expected to affect recombination near this gene, so that the inversions present in Kis should have been introgressed in AceRKis. If Kis was polymorphic at the beginning of the backcrosses, a unique variant of some genes could have been fixed in AceRKis, due to founder effect, while remaining polymorphic in Kis, thus inducing differences between the two strains. However, as each generation required a new backcross on Kis, this effect would be relatively limited, as most of the Kis polymorphism would have passed in AceRKis. We cannot however rule out the possibility that other genes, such as regulatory regions and loci physically linked to *ace-1*$^R$ could have hitch-hiked with it during the introgression process. Indeed, at the risk level of 0.05, a genetic distance of 15 cM around *ace1*$^R$ would not have been replaced during the successive backcrosses [5,50]. According to the average relationship between genetic distance and physical distance on the chromosome X [5,50], that would correspond to approximately 11 Mb of DNA that could differ between the two strains, depending on the initial polymorphism. In conclusion, we cannot exclude that part of the differences in salivary protein expression observed in the present work between Kis and AceRKis could be due to cis-regulatory elements or other loci from the genome in which the *ace-1*$^R$ mutation originally occured, rather than to the mutation itself. The analysis of the non-synonymous polymorphism between the two strains, particularly in the flanking regions of the *ace-1* locus, could help in quantifying the real impact of the introgression process. Finally, additional crossing experiments between the two strains followed by the analysis of the co-segregation of salivary protein expression and the resistance gene are required to formally establish the causal relationship between the regulation of the salivary protein genes and the resistance allele.

## Conclusion

The present work is the first attempt to determine whether insecticide-resistance in *An. gambiae* mosquitoes could induce a modulation of protein expression in the salivary glands and have some consequences on blood feeding, trophic behaviour and on *Plasmodium* transmission. We describe for the first time modifications in the sialome composition of insecticide-resistant *An. gambiae*. The known or putative functions inferred from sequence similarity analyses revealed that the differentially regulated proteins identified in this study are involved in pathogen invasion, blood feeding process, or protection against oxidation. These proteins deserve functional studies to understand the relevance of the differences in expression recorded in the insensitive acetylcholinesterase resistant mosquitoes. This data confirms the profound alterations linked to the resistance in the anopheline mosquito biology but additional genetic studies on the polymorphism between the two strains are needed to confirm whether *ace-1*$^R$ allele is responsible for the variation in protein expression and to determine whether the expression levels influence phenotypes relevant for transmission. Experiments on field mosquitoes bearing the *ace-1*$^R$ allele or the *ace-1* gene duplication [6] could also help to specify the link between *ace-1* alleles and phenotypes. Further experiments on the behaviour of these mosquitoes during blood feeding will contribute to the understanding of the impact of resistance on blood meal intake and on the transmission of the pathogen. This work opens the way for studying the profound biological alterations in the mosquito linked to the different insecticide resistance mechanisms, alone or in combination.

## Supporting Information

**Table S1 Details of the nanoLC ESI MS/MS based identification of the differentially expressed salivary gland proteins in insecticide-resistant AcerKis strain.** Protein identification was performed using nanoLC ESI MS/MS analysis. Insecta entries of Swiss-Prot and TrEMBL databases (release 2010_11) were searched by using the Proteome Discoverer software. Spot # refers to the SameSpots Analysis spot number (see Figure 1A); Accession number: accession number or entry name in UniProtKB/Swiss-Prot or UniProtKB/TrEMBL; In the peptide sequence column, m means methionine oxydation.

## Acknowledgments

We acknowledge Marie-Noëlle Lacroix and Blandine Luce for mosquito rearing in insect facility (UMR 224/LIN-IRD, Montpellier), Pierrick Labbé (ISEM, Université Montpellier 2) for helpful discussions and advice on introgression and genetics, and Fingani Annie Mphande (IRD) for editing the manuscript.

## Author Contributions

Conceived and designed the experiments: SC VC FR FMD. Performed the experiments: SC MR MS FM FMD. Analyzed the data: SC MS ED FMD. Contributed reagents/materials/analysis tools: MS ED ID PM FC FR. Wrote the paper: SC FMD.

## References

1. Pates HV, Takken W, Stuke K, Curtis CF (2001) Differential behaviour of Anopheles gambiae sensu stricto (Diptera: Culicidae) to human and cow odours in the laboratory. Bull Entomol Res 91: 289–296.
2. Hemingway J, Ranson H (2000) Insecticide resistance in insect vectors of human disease. Annu Rev Entomol 45: 371–391.
3. Weill M, Berthomieu A, Berticat C, Lutfalla G, Negre V, et al. (2004) Insecticide resistance: a silent base prediction. Curr Biol 14: R552–553.
4. N'Guessan R, Darriet F, Guillet P, Carnevale P, Traore-Lamizana M, et al. (2003) Resistance to carbosulfan in Anopheles gambiae from Ivory Coast, based on reduced sensitivity of acetylcholinesterase. Med Vet Entomol 17: 19–25.
5. Djogbenou L, Weill M, Hougard JM, Raymond M, Akogbeto M, et al. (2007) Characterization of insensitive acetylcholinesterase (ace-1R) in Anopheles gambiae (Diptera: Culicidae): resistance levels and dominance. J Med Entomol 44: 805–810.
6. Djogbenou L, Chandre F, Berthomieu A, Dabire R, Koffi A, et al. (2008) Evidence of introgression of the ace-1(R) mutation and of the ace-1 duplication in West African Anopheles gambiae s. s. PLoS One 3: e2172.
7. Lenormand T, Bourguet D, Guillemaud T, Raymond M (1999) Tracking the evolution of insecticide resistance in the mosquito Culex pipiens. Nature 400: 861–864.
8. Gazave E, Chevillon C, Lenormand T, Marquine M, Raymond M (2001) Dissecting the cost of insecticide resistance genes during the overwintering period of the mosquito Culex pipiens. Heredity (Edinb) 87: 441–448.
9. Bourguet D, Guillemaud T, Chevillon C, Raymond M (2004) Fitness costs of insecticide resistance in natural breeding sites of the mosquito Culex pipiens. Evolution 58: 128–135.
10. Berticat C, Boquien G, Raymond M, Chevillon C (2002) Insecticide resistance genes induce a mating competition cost in Culex pipiens mosquitoes. Genet Res 79: 41–47.
11. Berticat C, Bonnet J, Duchon S, Agnew P, Weill M, et al. (2008) Costs and benefits of multiple resistance to insecticides for Culex quinquefasciatus mosquitoes. BMC Evol Biol 8: 104.
12. Djegbe I, Cornelie S, Rossignol M, Demettre E, Seveno M, et al. (2011) Differential expression of salivary proteins between susceptible and insecticide-resistant mosquitoes of Culex quinquefasciatus. PLoS One 6: e17496.
13. Calvo E, Mans BJ, Andersen JF, Ribeiro JM (2006) Function and evolution of a mosquito salivary protein family. J Biol Chem 281: 1935–1942.
14. Alout H, Djogbenou L, Berticat C, Chandre F, Weill M (2008) Comparison of Anopheles gambiae and Culex pipiens acetycholinesterase 1 biochemical properties. Comp Biochem Physiol B Biochem Mol Biol 150: 271–277.
15. Djogbenou L, Noel V, Agnew P (2010) Costs of insensitive acetylcholinesterase insecticide resistance for the malaria vector Anopheles gambiae homozygous for the G119S mutation. Malar J 9: 12.
16. Alout H, Ndam NT, Sandeu MM, Djegbe I, Chandre F, et al. (2013) Insecticide resistance alleles affect vector competence of Anopheles gambiae s.s. for Plasmodium falciparum field isolates. PLoS One 8: e63849.
17. Fontaine A, Diouf I, Bakkali N, Misse D, Pages F, et al. (2011) Implication of haematophagous arthropod salivary proteins in host-vector interactions. Parasit Vectors 4: 187.
18. Choumet V, Carmi-Leroy A, Laurent C, Lenormand P, Rousselle JC, et al. (2007) The salivary glands and saliva of Anopheles gambiae as an essential step in the Plasmodium life cycle: a global proteomic study. Proteomics 7: 3384–3394.
19. Donovan MJ, Messmore AS, Scrafford DA, Sacks DL, Kamhawi S, et al. (2007) Uninfected mosquito bites confer protection against infection with malaria parasites. Infect Immun 75: 2523–2530.
20. Schneider BS, Higgs S (2008) The enhancement of arbovirus transmission and disease by mosquito saliva is associated with modulation of the host immune response. Trans R Soc Trop Med Hyg 102: 400–408.
21. Shute GT (1956) A method of maintaining colonies of East African strains of Anopheles gambiae. Ann Trop Med Parasitol 50: 92–94.
22. Wilm M, Shevchenko A, Houthaeve T, Breit S, Schweigerer L, et al. (1996) Femtomole sequencing of proteins from polyacrylamide gels by nano-electrospray mass spectrometry. Nature 379: 466–469.
23. Karas M, Hillenkamp F (1988) Laser desorption ionization of proteins with molecular masses exceeding 10,000 daltons. Anal Chem 60: 2299–2301.
24. Wilkins MR, Williams KL (1997) Cross-species protein identification using amino acid composition, peptide mass fingerprinting, isoelectric point and molecular mass: a theoretical evaluation. J Theor Biol 186: 7–15.
25. Olsen JV, de Godoy LM, Li G, Macek B, Mortensen P, et al. (2005) Parts per million mass accuracy on an Orbitrap mass spectrometer via lock mass injection into a C-trap. Mol Cell Proteomics 4: 2010–2021.
26. Petersen TN, Brunak S, von Heijne G, Nielsen H (2011) SignalP 4.0: discriminating signal peptides from transmembrane regions. Nat Methods 8: 785–786.
27. Okulate MA, Kalume DE, Reddy R, Kristiansen T, Bhattacharyya M, et al. (2007) Identification and molecular characterization of a novel protein Saglin as a target of monoclonal antibodies affecting salivary gland infectivity of Plasmodium sporozoites. Insect Mol Biol 16: 711–722.
28. Ghosh AK, Devenport M, Jethwaney D, Kalume DE, Pandey A, et al. (2009) Malaria parasite invasion of the mosquito salivary gland requires interaction between the Plasmodium TRAP and the Anopheles saglin proteins. PLoS Pathog 5: e1000265.
29. Sultan AA, Thathy V, Frevert U, Robson KJ, Crisanti A, et al. (1997) TRAP is necessary for gliding motility and infectivity of plasmodium sporozoites. Cell 90: 511–522.
30. Crawford JE, Rottschaefer SM, Coulibaly B, Sacko M, Niare O, et al. (2013) No evidence for positive selection at two potential targets for malaria transmission-blocking vaccines in Anopheles gambiae s.s. Infect Genet Evol 16: 87–92.
31. Marinotti O, Nguyen QK, Calvo E, James AA, Ribeiro JM (2005) Microarray analysis of genes showing variable expression following a blood meal in Anopheles gambiae. Insect Mol Biol 14: 365–373.
32. Arca B, Lombardo F, de Lara Capurro M, della Torre A, Dimopoulos G, et al. (1999) Trapping cDNAs encoding secreted proteins from the salivary glands of the malaria vector Anopheles gambiae. Proc Natl Acad Sci U S A 96: 1516–1521.
33. Valenzuela JG, Charlab R, Gonzalez EC, de Miranda-Santos IK, Marinotti O, et al. (2002) The D7 family of salivary proteins in blood sucking diptera. Insect Mol Biol 11: 149–155.
34. Brennan JD, Kent M, Dhar R, Fujioka H, Kumar N (2000) Anopheles gambiae salivary gland proteins as putative targets for blocking transmission of malaria parasites. Proc Natl Acad Sci U S A 97: 13859–13864.
35. Francischetti IM, Valenzuela JG, Pham VM, Garfield MK, Ribeiro JM (2002) Toward a catalog for the transcripts and proteins (sialome) from the salivary gland of the malaria vector Anopheles gambiae. J Exp Biol 205: 2429–2451.
36. Baker DA, Nolan T, Fischer B, Pinder A, Crisanti A, et al. (2011) A comprehensive gene expression atlas of sex- and tissue-specificity in the malaria vector, Anopheles gambiae. BMC Genomics 12: 296.
37. Hemingway J, Hawkes NJ, McCarroll L, Ranson H (2004) The molecular basis of insecticide resistance in mosquitoes. Insect Biochem Mol Biol 34: 653–665.
38. Prapanthadara LA, Ketterman AJ (1993) Qualitative and quantitative changes in glutathione S-transferases in the mosquito Anopheles gambiae confer DDT-resistance. Biochem Soc Trans 21 (Pt 3): 304S.
39. Ranson H, Rossiter L, Ortelli F, Jensen B, Wang X, et al. (2001) Identification of a novel class of insect glutathione S-transferases involved in resistance to DDT in the malaria vector Anopheles gambiae. Biochem J 359: 295–304.
40. Ranson H, Prapanthadara L, Hemingway J (1997) Cloning and characterization of two glutathione S-transferases from a DDT-resistant strain of Anopheles gambiae. Biochem J 324 (Pt 1): 97–102.
41. Chen L, Hall PR, Zhou XE, Ranson H, Hemingway J, et al. (2003) Structure of an insect delta-class glutathione S-transferase from a DDT-resistant strain of the malaria vector Anopheles gambiae. Acta Crystallogr D Biol Crystallogr 59: 2211–2217.
42. Wang Y, Qiu L, Ranson H, Lumjuan N, Hemingway J, et al. (2008) Structure of an insect epsilon class glutathione S-transferase from the malaria vector Anopheles gambiae provides an explanation for the high DDT-detoxifying activity. J Struct Biol 164: 228–235.
43. Wilce MC, Parker MW (1994) Structure and function of glutathione S-transferases. Biochim Biophys Acta 1205: 1–18.
44. Arca B, Lombardo F, Lanfrancotti A, Spanos L, Veneri M, et al. (2002) A cluster of four D7-related genes is expressed in the salivary glands of the African malaria vector Anopheles gambiae. Insect Mol Biol 11: 47–55.
45. Mans BJ, Calvo E, Ribeiro JM, Andersen JF (2007) The crystal structure of D7r4, a salivary biogenic amine-binding protein from the malaria mosquito Anopheles gambiae. J Biol Chem 282: 36626–36633.
46. Alvarenga PH, Francischetti IM, Calvo E, Sa-Nunes A, Ribeiro JM, et al. (2010) The function and three-dimensional structure of a thromboxane A2/cysteinyl leukotriene-binding protein from the saliva of a mosquito vector of the malaria parasite. PLoS Biol 8: e1000547.
47. Hoffmann AA, Rieseberg LH (2008) Revisiting the Impact of Inversions in Evolution: From Population Genetic Markers to Drivers of Adaptive Shifts and Speciation? Annu Rev Ecol Evol Syst 39: 21–42.
48. Petrarca V, Beier JC (1992) Intraspecific chromosomal polymorphism in the Anopheles gambiae complex as a factor affecting malaria transmission in the Kisumu area of Kenya. Am J Trop Med Hyg 46: 229–237.

49. Costantini C, Ayala D, Guelbeogo WM, Pombi M, Some CY, et al. (2009) Living at the edge: biogeographic patterns of habitat segregation conform to speciation by niche expansion in Anopheles gambiae. BMC Ecol 9: 16.

50. Pombi M, Stump AD, Della Torre A, Besansky NJ (2006) Variation in recombination rate across the X chromosome of Anopheles gambiae. Am J Trop Med Hyg 75: 901–903.

# 16

# Honeybee Colony Disorder in Crop Areas: The Role of Pesticides and Viruses

**Noa Simon-Delso[1,2]*, Gilles San Martin[3], Etienne Bruneau[1], Laure-Anne Minsart[1], Coralie Mouret[1], Louis Hautier[3]**

**1** Beekeeping Research and Information Centre, Louvain la Neuve, Belgium, **2** Environmental Sciences, Copernicus Institute, Utrecht University, Utrecht, The Netherlands, **3** Plant Protection and Ecotoxicology Unit, Life Sciences Department, Walloon Agricultural Research Centre, Gembloux, Belgium

## Abstract

As in many other locations in the world, honeybee colony losses and disorders have increased in Belgium. Some of the symptoms observed rest unspecific and their causes remain unknown. The present study aims to determine the role of both pesticide exposure and virus load on the appraisal of unexplained honeybee colony disorders in field conditions. From July 2011 to May 2012, 330 colonies were monitored. Honeybees, wax, beebread and honey samples were collected. Morbidity and mortality information provided by beekeepers, colony clinical visits and availability of analytical matrix were used to form 2 groups: healthy colonies and colonies with disorders (n = 29, n = 25, respectively). Disorders included: (1) dead colonies or colonies in which part of the colony appeared dead, or had disappeared; (2) weak colonies; (3) queen loss; (4) problems linked to brood and not related to any known disease. Five common viruses and 99 pesticides (41 fungicides, 39 insecticides and synergist, 14 herbicides, 5 acaricides and metabolites) were quantified in the samples.The main symptoms observed in the group with disorders are linked to brood and queens. The viruses most frequently found are Black Queen Cell Virus, Sac Brood Virus, Deformed Wing Virus. No significant difference in virus load was observed between the two groups. Three acaricides, 5 insecticides and 13 fungicides were detected in the analysed samples. A significant correlation was found between the presence of fungicide residues and honeybee colony disorders. A significant positive link could also be established between the observation of disorder and the abundance of crop surface around the beehive. According to our results, the role of fungicides as a potential stressor for honeybee colonies should be further studied, either by their direct and/or indirect impacts on bees and bee colonies.

**Editor:** Fabio S. Nascimento, Universidade de São Paulo, Faculdade de Filosofia Ciências e Letras de Ribeirão Preto, Brazil

**Funding:** The Walloon Region was the only funder of this study. The funders had no role in study design, data collection and analysis, or preparation of the manuscript.

**Competing Interests:** The authors have declared that no competing interests exist.

* Email: simon@cari.be

## Introduction

The evolution of pollinator populations has been the subject of an increasing number of studies, most of them showing worrying negative trends [1–4]. Furthermore, beekeepers have long been notifying enhanced bee winter losses and disorders [5–8]. Bees contribute to ecosystem services and their decline thus threatens pollination of both wild and cultured plants, endangering biodiversity, food and fibre production [9,10]. This decline may also have an impact on the production of other goods with pharmacological uses (e.g. honey, propolis) [11,12], and scientific and technological inspiration (e.g. development of visual guided flight robotics) [13]. Bees are also part of our culture (e.g. culinary, hobby occupation, etc.), contributing to the dynamism of rural and urban areas [14] and providing a source of inspiration and well-being for many [15].

Belgium shares the trends observed worldwide both in terms of wild and reared pollinators [8,16], with enhanced winter mortality observed from 1999 [8,17,18]. However, apart from colony mortality, Walloon beekeepers have reported a number of imprecise symptoms: colony weakness, mainly in spring; fast renewal of young queens; rapid loss of individuals in the colony, mainly foragers, or slow loss of individuals in the colony. In many cases brood and food remains in the colony. Sometime a small cluster of bees with the queen survives [17]. Some beekeepers described unspecific brood abnormalities not characteristic to any known disease (Baudoin and Lequeux, pers. com.). Other studies describe similar mortality trends [8,19,20] as well as unspecific symptoms: increased colony mortality and/or weakening [21–23], queen failure [21,23–25] or honey yield reduction [21].

Multifarious factors have been proposed to provide a cause to such a phenomenon. Climate change is proposed as one of them, together with a decrease of genetic variability of the bee colonies, electromagnetic radiation, pathogens and parasites or the impact of intensive agricultural systems (nutritional lack, GM plants or pesticides) [26]. Previous studies developed in the region discard some of these factors as causes of bee losses, specifically *Nosema* spp and American Foulbrood [27]. However, pesticide residues and certain viruses were detected in bee colonies [28,27]. These elements were the most relevant to us.

Countless studies have shown lethal and sub-lethal effects of pesticides on bees [29–31]. Insecticides are often the most studied for obvious reasons. However other substances (fungicides, herbicides) deserve analysis for their specific toxicity, individually or in synergy with other substances [32,33] or pathogens [34–36], or their extensive exposure given their large scale and/or repeated use. Indeed, residue analyses show that these types of substance are found in the hives, even though the crops in which they were applied would suggest no bee exposure [37–40].

Beside pesticides, bee viruses are often mentioned as a putative cause of decline of the colonies, or at least to reside among the presumed multi-factorial causes [41]. In Europe, at least 12 viruses infecting bees have been compiled [21]. Many honeybee viruses commonly occur in seemingly healthy populations that continue to run well. Viruses may remain latent and confined within certain cells or tissues with no active replication and no disruption of cellular function. Likewise, they may be replicating at low level in permissive cells but in non-vital sites or in honeybee life stages that do not exhibit any symptoms or obvious pathology [21,42,43]. Nevertheless, two viruses, ABPV and DWV, are able of inducing serious disorders to honeybees and have been shown to cause -in association with *V. destructor* - winter losses in Germany [23].

In this paper, we first studied the relationship of in-hive viral prevalence as well as the presence of pesticide residues on honeybee colonies' health. As a next step, we focused on the relationship between the environment surrounding the monitored apiaries and the health condition of their colonies.

## Materials and Methods

### Field work – colony follow up

A group of voluntary beekeepers were requested to participate in the study. All of them shared the following criteria: (1) have a minimum of 5 production colonies per apiary in June 2011; (2) regularly follow up the health status and development of their colonies; and (3) monitor the varroa infestation level and carry out officially recommended varroa treatments (treatment in July-August with veterinary medicaments based on thymol and winter treatment with veterinary medicaments based on oxalic acid). A total of 330 colonies distributed among 66 apiaries (5 colonies/apiary) in Walloon and Brussels regions (Belgium) were followed. In those apiaries composed of 5 or more colonies, the 5 well-developed colonies at the beginning of the study were selected.

Colonies received three visits along the study, the first one from mid-July to mid-August 2011, the second one from mid-September to mid-October 2011 and the third one from March to April 2012. State official beekeeping technicians specialised in bee health were trained in the framework of the project in order to minimise as far as possible the variability of the results due to handling and observations. For each of the visits, beekeepers were requested to fulfil a form asking information about the health history of the apiary and of the followed-up colonies, as well as about their beekeeping practices. Honeybee colony disorder symptoms were reported in the questionnaire. These include the following symptoms for which no causal agent could obviously be identified:

(1) dead and disappeared colonies : (1a) death of part or the whole colony, where dead bees can be found close to the hive or inside it. Beekeepers also describe the (1b) disappearance of part or the whole colony, leaving behind food reserves and brood, a phenomenon similar to the one described in North-American apiaries (Colony Collapse Disorder (CCD) [24];
(2) weak colonies : weakening of the colony, showing in occasions a slow development in spring under optimal conditions (e.g. optimal weather, low varroa pressure, etc) with as consequence the loss of the spring production, but in occasions showing abnormally small colonies or colonies with low activity;
(3) queen loss : replacement of young queens sometimes leading to queen-less colonies or interruption of the egg-laying activity of the queen [17];
(4) problems linked to brood and not related to any known disease.

The form also included requests about other symptoms typically linked to known diseases (e.g. diarrhoea, mummified larvae, varroa presence, presence of deformed wing bees, etc). Each of the questions requested further characterisation of the symptoms (e.g. population size, population dynamics, bees behaviour, etc.).

In addition, a thorough clinical inspection was carried out for each colony in the mentioned period. Special attention was given to the strength of the colony in terms of bee numbers, brood surface and reserves content, the presence of the queen and the varroa infestation level. Finally, two different samples were collected before and after the winter: (1) in-hive bees (minimum of 100 bees); (2) a section of the frame containing beebread and honey (about 100 $cm^2$). Samples were collected in hermetic plastic bags, cooled after collection and stored at $-20^{\circ}C$.

### Case choice – hierarchical sample clustering

Information available from bee colonies comprised field observations, beekeeper answers to the questionnaire and results from the analyses of the samples of beekeeping matrices (honey, beebread, wax and bees). Colonies for which this information was available were considered for our analyses. Colonies with well identified problems (heavy varroa infestation, lack of food or drone-laying queens) were discarded. After that, two groups were made: a group with disorders and another with healthy colonies. In order to constitute these groups and to limit variations due to potential different bee management, a hierarchical classification was made. The criteria used were the amount of food stores before the winter, the year of colony creation, the subspecies and the age of the queen. This classification was made by using Ward aggregation method. This method allows to build group with the lower variation within the group and the higher variation between groups [44].

### Virus analyses

The viruses under investigation were the Black Queen Cell Virus (BQCV), the Chronic Bee Paralysis Virus (CBPV), the Acute Bee Paralysis Virus (ABPV), the Deformed Wings Virus (DWV), and the Sac Brood Virus (SBV). Viral analyses were conducted on the honeybee workers collected before the winter, during the first and second periods (July–August and September–October). The samples were analysed with a quantitative RT-PCR by the National Bee Unit laboratory, Food and Environment Research Agency (Sand Hutton, York, United-Kingdom).

Total Nucleic acid (TNA) was extracted from 60 bees homogenised for 12 minutes with 20 ml GITC Lysis Buffer (5 M Guandine Thiocyanate, 0.05 M Tris base, 0.02 M EDTA, pH 8.0) in a 30 ml bottle containing 3, 7/16″ ball bearings. GITC Lysis buffer also contained 17.3 mM SDS buffer (173 mM Sodium dodecyl sulphate (SDS) in 100 ml MGW). The SDS buffer is added prior to use in warmed GITC Lysis Buffer. The homogenate was then incubated at 65°C for 40 minutes and then spun at 6189 g for 5 minutes. Polypropylene 96 -deep well plates (DT850301 Elkay Laboratory Products Ltd) were prepared as follows (1 well/extract); plate A: 800 µl extract, and 100 µl MagneSil$^{TM}$ beads (MD1441, Promega); plate B: 1 ml of GITC wash buffer (5 M Guandine Thiocynate, 0.05 M Tris base); plates C, D, E: 1 ml 70%v/v ethanol (E/00665DF/17, Fisher Scientific); plate F: 1 ml 5 M Betaine (B2629, Sigma), plate G, 300 ul 1×TE

Buffer (stock 100X TE (EC-862 National Diagnostics). Plates were loaded onto the Kingfisher Flex and processed as follows: Plate A – Bind 5 mins (fast dual mix), Plate B – Wash 3 mins (fast dual mix), Plate C, D, E – Wash 2 mins (fast dual mix), collect beads at 1 min intervals, Plate F – Wash 20 secs (medium), without releasing beads, Plate G – Mix 1 min then incubate at 65°C for 5 mins with mixing. All steps of the process are looped through twice. TNA was collected from plate G of each reaction and stored at −80°C prior to use in real-time PCR.

Reactions were set up in 96 well reaction plates using Absolute Blue QPCR ROX mix (AB-4139, Thermo Scientific) following the protocols supplied. 0.1 mM of 0.1 M Dithiothreitol (165680250, Arcos Organics) and 0.33 unit of MMLV (EPO441: Fermentas) were added to each reaction. The primers (Table 1) were all used at 400 nM and probes at 200 nM final concentration. Total nucleic acid (5 µl) was added to each reaction, giving a final reaction volume of 20 µl. Plates were cycled for 48°C/30 min, 95°C/10 min and 40 cycles of 60°C/1 min, 95°C/15 sec within the 7900 HT Sequence Detection System (Applied Biosystems), using real time data collection. The results were recorded as the cycle threshold (Ct) or cycle number after which a significant accumulation of florescence over the baseline was observed; an average (of duplicate wells) Ct value below 40 was regarded as a positive result with a threshold ΔRn setting of 0.2. Given the absence of internal standard, we assumed that the extraction method had the same efficiency towards bee DNA/RNA and viral RNA.

Virus primers and probes used for all pathogens tested have been previously described in Chantawannakul et al., 2006 [45]. Additional APBV primers are described in Martin et al., 2012. [46].

## Pesticide analyses

Samples collected before the winter were sent to Eurofins Chemiphar NV, Brugge, Belgium and analysed by SOFIA GmbH Chemisches Labor für Softwareentwicklung und Intelligente Analytik, Berlin, Germany. Two multi-residues methods (SF146 and SF150) were used searching for 99, 93 and 96 pesticides residues in wax (54 samples), beebread (108 samples) and honey (107 samples – one sample did not contain enough matrix) respectively (LOQ in Table 2).

For SF146 method, 10 ml of water was added to 10 g of samples. Methanol was added and the preparation was mixed. The mixture was filtered and centrifuged. Next, for analysis by gas chromatography coupled with mass spectrometry (GC-MS), the filtrate was mixed with sodium chloride and ethyl acetate solution 1:1 (v/v). All was dried with sodium sulfate and filtrated. This filtrate was then concentrated and analysed by GC-MS. For analysis with liquid chromatography coupled with tandem mass spectrometry (LC-MS/MS), 5 ml of sample were transferred in ChemElut column and eluted with dichloromethane. After concentration, 1 ml of water-methanol 1:1 (v/v) was added and this solution was analysed by LC-MS/MS. For the wax, some modifications were made. For GC-MS, sodium chloride was replaced by hexane to dissolve wax and the mixture was frozen until a precipitate appears. For LC-MS/MS, 0.2 g of sample were extracted with 30 ml of mixture of naphtha:acetonitrile 1:2 (v/v). 10 ml of acetonitrile phase was isolated and concentrated before the analysis by LC-MS/MS.

For SF150 method, 10 ml of water was added to 10 g of samples as well as methanol and hydrochloric acid. All was mixed, filtrated and centrifuged. A solution of sodium chloride (20%) was added and 5 ml of sample were transferred in ChemElut column and eluted with dichloromethane. After concentration, 1 ml of water-methanol 1:1 (v/v) was added and this solution was analysed by LC-MS/MS.

## Descriptive analyses

A descriptive analysis of virus prevalence and pesticide residues was first carried out. The relationship between disorder and these stressing factors was then assessed. For each virus type, a linear model (two way ANOVA) was used, with the cycle threshold value as dependent variable and the visit (first or second), group (with disorder or healthy) and their interaction as explanatory variables. For the pesticides residues, we used a similar model with the total number of residues as dependent variable and type (fungicide or insecticide/acaricide), group (with disorder or healthy) and their interaction as explanatory variable. Based on these models, we constructed a contrast matrix to test explicit post-hoc hypotheses (e.g. compare virus load between groups within each visit…). As several dependent variables had a strongly asymmetric distribution (i.e. non-normal), we computed all p-values by permutation (n = 1000). We applied a Bonferoni correction on the post-hoc tests p-values to take into account the multiplicity of the tests.

## Relationship between bee colony disorders and potential stressors or surrounding environment of the apiary

We used two separate generalized linear mixed models with a binomial distribution and logit link function. In both models, the "group" (with disorder or healthy) was used as dependent binary variable and the apiary was used as random effect (grouping

**Table 1.** List of primers used for virus analyses.

| Target | Primer Name | Sequence (5′-3′) |
|---|---|---|
| BQCV | BQCV 9195F | GGT GCG GGA GAT GAT ATG GA |
| | BQCV 8265R | GCC GTC TGA GAT GCA TGA ATA C |
| | BQCV 8217T | TTT CCA TCT TTA TCG GTA CGC CGC C |
| SBV | SBV 311F | AAG TTG GAG GCG CGy AAT TG |
| | SBV 380R | CAA ATG TCT TCT TAC dAG AGG yAA GGA TTG |
| | SBV 331T (MGB) | CGG AGT GGA AAG AT |
| CPV | CPV 304F | TCT GGC TCT GTC TTC GCA AA |
| | CPV 371R | GAT ACC GTC GTC ACC CTC ATG |
| | CPV 325T | TGC CCA CCA ATA GTT GGC AGT CTG C |

**Table 2.** Active ingredients or metabolite included in the multi-residue analyses by beekeeping matrix (A = Acaricide; F = Fungicide; H = Herbicide; I = Insecticide; S = Synergist).

| Active ingredient or metabolite* | Class | LOQ (mg/kg) | | |
|---|---|---|---|---|
| | | Wax | Beebread | Honey |
| 2,4-D | H | 0.1 | – | 0.01 |
| Abamectin | I | 0.1 | 0.1 | 0.005 |
| Acetamiprid | I | 0.1 | 0.1 | 0.005 |
| Aldicarb | I | 0.1 | 0.1 | 0.005 |
| Alpha-cypermethrin | I | 0.1 | 0.1 | 0.005 |
| Amitraz | A | 0.1 | 0.1 | 0.01 |
| Azoxystrobin | F | 0.1 | 0.1 | 0.005 |
| Bentazone | H | 0.2 | – | 0.02 |
| Benthiavalicarb | F | 0.1 | 0.1 | 0.005 |
| Beta-cyfluthrin | I | 0.1 | 0.1 | 0.01 |
| Bifenthrin | I | 0.1 | 0.1 | 0.005 |
| Boscalid | F | 0.1 | 0.1 | 0.005 |
| Captan | F | 0.1 | 0.1 | 0.01 |
| Carbaryl | I | 0.1 | 0.1 | 0.005 |
| Chlorpyriphos | I | 0.1 | 0.1 | 0.005 |
| Chlorpyriphos-methyl | I | 0.1 | 0.1 | 0.005 |
| Chlorothalonil | F | 0.1 | 0.1 | 0.005 |
| Clothianidin | I | 0.1 | 0.1 | 0.01 |
| Coumaphos | A | 0.1 | 0.1 | 0.05 |
| Coumaphos oxon * | A | 0.1 | 0.1 | – |
| Coumaphos phenolic* | A | 0.1 | 0.1 | – |
| Cyazofamid | F | 0.2 | 0.2 | 0.02 |
| Cyfluthrin | I | 0.1 | 0.1 | 0.01 |
| Cymoxanil | F | 0.1 | 0.1 | 0.01 |
| Cypermethrin | I | 0.1 | 0.1 | 0.005 |
| Cyproconazole | F | 0.1 | 0.1 | 0.005 |
| Cyprodinil | F | 0.1 | 0.1 | 0.01 |
| DDT | I | 0.1 | 0.1 | 0.005 |
| Deltamethrine | I | 0.1 | 0.1 | 0.005 |
| Dichlorprop-P | H | 0.1 | – | 0.01 |
| Difenoconazole | F | 0.1 | 0.1 | 0.005 |
| Diflubenzuron | I | 0.2 | 0.2 | 0.02 |
| Dimethenamid-P | H | 0.1 | 0.1 | 0.005 |
| Dimethoate | I | 0.1 | 0.1 | 0.005 |
| Dimethomorph | F | 0.1 | 0.1 | 0.01 |
| Dimoxystrobin | F | 0.1 | 0.1 | 0.005 |
| Epoxiconazole | F | 0.1 | 0.1 | 0.005 |
| Esfenvalerate | I | 0.1 | 0.1 | 0.005 |
| Ethofumesate | H | 0.5 | 0.5 | 0.05 |
| Famoxadone | F | 0.1 | 0.1 | 0.01 |
| Fenhexamid | F | 0.1 | 0.1 | 0.005 |
| Fenoxycarb | F | 0.1 | 0.1 | 0.005 |
| Fenpropidin | F | 0.1 | 0.1 | 0.01 |
| Fenpropimorph | F | 0.1 | 0.1 | 0.005 |
| Fipronil | I | 0.1 | 0.1 | 0.01 |
| Flonicamid | I | 2 | 2 | 0.2 |
| Fluazinam | F | 0.1 | 0.1 | 0.01 |
| Flufenacet | H | 0.1 | 0.1 | 0.01 |

**Table 2.** Cont.

| Active ingredient or metabolite* | Class | LOQ (mg/kg) | | |
|---|---|---|---|---|
| | | Wax | Beebread | Honey |
| Fluopicolide | F | 0.1 | 0.1 | 0.01 |
| Flusilazole | F | 0.1 | 0.1 | 0.005 |
| Heptenophos | I | 0.1 | 0.1 | 0.01 |
| Imidacloprid | I | 0.1 | 0.1 | 0.005 |
| Indoxacarb | I | 0.1 | 0.1 | 0.005 |
| Iprodione | F | 0.1 | 0.1 | 0.005 |
| Kresoxim-methyl | I | 0.1 | 0.1 | 0.005 |
| Lambda-cyhalothrin | I | 0.1 | 0.1 | 0.005 |
| Lindane | I | 0.1 | 0.1 | 0.005 |
| Linuron | H | 0.1 | 0.1 | 0.005 |
| MCPA | H | 0.1 | – | 0.01 |
| Mecoprop-P | H | 0.1 | – | 0.01 |
| Metalaxyl-M | F | 0.1 | 0.1 | 0.005 |
| Metamitron | H | 0.1 | 0.1 | 0.005 |
| Metconazole | F | 0.2 | 0.2 | 0.02 |
| Methiocarb | I | 0.1 | 0.1 | 0.005 |
| Methoxyfenozide | F | 0.1 | 0.1 | 0.005 |
| Metribuzin | H | 0.1 | 0.1 | 0.005 |
| Penconazole | F | 0.1 | 0.1 | 0.005 |
| Pendimethalin | H | 0.1 | 0.1 | 0.005 |
| Permethrin | I | 0.2 | 0.2 | 0.02 |
| Picoxystrobin | F | 0.1 | 0.1 | 0.005 |
| Piperonyl butoxide | S | 0.1 | 0.1 | 0.01 |
| Pirimicarb | I | 0.1 | 0.1 | 0.005 |
| Prochloraz | F | 0.1 | 0.1 | 0.005 |
| Propamocarb | F | 0.1 | 0.1 | 0.005 |
| Propiconazole | F | 0.1 | 0.1 | 0.01 |
| Propyzamide | F | 0.1 | 0.1 | 0.01 |
| Pymetrozine | I | 0.1 | 0.1 | 0.005 |
| Pyraclostrobin | F | 0.1 | 0.1 | 0.005 |
| Pyrethrin | I | 0.2 | 0.2 | 0.02 |
| Pyrimethanil | F | 0.1 | 0.1 | 0.005 |
| Rotenone | I | 0.1 | 0.1 | 0.01 |
| Spinosad | I | 0.1 | 0.1 | – |
| Spirodiclofen | I | 0.1 | 0.1 | 0.01 |
| Spirotetramat | F | 0.1 | 0.1 | 0.01 |
| Spiroxamine | F | 0.1 | 0.1 | 0.005 |
| Sulcotrione | H | 0.1 | – | 0.01 |
| Tau-fluvalinate | IA | 0.1 | 0.1 | 0.01 |
| Tebuconazole | F | 0.1 | 0.1 | 0.005 |
| Tebufenozide | I | 0.1 | 0.1 | 0.005 |
| Tefluthrine | I | 0.1 | 0.1 | 0.005 |
| Terbuthylazine | H | 0.1 | 0.1 | 0.01 |
| Tetraconazole | F | 0.1 | 0.1 | 0.01 |
| Thiabendazole | F | 0.1 | 0.1 | 0.005 |
| Thiacloprid | I | 0.1 | 0.1 | 0.005 |
| Thiamethoxam | I | 0.1 | 0.1 | 0.005 |
| Thiophanate-methyl | F | 0.1 | 0.1 | 0.005 |
| Trifloxystrobin | F | 0.1 | 0.1 | 0.01 |

**Table 2.** Cont.

| Active ingredient or metabolite* | Class | LOQ (mg/kg) | | |
|---|---|---|---|---|
| | | Wax | Beebread | Honey |
| Zeta-cypermethrin | I | 0.1 | 0.1 | 0.005 |
| Zoxamide | F | 0.1 | 0.1 | 0.01 |

factor) to take into account the non-independence between colonies from the same apiary and allow the intercept to vary between apiaries [47].

With the first model, we explored the relationship between the probability of disorder in a colony and potential stress factors, i.e. pesticides or viruses total load into the colonies. We used as explanatory variables the total number of fungicides residues, the total number of insecticides or acaricides (pooled together) residues, the total number of viruses detected for both visits and all first level interactions between these three explanatory variables. The insecticides and acaricides were pooled together because the most frequently found active ingredient, namely tau-fluvalinate, is authorized in Belgium for both purposes.

With the second model, we investigated the relationship between the probability of disorder in a colony and the structure of the agricultural landscape around the beehive. The surfaces of all different kinds of agricultural soil occupancy in a circle with a radius of 1500 m around each apiary were calculated according to the information provided by the farmers to the Walloon administration (Land Parcel Identification System). The different soil occupancy categories were pooled into two categories defined according to the potential frequency of plant protection product use: (1) grasslands (low pesticides use), (2) crops *sensu lato* (potentially higher pesticides use), including major crops (mainly: cereals, potatoes, beet, oilseed rape, maize, flax, …) but also surfaces dedicated to fruit or vegetables production, fodder production (mainly legumes) and to horticulture. The surfaces of these two land use options, grasslands and crops (*sensu lato*), were used as explanatory variables in the model. The 1500 m radius was chosen accordingly to the mean pollen and nectar collecting distance for the honeybee [48,49]. As the agricultural soil occupancy data was available only for Wallonia, we eliminated from the analysis the apiaries whose buffer had less than 70% of their surface within the administrative boundaries of Wallonia (n = 2).

We used Likelihood Ratio (LR) Tests to evaluate the significance of the explanatory variables [47,50]. We respected the marginality rules, i.e. all main effects were tested after removing from the model the interactions in which they are involved [51].

All analyses were performed with R (R Core Team 2013) and the mixed models were fitted with the package lme4 [52].

## Results

### Description of honeybee colony disorders

We gathered all sources of information, i.e. questionnaires, analytical matrices and information from clinical visits, for 173 colonies. After the clustering, 54 colonies coming from 21 apiaries (Figure 1) were considered for the study: 25 presenting bee disorders and 29 not presenting them, defined as healthy group. Their data and samples were collected by 8 beekeeping technicians.

In the group presenting disorders, 6 colonies were dead or unviable -with only a handful of bees remaining with the queen- at the spring visit (Table 3). Pre-wintering weakness, winter worker losses, late-season re-queening were reported in these cases. One colony was dead due to queen failure without re-queening in spring. Furthermore, a number of colonies were weak in spring. They were characterised by a low number of individuals and a slow development. In total, nine colonies lost their queens during the study, five of them leading to queenless colonies. These were considered as "disorder colonies" in the framework of the study. Finally, brood abnormalities not linked to known disease were observed in 10 colonies, most of them both before and after the winter.

Symptoms of other diseases or bee parasite observation remained low within the selected colonies. Wax moths were observed in two cases, one in each group. *Varroa destructor* was present in all colonies of the study. All of them were treated according to official veterinary advice. Diarrhoea was observed in one of the colonies showing brood abnormalities. Anatomical modifications (i.e. deformed or undeveloped wings, small abdomens) were observed in 6 colonies, half of them being healthy and half of them showing disorders. No symptoms of any of the foulbroods were observed.

### Virus content

One bees sample from the "disorder" group could not be analysed due to a lack of sufficient matrix. The three most abundantly found viruses were the Deformed Wings Virus (DWV), the Black Queen Cell Virus (BQCV) and the Sac Brood Virus (SBV) (Figure 2). The Acute Bee Paralysis Virus (ABPV) and the Chronic Bee Paralysis Virus (CBPV) were found only in a very

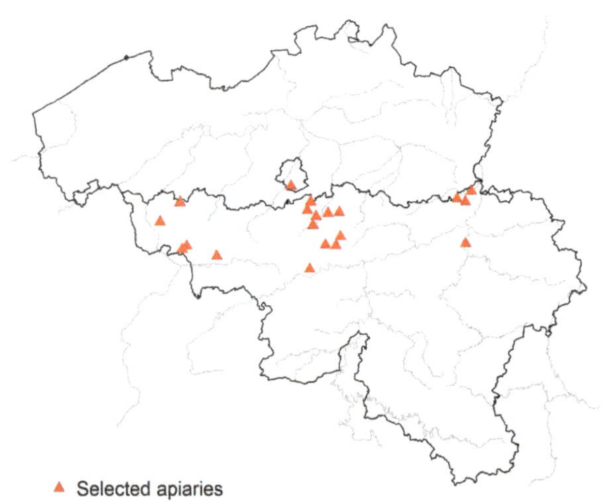


**Figure 1. Spatial distribution of selected apiaries.**

**Table 3.** Symptoms observed in the group with disorders.

| Symptoms of disorder | Frequency |
|---|---|
| Mortality | 2* |
| Weakening | 3 |
| Queen failure | 5 |
| Brood problems | 9 |
| Mortality+Weakening | 2 |
| Mortality+Weakening+Queen failure | 2 |
| Mortality+Queen failure | 1 |
| Queen failure+Brood problems | 1 |
| Total | 25 |

* One of these colonies not considered in the model due to a lack of viral results.

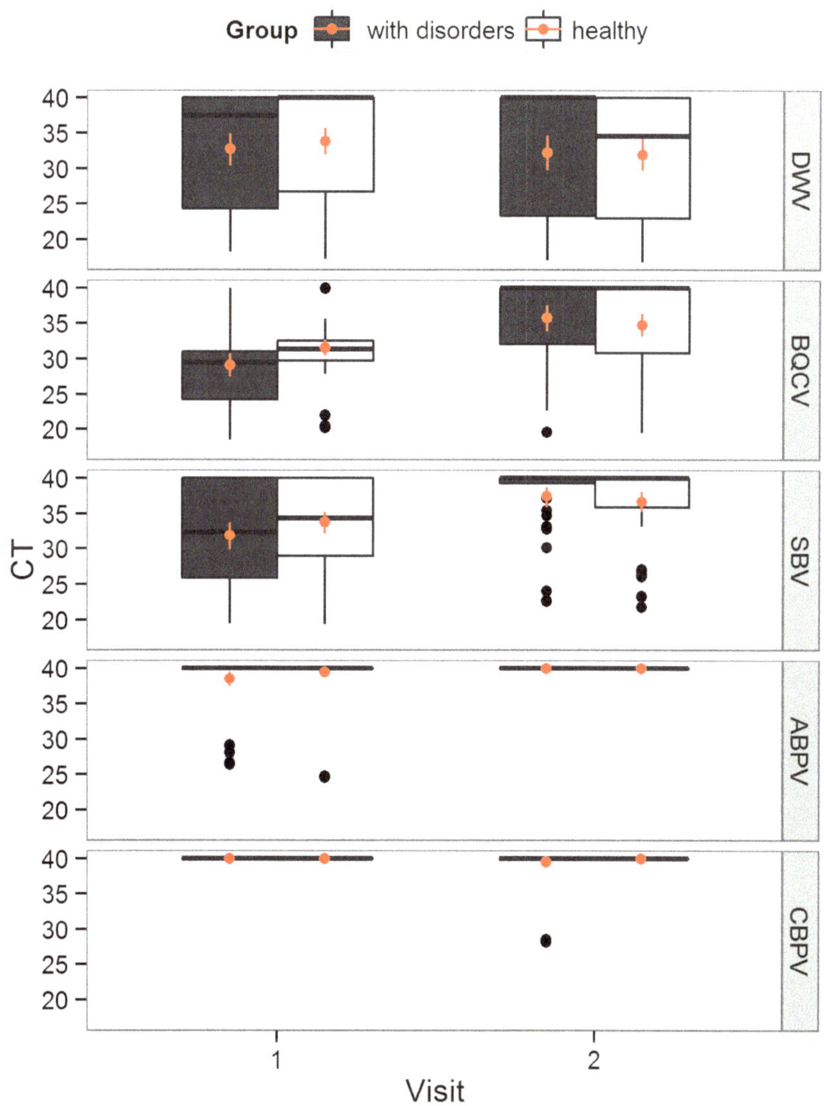


**Figure 2. Virus content according to the Cycle Threshold (CT) for the groups "with disorder" and "healthy".** Boxplot of Cycle Threshold for the first and second visits (visit 1 - mid-July to mid-august and visit 2 - mid-September to mid-October) and the group with disorders (grey, n = 24 colonies) and the healthy one (white, n = 29 colonies). Deformed Wings Virus (DWV), Black Queen Cell Virus (BQCV), SacBrood Virus (SBV), Acute Bee Paralysis Virus (ABPV), Chronic Bee Paralysis Virus (CBPV). In red, mean with confidence interval estimated by boostrap method. Note: CT values below 40 were regarded as positive results and the lowest CT values correspond to the higher virus contents.

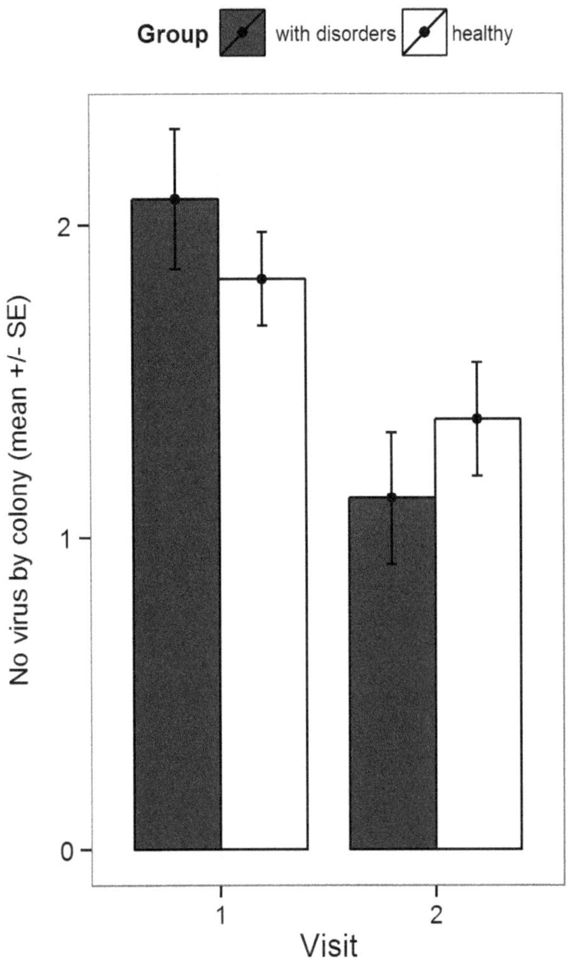


**Figure 3. Average number of different viruses per colony.** Data shown for the first and second visits (visit 1 - mid-July to mid-august and visit 2 - mid-September to mid-October) and the group with disorders (grey, n = 24 colonies) and the healthy one (white, n = 29 colonies). Whiskers show the standard error (SE).

limited number of samples and did not allowed particular statistical analyses on these two viruses.

There was no significant difference in viral content between the group with disorders or the healthy one, and independently of the visit for any of the three most abundant viruses (2-way ANOVA tested by permutations, DWV: $p = 0.731$, BQCV: $p = 0.373$, SBV: $p = 0.54$). We observed a decrease of virus content from summer to fall 2011, independently of the group, for BQCV ($p < 0.001$) and SBV ($p < 0.001$) but not for DWV ($p = 0.271$). For BQCV only, we observed a significant ($p = 0.036$) group×visit interaction indicating that the decrease of the virus abundance from visit 1 to visit 2 was higher in the group "with disorder" than in the group "healthy". Overall, the number of virus changed between visit 1 and visit 2 and a significant decrease was observed in the group with disorders ($p < 0.001$) (Figure 3).

## Pesticide analysis

172 agrochemical residues of 23 different active ingredients were detected in 94 out of 269 samples. Wax was the most contaminated matrix: 109 residues of 15 different active ingredients; while 39 and 24 residues of 10 and 8 substances were detected in beebread and honey, respectively (Figures 4 & 5).

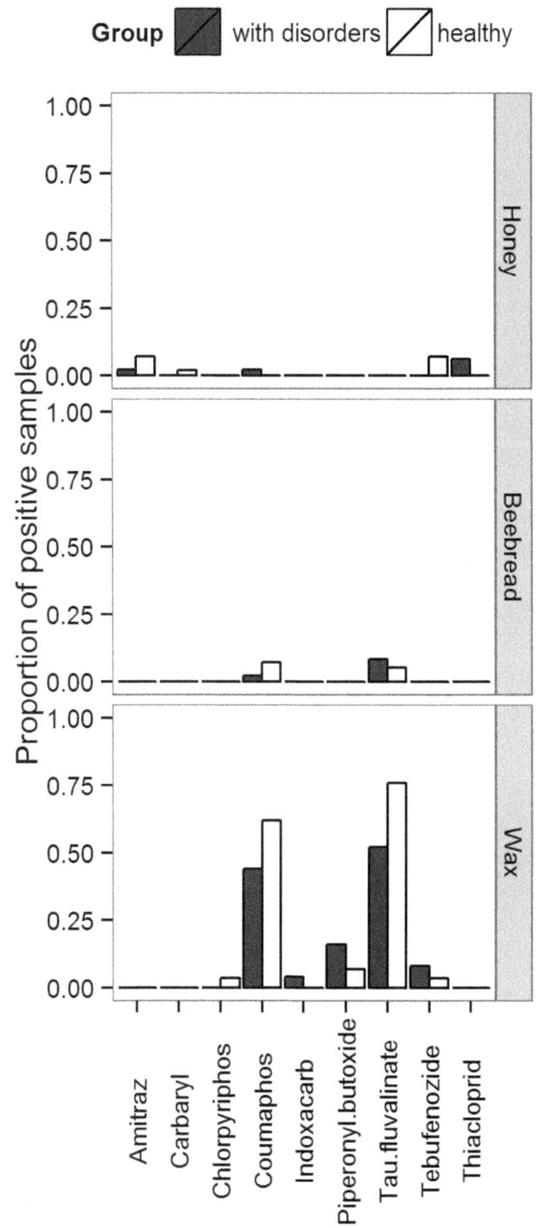


**Figure 4. Proportion of samples containing residues of acaricides/insecticides in the different beekeeping matrices (honey, beebread and wax).** Data shown for the group with disorders (grey, n = 25 colonies) and the healthy one (white, n = 29 colonies).

Residue levels contained in wax and beebread were higher (0.21–3.1 mg/kg) than those in honey (0.001–0.058 mg/kg).

For the subsequent statistical analysis, the results obtained for insecticide and acaricides residues were pooled together because the most frequently found active ingredient, tau-fluvalinate (n = 46), can be used in Belgium as an acaricide against varroa mite and as an insecticide to control *Meligethes aeneus* in rape. The second most frequently found residue was the coumaphos (n = 35), followed by two fungicides, boscalid (n = 33), iprodione (n = 13) (Table 4). Some residues of neonicotinoid insecticides, synergist and herbicide were also detected: thiacloprid (n = 3), piperonyl butoxide (n = 6), terbuthylazine (n = 1). The highest

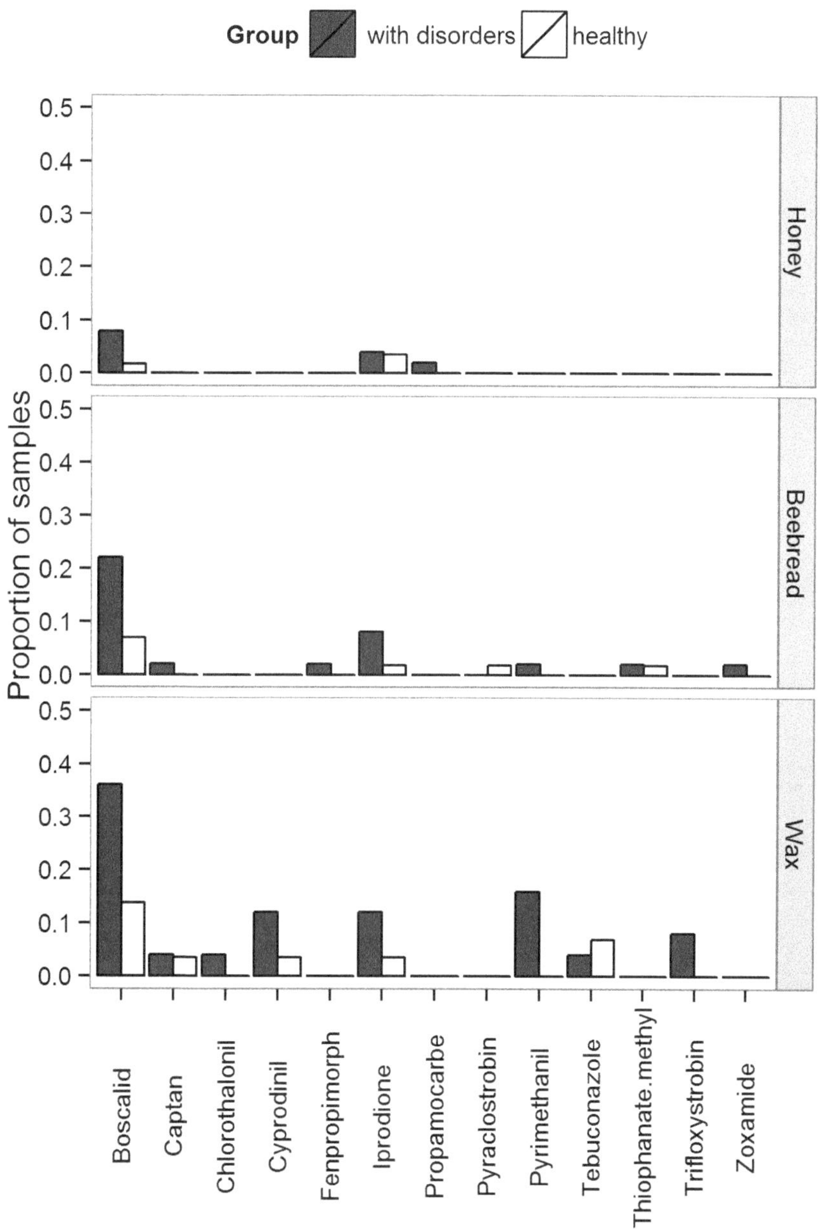


**Figure 5. Proportion of samples containing residues of fungicides in the different beekeeping matrices (honey, beebread and wax).** Data shown for the group with disorders (grey, n = 25 colonies) and the healthy one (white, n = 29 colonies).

residue level concerned captan with 3.1 and 1.9 mg/kg in the wax and bee bread of the same colony, respectively. Despite of being the most frequently found in matrices, tau-fluvalinate and coumaphos residues never exceeded 0.71 and 0.58 mg/kg, respectively. Boscalid, the most commonly found fungicide, ranged from 0.005 to 1.3 mg/kg.

There was a significant difference ($p = 0.01$) in terms of number of fungicide substances found between the "disorder" group (mean = 2.0) and the "healthy" one (mean = 0.7) as presented in Figure 6. The total number of insecticides/acaricides residues was slightly higher in the disorder group (mean = 2.1) than in the healthy group (mean = 1.64) but this difference is not statistically significant ($p = 0.79$). The most frequent fungicides in the group with disorders were boscalid and iprodione, detected in the three investigated matrices (Figure 5). However, for single active substances, no significant difference of number of residues was observed between the groups "with disorder" and "healthy".

## Probability of disorders in relation to potential stressors

There is clearly a significant positive relationship between the probability of a colony showing disorders and the total number of fungicides (Figure 7, LR = 7.128, df = 1, $p = 0.008$). The estimated probability for a colony to be in the "with disorder" group is 0.26 without fungicides residues, 0.60 with 2 fungicides residues and 0.88 with 4 fungicides residues when the insecticides-acaricides residues number and virus load are fixed to their observed mean value (Figure 7). None of the other variables (total number of viruses, total number of insecticides-acaricides) and none of the first level interactions seem to have any explanatory power on the probability of disorders in a colony (Table 5). No direct link could

**Table 4.** Residues of active ingredients found in wax, beebread and honey samples from colonies with disorders (Group D, n = 25 colonies) and healthy ones (Group H, n = 29 colonies).

| Active ingredient | Group | Wax LOQ (mg/kg) | Wax Number of samples (n = 54) | | | Wax Residues amount (mg/kg) | | | Bee bread LOQ (mg/kg) | Bee bread Number of samples (n = 108) | | | Bee bread Residues amount (mg/kg) | | | Honey LOQ (mg/kg) | Honey Number of samples (n = 107) | | | Honey Residues amount (mg/kg) | | |
|---|---|---|---|---|---|---|---|---|---|---|---|---|---|---|---|---|---|---|---|---|---|---|
| | | | <LOD | Detected | Quantified | Range | Mean | s.d. | | <LOD | Detected | Quantified | Range | Mean | s.d. | | <LOD | Detected | Quantified | Range | Mean | s.d. |
| Amitraz | D | 0.1 | – | – | – | | | | 0.1 | – | – | – | | | | 0.01 | 49 | – | 1 | 0.022 | 0.02 | |
| | H | | – | – | – | | | | | – | – | – | | | | | 53 | – | 4 | 0.01–0.028 | 0.02 | 0.01 |
| Boscalid | D | 0.1 | 16 | 8 | 1 | 0.29 | 0.29 | – | 0.1 | 39 | 7 | 4 | 0.4–1.3 | 0.68 | 0.42 | 0.005 | 46 | – | 4 | 0.005–0.026 | 0.01 | 0.01 |
| | H | | 25 | 4 | – | – | – | – | | 54 | 4 | – | – | – | – | | 56 | – | 1 | 0.058 | 0.06 | – |
| Captan | D | 0.1 | 24 | – | 1 | 3.1 | 3.1 | – | 0.1 | 48 | – | 1 | 1.9 | 1.90 | – | 0.01 | – | – | – | | | |
| | H | | 28 | 1 | – | – | – | – | | 58 | – | – | – | – | – | | – | – | – | | | |
| Carbaryl | D | 0.1 | – | – | – | | | | 0.1 | – | – | – | | | | 0.005 | 50 | – | | | | |
| | H | | – | – | – | | | | | – | – | – | | | | | 56 | – | 1 | 0.02 | 0.02 | – |
| Chlorpyriphos | D | 0.1 | 25 | – | – | | | | 0.1 | – | – | – | | | | 0.005 | – | – | – | | | |
| | H | | 28 | 1 | – | | | | | – | – | – | | | | | – | – | – | | | |
| Chlorothalonil | D | 0.1 | 24 | 1 | – | | | | 0.1 | – | – | – | | | | 0.005 | – | – | – | | | |
| | H | | 29 | – | – | | | | | – | – | – | | | | | – | – | – | | | |
| Coumaphos | D | 0.1 | 14 | 9 | 2 | 0.32–0.34 | 0.33 | 0.01 | 0.1 | 48 | 1 | – | | | | 0.05 | 49 | – | 1 | 0.012 | 0.01 | – |
| | H | | 11 | 10 | 8 | 0.23–0.58 | 0.37 | 0.01 | | 54 | 4 | – | | | | | 57 | – | – | | | |
| Cyprodinil | D | 0.1 | 22 | 3 | – | | | | 0.1 | – | – | – | | | | 0.01 | – | – | – | | | |
| | H | | 28 | 1 | – | | | | | – | – | – | | | | | – | – | – | | | |
| Fenpropimorph | D | 0.1 | – | – | – | | | | 0.1 | 49 | 1 | – | | | | 0.005 | – | – | – | | | |
| | H | | – | – | – | | | | | 58 | – | – | | | | | – | – | – | | | |
| Indoxacarb | D | 0.1 | 24 | 1 | – | | | | 0.1 | – | – | – | | | | 0.005 | – | – | – | | | |
| | H | | 29 | – | – | | | | | – | – | – | | | | | – | – | – | | | |
| Iprodione | D | 0.1 | 22 | 1 | 2 | 0.24–1.5 | 0.87 | 0.89 | 0.1 | 45 | – | 4 | 0.34–1.5 | 0.90 | 0.48 | 0.005 | 48 | – | 2 | 0.017–0.022 | 0.02 | 0.00 |
| | H | | 28 | 1 | – | – | – | – | | 57 | 1 | – | – | – | – | | 55 | – | 2 | 0.022–0.04 | 0.03 | 0.01 |
| Piperonyl butoxide | D | 0.1 | 21 | 4 | – | | | | 0.1 | – | – | – | | | | 0.01 | – | – | – | | | |
| | H | | 27 | 2 | – | | | | | – | – | – | | | | | – | – | – | | | |
| Propamocarbe | D | 0.1 | – | – | – | | | | 0.1 | – | – | – | | | | 0.005 | 49 | – | 1 | 0.008 | 0.01 | 0.00 |

**Table 4.** Cont.

| Active ingredient | Group | Wax | | | | | | | Bee bread | | | | | | | Honey | | | | | | |
|---|---|---|---|---|---|---|---|---|---|---|---|---|---|---|---|---|---|---|---|---|---|---|
| | | LOQ (mg/kg) | Number of samples (n = 54) | | | Residues amount (mg/kg) | | | LOQ (mg/kg) | Number of samples (n = 108) | | | Residues amount (mg/kg) | | | LOQ (mg/kg) | Number of samples (n = 107) | | | Residues amount (mg/kg) | | |
| | | | <LOD | Detected | Quantified | Range | Mean | s.d. | | <LOD | Detected | Quantified | Range | Mean | s.d. | | <LOD | Detected | Quantified | Range | Mean | s.d. |
| | H | | – | – | – | | | | | – | – | – | | | | | 57 | – | – | – | – | – |
| Pyraclostrobin | D | 0.1 | – | – | – | | | | 0.1 | 50 | – | – | | | | 0.005 | – | – | – | | | |
| | H | | – | – | – | | | | | 57 | 1 | – | | | | | – | – | – | | | |
| Pyrimethanil | D | 0.1 | 21 | 4 | – | | | | 0.1 | 49 | 1 | – | | | | 0.005 | – | – | – | | | |
| | H | | 29 | – | – | | | | | 58 | – | – | | | | | – | – | – | | | |
| Tau-fluvalinate | D | 0.1 | 12 | 9 | 4 | 0.29–0.46 | 0.4 | 0.08 | 0.1 | 45 | 4 | – | | | | 0.01 | 50 | – | – | – | – | – |
| | H | | 7 | 13 | 9 | 0.21–0.71 | 0.5 | 0.08 | | 55 | 3 | – | | | | | 53 | – | 4 | 0.011–0.02 | 0.02 | 0.00 |
| Tebuconazole | D | 0.1 | 24 | 1 | – | | | | 0.1 | – | – | – | | | | 0.005 | – | – | – | | | |
| | H | | 27 | 2 | – | | | | | – | – | – | | | | | – | – | – | | | |
| Tebufenozide | D | 0.1 | 23 | 2 | – | | | | 0.1 | – | – | – | | | | 0.005 | – | – | – | | | |
| | H | | 28 | 1 | – | | | | | – | – | – | | | | | – | – | – | | | |
| Terbuthylazine | D | 0.1 | 24 | 1 | – | | | | 0.1 | – | – | – | | | | 0.01 | – | – | – | | | |
| | H | | 29 | – | – | | | | | – | – | – | | | | | – | – | – | | | |
| Thiacloprid | D | 0.1 | – | – | – | | | | 0.1 | – | – | – | | | | 0.005 | 47 | – | 3 | 0.009–0.013 | 0.01 | 0.00 |
| | H | | – | – | – | | | | | – | – | – | | | | | 57 | – | – | – | – | – |
| Thiophanate-methyl | D | 0.1 | – | – | – | | | | 0.1 | 49 | 1 | – | | | | 0.005 | – | – | – | | | |
| | H | | – | – | – | | | | | 57 | – | 1 | 0.38 | 0.38 | – | | – | – | – | | | |
| Trifloxystrobin | D | 0.1 | 23 | 2 | – | | | | 0.1 | – | – | – | – | – | – | 0.01 | – | – | – | | | |
| | H | | 29 | – | – | | | | | – | – | – | | | | | – | – | – | | | |
| Zoxamide | D | 0.1 | – | – | – | | | | 0.1 | 49 | 1 | – | | | | 0.01 | – | – | – | | | |
| | H | | – | – | – | | | | | 58 | – | – | | | | | – | – | – | | | |

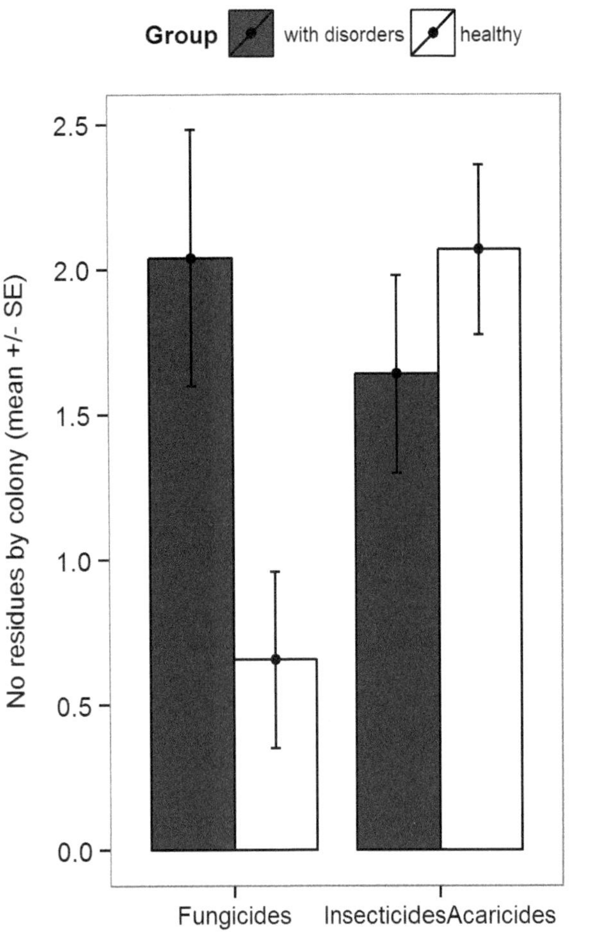


**Figure 6. Average number of residues per colony.** Data shown for the group with disorders (grey, n = 25 colonies) and the healthy ones (white, n = 29 colonies). Whiskers show the standard error (SE).

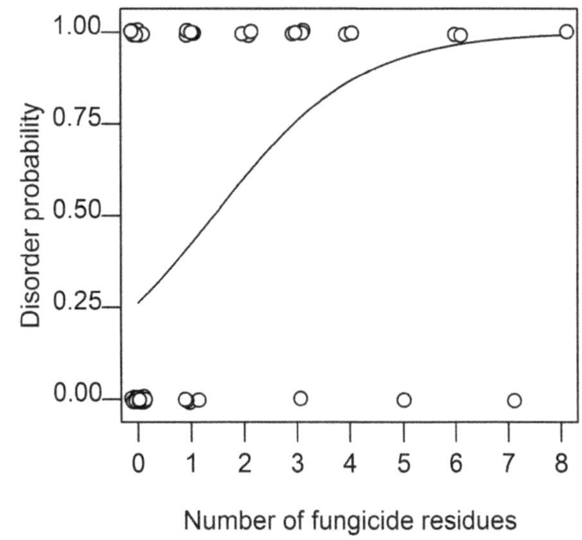


**Figure 7. Probability of honeybee colony disorders depending on the number of fungicide residues detected.** Model based on averaged coefficients and median value both for the number of insecticides-acaricides residues and total number of virus detected for both visits (n = 53 colonies).

be established between bee colony disorders and the amount (in µg/kg) of residues found in the matrices.

### Probability of disorders and crop/grassland surface

Our data clearly show a significant increase in disorder probability with the increase of crop surfaces (*sensu lato*, i.e. including fruit, vegetables, fodder production and horticulture) in the surrounding of the apiary (Figure 8, LR = 8.052, df = 1, $p = 0.0045$). The predicted probability of disorders is close to 0.1 for a crop surface of 0 ha in the radius of 1500 m and increases up to 0.8 for a surface of 500 ha of crops when fixing the grasslands surface to its observed mean.

On the contrary, the probability of disorders strongly decreases when the grassland surfaces increases (Figure 8, LR = 14.527, df = 1, $p = 0.0001$) after controlling for the crop surface. The predicted probability of disorders is close to 1 for a grassland surface of 0 ha in a radius of 1500 m and drops to ~0.1 for a surface of 150 ha of grasslands when fixing the crops surface to its observed mean.

Very similar results were found with a 3000 m buffer around the apiary (results not shown).

## Discussion

The bee disorders observed after the winter in the Walloon region happened despite of normal climatic conditions. The autumn of 2011 was dry (140.4 mm rainfall water from October to December, average being 219.9 mm), sunny and warm for Belgian conditions (12.4°C on average, which is normally 10.9°C for this period), followed by average winter and spring 2012 in terms of temperature, rainfall, with the exception of a cold wave lasting twelve days in February [53]. These conditions would not induce, *a priori,* a risk for honeybee colonies. Furthermore, based on studies carried out on the palynological diversity of the pollens collected by these colonies before the winter (Table S1), the hypothesis of nutritional lack is unlikely. All samples analysed contained at least 8 different botanic sources of pollen, including rich protein content as oilseed rape, ivy and *Phacelia*.

Viruses infections has often been mentioned as a source of stress for honeybee colonies [7]. However no significant difference in quantity or occurrence was observed between the healthy and the "with disorder" groups. DWV is one of the most commonly detected virus in *A. mellifera*. The prevalence of this virus is even more important at colonies infested by *Varroa destructor*, a well-established vector of this virus [7]. In accordance with the present study, further studies run on Belgian apiaries show DWV, BQCV and SBV as the most frequently found viruses [54]. Unlike DWV, we observed that the amount of SBV and BQCV has dropped significantly between the first and second visits for each group. In the case of BQCV, this is of no surprise as the cycle of incidence of this virus has been shown to increase in late winter, with a peak in May or June followed by a rapid decline [55]. The observed decrease of SBV has also been reported in other studies [42] in which the authors suggested that bees could develop a molecular defensive mechanism to reduce virus multiplication, or that the change in bee susceptibility to the virus could result from environmental conditions such as the quality of pollen.

We cannot prove any causal relationship between any of these viruses analysed and bee disorders, nor does the interaction

**Table 5.** Analysis of deviance table for generalized linear binomial models describing the relationships between the colony disorder probability and three variables: the total number of (1) fungicide residues (totfungicides), (2) insecticide-acaricide residues (totinsaca) and (3) virus detected for both visits (totvirus).

| | LR | df | p(>Chisq) |
|---|---|---|---|
| totfungicides | 7.13 | 1 | 0.008 |
| totinsaca | 0.005 | 1 | 0.943 |
| totvirus | 0.136 | 1 | 0.712 |
| totfungicides:totvirus | 2.222 | 1 | 0.136 |
| totinsaca:totvirus | 0.975 | 1 | 0.323 |
| totfungicides:totinsaca | 0.901 | 1 | 0.342 |

LR = likelihood ratio.

between these two factors. Cox-Foster et al., 2007 [41] found no clear correlation between a variety of pathogens, including *Nosema* spp., DWV, CBPV, ABPV, BQCV, *Mellisococcus pluton* and *Paenibacillus larvae* ssp and CCD. No specific spore counts were carried out in our study. However, no signs of nosemosis or foulbrood could be linked to colonies presenting bee disorders. It is noteworthy that a monitoring run at Belgian level found *Nosema* spp spores in one out of four colonies [56]. Nevertheless, Cox-Foster et al., 2007 [41], show a positive correlation between IAPV and CCD, which *a priori* would not be relevant in our conditions given the low prevalence of IAPV in this country [27,54]. A recent publication [36] shows a positive correlation between the presence of fungicides in pollen loads and the presence of spores of *Nosema ceranae*. We do not exclude the potential involvement of *Nosema* spp. in the case of bee disorders. However, in the framework of our study, *N. ceranae* seems either to play a role in the development of this weakening, while remaining asymptomatic, or not to play a decisive role in it.

When considering pesticide residues, no direct link could be established between bee colony disorders and the amount (in µg/kg) of residues found in the matrices. Neither could we identify any specific molecule as cause of bee disorders. Nevertheless, the study of the residue load of pesticides, specifically fungicides, opens new avenues for a better understanding of honeybee colony disorders.

Even if insecticides/acaricides were the most numerous residues detected in hives mainly in wax, no significant difference in the number of accumulated residues was observed between colonies with disorders and the healthy ones. Indeed the two most abundant active ingredients, tau-fluvalinate and coumaphos, came most probably from varroa control measures even if tau-fluvalinate is used as an insecticide (Mavrik 2F) against *Meligethes aeneus* in oilseed rape. These active ingredients seem to be a frequent outcome of residue analyses studies [57,58,39,38,40,59]. Synergistic effects have been shown between acaricides and other molecules [33,60]. Nevertheless, our modelling did not suggest any synergistic effects in the appearance of bee disorders occurring between residues of fungicides and insecticides-acaricides. Residues of synergist, piperonyl butoxide, were also detected that indicates a prior exposure to synthetic or natural pyrethroids even though residues were not found in the analysed matrix, probably due to a fast degradation of these active ingredients [61]. A non-authorised active ingredient was also detected: carbaryl, forbidden in Belgium since 2007, indicating an illegal agricultural or gardening use.

The only neonicotinoid found, thiacloprid, was detected in honey during the sampling period of July-August in two colonies. However, the limits of detection achieved in our study do not allow to determine the presence of neonicotinoids -with the exception of acetamiprid- or fipronil at levels in the range of the acute toxicity of these substances (30–40 µg/Kg). Residues of these substances could be present at lower levels and thus an exposure to these substances cannot be excluded.

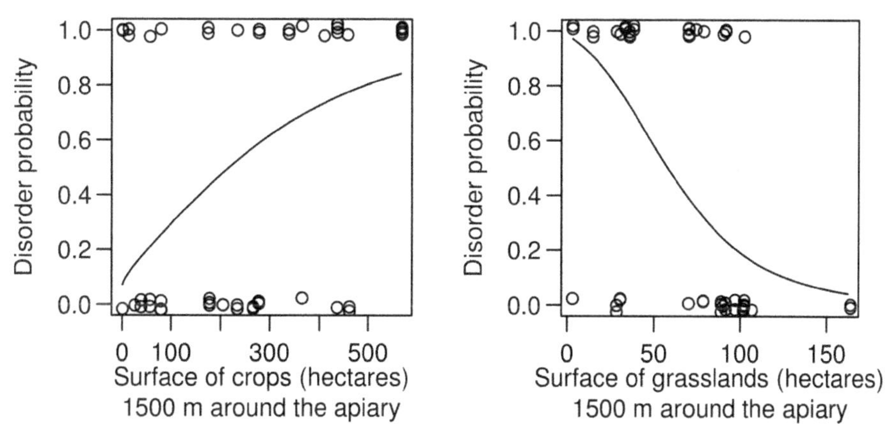


**Figure 8. Probability of honeybee colony disorders depending on the apiary's environment.** Consideration of crop surface *vs.* grassland surface in a radius of 1500 m around the apiary. Crops include fruit, vegetables, fodder production and horticulture. For each graph, the value of the variable not displayed is fixed to its observed mean (n = 53 colonies).

According to our results, the number of fungicide residues seems to plays a role in the appearance of honeybee colony disorder.

Significantly, more fungicides residues were detected in colonies with disorders than in the healthy ones. Mainly four active ingredients, frequently used in plant protection, were detected: boscalid, iprodione, pyrimethanil and cyprodinil. Fungicides are often considered safe for honeybees based on their acute toxicity. However some direct toxic effects on bees either by the mother compounds or their metabolites have been reported in the past [62]. For example, boscalid shows low acute toxicity to bees [63], despite the fact that beekeepers in the USA have already reported losses and adverse effects on bee brood development related to the foliar application of this systemic active ingredient [64]. Incidents were often linked to the use of a co-formulation boscalid – pyraclostrobin. However, both molecules appeared above detection levels in only two colonies of our study and could not explain the general trend. Interestingly, one metabolite of boscalid, the 2-chloronicotinic acid, is similar to the 6-chloronicotinic acid, a metabolite of imidacloprid. The latter has proved to be lethal to bees at low concentrations (0.1 μg/L) following chronic exposure [65]. Further research should be carried out in order to clarify the mechanism of bee toxicity of boscalid. Indeed, boscalid has proved toxic to other aquatic invertebrates, reducing daphnid fecundity and *Chironomid* emergence [64]. In addition, synergism with other active ingredients like insecticides are possible and increase the toxicity for honeybee [66,32,67]. Two other fungicides were also detected at very high levels in wax and beebread: iprodione and captan. The former is known for its synergistic effects in collaboration with insecticidal compounds [33]. The latter has been shown to induce effects on growth and development of larval honeybees [68,69]. Chronic and larval toxicity studies would be interesting at this stage in order to evaluate possible direct toxic effects on bee individuals and their behaviour. Indeed, a recent study showed increased larval mortality following chronic exposure to tau-fluvalinate, coumaphos, chlorpyriphos and chlorotalonyl or some of their combinations [70]. All these substances were found in our study. Increased mortality rates in the fall may compromise the size and age structure of the wintering cluster, which could lead to winter losses.

The indirect effects of fungicides on bees or on bee colonies are relatively little-known to date.

Fungicides may have an impact on the colony by modifying the existing microflora present in the food stores or in the bee intestinal tract [71]. Studies have already shown the possible modification of microbial composition both at beebread level [62,72] and at intestinal level [73]. This modification in the composition of microbiota may lead to dysbiosis [74]. The impact that such an unbalance in the bee gut microflora may have on bee health has already been considered. The link between the unspecific symptoms observed in our study and a possible microbial alteration could be subject of further research.

In parallel, the potential impact of microbial modification on digestibility and availability of nutrients should be a target for further research. Indeed, the content of essential amino acids might be altered when beebread is contaminated with fungicides (DeGrandi-Hoffman, 2013, pers com.). Given the importance of nutrition, especially pollen, in the good development of the colony [75] alterations in composition or lack of essential nutrients would put the homeostasis of the colony at stake. Some studies have already shown the impact of nutritional lack on bee development and health [76]. Provided that pollen is the unique source of amino acids for honeybees, royal jelly production could also be affected [77,78] with unexpected potential consequences for its main consumers, larvae and the queen. A poor nutrition of the queen, could have as a result an impact in its activity. Likewise, a poor nutrition of the larvae has been shown to impact their development [76]. As a result, the presence of fungicides on beebread and honey may have both a direct effect on their health, but also an indirect one on the colony development.

Fungicides are widely used in agriculture and are broadly present in bee matrices, sometimes at high concentrations at levels of mg/kg [79,80,38,59]. Boscalid, cyprodinil, iprodione are used to control a broad range of fungal pathogens including *Botrytis* spp., *Alternaria* spp. and *Sclerotinia* spp. on a wide range of crops including fruits, vegetables and ornamentals. Pyrimethanil is more specifically used to control grey mould on fruits, vegetables and ornamentals, and leaf scab on apple trees [81]. These active ingredients were already reported in bee matrices in Europe and in the USA [23,39,62]. Their frequent presence in bee matrices might be an indication of chemical plant protection intensity in all agricultural landscapes. Fungicides could also be markers of exposure to other active ingredients with higher toxicity to bees. Mixes of products like fungicides – insecticides are often applied to reduce the number of spray applications. As a result, other pesticides often used in combination with the fungicides found or applied in the same crops could have been at the origin of the effects observed in this study. However, the sensibility of the residue analyses used in our study might explain the lack of detection of such components. Further intensive monitoring and a thorough record of the agricultural practices on pesticide application in tank mixes would help clarifying this alternative explanation.

Factors different from fungicides are most likely involved in the development of bee disorders. According to our model, the disorder probability was not absent when fungicides residues were not detected in presence of insecticide residues and virus. Non identified pathogens, chemicals or factors of different nature could be operating as silent stressors. Ravoet and colleagues (2013) [82] reported the presence of new pathogens in Belgium that were not taken into consideration in our analyses (i.e. *Crithidia mellificae* or the Lake Sinai Virus (LSV)). Other stressing factors could be also linked to the intensive agriculture. In fact, we observed an increase of colony disorders in the area with high density of crops in comparison with areas with grassland. Some studies and beekeeper claims go in line with our outcome regarding the concentration of bee problems in areas with intensive agriculture [83–86]. Furthermore, the same negative trends on pollinators and biodiversity in areas with intensive agriculture have already been described as the result of as increased pesticide use, decreased landscape heterogeneity, loss and fragmentation of natural habitat [87,88].

In conclusion, the five virus studied (ABPV, CBPV, QBCV, SBV, DWV) do not seem determinant in the appearance of bee disorders in our study. These disorders seem clearly linked to the environment of the apiaries and were observed mainly in agricultural crop areas. We observed also that the number of fungicide residues appears as the main potential stress factor linked to bee disorder. However other stressing factors could be acting or interacting at the same time: insecticides exposure, a lack of amino acids and oligo-elements, etc. Our results open new avenues for future research in order to better understand the side effects of fungicides on the bee colony and questions the sustainability of intensive agriculture model and its impact on bees. Specific toxicological studies on both adult bees and larvae would be recommendable in order to better characterise the toxicity of fungicides.

## Acknowledgments

We would like to salute and to thank the beekeepers and beekeeping technicians for their dedication and active contribution to our study. Likewise, we thank Gloria DiGrandi-Hoffman, Gerard Arnold, Szaniszlo Szöke and Martin Dermine for their critical and constructive review of the article. We would like also to thank the Direction Générale Opérationnelle Agriculture, Ressources Naturelles et Environnement (DGO3), Département des Aides (D4), Direction des Surfaces (D42), Service 42/3 - LPIS (Land Parcel Identification System) - Service Public de Wallonie for the spatial data. We would also like to thank the Molecular Technology Unit at The Food and Environment Research Agency, Sand Hutton, York, YO41 1LZ for Bee Pathogen Screening Services. Finally, we would like to thank the reviewers for their helpful comments.

## Author Contributions

Conceived and designed the experiments: NS LH EB. Performed the experiments: NS LH CM. Analyzed the data: GSM LH. Wrote the paper: LH NS LAM GSM.

## References

1. Biesmeijer JC, Roberts SPM, Reemer M, Ohlemüller R, Edwards M, et al. (2006) Parallel declines in pollinators and insect-pollinated plants in Britain and the Netherlands. Science 313: 351–354.
2. Potts SG, Biesmeijer JC, Kremen C, Neumann P, Schweiger O, et al. (2010) Global pollinator declines: trends, impacts and drivers. Trends Ecol Evol 25: 345–353.
3. Goulson D, Lye GC, Darvill B (2008) Decline and conservation of bumble bees. Annu Rev Entomol 53: 191–208.
4. Carvalheiro LG, Kunin WE, Keil P, Aguirre-Gutiérrez J, Ellis WN, et al. (2013) Species richness declines and biotic homogenisation have slowed down for NW-European pollinators and plants. Ecol Lett 16: 870–878. doi:10.1111/ele.12121.
5. Kluser S, Peduzzi P (2007) Global pollinator decline: a literature review. UNEP/GRID- Europe. © UNEP 2007. Switzerland.
6. Maxim L, van der Sluijs JP (2010) Expert explanations of honeybee losses in areas of extensive agriculture in France: Gaucho ® compared with other supposed causal factors. Environ Res Lett 5: 014006. doi:10.1088/1748-9326/5/1/014006.
7. vanEngelsdorp D, Meixner MD (2010) A historical review of managed honey bee populations in Europe and the United States and the factors that may affect them. J Invertebr Pathol 103: S80–S95.
8. Van der Zee R, Pisa L, Andonov S, Brodschneider R, Charriere J-D, et al. (2012) Managed honey bee colony losses in Canada, China, Europe, Israel and Turkey, for the winters of 2008–9 and 1009–10. J Apic Res Bee World 51: 100–114.
9. Klein A-M, Müller C, Hoehn P, Kremen C (2009) Understanding the role of species richness for crop pollination services. Biodiversity, ecosystem function and human wellbeing. New York. 195–208.
10. Lautenbach S, Seppelt R, Liebscher J, Dormann CF (2012) Spatial and temporal trends of global pollination benefit. PloS One 7: e35954.
11. Banskota AH, Tezuka Y, Kadota S (2001) Recent progress in pharmacological research of propolis. Phytother Res 15: 561–571. doi:10.1002/ptr.1029.
12. Jull AB, Rodgers A, Walker N (2008) Honey as a topical treatment for wounds. Cochrane Database of Systematic Reviews. Chichester, UK: John Wiley & Sons, Ltd. 1–43.
13. Srinivasan MV (2011) Honeybees as a Model for the Study of Visually Guided Flight, Navigation, and Biologically Inspired Robotics. Physiol Rev 91: 413–460. doi:10.1152/physrev.00005.2010.
14. Bradber N (1990) Beekeeping in rural development. International Bee Research Association. 15 pp.
15. UNEP (2005) Ecosystems and Human Well-Being: Our Human Planet: Summary for Decision Makers. Island Press. Washington: Island Press.
16. Rasmont P, Pauly A, Terzo M, Patiny S, Michez D, et al. (2005) The survey of wild bees (Hymenoptera, Apoidea) in Belgium and France. Food Agric Organ Rome: 18.
17. Lefevbre M, Bruneau E (2004) État des lieux du phénomène de dépérissement des ruchers en Région wallonne. Louvain la Neuve.
18. Nguyen BK, Mignon J, Laget D, de Graaf D, Jacobs F, et al. (2010) Honey bee colony losses in Belgium during the 2008–9 winter. J Apic Res 49: 337–339. doi:10.3896/IBRA.1.49.4.07.
19. Neumann P, Carreck NL (2010) Honey bee colony losses. J Apic Res 49: 1–6.
20. Vanengelsdorp D, Hayes J, Underwood RM, Caron D, Pettis J (2011) A survey of managed honey bee colony losses in the USA, fall 2009 to winter 2010. J Apic Res 50: 1–10.
21. Aubert M, Faucon J-P, Chauzat M-P (2008) Enquête prospective multi-factorielle: influence des agents microbiens et parasitaires, et des résidus de pesticides sur le devenir de colonies d'abeilles domestiques en conditions naturelles. AFSSA.
22. Higes M, Martín-Hernández R, Garrido-Bailón E, González-Porto AV, García-Palencia P, et al. (2009) Honeybee colony collapse due to *Nosema ceranae* in professional apiaries. Environ Microbiol Rep 1: 110–113. doi:10.1111/j.1758-2229.2009.00014.x.
23. Genersch E, von der Ohe W, Kaatz H, Schroeder A, Otten C, et al. (2010) The German bee monitoring project: a long term study to understand periodically high winter losses of honey bee colonies. Apidologie 41: 332–352.
24. vanEngelsdorp D, Speybroeck N, Evans JD, Nguyen BK, Mullin C, et al. (2010) Weighing Risk Factors Associated With Bee Colony Collapse Disorder by Classification and Regression Tree Analysis. J Econ Entomol 103: 1517–1523. doi:10.1603/EC09429.
25. Pettis JS (2013) The Role Of Pesticides In Queen Health & Sperm Viability. Apimondia 2014, Kiev.
26. Farooqui T (2013) A potential link among biogenic amines-based pesticides, learning and memory, and colony collapse disorder: A unique hypothesis. Neurochem Int 62: 122–136. doi:10.1016/j.neuint.2012.09.020.
27. Nguyen BK, Ribiere M, Vanengelsdorp D, Snoeck C, Saegerman C, et al. (2011) Effects of honey bee virus prevalence, Varroa destructor load and queen condition on honey bee colony survival over the winter in Belgium. J Apic Res 50: 195–202.
28. Nguyen BK, Saegerman C, Pirard C, Mignon J, Widart J, et al. (2009) Does imidacloprid seed-treated maize have an impact on honey bee mortality? J Econ Entomol 102: 616–623.
29. Desneux N, Decourtye A, Delpuech J-M (2007) The Sublethal Effects of Pesticides on Beneficial Arthropods. Annu Rev Entomol 52: 81–106. doi:10.1146/annurev.ento.52.110405.091440.
30. Belzunces LP, Tchamitchian S, Brunet J-L (2012) Neural effects of insecticides in the honey bee. Apidologie 43: 348–370. doi:10.1007/s13592-012-0134-0.
31. Van der Sluijs JP, Simon-Delso N, Goulson D, Maxim L, Bonmatin J-M, et al. (2013) Neonicotinoids, bee disorders and the sustainability of pollinator services. Curr Opin Environ Sustain 5: 293–305. doi:10.1016/j.cosust.2013.05.007.
32. Pilling ED, Jepson PC (1993) Synergism between EBI fungicides and a pyrethroid insecticide in the honeybee (Apis mellifera). Pestic Sci 39: 293–297.
33. Fischer R, Wachendorff-Neumann U (2003) Active ingredient combinations with insecticidal, fungicidal and acaricidal properties.
34. Alaux C, Brunet J-L, Dussaubat C, Mondet F, Tchamitchan S, et al. (2010) Interactions between *Nosema* microspores and a neonicotinoid weaken honeybees (*Apis mellifera* ). Environ Microbiol 12: 774–782. doi:10.1111/j.1462-2920.2009.02123.x.
35. Vidau C, González-Polo RA, Niso-Santano M, Gómez-Sánchez R, Bravo-San Pedro JM, et al. (2011) Fipronil is a powerful uncoupler of oxidative phosphorylation that triggers apoptosis in human neuronal cell line SHSY5Y. NeuroToxicology 32: 935–943. doi:10.1016/j.neuro.2011.04.006.
36. Pettis JS, Lichtenberg EM, Andree M, Stitzinger J, Rose R (2013) Crop pollination exposes honey bees to pesticides which alters their susceptibility to the gut pathogen Nosema ceranae. PloS One 8: e70182.
37. Bogdanov S (2005) Contaminants of bee products. Apidologie 37: 1–18. doi:10.1051/apido:2005043.
38. Johnson RM, Ellis MD, Mullin CA, Frazier M (2010) Pesticides and honey bee toxicity – USA. Apidologie 41: 312–331. doi:10.1051/apido/2010018.
39. Mullin CA, Frazier M, Frazier JL, Ashcraft S, Simonds R, et al. (2010) High Levels of Miticides and Agrochemicals in North American Apiaries: Implications for Honey Bee Health. PLoS ONE 5: e9754. doi:10.1371/journal.pone.0009754.
40. Chauzat M-P, Martel A-C, Cougoule N, Porta P, Lachaize J, et al. (2011) An assessment of honeybee colony matrices, Apis mellifera (Hymenoptera: Apidae) to monitor pesticide presence in continental France. Environ Toxicol Chem 30: 103–111. doi:10.1002/etc.361.
41. Cox-Foster DL, Conlan S, Holmes EC, Palacios G, Evans JD, et al. (2007) A metagenomic survey of microbes in honey bee colony collapse disorder. Science 318: 283–287.
42. Tentcheva D, Gauthier L, Zappulla N, Dainat B, Cousserans F, et al. (2004) Prevalence and seasonal variations of six bee viruses in Apis mellifera L. and Varroa destructor mite populations in France. Appl Environ Microbiol 70: 7185–7191.
43. Gauthier L, Tentcheva D, Tournaire M, Dainat B, Cousserans F, et al. (2007) Viral load estimation in asymptomatic honey bee colonies using the quantitative RT-PCR technique. Apidologie 38: 426–435.
44. Zuur AF, Ieno EN, Smith GM (2007) The Analysis of Ecological Data. Springer-Verlag.
45. Chantawannakul P, Ward L, Boonham N, Brown M (2006) A Scientific Note on the Detection of Honeybee Viruses Using real-time PCR (TaqMan®) in Varroa

Mites Collected from a Thai Honeybee (Apis mellifera) Apiary. Journal of Invertebrate Pathology: 91: 69–73.
46. Martin SJ, Highfield AC, Brettell L, Villalobos EM, Budge GE, et al. (2012) Global honey bee viral landscape altered by a parasitic mite. Science 336: 1304–1306.
47. Zuur AF, Ieno EN, Elphick CS (2010) A protocol for data exploration to avoid common statistical problems. Methods Ecol Evol 1: 3–14.
48. Beekman M, Ratnieks FLW (2000) Long-range foraging by the honey-bee, Apis mellifera L. Funct Ecol 14: 490–496.
49. Steffan-dewenter I, Kuhn A (2003) Honeybee foraging in differentially structured landscapes. Proc R Soc Lond B Biol Sci 270: 569–575.
50. Pinheiro JC, Bates DM (2000) Mixed-Effects Models in S and S-PLUS. Springer. 560 p.
51. Fox J (2002) An R and S-Plus companion to applied regression. Sage Publications, Inc. Available: http://books.google.be/books?hl = en&lr = &id = xWS8kgRjGcAC &oi = fnd&pg = PR9&dq = Fox+2002+An+R+and+S+Plus+&ots = o5qCjIMtOb &sig = lvi2jQFv4iDIBrv0QtaWiRIlRBM. Accessed 2012 Sep 18.
52. Bates D (2013) Linear mixed model implementation in lme4. Ms Univ Wis. Available: http://debianmirror.wwi.dk/cran/web/packages/lme4/vignettes/Implementation.pdf. Accessed 2014 Feb 3.
53. IRM IR de M (2013) Bilan climatologique annuel. Available: http://www.meteo.be/meteo/view/fr/1317239-Bilan+climatologique+annuel.html. Accessed 2013 Dec 15.
54. De Smet L, Ravoet J, de Miranda JR, Wenseleers T, Mueller MY, et al. (2012) BeeDoctor, a versatile MLPA-based diagnostic tool for screening bee viruses. PloS One 7: e47953.
55. Bailey L, Ball BV, Perry JN (1981) The prevalence of viruses of honey bees in Britain. Ann Appl Biol 97: 109–118.
56. Verhoeven B (2014) Projet pilote de surveillance de la mortalité des abeilles 2012–2013 Premiers résultats. AFSCA. Journé apicole 26 janvier 2014, Namur.
57. Chauzat M-P, Faucon J-P, Martel A-C, Lachaize J, Cougoule N, et al. (2006) A survey of pesticide residues in pollen loads collected by honey bees in France. J Econ Entomol 99: 253–262.
58. Chauzat M-P, Faucon J-P (2007) Pesticide residues in beeswax samples collected from honey bee colonies (Apis mellifera L.) in France. Pest Manag Sci 63: 1100–1106. doi:10.1002/ps.1451.
59. Wiest L, Buleté A, Giroud B, Fratta C, Amic S, et al. (2011) Multi-residue analysis of 80 environmental contaminants in honeys, honeybees and pollens by one extraction procedure followed by liquid and gas chromatography coupled with mass spectrometric detection. J Chromatogr A 1218: 5743–5756. doi:10.1016/j.chroma.2011.06.079.
60. Johnson RM, Pollock HS, Berenbaum MR (2009) Synergistic interactions between in-hive miticides in Apis mellifera. J Econ Entomol 102: 474–479.
61. Yu SJ (2008) The toxicology and biochemistry of insecticides. CRC N Y NY USA. Available: http://www.lavoisier.fr/livre/notice.asp?ouvrage = 1795057. Accessed 2014 Jan 31.
62. DeGrandi-Hoffman G, Chen Y, Simonds R (2013) The Effects of Pesticides on Queen Rearing and Virus Titers in Honey Bees (Apis mellifera L.). Insects 4: 71–89. doi:10.3390/insects4010071.
63. Pesticide Properties Database (PPDB) (2014) Boscalid (Ref: BAS 510F). Available: http://sitem.herts.ac.uk/aeru/ppdb/en/86.htm. Accessed 2014 Jan 3.
64. Aubee C, Lieu D (2010) Environmental Fate and Ecological Risk Assessment for Boscalid New Use on Rapeseed, Including Canola (Seed Treatment). U.S. EPA - Washington.
65. Suchail S, Guez D, Belzunces LP (2001) Discrepancy between acute and chronic toxicity induced by imidacloprid and its metabolites in Apis mellifera. Environ Toxicol Chem SETAC 20: 2482–2486.
66. Belzunces LP, Colin ME (1992) Les phenoménes d'affaiblissement de ruchers.
67. Meled M, Thrasyvoulou A, Belzunces LP (1998) Seasonal variations in susceptibility of Apis mellifera to the synergistic action of prochloraz and deltamethrin. Environ Toxicol Chem 17: 2517–2520.
68. Thompson HM (2003) Behavioural effects of pesticides in bees–their potential for use in risk assessment. Ecotoxicology 12: 317–330.
69. Mussen EC, Lopez JE, Peng CY (2004) Effects of selected fungicides on growth and development of larval honey bees, Apis mellifera L.(Hymenoptera: Apidae). Environ Entomol 33: 1151–1154.
70. Zhu W, Schmehl DR, Mullin CA, Frazier JL (2014) Four Common Pesticides, Their Mixtures and a Formulation Solvent in the Hive Environment Have High Oral Toxicity to Honey Bee Larvae. PloS One 9: e77547.
71. Batra LR, Batra SWT, Bohart GE (1973) The mycoflora of domesticated and wild bees (Apoidea). Mycopathol Mycol Appl 49: 13–44.
72. Yoder JA, Jajack AJ, Rosselot AE, Smith TJ, Yerke MC, et al. (2013) Fungicide Contamination Reduces Beneficial Fungi in Bee Bread Based on an Area-Wide Field Study in Honey Bee, *Apis mellifera*, Colonies. J Toxicol Environ Health A 76: 587–600. doi:10.1080/15287394.2013.798846.
73. Anderson KE, Sheehan TH, Eckholm BJ, Mott BM, DeGrandi-Hoffman G (2011) An emerging paradigm of colony health: microbial balance of the honey bee and hive (Apis mellifera). Insectes Sociaux 58: 431–444. doi:10.1007/s00040-011-0194-6.
74. Sartor RB (2008) Therapeutic correction of bacterial dysbiosis discovered by molecular techniques. Proc Natl Acad Sci 105: 16413–16414.
75. Di Pasquale G, Salignon M, Le Conte Y, Belzunces LP, Decourtye A, et al. (2013) Influence of Pollen Nutrition on Honey Bee Health: Do Pollen Quality and Diversity Matter? PLoS ONE 8: e72016. doi:10.1371/journal.pone.0072016.
76. Brodschneider R, Crailsheim K (2010) Nutrition and health in honey bees. Apidologie 41: 278–294.
77. Standifer LN (1967) A comparison of the protein quality of pollens for growth-stimulation of the hypopharyngeal glands and longevity of honey bees, Apis mellifera L. (Hymenoptera: Apidae). Insectes Sociaux 14: 415–425. doi:10.1007/BF02223687.
78. Liming W, Jinhui Z, Xiaofeng X, Yi L, Jing Z (2009) Fast determination of 26 amino acids and their content changes in royal jelly during storage using ultra-performance liquid chromatography. J Food Compos Anal 22: 242–249.
79. Kubik M, Nowacki J, Pidek A, Warakomska Z, Michalczuk L, et al. (1999) Pesticide residues in bee products collected from cherry trees protected during blooming period with contact and systemic fungicides. Apidologie 30: 12. doi:10.1051/apido:19990607.
80. Kubik M, Nowacki J, Pidek A, Warakomska Z, Michalczuk L, et al. (2000) Residues of captan (contact) and difenoconazole (systemic) fungicides in bee products from an apple orchard. Apidologie 31: 531–542.
81. Pesticide Properties Database (PPDB) (2014) Pyrimethanil (Ref: SN 100309). Available: http://sitem.herts.ac.uk/aeru/ppdb/en/573.htm. Accessed 2014 Jan 3.
82. Ravoet J, Maharramov J, Meeus I, De Smet L, Wenseleers T, et al. (2013) Comprehensive bee pathogen screening in Belgium reveals Crithidia mellificae as a new contributory factor to winter mortality. PloS One 8: e72443.
83. Oomen PA, Belzunces L, Pélissier C, Lewis G (2001) Honey bee poisoning incidents over the last ten years, as reported by bee keepers in The Netherlands. Les Colloques 98: 129–136.
84. Brittain CA, Vighi M, Bommarco R, Settele J, Potts SG (2010) Impacts of a pesticide on pollinator species richness at different spatial scales. Basic Appl Ecol 11: 106–115.
85. CRA-API (2011) Effects of coated maize seed on honey bees. Report based on results obtained from the first year of activity of the APENET project. Italy.
86. Maxim L, van der Sluijs J (2013) Seed-dressing systemic insecticides and honeybees. Late Lessons from Early Warnings: Science, Precaution, Innovation. Copenhagen. 401–438. Available: http://www.beekeeping.com/articles/us/late_lessons_from_early_warnings_2.pdf. Accessed 2014 Feb 10.
87. Holzschuh A, Steffan-Dewenter I, Tscharntke T (2008) Agricultural landscapes with organic crops support higher pollinator diversity. Oikos 117: 354–361.
88. Kleijn D, Kohler F, Báldi A, Batáry P, Concepcion ED, et al. (2009) On the relationship between farmland biodiversity and land-use intensity in Europe. Proc R Soc B Biol Sci 276: 903–909.

# 17

# Effects of Caste on the Expression of Genes Associated with Septic Injury and Xenobiotic Exposure in the Formosan Subterranean Termite

**Claudia Husseneder***, **Dawn M. Simms**

Department of Entomology, Louisiana State University Agricultural Center, Baton Rouge, Louisiana, United States of America

## Abstract

As social insects, termites live in densely populated colonies with specialized castes under conditions conducive to microbial growth and transmission. Furthermore, termites are exposed to xenobiotics in soil and their lignocellulose diet. Therefore, termites are valuable models for studying gene expression involved in response to septic injury, immunity and detoxification in relation to caste membership. In this study, workers and soldiers of the Formosan subterranean termite, *Coptotermes formosanus,* were challenged by bacterial injection or by no-choice feeding with a sublethal concentration (0.5%) of phenobarbital. Constitutive and induced expression of six putative immune response genes (two encoding for lectin-like proteins, one for a ficolin-precursor, one for the Down syndrome cell adhesion molecule, one for a chitin binding protein, and one for the gram-negative binding protein 2) and four putative detoxification genes (two encoding for cytochrome P450s, one for glutathione S-transferase, and one for the multi antimicrobial extrusion protein), were measured via quantitative real time polymerase chain reaction and compared within and among 1) colonies, 2) treatment types and 3) castes via ANOVA. Eight genes were inducible by septic injury, feeding with phenobarbital or both. Colony origin had no effect on inducibility or differential gene expression. However, treatment type showed significant effects on the expression of the eight inducible genes. Caste effects on expression levels were significant in five of the eight inducible genes with constitutive and induced expression of most target genes being higher in workers than in soldiers.

**Editor:** Omprakash Mittapalli, The Ohio State University/OARDC, United States of America

**Funding:** This study was funded by USDA-ARS Specific Cooperative Agreements and grants by the state of Louisiana to CH. The funders had no role in study design, data collection and analysis, decision to publish, or preparation of the manuscript.

**Competing Interests:** The authors have declared that no competing interests exist.

* Email: chusseneder@agcenter.lsu.edu

These authors contributed equally to this work.

## Introduction

As social insects, termites (order Blattaria, formerly Isoptera [1]) live in densely populated colonies with a specialized caste system characterized by morphological differentiation and division of labor between worker, soldier and reproductive castes. Relatedness within termite colonies is usually high since members of all castes are the offspring of a limited number of reproductives (kings and queens [2]). Morphological and behavioral differentiations among castes are therefore based on gene expression rather than genotype [3].

Termites must have evolved strong and versatile defense mechanisms against toxic xenobiotics and pathogens in adaptation to living in densely populated colonies of closely related individuals, with intense social interactions in nest conditions conducive to microbial growth and transmission as well as being exposed to contaminants in the soil, insecticides and a lignocellulosic diet containing secondary metabolites produced by plants. Detoxification enzymes to metabolize xenobiotics are therefore essential. Previous studies using purification and biochemical characterization showed that termites possess multiple cytochrome P450s [4], glutathione S-transferases (GSTs [4]) and microsomal esterases [5] and show different levels of susceptibility to termiticides [6,7,8]. However, the relationship between constitutive or inducible gene expression and enzyme activity/insecticide susceptibility has not yet been established.

While detoxification enzymes protect insects from toxic effects of xenobiotics, the innate immune system is the first line of defense against microbial infections. The innate immune system is primarily non-specific and serves to recognize invading pathogens, activate humoral and cellular immune responses leading to inactivation and clearance of pathogens.

The insect innate immune system recognizes invading pathogens through an array of pattern recognition receptors (PRR) that bind to conserved microbial cell membrane or wall epitopes serving as immune elicitors, such as lipopolysaccharides (Gram-negative bacteria), peptidoglycans (Gram-positive bacteria) and β-1,3-glucan (fungi). Insect PRRs include C-type lectins, gram-negative bacteria binding proteins, peptidoglycan-, lipopolyssaccharide-and β-1,3-glucan binding protein, and β-1,3-glucan recognition protein [9,10,11]. Binding of microbial immune elicitors by PRRs triggers cellular and humoral immune responses.

Cellular defense consists of a variety of hemocytes responsible for sealing wounds via aggregation [12], phagocytosis, encapsulation of invading microbes and nodulation, which leads to melanization by the prophenol oxidase (PPO) pathway and sequestration [13,14]. Humoral response includes the release of antimicrobial peptides (AMPs) into the hemolymph mainly from the fat body and the activation of proteases, which trigger melanization via the PPO pathway ([15], and references therein). The PPO pathway also produces intermediates that are directly toxic to microorganisms (e.g., quinones).

While the PPO pathway initiates melanization and produces antimicrobial compounds within minutes, recognition of foreign cell wall components also triggers intracellular signaling pathways that work within a couple of hours to days. The Toll pathway responds mainly to Gram-positive bacteria and fungi, while the immune deficiency (IMD) pathway reacts predominantly upon infection with Gram-negative bacteria. Both pathways result in expression of PRRs and antifungal peptides and/or AMPs. The Toll pathway also triggers activation of serine proteases involved in melanization. Other signaling pathways (JNK- and JAK/STAT) are triggered by exposure to lipopolysaccharides of Gram-negative bacteria, but their contribution to insect immunity is less well researched [16].

Termites possess both humoral and cellular immunity (reviewed in [17]). Pioneering work by, e.g., [18,19] showed that immunization via sub-lethal dosages of pathogens induces the production of AMPs in the termite hemolymph. A number of genes and peptides related to this kind of humoral immunity have been described in termites. Among them are a transferrin gene which was up-regulated following exposure to an entomopathogenic fungus [20], termicin, spinigerin [18] and gram-negative binding proteins (GNBPs) which can act as both PRRs and AMPs in several termite species [21]. Production of AMPs appears to differ not only between species [22], but also between termites belonging to different colonies or castes [17,19,23]. The components of cellular immunity in termites are less well studied although melanization and encapsulation of foreign bodies have been frequently used as bioassay for immune competence (e.g., [24]).

While both, detoxification enzymes as well as immune proteins have been detected in termites, full genome screening and gene expression studies are still in their infancy where termites are concerned. We chose the Formosan subterranean termite (FST), *Coptotermes formosanus* Shiraki (Blattodea: Rhinotermitidae), as model to investigate constitutive and induced gene expression related to xenobiotic metabolism and immunity in termites. This termite represents one of the most successful and economically important invasive pest species worldwide [25,26]. As a subterranean termite, *C. formosanus* forages and nests mainly in the soil and is associated with urban environments and as such exposed to xenobiotics not only through their natural diet (wood) but also through soil contaminants and insecticides. Colonies are densely populated and can contain over a million individuals [27]. Such a life style should be conducive for growth and spread of pathogens, however, 50 years of failed biocontrol efforts [28] suggests the existence of strong synergistic immune and detoxification defense mechanisms of subterranean termite colonies.

Annotation of our previously constructed expressed sequence tag library of the FST revealed the presence of putative detoxification and immune-related genes [29]. Out of those, we selected the 10 genes with the most similarity to known detoxification genes (e.g., those encoding cytochrome P450s, glutathione S-transferase) and immune genes (e.g., those encoding for gram-negative binding proteins, lectin- and fibrinogen-like peptides as well as the Down syndrome cell adhesion molecule [29]). A recent study aiming at a more complete description of the immune related transcriptome in *C. formosanus* [30] confirmed the presence of those genes among hundreds of putatively immune-related genes encoding for PRRs, signal modulators, transductors and effector molecules (such as AMPs) whose expression was upregulated after fungal and/or bacterial challenge. However, colony or caste specific gene expression, inducible detoxification or possible crosslinks between genes involved in immunity and xenobiotic metabolism were not considered. In fact, very few studies, have considered differential gene expression of immunity and detoxification genes among castes of termites [31,32]. Therefore, the principal goals of our study were to identify genes associated with immunity and/or detoxification in FSTs, test whether expression can be induced by septic injury with bacteria and/or sublethal xenobiotic challenge with phenobarbital and to test if expression of immune response and/or xenobiotic metabolism genes is colony or caste dependent.

## Results

### 1. Constitutive and induced gene expression

Examined in this study were a total of 10 genes putatively involved in immunity, and/or detoxification in termites (Table 1). Constitutive (untreated controls) expression of target gene transcripts was rather low, overall (mean $R = 0.014$, $SD = 0.163$, Table S1). All but two target genes were inducible, either with a sublethal dose (0.5%) of the xenobiotic phenobarbital (PB) (mean $R = 2.014$, $SD = 1.379$) or septic injury with non-pathogenic bacteria (*Escherichia coli/Pilibacter termitis*) (mean $R = 2.668$, $SD = 1.988$) or both treatments (Table S1, Figure 1). Only one of the cytochrome P450 genes (CYP450) and the gene for Down syndrome cell adhesion molecule (Dscam) were not induced by either treatment. Most genes were inducible by both treatments. The genes inducible by xenobiotic and septic injury/immune challenge encoded for CYP15A1, Chitin-binding protein (CBP), two lectin-related proteins (LECT-like, CLECT), the Ficolin-2 precursor (FICO-2) and the gram-negative binding protein 2 (GNBP-2). The gene encoding for glutathione S-transferase (GST) was exclusively induced by 0.5% PB treatment, while the gene for the Multi antimicrobial extrusion protein (MatE) was induced predominantly with the *E. coli/P. termitis* treatment.

### 2. Effects of colony origin

Tests for variance between colonies and for interactions between colony and treatment or between colony and caste showed no significant effect of colony origin at any of the 10 target genes. Probabilities of colony variability and interaction ranked from 0.127 to 0.783. Thus, regardless of treatment or caste, R values were not significantly different among the three termite colonies.

### 3. Effects of treatment

In contrast to colony origin, treatment type (0.5% PB vs. *E. coli/P. termitis*) had a significant effect on the expression of each of the eight inducible target genes (Figure 1, Table S1). CYP15A1 was induced in both types of treatment; however, relative expression was significantly higher ($F = 88.03$, $P < 0.0001$: PROC GLM) for the *E. coli/P. termitis* treatment as compared to the 0.5% PB treatment due to a strong upregulation of gene expression in workers after bacteria treatment. Similarly, GNBP-2 showed induced expression in both treatment types but with significantly higher expression ($F = 28.08$, $P < 0.0001$) for termite workers injected with *E. coli/P. termitis* than for termites fed with 0.5% PB. Also, CBP was significantly more induced ($F = 31.19$, P<

**Table 1.** Descriptions of target and reference genes.

| Gene | Abbrev. | Species of Origin (Accession No.) | Top Matches in GenBank, Species of Origin (Accession No. for Nucleotide or Peptide Sequence) | Putative Function | Primer Sequence (5′-3′) |
|---|---|---|---|---|---|
| **Putative xenobiotic metabolism related genes** | | | | | |
| Cytochrome P450 15A1 | CYP15A1 | *Coptotermes formosanus* | CYP151A, *Reticulitermes flavipes*(ACN93795) | Epoxidase, | ATTTGTCACTCGCCTTGGTC |
| | | FK835449 | CYP151A, *Diploptera punctata* (AAS13464) | juvenile hormone synthesis | AAAGTTGCCATACGTGGAGG |
| Cytochrome P450 | CYP450 | *Coptotermes formosanus* | CYP314a1, *Drosophila melanogaster* (Q9VUF8.3) | Ecdysone 20-monooxygenase | ACTGGTTTGTAGTCGGTGCC |
| | | FK833823 | CYP12A2, Musca domestica (O18635.1) | Monooxygenase, oxidoreductase | TTAGCACCGAGAGACGTTCA |
| Glutathion-S-Transferase | GST | *Hodotermopsis sjoestedti* | Glutathione S transferase, *Blattella germanica* | Metabolism of lipophilic toxins, | GTTTCAAGCCTCGCTTTTGAC |
| | | DC239424 | (FJ855500.1) | excretion of xenobiotics | CGTCACTGTGAGACTGCCAT |
| Multi antimicrobial extrusion | MatE | *Coptotermes formosanus* | MatE, *Trichonomonas vaginalis* (XP_001314814.1) | Sodium antiporter, cationic efflux | TTGACACAACACCTTTCGGA |
| protein | | FK833694 | | transport pump, excretion of organic cations | GCCGTGTTTGAGGATGAAAT |
| **Putative Immune related genes** | | | | | |
| Chitin binding protein | CBP | *Coptotermes formosanus* | chitin-binding protein, *Holotrichia oblita* (AFD28282.1) | Peritrophin A-type chitin-binding protein | CGGGAGGACTGCTAGAACTG |
| | | FK832766 | | | TGAGGGAACTCCCTGTTCAC |
| Down syndrome | DSCAM | *Coptotermes formosanus* | Downsyndrome cell adhesion molecule | immunglobulin-like fibronectin, | CCGTACCATTGACGGAGAGT |
| cell adhesion molecule | | FK835436 | *Harpegnathos saltator* (EFN81146.1) | cell recognition and adhesion, | CACTTGTGTTGCCCGTAATG |
| | | | | axon guidance receptor in insects | |
| Ficolin-2 precursor | FICO-2 | *Coptotermes formosanus* | Ficolin-2 precursor, *Pediculus humanus corporis*, (XP_002430827.1) | Antigen recognition, carbohydrate binding, | CTCCGTCGGACTCTCATCTC |
| | | FK836944 | scabrous protein-like, *Tribolium castaneum*, (XP_972571.1) | Enhances phagocytosis and opsonization | GCGCTATAACCTGGGTCTCA |
| Lectin-related hemolymph | LECT-like | *Coptotermes formosanus* | Hemolymph lipopolysaccharide-binding protein, *Periplaneta americana* (P26305.1) | *E. coli* Lipopolysaccharide binding, | CTTCCCAGTGTCGAAAGCTC |
| Lipospolysaccharide binding protein | | FK834461 | Lectin-related regenectin, *P. americana* (BAA82265.1) | Tissue regeneration | TGGGTACCCCATAGGAGATG |
| C-type lectin | CLECT | *Coptotermes formosanus* | conserved hypothetical C-type lectin, *P. humanus corporis* (XP_002424769.1) | carbohydrate-binding, pathogen recognition | AGGGTGAAAGAGGAGCAACA |
| | | FK835521 | | innate immunity | TAAAAATCCCAACAACCCCA |
| Gram-negative binding protein 2 | GNBP-2 | *Drepanotermes rubriceps* | gram negative bacteria binding protein 2 from different termite species | insect pathogen recognition protein | CTTCCCAGTGTCGAAAGCTC |
| | | DQ058934 | (e.g., AAZ0893.1-AAZ08504.1, AEK64801.1, ADJ19004.1) | beta(1,3)-glucanase effector | TGGGTACCCCATAGGAGATG |
| **Reference genes** | | | | | |
| Cytoplasmic heat shock protein 70 | HSP | *Coptotermes formosanus* | | ATP and nucleotide binding | AGCCTTGGCCACAACAGTGCAA |

**Table 1.** Cont.

| Gene | Abbrev. | Species of Origin (Accession No.) | Top Matches in GenBank, Species of Origin (Accession No. for Nucleotide or Peptide Sequence) | Putative Function | Primer Sequence (5′-3′) |
|---|---|---|---|---|---|
| | | FK835495 | | | GGAGCAGGAGCAGGACCGACT |
| NADH Dehydrogenase subunit 4 | NADH | *Coptotermes formosanus* | | Mitochondrial electron, sodium ion and | ACGAAGCAACCCATAACCACCAAGC |
| | | FK833785 | | proton transport, ubiquinone activity | GGGCTCATGTTGAGGCTCCTGTT |
| Elongation factor1-alpha | EFA | *Coptotermes formosanus* | | GTP binding, | GGCGGTCACATTTCTCCTTA |
| | | FK834645 | | regulation of translational elongation | CTGAACCACCCTGGTCAGAT |

0.0001) in *E. coli/P. termitis* injected termites than in 0.5% PB fed termites. MatE was only induced (F = 32.37, P<0.001) in the *E. coli/P. termitis* treatment group, while gene expression after PB treatment was not different from that of the controls.

In contrast, GST was exclusively induced after 0.5% PB treatment (F = 3.53, P = 0.0312), while gene expression did not differ from controls after *E. coli/P. termitis* treatment. Although induced by both treatments, relative expression of LECT-like, CLECT and FICO-2 was significantly higher (F = 33.03, P< 0.001; F = 61.72, P<0.0001; and F = 23.78, P<0.001, respectively) for termites fed with 0.5% PB than for termites injected with *E. coli/P. termitis*.

## 4. Effects of caste

Tests for differential expression showed that caste (worker or soldier) had a significant (P<0.05) effect on levels of expression in five of the eight inducible target genes (CYP15A1, MatE, GST, GNBP, and CLEC, Figure 1, Table S1). The gene expressing CYP15A1 was induced in both workers and soldiers, with a significant interaction (P = 0.032) of treatment and caste, due to the significantly higher (F = 34.79, P<0.0001) expression in workers as compared to in soldiers. Constitutive expression of CYP15A1 was also significantly increased (F = 213.65, P<0.001, PROC GLM) for untreated workers as compared to untreated soldiers.

Constitutive expression of MatE was significantly higher (F = 40.16, P<0.0001) in untreated workers than in untreated soldiers as well. Furthermore, MatE was only induced in workers by septic injury. The Gram-negative bacterial binding protein (GNBP-2) was, similarly, almost exclusively induced in workers in both treatments (F = 121.23, P<0.001). GST was also exclusively induced in workers, but only after challenge with 0.5% PB. The C-type lectin gene (CLECT), was the only target gene which showed significantly higher (F = 47.53, P<0.001) expression in soldiers than in workers.

Caste had a marginal effect on the inducibility of Ficolin-2 (FICO-2) and chitin binding protein (CBP). These genes were induced both in workers and in soldiers, but FICO-2 was expressed marginally higher in the worker caste (F = 7.31, P = 0.058), while CBP's expression was marginally biased towards soldiers (F = 8.10, P = 0.049). The second lectin related gene (LECT-like) also was induced in both castes, but showed no significant difference in expression levels as an effect of caste (F = 0.16, P = 0.692).

# Discussion

## 1. Effect of xenobiotic exposure and septic injury on expression of target genes

Constitutive expression of all the investigated genes in FST workers and soldiers was low; however, eight out of the ten genes were inducible by one or both treatment types in one or both castes. The initial categorization based on the putative function of the ten target genes into genes associated with detoxification/xenobiotic metabolism and septic injury/immunity response was somewhat supported by the induction experiments using phenobarbital (PB) treatments and septic injury with bacteria, however, there were also some unexpected results (Figure 2).

The expression of the glutathione S-transferase (GST) and one of the cytochrome P450 genes (CYP15A1) as expected was upregulated in response to exposure with the xenobiotic PB. The barbiturate PB is used to treat epilepsy in humans and acts on GABA-gated chloride channels in the central nervous system to reduce excitatory synaptic response similar in activity to some termiticides, i.e., Fipronil [33].

Response elements and nuclear receptors responsible for induction of many cytochrome P450s and GSTs have been described in insects and shown to respond to several groups of xenobiotics, including PB [34]. In insects the GSTs are known to metabolize plant allelochemicals and insecticides, and high levels of GSTs are related to insecticide resistance. Exposure to PB has been previously shown to induce GST activity in cockroaches [35], *Drosophila* [36,37,38], *Aedes aegypti* [39], among others (reviewed in [40]). Upregulation of the expression of the GST gene in FST workers upon exposure to PB was therefore expected. The lack of upregulation in soldiers and the effect of caste on differential gene expression are discussed below.

Although, cytochrome P450 enzymes are largely ubiquitous and have common catalytic mechanisms, they are translated from a large family of genes with high sequence diversity, which explains their numerous functions. In insects, cytochrome P450s are known to play a role in detoxification of insecticides and xenobiotics (reviewed in [40]), but also in regulation of caste development via metabolism of hormones, like the juvenile hormone JH [41,42,43].

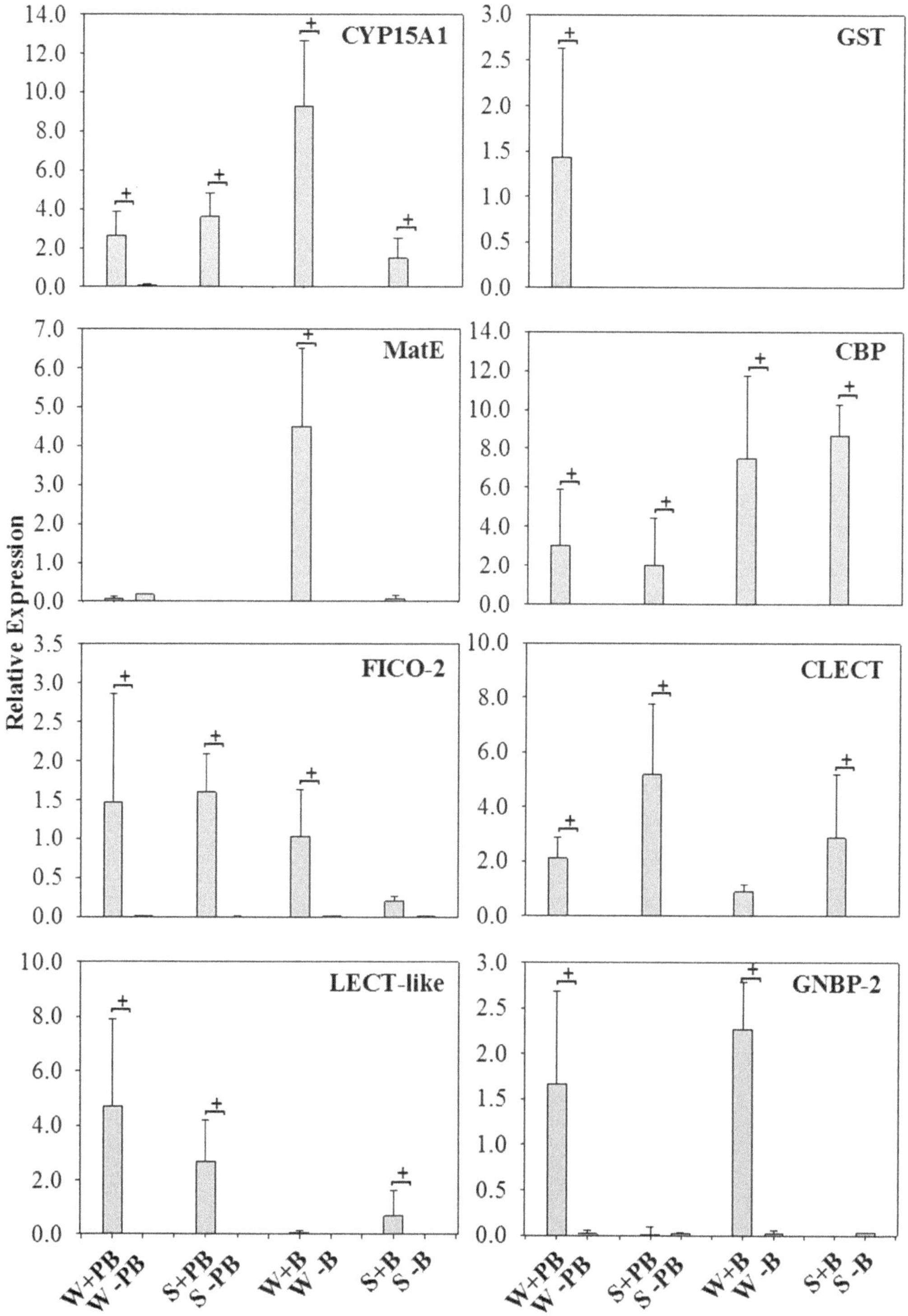


**Figure 1. Constitutive and induced expression of target gene amplicons.** The y-axes denote relative expression in R units averaged over three colonies (note that the ranges of y-values vary across graphs). The x-axes denote caste (W = worker, S = soldier) and treatment type (+PB = fed with 0.5% Phenobarbital, +B = septic injury with *E. coli/P. termitis* and -PB/-B = untreated control showing constitutive expression). 'Missing' columns have a relative expression of<0.01. + marks significant induction. There was no induced expression observed for either CYP450 or DSCAM (not shown).

The closest matches for the gene product of one of the CYP450s in our study were a CYP15A1 of the cockroach *Diploptera punctata* (AAS13464.1, [41]) and a CYP15A1 pseudogene found in *R. flavipes* (ACN93795.1, [43]) with an identity of 51–54%, which lies within the range of cross-species alignments of translated CYP15s (41.5–72.5%, [43]). The product of CYP15A1 epoxidizes methyl farnesoate to juvenile hormone III in cockroaches [41,44] and a similar function in JH biosynthesis was suggested in termites [43,45]. Tarver et al. [43] showed that CYP15F1, which is closely related to the truncated CYP15A1 pseudogene, is involved in JH dependent caste differentiation. However, this study [43] did not investigate whether these two genes were functionally related.

Unlike other prominent cytochrome P450s, which metabolize a variety of pesticides due to their broad substrate and product specificity, CYP15A1 was shown to be highly specific in the silkworm [41]. The CYP15A1 gene in our study was the only target gene that was constitutively expressed and induced by both treatment types in workers and soldiers. The fundamental role in JH biosynthesis and caste regulation of similar genes in termites [43] may explain the level of constitutive expression observed in our study.

The inducibility by PB and septic injury, however, suggests that the CYP15A1 gene product has expanded functionality in terms of xenobiotic metabolism and immunity beyond its primary function as epoxidase catalyzing the last step in JH biosynthesis. While PB inducibility of CYP450s of the families 4, 6, 9, 12 and 28 in insects is well known, the only inducers for CYP15 family members so far reported were JH and terpene components from soldier head extracts suspected to act as caste regulatory primer pheromones [43]. Our study is the first to demonstrate inducibility of a member of the CYP15 family by a xenobiotic. However, the involvement of the CYP15 enzyme in actual xenobiotic metabolism and detoxification has to be confirmed in enzyme and bioassays. Similar to CYP15A1, another enzyme involved in JH metabolism (JH III epoxide hydrolase responsible for JH degradation) has been shown to be upregulated after PB exposure in several insect species ([46], and references therein).

A link between JH levels and innate immunity has been discovered in the silkworm *Bombus mori* [47], when JH treatment acted as an immune activator and increased the transcription products of six AMPs. We suggest that septic injury in our study upregulated CYP15A1 expression in termite workers and soldiers, which results in JH biosynthesis to provide increased JH levels as an immune activator. This route would constitute an immune pathway different but complementary to the Toll- and Imd pathways.

Overexpression of CYP450s and GSTs is often associated with insecticide resistance ([37], and references therein). However, the investigated CYP450s and GST genes were not constitutively expressed at elevated levels in Formosan termites and induction with PB led to a rather moderate spike in gene expression (less than 4 fold), which is in the range of susceptible insects [48]. Insecticide resistant strains usually show a much higher fold-change [49]. Although there is variability in tolerance to

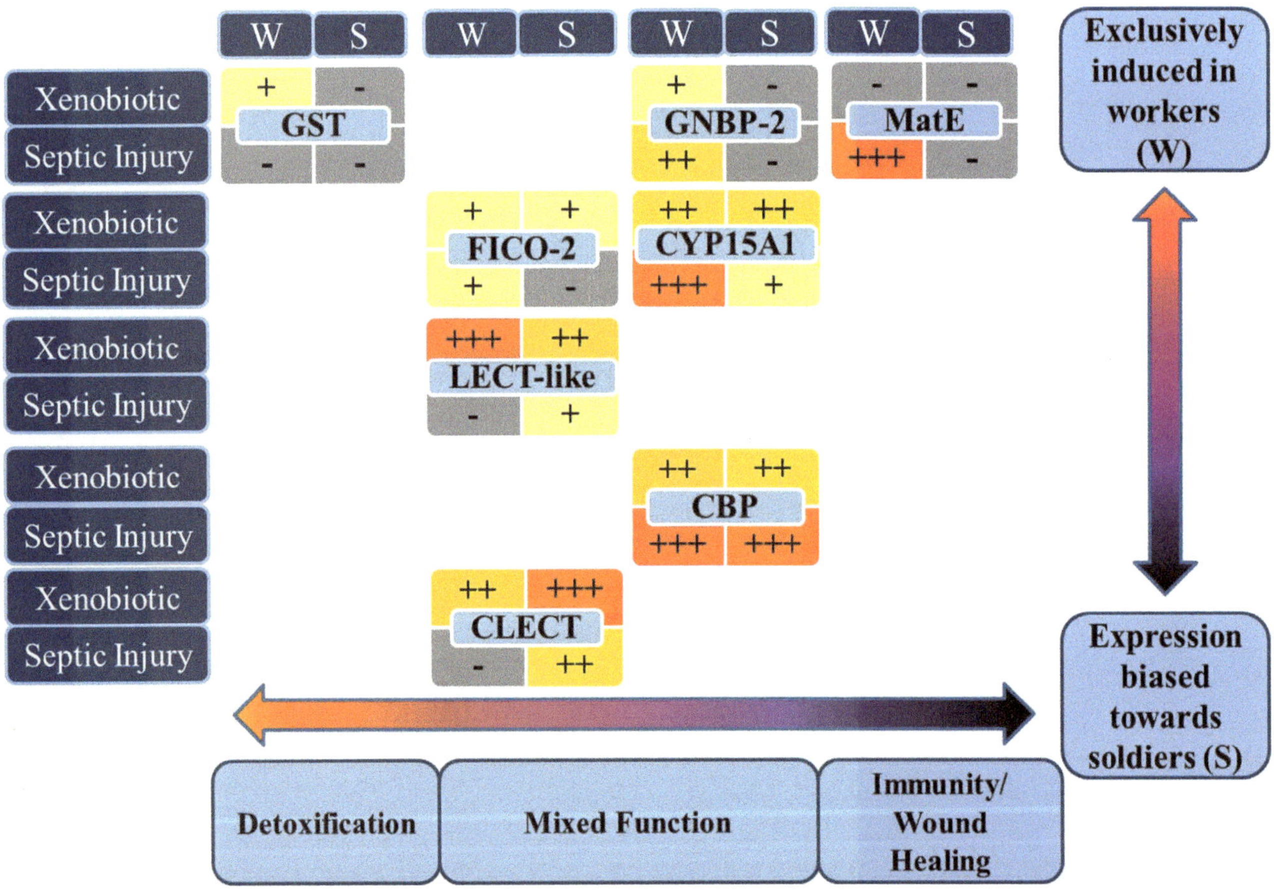


**Figure 2. Graphic summary of caste- and treatment dependent inducibility and function of eight detoxification and immune-related genes of the Formosan subterranean termite.** Treatment consisted of exposure to a xenobiotic substance (0.5% phenobarbital) or septic injury via bacteria injection. Relative levels of caste (workers vs. soldiers) dependent induced gene expression are indicated with +++ in red (>4 fold expression change compared to control), ++ in orange (2–4 fold expression change), + in yellow (<2 fold expression change) or – in grey (no significant induction). Genes are arranged from left to right according their putative function ranging from detoxification to immunity and from top to bottom according to their caste dependent expression.

insecticides among termite colonies and species, to date, no resistance or treatment failure has been recorded.

The second cytochrome P450 gene in the study (CYP450) did not respond to treatment at all. CYP450 in our study was most closely related (48% identity) to *shade* (CYP314a1) from the Halloween gene family, an ecdysone 20-monooxygenase ecdysone hydroxylase involved in 20-hydroecdysone (molting hormone) synthesis in *D. melanogaster* [50]. CYP450 also showed considerable similarity to CYP12A2 from the house fly *Musca domestica*. A similar gene (CYP12A1) in the house fly was shown to be inducible by PB and to metabolize different insecticides [51]. Thus, we initially hypothesized that this gene would be most likely up-regulated in response to PB treatment and possibly involved in xenobiotic metabolism. However, this gene apparently lacks the response elements and/or transcription factors needed for PB inducibility [34]. Genome-wide studies in other insects showed that only a third of CYP genes are inducible by xenobiotics [52].

Similar to GST and CYP450 genes, we initially expected the gene for MatE to be inducible by the xenobiotic PB as MatE proteins have been described as multi drug resistance efflux proteins. They function as transmembrane pumps and are involved in the excretion of xenobiotics in vertebrates, detoxification of metals and secondary metabolites in plants [53] and mediating resistance to antibiotics in bacteria [54,55]. It was somewhat unexpected that MatE expression was upregulated by septic injury and not by xenobiotic challenge in our study. The only evidence to date that MatE might be involved in antimicrobial defense is found in plant studies. Nawrath et al. [56] investigated a MatE gene homologue in *Arabidopsis* plants: expression was strongly induced by inoculation with an avirulent strain of *Pseudomonas syringae* and the authors concluded that this gene was likely involved in antimicrobial plant defense via the salicylic acid-dependent signaling pathway.

To our knowledge, MatE proteins and their function in insects have not yet been investigated. It is possible that the MatE gene in our study was directly involved in immune response in termites as indicated by the induced expression in FST workers injected with *E. coli* and *P. termitis*. However, it is more likely that MatE was upregulated as a consequence of the production of cytotoxic metabolites resulting from the activation of immune pathways by bacteria injection. For example, septic injury in insects activates the PPO pathway, which produces melanin, a phenolic biopolymer. Some melanin precursors are cytotoxic, which is advantageous for killing opportunistic invaders at the wound site [57]. However, metabolites might be inadvertently cytotoxic to the insect itself, especially since the PPO cascade is taking place in the open circulatory system and thus byproducts may accumulate in the hemolymph. MatE proteins play an important role in the removal of metabolic waste products and are believed to be a transport system for phenolic compounds [58]. Excessive melanin or its intermediaries from the PPO pathway might thus be extruded or sequestered by MatE in termites.

We initially assumed that the chitin binding protein (CBP) belongs into the category of immune related genes. The CBP in our study matched closest (31%) to a putative peritrophic membrane protein in a scarab beetle (*Holotrichia oblita*.), which contains carbohydrate and chitin binding domains. Invertebrates, such as shrimp, possess AMPs with chitin-binding capacity thought to be involved in bacteria phagocytosis, and perhaps antifungal activity and wound healing via chitin assembly [59]. Upregulation of CBP upon septic injury in workers and soldiers of Formosan termites suggests a similar involvement in immunity and/or wound healing via restoration of the cuticle.

However, PB treatment also induced CBP expression albeit at lower expression levels than bacteria challenge. A genome-wide analysis in the flour beetle, *Tribolium castaneum*, revealed that some proteins with peritrophin A-type chitin-binding domains were expressed in cuticle-forming tissues while others were associated with the peritrophic membrane and expressed in the midgut during feeding [60]. The peritrophic membrane is a barrier against pathogens and a sink for toxic substances [61], which would explain upregulation of certain peritrophins after ingestion of a xenobiotic like PB.

Similarly, the Ficolin 2 precursor (FICO-2) and the two lectins (CLECT, Lectin-like) in our study were assumed to be involved in immune defense enhancing phagocytosis and opsonization [62]. Ficolins are fibrinogen related proteins that act as PRRs by virtue of binding to carbohydrates on the surface of gram-negative and gram-positive bacteria [63]. This leads -at least in vertebrates- to the activation of the lectin complement pathway resulting in bacteria lysis. In insects, ficolins are not that well studied but the fact that their fibrinogen-like domain is conserved throughout the animal kingdom suggests that they are involved in ancestral forms of the innate immune system in invertebrates [64–66]. Genome-wide analyses and microarrays in *Drosophila* spp [15,66] confirmed that fibrinogen-related domains are involved in immune response in insects. The Ficolin precursor in our study showed high similarity (78%) to a putative Ficolin-2 precursor of the human louse (*Pediculus humanus corporis*) and scabrous protein of *T. castaneum* (72%) and was upregulated by septic injury.

Both lectin related gene products in our study belong to the C-type lectin superfamily of calcium dependent carbohydrate-recognition proteins. C-type lectins play a variety of roles in adaptive and innate immunity as PRR, cell adhesion molecules and activators of the complement pathway [67]. Lectins in the hemolymph of cockroaches and lepidopterans participate in innate immune response via phagocytosis, encapsulation, nodulation and activation of the PPO pathway ([9], and references therein). Immunelectins vary in their specificity towards microbial epitopes as some show induction after Gram-negative bacteria injection only, while others bind to Gram-negative and Gram-positive bacteria as well as fungi [9].

The product of the LECT-like gene in our study showed 32% identity to both a hemolymph lipopolysaccharide (LPS)-binding protein [68] and regenectin [69] from the cockroach *Periplaneta americana*. The LPS binding protein had affinity for lipopolysaccharides on the cell membrane of some Gram-negative bacteria, but not others, suggesting high selectivity to certain oligosaccharide structures and thus specific immune recognition of selective pathogens [68]. Regenectin is involved in muscle formation and leg regeneration in cockroaches, but in contrast to LPS-binding protein, the transcription of its gene is not enhanced by bacterial infection [69]. Transcription of the LECT-like gene was upregulated in FST soldiers upon septic injury, which most likely reflects bacteria binding function and/or participation in tissue regeneration for wound healing.

The product of the second lectin related gene (CLECT) in our study was highly related to a conserved hypothetical protein in the human body louse (90% identity) and C-type lectin-like precursors from several insect species such as the flour beetle (89%), several mosquito species (≥86%), *Drosophila* spp. (≥86%), and social insects, i. e., ants, honey and bumble bees (≥88%). CLECT belongs thus to the immunelectins acting as PRR [9] that activate the PPO pathway. C-type lectins are also involved in sclerotization, i.e., exoskeleton hardening in arthropods [70]. Sclerotization is catalyzed by phenoloxidases and thus linked to activation of the

PPO pathway. Furthermore, the calcium binding properties of C-type lectins are assumed to contribute to biomineralization of the cuticle by depositing calcium salts [70]. Thus, it was not surprising that CLECT was upregulated after septic injury similar to the LECT-like gene.

We were surprised, however, that the expression of the Ficolin precursor and both lectins was also induced by PB treatment. The majority of antiepileptic drugs such as PB show immunosuppressive effects in humans; however, under certain conditions, they can also stimulate the immune system [71]. Microarray analysis of gene expression revealed that *Drosophila* flies execute massive transcriptional response to PB treatment [72]. Not only were the detoxificication genes upregulated, but also genes involved in several metabolic and stress response pathways were affected. A nuclear receptor (DHR96) played a central role in regulating the response to PB. Middha and Wang [66] subsequently showed that some fibrinogen-related proteins are regulated by DHR96, thus linking expression of fibrinogen-like genes to PB exposure. This suggests a broader than originally assumed network of functions beyond innate immunity mediated by fibrinogen-like proteins in insects.

Gram-negative binding proteins are known in insects to sense pathogen invasion and trigger immune responses, i.e., the Toll and Imd pathways [73], and references therein) as well as the PPO cascade [74]. In termites, GNBP-2 has a double function as PRR that recognizes LPS (like lectins) and beta-1,3-glucans and as antifungal effector due to its beta-1,3-glucanase activity [21]. GNBP-2 is expressed in the salivary gland and by hemocytes and secreted via allogroming onto the cuticle and incorporated into nest and tunnel material presumably to suppress growth of fungal pathogens [21,73].

Prior to our study, the role of GNBP in termites has been explored mostly in terms of antifungal but not antibacterial activity [73,75,76]. Our study showed involvement of GNBP-2 in defense against bacterial septic injury deduced from the fact that expression in workers doubled after bacteria injection. This result supports Bulmer et al.'s [21] *in vitro* binding assays that showed that GNBP-2 had affinity to Gram-negative bacteria.

While the upregulation of GNBP-2 upon immune challenge was expected, it was surprising that feeding on PB also induced GNBP-2 expression in workers. There is no straightforward explanation, but we may speculate that workers increase grooming of colony mates showing signs of PB exposure, which might indirectly lead to increased production of GNBP-2 in the salivary gland as protective secretion [21,73]. Furthermore, PB acts as universal inducer for many genes in insects, not only CYP450s and GSTs but also for juvenile hormone epoxide hydrolases and for genes involved in carbohydrate metabolism, such as amylases [37]. GNBP-2 has a glucanase site that performs hydrolysis on glucosidic bonds and thus is a hydrolase involved in carbohydrate metabolisms. Like the increased transcription of other PRR immune genes upon xenobiotic exposure (see lectins and ficolin above) the regulatory and functional link between detoxification and immunity needs to be further investigated.

We initially hypothesized that the Down syndrome cell adhesion molecule (Dscam) was immune related. The gene in our study showed several putative conserved domains of the Immuneglobulin superfamily and fibronectin Type III similar to those in Hymenoptera and Diptera [77], which play a role in cell adhesion and wound healing in vertebrates. Insect studies showed that the Dscam receptor is involved in neural wiring and innate immunity [78]. Reduction of Dscam expression using mutants or RNA interference impaired phagocytosis of bacteria presumably due to reduced binding and opsonization of bacteria [79,80].

Surprisingly, Dscam was not induced by septic injury (or by PB treatment) in our study. Dscam in the insect genome has many alternative exons and encodes tens of thousands of isoforms resulting from alternative splicing [77,78]. Watson et al. [80] showed that some Dscam isoforms bound to *E. coli*, while others did not. There might be bacteria binding and non-binding isoforms in general, and/or binding is somewhat specific to epitopes of non-resident bacteria species, e. g., pathogens. The bacteria in our assay were native symbionts cultured from the termite gut, so the Dscam in our study might not be triggered.

## 2. Gene expression levels are mostly biased in favor of workers

Expression of five of the eight inducible target genes (CYP15A1, GST, MatE, GNBP-2, CBP and CLECT) was significantly and that of two additional genes (FICO-2, CBP) was marginally influenced by caste. Constitutive and/or induced gene expression was higher in workers in most of those differentially expressed genes except for CLECT, which was predominantly expressed in the soldier caste. In fact, three of the target genes (GST, MatE, and GNBP-2) were almost exclusively induced in workers.

In our study, higher expression of the "classic" detoxification genes such as members of the CYP450s (CYP15A1) and GST in the worker caste supports previous findings of higher enzyme levels in workers and higher tolerance against insecticides. Gatti and Henderson [6] and Gatti et al. [81] showed higher tolerance against organophosphates in workers attributed to a higher acetylcholinesterase activity in workers. Valles and Woodson [8] discuss their unpublished data which showed that soldiers of *C. formosanus* have only a fraction of detoxification enzyme activity compared to workers. Nevertheless, in a study of Osbrink et al. [7] soldiers of *C. formosanus* and *R. virginicus* in several colonies were more tolerant than workers to the majority of insecticides tested. This obvious variance among studies, species, enzymes and insecticides underlines the need for continued investigation of the relationship among gene expression, inducibility, metabolic enzyme production and activity, and insecticide susceptibility.

The major immune-related genes (GNBP-2, FICO-2) were induced and upregulated at higher levels after septic injury in workers than in soldiers. Soldiers might not rely primarily on the production of their own immune peptides for protection against pathogens. Soldiers have defense secretions, which in some species may have antifungal properties [23] and a higher degree of cuticle sclerotization than workers, which may aid as a mechanical barrier. Soldiers also may have antibacterial and/or antifungal activity via chitin assembly, wound healing and systemic protection related to increased CBP expression, which has been suggested for other highly sclerotized arthropods [59]. Sclerotization results from melanization and melanin plays an important role in innate immunity via wound healing and encapsulation of invaders. Melanin production during the PPO pathway also leads to toxic metabolites, such as quinones that could kill invaders [82]. Thus soldiers might have more non-protein related mechanical and biochemical defenses in form of sclerotization, melanin and its byproducts in addition to antimicrobial gland secretions. Soldiers could also receive passive immunization via secretions from workers. It has been shown that GNBP-2 is produced in the workers' salivary gland and secreted to coat nest material and the cuticle [21] and that social transfer of immunity can occur [83]. Workers might overproduce GNBP-2 and other immune peptides and transfer them to soldiers via grooming.

CLECT was the only gene that was expressed at significantly higher levels in soldiers. C-type lectins are involved in mineralization and sclerotization of the cuticle [70]. Soldiers are more sclerotized than workers and thus upregulation of CLECT expression might be naturally facilitated and high levels of CLECT can be produced on demand. Although soldiers are likely to frequently get injured during colony defense, they did not upregulate the other immune proteins investigated in this study (GNBP-2 and Ficolin) to the same degree as workers. Consequently, the soldier caste might primarily rely on lectins, which they use naturally for exoskeleton hardening to support their innate immunity and complement passive transfer of immune proteins from workers secretions. Research into caste specific differences, concerning not only the level of immune defense but also the mechanisms and pathways used, might be of interest in future studies on social insects. The present study provides the foundation for this endeavor.

## Materials and Methods

### 1. Sample collection

Worker and soldier FSTs were collected in May 2009 from three field colonies in New Orleans, Louisiana, USA, an urban habitat in which this invasive species thrives: Colony 1 - City Park, adjacent to Tad Gromley Stadium at 1 Palm Drive; colony 2 - Louis Armstrong Park at 901 North Rampart Street and colony 3 - French Quarter at 300 Canal Street. No specific permissions were required for accessing these locations for sampling activities and no endangered or protected species were involved in the studies. Voucher specimens are deposited in the Louisiana State Arthropod Museum at the LSU Agricultural Center. Based on previous studies of FST colony size and flight distance [84–86], samples from these locations represented three independent colonies from the New Orleans population. Termites (~500 workers and 500 soldiers) were collected from untreated inground monitoring station (Sentricon, Dow AgroSciences LLC: Indianapolis, IN) at each location and then transferred to plastic tubs (50×30×20 cm) lined with moist corrugated cardboard and maintained overnight in the laboratory at 25°C and 30% relative humidity (Intellus Environmental Controller, Percival: Perry, IA).

### 2. Determining sublethal treatment conditions for septic injury and phenobarbital challenge

Since bacteria invasion is generally associated with injury, and the responses to injury and infection are likely to be intricately connected, we decided to test the response to septic injury. To determine the appropriate sublethal treatment for inducing gene expression related to septic injury non-pathogenic bacteria, that were previously cultured from the guts of Formosan termite workers (gram-negative *Escherichia coli* [87] and gram-positive *Pilibacter termitis* [88] ATCC BAA-1030), were revived from glycerol stocks. *E. coli* was cultured under aerobic conditions in 1 mL of BHI broth for 1 day on a shaking incubator at 30°C, while *P. termitis* was grown anaerobically, using $H_2/CO_2$ BBL GasPaks (Becton Dickinson and Company: Franklin Lakes, NJ), in 1 mL of BHI broth at 30°C for 3 days. Sterile 28 gauge needles (Comfort Point: Nanaimo, British Columbia) were each soaked for approximately 60 seconds in a mixture of equal volumes of exponential growth phase *E. coli* and *P. termitis* (200 µL supernatant of each, placed in a 1.5 mL micro-centrifuge tube). Twenty workers and 20 soldiers from colony 1 were cold immobilized on a thermoelectric laboratory chill table (Bioquip: Gardenia, CA) and placed with their ventral surface exposed on a dissecting microscope stage (Leica MZ7.5 Stereomicroscope, Meyer Instruments: Houston, Texas). The abdomen was swabbed with 70% ethanol before being pierced through the intersegmental membrane of the $4^{th}$ and $5^{th}$ segments with the septic needle [18]. Following inoculation with bacteria, individual termites were isolated in 47 mm Petri dishes and kept at 25°C and 30% relative humidity (Intellus Environmental Controller, Percival: Perry, IA) for 48 hours. Twenty workers and 20 soldiers, which were not inoculated with bacteria, were held individually in separate dishes under the same conditions as controls. Mortality was determined as the inability for the termite to right itself after 30 seconds and was assessed for treated and untreated termites at 12, 24 and 48 hrs. Since no mortality was recorded at 24 hrs but 50% and 70% mortality in workers and soldiers, respectively, was recorded at 48 hrs, the 24 hr treatment was chosen as the sublethal treatment condition for target gene induction (Table 2). The 24 hr sublethal exposure time was consistent with rapid induction of innate immunity in insects (i.e., peptide synthesis in *Drosophila* [89] and in *P. spiniger* [18]).

For the xenobiotic challenge we chose phenobarbital (PB, Sigma-Aldrich: St. Louis, MO), a universal inducer of a broad variety of detoxification genes in insects [36,40]. Most studies test reactions to xenobiotics by topical application or ingestion. Since those would be the natural pathways of exposure in termites we used droplet feeding, which likely leads to ingestion and topical exposure via contact with the droplets and possibly transfer by grooming. To determine sublethal dose and time for PB treatments, 20 workers and 20 soldiers from colony 1 were each placed individually in a 47 mm Petri dish (Millipore: Billerica, MA). Five 2 µL droplets of 1% PB were pipetted directly onto the inside surface of each Petri dish. Droplets of 1% PB also were added to 20 Petri dishes which did not contain a termite as an evaporation control. All 60 dishes were sealed with parafilm and kept at 25°C and 30% relative humidity (Intellus Environmental Controller, Percival: Perry, IA) for 48 hrs. The PB-infused water droplets were monitored at 12 hr intervals. Level of evaporation was negligible over the 48 hr incubation period, and the PB droplets were consumed by workers and soldiers at approximately equal rate, confirming uptake.

We then used the droplet feeding assay to determine the maximum sublethal time and dose of PB treatment. Phenobarbital was serially diluted with distilled water (1%, 0.5%, 0.25% and 0.125%), and each concentration of PB was fed to 20 workers and 20 soldiers in 2 µl droplets as described above. Controls were fed distilled water only. Mortality was assessed as described above. Based on the results of this experiment (Table 2), the 0.5% PB treatment at 24 hrs was chosen as the sublethal treatment condition for target gene induction in both workers and in soldiers, which lies within the published time frame recommended for measuring PB- induced gene expression in insects ranging from 4 hrs in *Drosophila* [38] to 5 days in cockroaches [35].

### 3. Experimental design for septic injury and PB challenge

Following the established sublethal treatment conditions, 20 workers and 20 soldiers from each of the three FST colonies were fed 0.5% PB, 20 workers and 20 soldiers per colony were inoculated with the *E. coli/P. termitis* mixture and 40 workers as well as 40 soldiers per colony were left untreated (controls). Termites were kept in individual Petri dishes at conditions described above and flash-frozen at −80°C (Ultra low VIP, Sanyo Scientific: Wood Dale, IL) after 24 hrs. From each of the three colonies (biological replicates), five individuals of each caste (worker/soldier) and treatment (0.5% PB, septic injury with *E. coli/P. termitis* untreated control) were randomly chosen as technical replicates for quantitative real time PCR (qRT-PCR) reactions.

**Table 2.** Mortality of FST workers and soldiers to determine sublethal treatment conditions.

| Time (h) | Treatment | Mortality (%) | |
|---|---|---|---|
| | | Workers | Soldiers |
| 48 | 1% PB | 100 | 100 |
| 24 | 1% PB | 70 | 70 |
| 12 | 1% PB | 50 | 70 |
| 48 | 0.5% PB | 50 | 50 |
| **24** | **0.5% PB** | **0** | **0** |
| 12 | 0.5% PB | 0 | 0 |
| 48 | 0.25% PB | 0 | 0 |
| 24 | 0.25% PB | 0 | 0 |
| 12 | 0.25% PB | 0 | 0 |
| 48 | 0.125% PB | 0 | 0 |
| 24 | 0.125% PB | 0 | 0 |
| 12 | 0.125% PB | 0 | 0 |
| 48 | *E.coli/P. termitis* | 50 | 70 |
| **24** | ***E.coli/P. termitis*** | **0** | **0** |
| 12 | *E.coli/P. termitis* | 0 | 0 |
| 48 | Control | 0 | 0 |
| 24 | Control | 0 | 0 |
| 12 | Control | 0 | 0 |

Termites were fed with specified concentrations of serially diluted phenobarbital (PB), injected with a 50:50 mixture of *E. coli* and *P. termitis* or left untreated (Control). Sublethal treatment conditions (maximum dose and/or exposure time prior to observed mortality) for target gene induction are in bold print.

## 4. RNA isolation and cDNA synthesis

Total RNA was extracted from whole bodies of individual termites using the RNA Purification Microkit (QIAGEN, Valencia, CA). RNA concentrations were assessed at the 260/280 nm ratio of absorbance using a ND-1000 spectrophotometer (Nanodrop Technologies: Wilmington, DE). RNA samples were normalized to 10 ng per µl by adding RNAse free water and were then used as template for complimentary DNA (cDNA) synthesis via the iScript Kit (Biorad: Hercules, CA, USA) using a PTC-200 DNA Engine gradient thermocycler (MJ Research: San Francisco, California). The cDNA was normalized to 1,000 ng per µl and used as template for qRT-PCR.

## 5. Primer design for target and reference genes

We selected ten target genes putatively involved in immunity and xenobiotic metabolism from our Expressed Sequence Tag (EST) library of the FST [29], the literature and the NCBI database. Two ESTs matching to genes from Cytochrome P450 families [CYP15A1 (FK835449), CYP450 (FK833823)], and ESTs encoding for glutathione S-transferase [GST (DC239424)] and the multi antimicrobial extrusion protein [MatE (FK833694)] were assumed to be mainly involved in xenobiotic metabolism. Two ESTs encoding for lectin-related peptides [CLECT (FK835521), LECT-like (FK834461)], and those encoding for a Ficolin-2 precursor [FICO-2, (FK836944)], the Down syndrome cell adhesion molecule [Dscam (FK835436)], a chitin binding protein [CBP (FK832766)] and the gram-negative binding protein 2 [GNBP-2 (DQ058934)] are likely to play a role in pathogen recognition due to their putative cell adhesion function and were thus assumed to be immune related.

Primers for GST and GNBP-2 were already developed for a previous study that examined differential gene expression among FST queens and precopulatory females [29]. Three primer pairs were designed for each of the remaining seven target genes (PRIMER-BLAST software suite, NCBI) to have between 17 and 28 bases, with ~50% to 60% GC content.

The EST library for FST [29] was first screened for reference genes used in previous studies (e.g., [90,91]) for quantifying gene expression, i.e., Glyceraldehyde-3-phosphate dehydrogenase (FK834351), TATA-binding protein (FK835420), Elongation factor-1-alpha (FK834645) and Ubiquitin specific peptidase (FK831079). Next, the library was screened for genes most likely to be involved in cell maintenance and/or basal cell metabolism, i.e., "housekeeping" genes. Cytoplasmic heat shock protein (FK835495) and NADH dehydrogenase (FK833785) were selected, bringing the total number of putative reference genes to six. Table S2 lists the two to three primer pairs designed for each of the reference genes, along with their melting points, sequences and GC content.

Each primer pair designed for target genes was tested using qRT-PCR with cDNA from three untreated, randomly chosen workers from colony 1 as template. Primer pairs designed for reference genes were tested with cDNA template synthesized from three colony 1 workers and soldiers that were randomly chosen from each of the 0.5% PB treatment group, the *E. coli/P. termitis* treatment group and the untreated control group.

Amplifications of target and reference genes were performed using the IQ5 iCycler (Biorad: Hercules, CA, USA) in reactions each containing 5 µl cDNA, 25 µl SYBR-green Supermix (BioRad), 150 nm forward and reverse primer for each candidate or reference gene (see below), and 17 µl $H_2O$ for a total volume of 50 µl. The thermal protocol consisted of one cycle of 95°C for 3 min. followed by 40 cycles of 1) 95°C for 10 sec and 2) 58°C for 30 sec. Relative fluorescence units and Ct values (fractional cycle at which amplification reaches a detection threshold) were recorded at the end of each cycle by the iCycler software.

For each target gene, 1 of the 3 primer pairs was selected based on the following criteria [92–94]: 1) consistency across technical replicates (different individuals run on different plates), 2) target amplicon at 50 to 100 bp, 3) melting curve with a single peak and 4) reaction efficiency (E) ≥1.8. Primers for reference genes needed to pass all of the same criteria while also being consistent across treatment types (0.5% PB and *E. coli/P. termitis,* control) and among castes (workers and soldiers). Table S2 lists the ten target and three reference genes with their corresponding primer pairs which passed all of the above criteria.

## 6. Relative expression and tests for differential expression

LinRegPCR software [93] was used to calculate the individual efficiencies (E) for reactions, which were then averaged across the five technical replicates for each gene and treatment type. $C_T$ values given by the iCycler qRT-PCR software were similarly averaged across technical replicates. For the three reference genes, geometric means (normalization factor: $NF_n$, $n = 3$, [95]) of $C_T$ and E values were calculated, and these geometric means were next used to calculate relative expression (R) of target genes. Relative gene expression levels (R) were calculated based on the methods described in [96], in which amplification efficiencies (E) for target and reference samples are not assumed to be equal $[(R) = E_{target}^{C_T,target\ (control-sample)} / E_{ref}^{C_T,ref\ (control-sample)}]$ where *target* refers to the target gene under investigation and *ref* denotes the geometric mean of $C_T$ and E values for the three reference genes. When calculating R values for constitutive expression, *control* refers to a non-template control, and *sample* refers to a biological control (i.e., untreated termites). However, when calculating R values for induced target gene expression, *control* refers to untreated controls, and *sample* refers to a treatment type (i.e., 0.5% PB or *E. coli/P. termitis*).

R values were compared between constitutively expressed and induced target amplicons within and among 1) three colonies, 2) three treatment groups (untreated control, 0.5% PB or *E. coli/P. termitis*), and 3) two castes (worker or soldier) via non-parametric, multivariate analysis of variance (PROC GLM, SAS/STAT Software 9.3, SAS Institute Inc., 2001). Furthermore, the GLIMMIX and RANK procedures were used to analyze non-normally distributed data points. The statistical model used was: gene = 1 to 10; expressed(gene) = treatment caste treatment*caste; random treatment*caste*colony, where "gene" refers to the 10 target genes, and "expressed(gene)" is the interaction of treatment, caste and colony. Differences in expression were considered significant when $P<0.05$.

## Acknowledgments

We thank the members of the New Orleans Mosquito, Termite and Rodent Control Board (especially Dr. C. Riegel, Dr. C. Cottone and B. Yokum) for providing termite colonies, J. Donaldson for technical assistance, A. Howells for assisting with the preparation of this manuscript and Dr. J. Geaghan (LSU Department of Experimental Statistics) for help with writing the SAS program code. We thank Drs. L. Foil and J. Ottea for providing valuable expertise in insect toxicology.

## Author Contributions

Conceived and designed the experiments: CH DS. Performed the experiments: DS. Analyzed the data: CH DS. Contributed reagents/materials/analysis tools: CH. Contributed to the writing of the manuscript: CH DS.

## References

1. Inward D, Beccaloni G, Eggleton P (2007) Death of an order: a comprehensive molecular phylogenetic study confirms that termites are eusocial cockroaches. Biol Lett 3: 331–335.
2. Vargo EL, Husseneder C (2009) Biology of subterranean termites: Insights from molecular studies of *Reticulitermes* and *Coptotermes*. Annu Rev Entomol 54: 379–403.
3. Ishikawa Y, Okada Y, Ishikawa A, Miyakawa H, Koshikawa S, et al. (2010) Gene expression changes during caste-specific neuronal development in the damp-wood termite *Hodotermopsis sjostedti*. BMC Genomics 11: 314–322.
4. Haritos VS, French JRJ, Ahokas JT (1994) Cytochrome P450 monooxigenase and glutathione S-transferase activity of two Australian termites: *Mastotermes darwiniensis* and *Coptotermes acinaciformis*. Insect Biochem Mol Biol 24: 929–932.
5. Valles SM, Oi FM, Strong CA (2001) Purification and characterization of *trans*-permethrin metabolizing microsomal esterases from workers of the eastern subterranean termite, *Reticulitermes flavipes* (Kollar). Insect Biochem Mol Biol 31: 715–725.
6. Gatti SS, Henderson G. (1996) Differential response of Formosan subterranean termite castes (Isoptera: Rhinotermitidae) to selected termiticides. Sociobiology 28: 23–32.
7. Osbrink WLA, Lax AR, Brenner RJ (2001) Insecticide susceptibility in *Coptotermes formosanus* and *Reticulitermes virginicus* (Isoptera: Rhinotermitidae). J Econ Entomol 94: 1217–1228.
8. Valles SM, Woodson WD (2002) Insecticide susceptibility and detoxication enzyme activities among *Coptotermes formosanus* Shiraki workers sampled from different locations in New Orleans. Comp Biochem Physiol Part C 131: 469–476.
9. Yu XQ, Zhu YF, Ma C, Fabrick JA, Kanost MR (2002) Pattern recognition proteins in *Manduca sexta* plasma. Insect Biochem Mol Biol 32: 1287–1293.
10. Bechkage NE (2008) Insect Immunology. First Edition. San Diego CA, Academic Press, p. 360.
11. Tsakas S, Marmara VJ (2010) Insect immunity and its signaling: an overview. Invertebrate Surviv J 7: 228–238.
12. Theopold U, Schmidt O, Söderhäll K, Dushay MS (2004) Coagulation in arthropods: defence, wound closure and healing. Trends Immunol 25: 289–294.
13. Lavine MD, Strand MR (2002) Insect hemocytes and their role in immunity. Insect Biochem Mol Biol 32: 1295–1309.
14. Cerenius L, Söderhäll K (2004) The prophenoloxidase-activating system in invertebrates. Immunol Rev 198: 116–126.
15. Johansson KC, Metzendorf C, Söderhäll K (2005) Microarray analysis of immune challenged Drosophila hemocytes. Exp Cell Res 305: 145–155.
16. Boutros M, Agaisse H, Perrimon N (2002) Sequential activation of signaling pathways during innate immune responses in Drosophila. Dev Cell 3: 711–722.
17. Rosengaus RB, Traniello JFA, Bulmer M (2011) Ecology, behavior and evolution of disease resistance in termites. In: Bignell DE, Roisin Y, Lo N, editors. Biology of Termites: A Modern Synthesis, Springer, Netherlands. 165–191. Doi:10.1007/978-90-481-3977-4_7.
18. Lamberty M, Zachary D, Lanot R, Bordereau C, Robert A, et al. (2001) Constitutive expression of a cysteine-rich antifungal and a linear antibacterial peptide in a termite insect. J Biol Chem 276: 4085–4092.
19. Rosengaus RB, Cornelisse T, Guschanski K, Traniello JFA (2007) Inducible immune proteins in the dampwood termite *Zootermopsis angusticollis*. Naturwissenschaften 94: 25–33.
20. Thompson GJ, Crozier YC, Crozier RH (2003) Isolation and characterization of a termite transferrin gene up-regulated on infection. Insect Mol Biol 12: 1–7.
21. Bulmer MS, Bachelet I, Raman R, Rosengaus RB, Sasisekharan R (2009) Targeting an antimicrobial effector function in insect immunity as a pest control strategy. Proc Natl Acad Sci USA 31: 12652–12657.
22. Xu P, Shi M, Lai R, Chen X-X (2012) Differences in numbers of termicins expressed in two termite species affected by fungal contamination of their environments. Genet Mol Res 11: 2247–2257.
23. Rosengaus RB, Levebre ML, Traniello JFA (2000) Inhibition of fungal spore germination by *Nasutitermes*: Evidence for a possible antiseptic role of soldier defense secretions. J Chem Ecol 26: 21–39.
24. Calleri DV, Reid EM, Rosengaus RB, Vargo EL, Traniello JFA (2006) Inbreeding and disease resistance in a social insect: effects of heterozygosity on

immunocompetence in the termite *Zootermopsis angusticollis*. Proc R Soc B 273: 2633–2640.

25. Husseneder C, Delatte JR, Grace JK, Simms DM, Wang C, et al. (2011) Genetic diversity and colony breeding structure in native and introduced ranges of the Formosan subterranean termite, *Coptotermes formosanus*. Biol Invasions 14: 419–437.
26. Global Invasive Species Database, 2013. 100 of the World's Worst Invasive Alien Species. Available: http://www.issg.org/database/species/search.asp?st = 100ss&fr = 1&str = &lang = EN. Accessed 2014 Jul 29.
27. Su N-Y, Scheffrahn RH (1988) Foraging population and territory of the Formosan subterranean termite (Isoptera: Rhinotermitidae) in an urban environment. Sociobiology 14: 353–359.
28. Chouvenc T, Su N-Y, Grace JK (2011) Fifty years of attempted biological control of termites – Analysis of a failure. J Biol Control 59: 69–82.
29. Husseneder C, McGregor C, Lang RP, Collier R, Delatte J (2012) Transcriptome profiling of female alates and egg-laying queens of the Formosan subterranean termite. Comp Biochem Physiol Part D. Genomics Proteomics 1: 14–27.
30. Hussain A, Li Y-F, Cheng Y, Liu Y, Chen C-C, et al. (2013) Immune-Related Transcriptome of *Coptotermes formosanus* Shiraki Workers: The Defense Mechanism. PLOS ONE 8(7): e69543. doi:10.1371/journal.pone.0069543.
31. Cornette R, Koshikawa S, Hojo M, Matsumoto T, Miura T (2006) Caste-specific cytochrome P450 in the damp-wood termite *Hodotermopsis sjostedti* (Isoptera, Termopsidae). Insect Mol Biol 15: 235–244.
32. Steller MM, Kambhampati S, Caragea D (2010) Comparative analysis of expressed sequence tags from three castes and two life stages of the termite *Reticulitermes flavipes*. BMC Genomics 11: 463–474.
33. Ratra GS, Casida JE (2001) GABA receptor subunit composition relative to insecticide potency and selectivity. Toxicol Lett 122: 215–222.
34. Sueyoshi T, Negishi M (2001) Phenobarbital response elements of cytochrome P450 genes and nuclear receptors. Annu Rev Pharmacol Toxicol 41: 123–143.
35. Brown D, Zhang L, Wen Z, Scott JG (2003) Induction of P450 monooxigenases in the German cockroach, *Blattella germanica*. L Arch Insect Biochem Physiol 53: 119–124.
36. Le Goff G, Hilliou F, Siegfried BD, Boundy S, Wajnberg E, et al. (2006) Xenobiotic response in *Drosophila melanogaster*: sex dependence of P450 and GST gene induction. Insect Biochem Mol Biol 36: 674–82.
37. Sun W, Margam VM, Sun L, Buczkowski G, Bennett GW, et al. (2006) Genome-wide analysis of Phenobarbital-inducible genes in *Drosophila melanogaster*. Insect Mol Biol 15: 455–464.
38. Willoughby L, Chung H, Lumb C, Robin C, Batterham P, et al. (2006) A comparison of *Drosophila melanogaster* detoxification gene induction responses for six insecticides, caffeine and phenobarbital. Insect Biochem Mol Biol 36: 934–942.
39. Lumjuan N, Stevenson BJ, Prapanthadara L-A, Somboon P, Brophy PM, et al. (2007) The *Aedes aegypti* glutathione transferase family. Insect Biochem Mol Biol 37: 1026–1035.
40. Feyereisen R (2005) Insect Cytochrome P450. In: Gilbert LI, Iatrou K, Gill SS' editors. Comprehensive Molecular Insect Science Vol. 4. Elsevier, Oxford, UK., 1–77.
41. Helvig C, Koener JF, Unnitan GC, Feyereisen R (2004) CYP15A1, the cytochrome P450 that catalyzes epoxidation of methyl farnesoate to juvenile hormone III in cockroach corpora allata. Proc Natl Acad Sci USA 101: 4024–4029.
42. Zhou X, Song C, Grzymala TL, Oi FM, Scharf ME (2006) Juvenile hormone and colony conditions differentially influence cytochrome P450 gene expression in the termite *Reticulitermes flavipes*. Insect Mol Biol 15: 749–61.
43. Tarver MR, Coy MR, Scharf ME (2012) *Cyp15F1*: A Novel Cytochrome P450 Gene Linked to Juvenile Hormone- Dependent Caste Differentiation in the Termite *Reticulitermes flavipes*. Arch Insect Biochem Physiol 80: 92–108.
44. Marchal E, Zhang YL, Badisco L, Velinden H, Hult EF, et al. (2011) Final steps in juvenile hormone biosynthesis in the desert locust, *Schistocerca gregaria*. Insect Biochem Mol Biol 41: 219–227.
45. Tarver MR, Zhou X, Scharf ME (2010) Socio-environmental and endocrine influences on developmental and caste-regulatory gene expression in the eusocial termite *Reticulitermes flavipes*. BMC Mol Biol 11: 28–43.
46. Anspaugh DD, Roe MR (2005) Regulation of JH epoxide hydrolase versus JH esterase activity in the cabbage looper, *Trichoplusia ni*, by juvenile hormone and xenobiotics. J Insect Physiol 51: 523–535.
47. Tian L, Guo E, Diao Y, Zhuo S, Peng Q, et al. (2012) Genome-wide regulation of innate immunity by juvenile hormone and 20-hydroxyectdysone in the *Bombyx* fat body. BMC Genomics, 11: 549.
48. Scott JG, Sridhar P, Liu N (1996) Adult specific expression and induction of cytochrome $P450_{lpr}$ in house flies. Arch Insect Biochem Physiol. 31: 313–323. doi:10.1002/(SICI)1520–6327(1996)31: 3<313::AID-ARCH6>3.0.CO;2-Y.
49. Kasai S, Scott JG (2001) Expression and regulation of *CYP6D3* in the house fly, *Musca domestica* (L.). Insect Biochem Mol Biol 32: 1–8.
50. Petryk A, Warren JT, Marqués G, Jarcho MP, Gilbert LI, et al. (2003) Shade is the Drosophila P450 enzyme that mediates the hydroxylation of ecdysone to the steroid insect molting hormone 20-hydroxyecdysone. Proc Natl Acad Sci USA 100: 13773–13778. doi:10.1073/pnas.2336088100.
51. Guzov VM, Unnitan GC, Chernogolov AA, Feyereisen R (1998) CYP12A1, a mitochondrial cytochrome P450 from the house fly. Arch Biochem Biophys 359: 231–240.
52. Giraudo M, Unnitan GC, Le Goff G, Feyereisen R (2009) Regulation of cytochrome P450 expression in *Drosophila*: Genomic insights. Pestic Biochem Physiol. 97(2): 115–122. doi:10.1016/j.pestb.2009.06.009.
53. Moriyama Y, Hiasa M, Matsumoto T, Omote H (2008) Multidrug and toxic compound extrusion (MATE)-type proteins as anchor transporters for the excretion of metabolic waste products and xenobiotics. Xenobiotica 38: 1107–1118.
54. Kaatz GW, McAleese F, Seo SM (2005) Multidrug Resistance in *Staphylococcus aureus* Due to Overexpression of a Novel Multidrug and Toxin Extrusion (MATE) Transport Protein. Antimicrob Agents Chemother 49: 1857–1864.
55. Tanaka Y, Hipolito CJ, Maturana AD, Ito K, Kuroda T, et al. (2013) Structural basis for the drug extrusion mechanism by a MATE multidrug transporter. Nature 496: 247251.
56. Nawrath C, Heck S, Parintawong N, Métraux J-P (2002) EDS5, an Essential Component of Salicylic Acid–Dependent Signaling for Disease Resistance in Arabidopsis, Is a Member of the MATE Transporter Family. The Plant Cell 14: 275–286.
57. Sugumaran M (2002) Comparative biochemistry of eumelanogenesis and the protective roles of phenoloxidase and melanin in insects. Pigment Cell Res 15: 2–9.
58. Omote H, Hiasa M, Matsumoto T, Otsuka M, Moriyama Y (2006) The MATE proteins as fundamental transporters of metabolic and xenobiotic organic cations. TRENDS Pharmacol Sci 27. doi:10.10/j.tips.2006.09.001.
59. Destoumieux D, Muñoz M, Cosseau C, Rodriguez J, Bulet P, et al. (2000) Penaeidins, antimicrobial peptides with chitin-binding activity, are produced and stored in shrimp granulocytes and released after microbial challenge. J Cell Sci 113: 461–469.
60. Jasrapuria S, Arakane Y, Osman G, Kramer KJ, Beeman RW, et al. (2010) Genes encoding proteins with peritrophin A-type chitin-binding domains in *Tribolium castaneum* are grouped into three distinct families based on phylogeny, expression and function. Insect Biochem Mol Biol 40: 214–227.
61. Wijffels G, Eisemann C, Riding G, Pearson R, Jones A, et al. (2001) A novel family of chitin-binding proteins from insect type 2 peritrophic matrix. J Biol Chem 276: 15527–15536.
62. Endo Y, Matsushita M, Fujita T (2007) Role of ficolin in innate immunity and its molecular basis. Immunobiology 212: 371–379.
63. Matsushita M (2010) Ficolins: Complement-activating lectins involved in innate immunity. J Innate Immun 2: 24–32.
64. Holmskov U, Thiel S, Jensenius JC (2003) Collectins and ficolins: Humoral lectins of the innate immune defense. Annu Rev Immunol 21: 547–578.
65. Wang X, Zhao Q, Christensen BM (2005) Identification and characterization of the fibrinogen-like domain of fibrinogen-related proteins in the mosquito, *Anopheles gambiae*, and the fruit fly, *Drosophila melanogaster*, genomes. BMC Genomic 6: 114.
66. Middha S, Wang X (2008) Evolution and potential function of fibrinogen-like domains across twelve *Drosophila* species. BMC Genomics 9: 260–268.
67. Yang J, Wang L, Zhang H, Wang H, Song L (2011) C-type lectin in *Chlamys farreri* (CfLec-1) mediating immune recognition and opsonization. PLOS ONE 2: e17089.
68. Jomori Y, Kubo T, Natori S (1990) Purification and characterization of lipopolysaccharide-binding protein from hemolymph of American cockroach *Periplaneta americana*. Eur J Biochem 190: 201–206.
69. Arai T, Kawasaki K, Kubo T, Natori S (1998) Cloning of cDNA for regenectin, a humoral C-type lectin of *Periplaneta americana*, and expression of the regenectin gene during leg regeneration. Insect Biochem Mol Biol 28: 987–994.
70. Kuballa AV, Elizur A (2008) Differential expression profiling of components associated with exoskeletal hardening in crustaceans. BMC Genomics 9: 575. doi:10.1186/1471-2164-9-575.
71. Basta-Kaim A, Budziszewska A, Lasori W (2008) Effects of antileptic drugs on immune system. Przegl Lek 65: 799–802 [in Polish].
72. King-Jones K, Horner MA, Lam G, Thummel CS (2006) The DHR96 nuclear receptor regulates xenobiotic responses in *Drosophila*. Cell Metab 4: 37–48.
73. Hamilton C, Bulmer MS (2011) Molecular antifungal defenses in subterranean termites: RNA interference reveals in vivo roles of termicins and GNBPs against a naturally encountered pathogen. Dev Comp Immunol 36(2): 372–377. doi:10.1016/j.dci.2011.07.008.
74. Ma C, Kanost MR (2000) A beta 1,3-glucan recognition protein from an insect, *Manduca sexta*, agglutinates microorganisms and activates the phenoloxidase cascade. J Biol Chem 275: 7505–7514.
75. Hamilton C, Lay F, Bulmer MS (2011) Subterranean termite prophylactic secretions and external antifungal defenses. J Insect Physiol 57: 1259–1266.
76. Gao Q, Tancredi SE, Thompson GJ (2012). Identification of mycosis-related genes in the Eastern subterranean termite by suppression subtractive hybridization. Arch Insect Biochem Physiol 80: 63–76. doi:10.1002/arch.21026.
77. Graveley BR, Kaur A, Gunning D, Zipurski SL, Rowen L, et al. (2004) The organization and evolution of Dipteran and Hymenopteran *Down syndrome cell adhesion molecule* (*Dscam*) genes. RNA 10: 1499–1506.
78. Schmucker D, Chen B (2009) Dscam and DSCAM: complex genes in simple animals, complex animals yet simple genes. Genes Dev 23: 147–156.
79. Dong Y, Taylor HE, Dimopoulos G (2006) AgDscam, a Hypervariable Immunoglobulin Domain-Containing Receptor of the *Anopheles gambiae* Innate Immune System. PLOS Biol 4: 1137–1147.

80. Watson FL, Püttmann-Holgado R, Thomas F, Lamar DL, Hughes M, et al. (2005). Extensive diversity of Ig-superfamily proteins in the immune system of insects. Science 309: 1874–1878.
81. Gatti SS, Henderson G, Abdel-aal YAI, Ibrahim SA (2002) Acetylcholinesterase mediated susceptibility of soldiers and workers of Formosan subterranean termite (Isoptera: Rhinotermitidae) to chloropyrifos. J Econ Entomol 95: 813–819.
82. Dunn PE (1986) Biochemical aspects of insect immunology. Annu Rev Entomol 31: 321–339.
83. Traniello JFA, Rosengaus RB, Savoie K (2002) The development of immunity in a social insect: Evidence for the group facilitation of disease resistance. Proc Natl Acad Sci USA 99: 6838–6842.
84. Husseneder C, Messenger MT, Su N-Y, Grace JK, Vargo EL (2005) Colony social organization and population genetic structure of an introduced population of the Formosan Subterranean Termite from New Orleans, Louisiana, U.S.A. J Econ Entomol 98: 1421–1434.
85. Messenger MT, Mullins AJ (2005) New flight distance recorded for *Coptotermes formosanus* (Isoptera: Rhinotermitidae). Fla Entomol 88: 99–100.
86. Simms DM, Husseneder C (2009) Assigning individual alates of the Formosan subterranean termite to their colonies of origin within the context of an area-wide management program. Sociobiology 52: 1–20.
87. Husseneder C, Berestecky JM, Grace JK (2009) Changes in composition of culturable bacteria community in the gut of the Formosan subterranean termite depending on rearing conditions of the host. Ann Entomol Soc Am 102: 498–507.
88. Higashiguchi DT, Husseneder C, Grace JK, Berestecky JM (2006) *Pilibacter termitis* gen. nov., sp. nov., a lactic acid bacterium from the hindgut of the Formosan subterranean termite (*Coptotermes formosanus*). Int J Syst Evol Microbiol 56: 15–20.
89. Fehlbaum P, Bulet P, Michaut L, Lagueux M, Broekaert WF, et al. (1994) Septic injury of *Drosophila* induces the synthesis of a potent antifungal peptide with sequence homology to plant antifungal peptides. J Biol Chem 269: 33159–33163.
90. Dheda K, Huggett JF, Bustin SA, Johnson MA, Rook G, et al. (2004) Validation of housekeeping genes for normalizing RNA expression in real-time PCR. BioTechniques 37: 112–119.
91. Infante C, Matsuoka MP, Asensio E, Canavate JP, Reith M, et al. (2008) Selection of housekeeping genes for gene expression studies in larvae from flatfish using real-time PCR. BMC Mol Biol 9: 28–39.
92. Bustin SA, Benes V, Garson JA, Hellemans J, Huggett J, et al. (2009) The MIQE Guidelines: Minimum Information for Publication of Quantitative Real-Time PCR Experiments. Clin Chem 55: 611–622.
93. Ramakers C, Ruijter JM, Lekanne Deprez RH, Moorman AFM (2003) Assumption-free analysis of quantitative real-time polymerase chain reaction (PCR) data. Neurosci Lett 339: 62–66.
94. Ruijter JM, Ramakers C, Hoogaars WMH, Karlen Y, Bakker O, et al. (2009) Amplification efficiency: linking baseline and bias in the analysis of quantitative PCR data. Nucleic Acid Res 37: 1–12.
95. Vandesompele J, De Preter K, Pattyn F, Poppe B, Van Roy N, et al. (2002) Accurate normalization of real-time quantitative RT-PCR data by geometric averaging of multiple internal control genes. Gen Biol 3: 1–12.
96. Pfaffl MW (2001) A new mathematical model for relative quantification in real-time RT-PCR. Nucl Acid Res 29: 2002–2007.

# Isolation, Identification and Characteristics of an Endophytic Quinclorac Degrading Bacterium *Bacillus megaterium* Q3

**Min Liu[1,2]❡, Kun Luo[1,2]❡, Yunsheng Wang[2,3], Aiping Zeng[1], Xiaomao Zhou[1], Feng Luo[2,3]*, Lianyang Bai[4]***

**1** Institute of Pesticide Science, Hunan Agricultural University, Changsha, China, **2** College of Plant Protection, Hunan Agricultural University, Changsha, China, **3** School of Computing, Clemson University, Clemson, South Carolina, United States of America, **4** Hunan academy of agricultural sciences, Changsha, China

## Abstract

In this study, we isolated an endophytic quinclorac-degrading bacterium strain Q3 from the root of tobacco grown in quinclorac contaminated soil. Based on morphological characteristics, Biolog identification, and 16S rDNA sequence analysis, we identified strain Q3 as *Bacillus megaterium*. We investigated the effects of temperature, pH, inoculation size, and initial quinclorac concentration on growth and degrading efficiency of Q3. Under the optimal degrading condition, Q3 could degrade 93% of quinclorac from the initial concentration of 20 mg/L in seven days. We analyzed the degradation products of quinclorac using liquid chromatography–tandem mass spectrometry (LC-MS/MS). The major degradation products by Q3 were different from those of previously identified quinclorac degrading strains, which suggests that Q3 may employ new pathways for quinclorac degradation. Our indoor pot experiments demonstrated that Q3 can effectively alleviate the quinclorac phytotoxicity in tobacco. As the first endophytic microbial that is capable of degrading quinclorac, Q3 can be a good bioremediation bacterium for quinclorac phytotoxicity.

**Editor:** Youjun Zhang, Institute of Vegetables and Flowers, Chinese Academy of Agricultural Science, China

**Funding:** This work was partially supported by key project funds of State Tobacco Monopoly Bureau of China (SR0558), Special national public welfare industry research (201303031) and funds from Hunan Agricultural University. The funders had no role in study design, data collection and analysis, decision to publish, or preparation of the manuscript.

**Competing Interests:** The authors have declared that no competing interests exist.

* Email: luofeng@clemson.edu (FL); bailianyang2005@aliyun.com (LB)

❡ These authors contributed equally to this work.

## Introduction

Quinclorac (3,7-dichloro-8-quinoline-carboxylic) is a highly selective auxin herbicide developed by the BASF Corporation. It is widely used for controlling barnyard grasses and certain dicot grasses in rice fields [1]. Quinclorac can be used as both a pre-emergence and post-emergence herbicide. Quinclorac stimulates the induction of 1-aminocyclopropane-1-carboxylic acid (ACC) synthase, a key enzyme in ethylene formation, which leads to the accumulation of cyanide in the tissues of plants and ultimately—to their death [2,3].

Meanwhile, quinclorac is highly stable, and degradation of quinclorac in nature is very slow [4]. The quinclorac residues are phytotoxic to many crops and vegetables, which has become one of the major problems in their rotation with rice [5]. Especially in a rice-tobacco rotation field, residues of quinclorac can cause serious phytotoxicity to sensitive tobacco, which has a strong effect on the yield and quality of tobacco. The quinclorac residues may also affect microbial activity and bring potential hazards to the environment through soil contamination [6]. Currently, two major methods are used to reduce quinclorac contamination. They are photodegradation and microbial degradation [7,8]. Photodegradation is only useful when quinclorac exists in the soil's surface. On the other hand, microbial degradation is not limited by this constraint. Bioremediation using microbial degradation is an effective way to control pesticide residue [9]. Several quinclorac degrading strains were isolated from quinclorac-contaminated soil. Lü et al. isolated a WZ1 strain from pesticide manufactory soil, which was capable of degrading quinclorac [10]. WZ1 was identified as *Burkholderia cepacia* [10]. At optimal conditions, WZ1 degraded 90% of quinclorac from an initial concentration of 1000 mg/L in 11 days [10,11,12]. In 2012, Xu et al. also isolated a *Bordetella sp.* strain HN36 [13] from a pesticide manufactory soil. HN36 strain degraded 96% of quinclorac from an initial concentration of 400 mg/L in 48 h [13]. In 2013, Dong et al. isolated a quinclorac degrading J3 bacterium strain from soil with long-term quinclorac usage. The J3 was identified as *Alcaligenes sp* [14]. The J3 strain degraded 70% of quinclorac from initial concentration of 100 mg/L in 7 days at optimal conditions [14]. Fan et al. isolated a *Pantoea* sp. QC06 strain from soil with long-term quinclorac usage in 2013. The strain QC06 degraded 95.31% of quinclorac from initial concentration of 50 mg/L in 7 days at optimal degrading conditions [15].

Currently, no study has reported on the capability of the endophytic microbial bacterium to degrade quinclorac. In this paper, we isolated the Q3, an endophytic quinclorac-degrading bacterium, from the root of tobacco grown in a quinclorac-contaminated field.

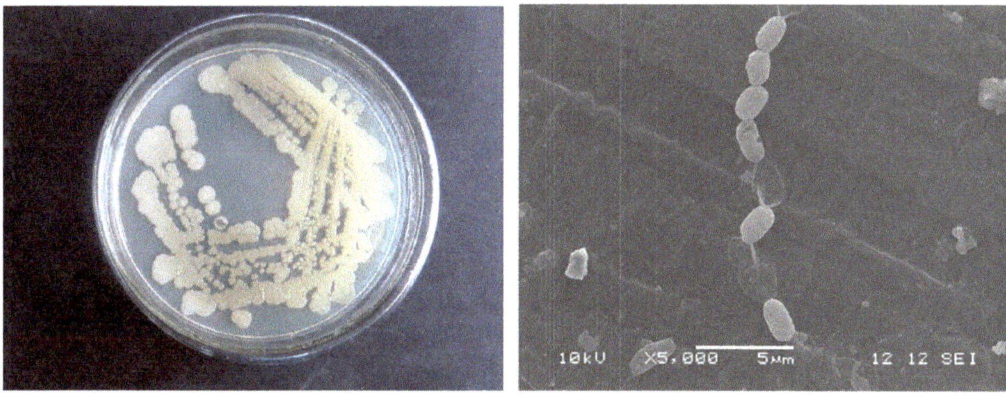


**Figure 1. Morphological characteristics of strain Q3.** Colony morphology on LB plate (left) and cell morphology under electron microscope (right).

Compared with microbial strains isolated from highly concentrated quinclorac soil from pesticide manufactory, endophytic microbial strains are capable of degrading quinclorac in lower concentration, such as the concentration of quinclorac commonly found in a rice field [16]. We investigated its morphological, physiological and biochemical characteristics. We sequenced the 16S rDNA of Q3 and identified it as *Bacillus megaterium*. We explored the degradation conditions of Q3 to find its optimal degradation condition. We analyzed the microbial degradation products of quinclorac using HPLC-MS/MS. The degradation products showed that Q3 may employ degradation pathways that differ from those of previously identified quinclorac degrading microbial. We also conducted indoor pot cultivation experiments to demonstrate the bioremediation function of Q3 on tobacco quinclorac phytotoxicity.

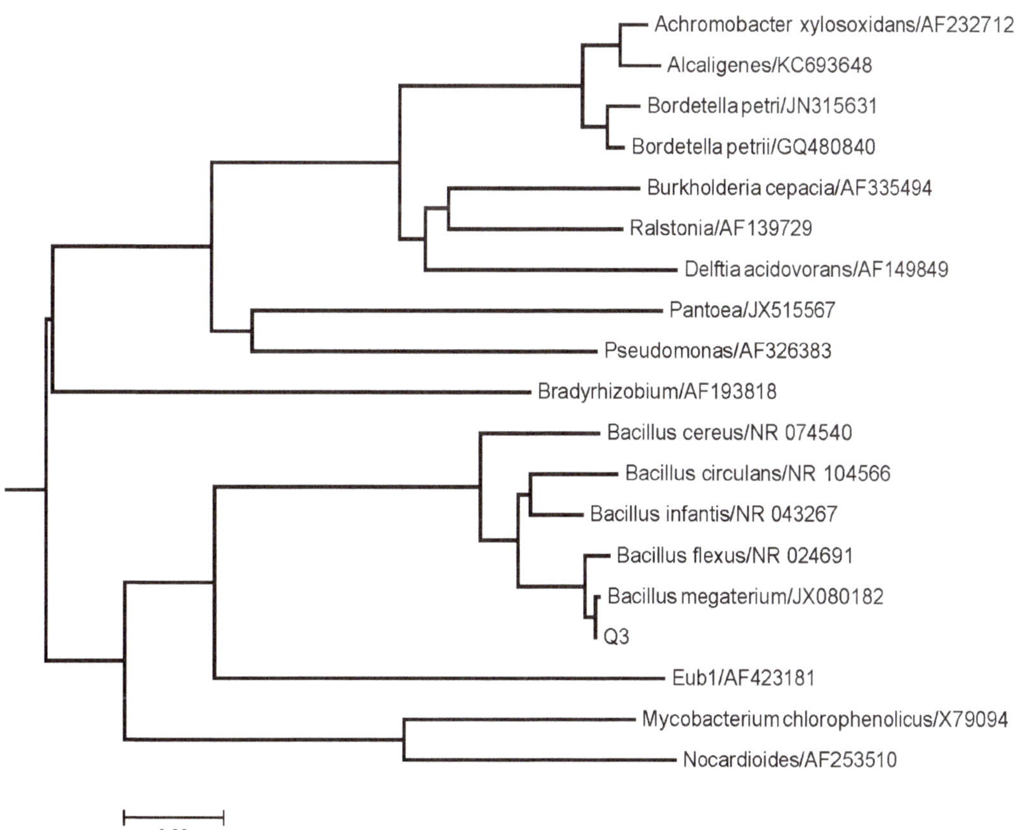


**Figure 2. The phylogenetic tree based on 16S rDNA of strain Q3 and those of herbicide-degrading species.** Q3 and other two quinclorac degrading bacteria, such as *Burkholderia cepacia* and *Bordetella petrii* are in different branches of the tree.

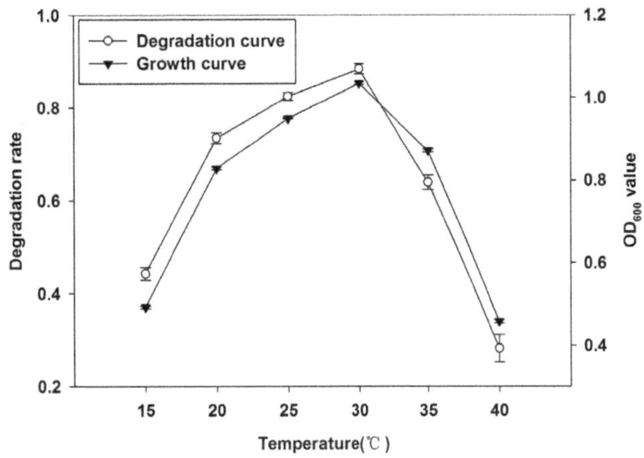


**Figure 3. Effect of temperature on degradation of quinclorac and growth of Q3.** Both growth and quinclorac degradation of Q3 reached peak values at temperature of 30°C.

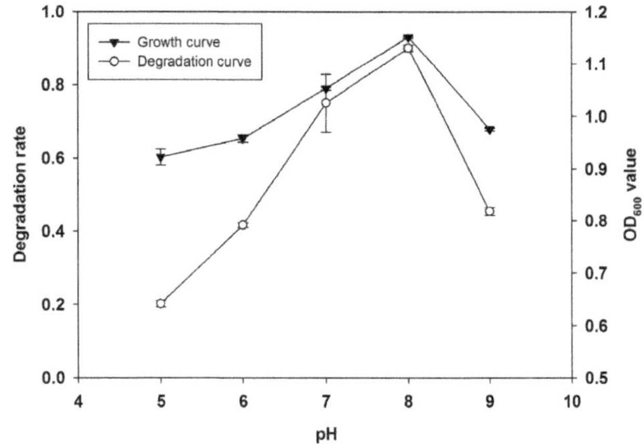


**Figure 4. Effect of pH on degradation of quinclorac and growth of Q3.** Both growth and quinclorac degradation of Q3 reached peak values at pH of 8.

## Materials and Methods

### Materials

Quinclorac standard (98.1%) was purchased from Dr. Ehrenstorfer GmbH (Germany. HPLC grade methanol was purchased from Tedia Company (USA). Other analytical reagents were bought from the China National Pharmaceutical Group Corporation, SINOPHARM.

The root of widely planted tobacco K326 was collected from the private land of Mr. Huibo Zhou in Shiyan village of Lougu Town, Liuyang City of Hunan Province. Mr. Zhou should be contacted for future permissions. The collection of root did not involve endangered or protected species and was safe to the environment. No specific permission is needed. The GPS coordinates of the field are E113.50′99.23″ and N28.36′ 34.24″.

The composition of mineral salt medium (MSM) was as follows ($g.L^{-1}$): $(NH_4)_2SO_4$ 1.0, $KH_2PO_4$ 2.0, $MgSO_4.7H_2O$ 0.5, NaCl 0.1, and $CaCO_3$ 0.5. For quinclorac degrading, 0.5 g sucrose and 0.5 g yeast powder were added to the mineral salt medium as an additional carbon source. The composition of Luria–Bertani's (LB) medium for purification and enlarged cultivation was as follows ($g.L^{-1}$): yeast powder 5, tryptone 10, NaCl 10. In solid LB medium, 15 $g.L^{-1}$ agar was added.

### Methods

**Isolation of quinclorac-degrading endophytic bacteria.** We collected the tobacco from a field with long-term usage of quinclorac. Then, the cleaned tobacco root was surface-sterilized sequentially using: 75% (v/v) ethanol for 2 min, ultrapure water three times with 1 min each and 1% mercuric chloride for 1 min. Finally the surface-sterilized tobacco root was washed by sterile ultrapure water three times again to clean sterilization agent residues. To confirm the success of surface sterilization, we inoculated the ultrapure water from the final rinse on petri dish with LB agar medium. The tobacco root was considered clean if no bacterial colony was found on petri dish after inoculation.

We grounded 0.3 g fresh root with 10 mL sterile ultrapure water using a sterile mortar. We spread the grinding fluid (50 ul) on a petri dish of mineral salt agar medium that contains 20 mg/L quinclorac. We cultured it in an incubator at 28°C for three days. Bacterial colonies with clear zones were selected as potential quinclorac-degrading bacteria.

**Characterization of quinclorac-degrading strain Q3.** The identification of strain Q3 was performed according to Bergey's Manual of Determinative Bacteriology. First, we studied the morphological characteristics of strain Q3. We streaked Q3 on solid LB media to observe its colonial morphology. We also cultured strain Q3 in liquid LB media for 28 h and scanned the strain morphology using a scanning electron microscope. Then, we assessed basic physiological and biochemical characteristics of strain Q3. We performed 71 carbon source utilization tests and 23 chemical-sensitivity tests of strain Q3 using a GNIII microplate (96 holes) from Biolog Corporation (USA).

**16S rDNA sequencing and phylogenetic analysis.** DNA of strain Q3 was extracted and purified with a commercial DNA extraction kit (TransGen Biotech, China). Then, the 16S rDNA was amplified by polymerase chain reaction (PCR) using general bacterial primer 27F (5′-AGAGTTTGATCCTGGCTCAG-3′) and 1492R (5′-GGTTACCTTGTTACGACTT-3′). The PCR mixture contained 2.0 μL 10×Ex Taq buffer, 1.6 μL $MgCl_2$(25 mM), 1.6 μL dNTP (2.5 mM), 1.0 μL of each primer, 0.5 μL DNA template, 0.2 μL Taq DNA polymerase (10000 $U.mL^{-1}$), and 12.1 μL $ddH_2O$ in a total volume of

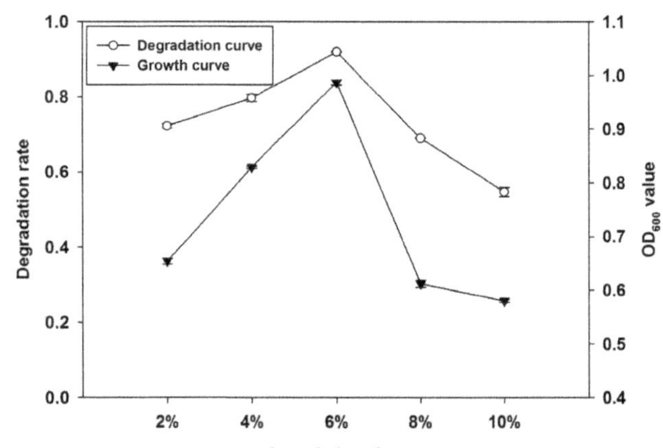


**Figure 5. Effect of inoculation size on degradation of quinclorac and growth of Q3.** Both growth and quinclorac degradation of Q3 reached peak values when inoculation size was 6%.

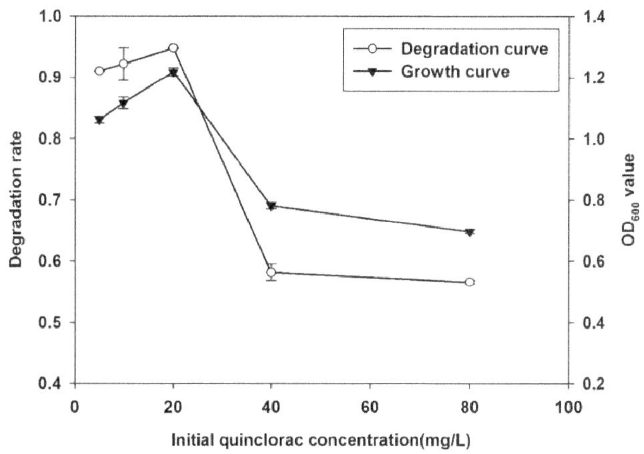


**Figure 6. Effect of initial quinclorac concentration on Q3 growth and degradation of quinclorac.** Both growth and quinclorac degradation of Q3 reached peak values when initial quinclorac concentration was 20 mg/L.

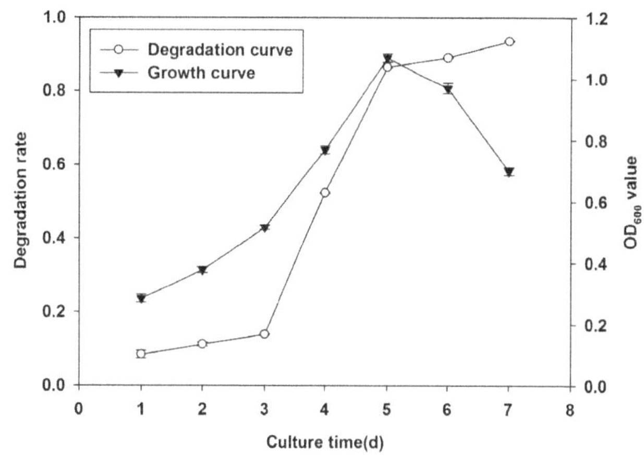


**Figure 7. The growth and quinclorac degradation curve of Q3 at optimal condition.** The degradation of quinclorac is highly correlated with the growth of Q3.

20 μL. The PCR amplification process consisted of one cycle of 5 min denaturation and Taq activation at 95°C, 40 cycles with each consisting of 30 s at 95°C, 30 s at 55°C, and 90 s at 72°C, and a final extension cycle of 10 min at 72°C. The PCR products were first verified by agarose gel electrophoresis. Then, purified PCR products were sequenced by Shanghai Majorbio Bio-pharm Technology Co., Ltd. The 16s rDNA sequence was aligned with published sequences in GenBank using BLAST program, and phylogenies was analyzed using MEGA (version 5.0) software.

**Measurement of *Bacillus megaterium* Q3 Growth.** *Bacillus megaterium* Q3 was cultivated in liquid LB culture for 24 h in advance. The culture liquid was then centrifuged at 2152×g for 3 min to collect bacteria cells. The collected cells were washed three times to remove remaining LB culture. Then, the cells were inoculated in a flask containing degrading culture for subsequent study.

Growth of Q3 was determined by the spectrophotometric method operated under 600 nm. First, 0.1 mL culture liquid was transferred into a centrifuge tube and centrifuged at 2152×g for 3 min. The cells were collected and washed with 0.3% sterile saline three times, and then fully resuspended in 1 mL 0.3% sterile normal saline. The $OD_{600}$ value was measured by a spectrophotometer (Shimadzu, Japan).

**Measurement of quinclorac degradation.** The degradation of quinclorac was monitored by high performance liquid chromatography (HPLC) (Shimadzu, Japan) equipped with a SPD-20A UV detector and an Athena $C_{18}$ column (CNW, 5 μm, 4.6 mm×250 mm) using a method developed by Wang et al. [17]. For each measurement, 1 mL degradation liquid was thoroughly mixed with 0.25 mL chloroform-n-butyl alcohol (volume ratio 4:1) for 30 min and centrifuged at 12396×g for 5 min. Then, the supernatant was filtered through a 0.22 μm filter membrane and detected by HPLC. Water (containing 0.2% acetic acid)/methanol (40/60, v/v) mixture was used as effluent with a flow rate of 0.8 mL/min for HPLC. The detections were performed at a wavelength of 240 nm with column temperature at 30°C. The injection volume was 20 μL.

## Measurement of quinclorac metabolites

**Extraction of degradation products.** 2 mL degradation liquid was transferred to a centrifuge tube and centrifuged at 12396×g for 5 min after deproteinization. We took the supernatant and added 4 mL ethyl acetate to extract the degradation products. Then, we added saturated sodium chloride to make the mixed liquid separate into two layers. The ethyl acetate layer was taken and dewatered by adding anhydrous sodium sulfate; then, the ethyl acetate layer was concentrated before examination by an ACQUITY/ZQ 4000 HPLC-MS/MS (Waters, America).

**Detection condition of HPLC.** ACQUITY UPLC BEH $C_{18}$ column (50×2.1 mm,1.7 μm)was used. 10 mmol/L ammonium acetate (containing 0.1% formic acid)-methanol was used as effluent with a flow rate of 0.2 mL/min. The sample room temperature was 10°C and the column temperature was set to 30°C. The injection volume was 10 μL.

**Detection condition of MS/MS.** The mass spectrometer was operated in the positive polarity mode. The MS/MS interface was performed under the following conditions: gas temperature of 350°C, cone voltage of 25 V, collision energy of 20 eV, and a capillary voltage of 3000 V. Full scans were conducted from 0 min to 8.0 min with scan time of 0.1 second. The scan scope was 100 to 290 m/z.

**Pot experiment.** The pot cultivation of tobacco were conducted under three conditions: 1) soil with quinclorac and Q3 added (bioremediation treatment); 2) soil with quinclorac but no Q3 added (phytotoxicity treatment and marked as YHCK); 3) soil without quinclorac and no Q3 added (blank control and marked as CK). Three replicate experiments were conducted in each case.

We collected quinclorac-contaminated soil from the tobacco field of Hunan Agricultural University. Rice had been planted in the field previously and quinclorac was used to control weeds with dose of 40 g (active ingredient)/667 $m^2$. The quinclorac concentration in collected soil was 0.04 mg/kg. After cultured in liquid LB medium for 24 h, 10 ml of Q3 fermentation broth was applied to soil in pot for seven days before tobacco was transplanted. Each pot had one tobacco seedling transplanted into its soil. The tobacco was cultured in a greenhouse after transplanting. Obvious phytotoxicity symptom was observed within approximately 30 days after transplanting. Then, the first investigation was conducted on leaf length, leaf width, and height of each tobacco plant. The second investigation was conducted about 15 days after the first one.

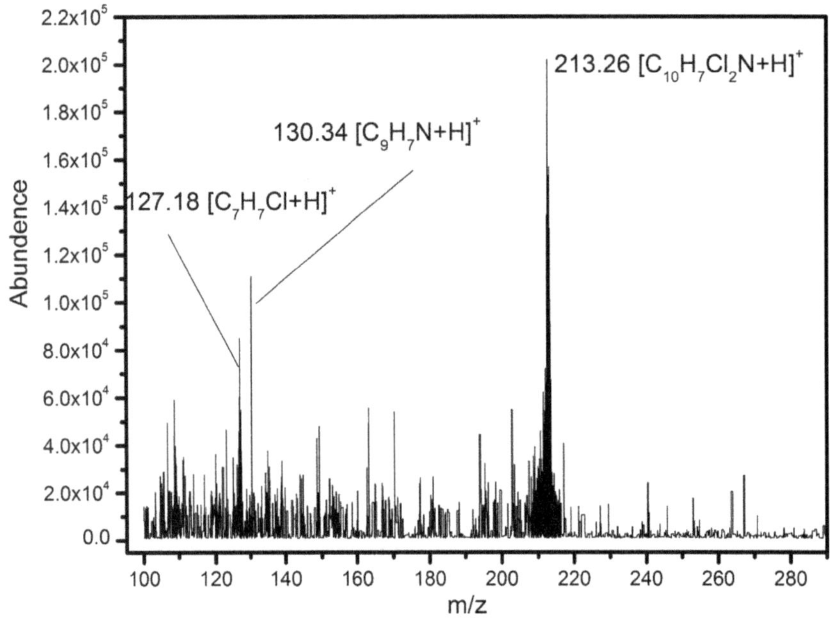


**Figure 8. Mass spectrum of quinclorac degradation products of Q3.** The molecular formula of quinclorac and three degradation products are labeled.

## Results

### Identification and Characterization of quinclorac degrading bacteria, *Bacillus megaterium* Q3

We cultured ground quinclorac-contaminated tobacco root for three days. We were able to isolate six bacteria strains that colonize with clear zones. Those six strains were denoted as strain Q1–Q6. We tested the quinclorac degradation capabilities of those six strains (data not shown). Among the six strains, strain Q3 had the highest quinclorac degradation rate compared to the other five strains. Thus, we focused our further research on strain Q3.

We first studied the morphological characteristics of strain Q3. The colonial morphology of strain Q3 on LB plates was thin flat, faint yellow, opaque, round with smooth edge and were about 1.2~3 mm in diameter (Figure 1 left). The cell morphology of strain Q3 under scanning electron microscope was short rod-like and 1.2~1.5×1.9~3.7 μm in size, which occurred singly or in short chains (Figure 1 right).

Then, we assessed the basic physiological and biochemical characteristics of strain Q3. Gram staining test showed that strain Q3 is a Gram-positive bacterium. Thus, we used Biolog GN III microplate (containing a 71 carbon source utilization test and 23 chemical-sensitivity test) to determine the Q3 strain's substrate utilization. After 18 h incubation, the Biolog GN action profile of Q3 was similar to that of *Bacillus megaterium* with a SIM index of 0.673, a PROB index of 0.977,and a DIST value of 5.774.

To further classify Q3, we sequenced part of its 16S rDNA sequence. We obtained a 1287 bp amplification fragment of 16s rDNA of strain Q3 by PCR. We compared this partial 16s rDNA sequence with sequences in the Genbank Database. The 16S rDNA sequence of Q3 was identical to that of *Bacillus megaterium* N1564-A29 (JX080182). The result was consistent with morphological characteristics, physiological and biochemical characteristics. We then aligned the 16 s rDNA sequences with those of several herbicide degrading bacteria and constructed a phylogenetic tree (Figure 2). The phylogenetic tree clearly showed that Q3 belongs to the same branch of *Bacillus megaterium* N1564-A29 and was obviously different from other quinclorac degrading bacteria, such as *Burkholderia cepacia* [10] and *Bordetella petrii* [13]. Therefore, we identified and designated isolated strain Q3 as *Bacillus megaterium*.

### Identification of optimal degradation conditions of Q3

**Effect of temperature on degradation of quinclorac.** We incubated the Q3 with degradation culture media at 15°C, 20°C, 25°C, 30°C, 35°C and 40°C separately. We measured the

**Table 1.** Bioremediation of strain Q3 on leaf length, leaf width and plant height of tobacco.

| | quinclorac+Q3/30days (mean±sd) | quinclorac/30days (mean±sd) | quinclorac+Q3/45days (mean±sd) | quinclorac/45days (mean±sd) |
|---|---|---|---|---|
| Leaf length | 89.49±2.71 | 83.18±0.68 | 93.40±0.64 | 79.25±0.78 |
| Leaf width | 92.14±3.24 | 64.75±2.46 | 91.70±1.71 | 40.09±0.59 |
| Plant height | 93.64±4.95 | 67.67±4.20 | 92.29±4.55 | 74.51±2.86 |

Note: Values are the means and standard deviation (sd) of three replicates. Values are percentage of leaf length, leaf width and plant height of treatments compared with controls without quinclorac phytotoxicity.

**Figure 9. Bioremediation of strain Q3 (left) and phytotoxicity of quinclorac (right) on tobacco.** The tobacco grew much better in Q3 added soil (left) than in soil without Q3, while soil in both containers were contaminated by quinclorac.

degradation rate of quinclorac and growth of Q3 after 7 days of incubation. Figure 3 plots the growth rate of Q3 and the degradation rate of quinclorac with temperatures. It showed that temperature had a significant effect on both Q3 growth and degradation of quinclorac. When the culture temperature increased from 15°C to 30°C, growth of Q3 and degradation of quinclorac also increased. At the temperature of 30°C, both $OD_{600}$ value and quinclorac degradation rate of Q3 reached peak values, which were 1.03 and 88.4%, respectively. When the temperature was increased above 30°C, both growth of Q3 and degradation of quinclorac decreased as temperature increased. So, the optimal temperature for the growth and quinclorac degradation of Q3 was 30°C.

**Effect of pH on degradation of quinclorac.** We set the pH of degradation culture medium to 5, 6, 7, 8 and 9, respectively, and incubated the Q3 at 30°C in incubator for seven days. Then, we measured the degradation rate of quinclorac and growth of Q3. Figure 4 plots the growth rate of Q3 and degradation rate of quinclorac with pH. It shows when the pH of culture increased from 5 to 8 as both growth of Q3 and degradation of quinclorac also increased. Q3 reached the highest growth rate and highest quinclorac degradation rate at pH 8 when the $OD_{600}$ value was 1.13 and quinclorac degradation rate was 93.1%, respectively. Growth of strain Q3 and degradation of quinclorac sharply decreased when culture medium was set to both low and high pH levels. So, the optimal pH for growth and quinclorac degradation of Q3 was 8.0.

**Effect of inoculation size on degradation of quinclorac.** To study the effect of inoculation size on growth and quinclorac degradation of Q3, we inoculated 2%, 4%, 6%, 8% and 10% (volume ratio) of Q3 into degradation medium, respectively. Then, we measured the quinclorac degradation rate and growth of Q3 after 7 days of incubation. Figure 5 shows the effect of inoculation size on Q3 growth and degradation of quinclorac. Figure 5 also illustrates the growth increase of both Q3 and the quinclorac degradation rate when inoculation size increased from 2% to 6%. The $OD_{600}$ value was.0.99 and quinclorac degradation rate was 91.9% at inoculation size of 6%, respectively. However, both growth and quinclorac degradation of Q3 decreased when the inoculation size was increased above 6%. So the optimal inoculation size for Q3 was 6%.

**Effect of initial quinclorac concentration on degradation of quinclorac.** To study the effect of the initial quinclorac concentration on degradation and Q3 growth, we incubated Q3 on different degradation mediums with initial quinclorac concentrations of 5, 10, 20, 40 and 80 mg/L. Then, we measured the degradation rate of quinclorac and growth rate of Q3 after 7 days. Figure 6 shows that Q3 can grow well when initial quinclorac concentration is less than 40 mg/L. However, Q3 cannot grow well when the initial quinclorac concentration reaches 40 mg/L. Meanwhile, the degradation rate of quinclorac correlated with the growth of Q3. Q3 had a high degradation rate when the initial quinclorac concentration was less than 40 mg/L. The degradation rate reached a peak value of 94.8% when initial quinclorac

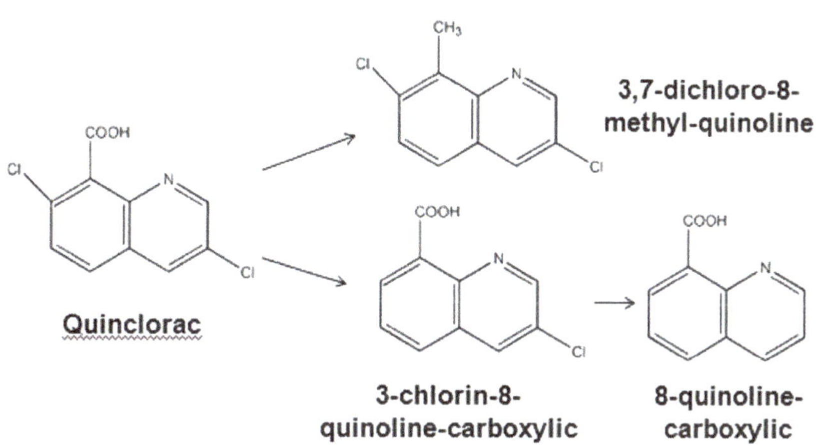


**Figure 10. Possible quinclorac degradation pathways of Q3.** The pathways were proposed based on the resolved quinclorac degradation products from mass spectrum.

concentration was 20 mg/L. The degradation rate reduced dramatically when the initial quinclorac concentration was beyond 40 mg/L.

**Effect of degradation time on quinclorac degradation.** We incubated Q3 under optimal conditions setting the temperature at 30°C, pH at 8, inoculation size at 6%, and initial quniclorac concentration at 20 mg/L. We measured the degradation rate of quinclorac and growth of Q3 for seven days. Figure 7 shows how the degradation rate of quinclorac increased slowly during the first three days. Then, the degradation rate increased sharply on the fourth and fifth day. After that, the degradation rate only slightly increased. The degradation rate on the seventh day was 93.6%. Meanwhile, $OD_{600}$ value of Q3 showed a gradual increase and reached its peak value by the fifth day whereupon the growth rate of Q3 began to decline.

## Analysis of quinclorac degradation products of Q3

We isolated the degradation products of quinclorac and analyzed those using LC-MS/MS. Three major degradation products were detected after Q3 was incubated in degradation medium for three days. Figure 8 shows the mass spectrum of degradation products on Day 3. The peak at 243.13 *m/z* is the original quinclorac. The peak at 213.29 *m/z* is resolved as 3, 7-dichloro-8-methyl-quinoline. The peak at 208.09 *m/z* is resolved as 3-chlorin-8-quinoline-carboxylic. And the peak at 191.10 *m/z* is resolved as 8-quinoline-carboxylic. The three degradation products were created separately with a retention time of about 2.00 min.

## Bioremediation of quinclorac phytotoxicity by Q3

To examine the potential usage of Q3 as a bioremediation for quinclorac phytotoxicity, we compared the growth of tobacco in pots with different kinds of soil: soil with quinclorac and Q3 added, soil with quinclorac but no Q3 added, and soil without quinclorac and no Q3 added as control. Table 1 compared the leaf length, leaf width, and plant height of tobacco in soils with and without Q3 bioremediation. The results showed that Q3 can significantly reduce tobacco phytotoxicity. After having been transplanted for 30 days in the soil with quinclorac (0.04 mg/L) but no Q3, the leaf length, leaf width and plant height of tobacco were all significantly inhibited, which were only 83.18%, 64.75% and 67.67% of the growth rate for those in the control group. In contrast, in soil with quinclorac and Q3, leaf length, leaf width and plant height of tobacco recovered to 89.49%, 92.14% and 93.64% of the growth rate for those plants in control group. After having been transplanted for 45 days, the recovery rate of leaf length, leaf width and plant height of tobacco in Q3 bioremediated quinclorac soil was 93.40%, 91.70% and 92.29, respectively. While without Q3 bioremediation, the leaf length, leaf width and plant height of tobacco were further affected by quinclorac and decreased to 79.25%, 40.09% and 74.51% of growth rates in control group. The results also showed that inhibition of quinclorac on leaf width of tobacco was more serious than inhibition on leaf length and plant height (Figure 9).

## Discussion

In this study, we isolated and identified an endophytic quinclorac-degrading bacterium strain Q3 that can effectively degrade quinclorac. All previously quinclorac degrading bacteria were isolated from soil. The initial degrading concentrations of those bacteria are much higher than the quinclorac concentration in rice fields [16]. On the other hand, the Q3 strain is capable of degrading quinclorac in low concentration. Furthermore, our indoor pot experiments showed that Q3 can be a good bioremediation for tobacco quinclorac phytotoxicity. However, it is unclear if the strain Q3 still had the same degradation and bioremediation efficiency when it is used in natural environment. We will investigate the intervention of Q3 in actual tobacco and rice fields in detail in the future.

Similar to previously identified four quinclorac degradation strains, Q3 reached peak degradation rate at temperature 30°C. However, the four strains identified before prefer the neutral condition at pH of 7, and the Q3 prefers the alkaline condition at pH of 8 for quinclorac degradation.

Endophytic strains usually not only exist in plants, but also conduct a stable breeding in soil, so endophytic strains could degrade pesticides in both soil and plants [18,19]. Meanwhile, endophytic strains could produce indole acetic acid (IAA) and siderophores to promote plant establishment and enhance plant growth and development, thus promoting phytoremediation of pesticide [18]. Therefore, endophytic degrading bacterium may be more efficient in pesticide degradation [18,19].

The major degradation products by Q3 were identified as 3, 7-dichloro-8-methyl-quinoline, 3-chlorin-8-quinoline-carboxylic and 8-quinoline-carboxylic. This is different from the results of previous study [10] where Lü et al. once reported that *Burkholderia cepacia* strain WZ1 could degrade quinclorac to 3,7-dichloro-8-quinoline and 2-choro,1,4- benzenedicarboxylic acid, respectively. Based on the results of mass spectrum, we hypothesized that the Q3 may employ multiple pathways to degrade quinclorac. One possible pathway is through the reduction of carboxyl, and the other pathway is through dechlorination of quinclorac (Figure 10). The degradation pathway of Q3 was different from previous reported by Lü et al. [10]. Due to lack of standard substance and failure to isolate a pure degradation product, we were not able to make qualitative and quantitative analysis of the degradation products. Furthermore, how strain Q3 degrades quinclorac and what the key enzyme is in the process of quinclorac degradation remains unknown. In our future work, we will sequence the whole genome of strain Q3 and elucidate the key enzymes for quinclorac degradation and transcription factors that control the quinclorac degradation.

## Author Contributions

Conceived and designed the experiments: AZ XZ FL LB. Performed the experiments: ML KL. Analyzed the data: YW. Contributed reagents/materials/analysis tools: ML KL. Wrote the paper: AZ XZ FL LB.

## References


1. Grossmann K (1998) Quinclorac belongs to a new class of highly selective auxin herbicides. Weed Science: 707–716.
2. Grossmann K (2010) Auxin herbicides: current status of mechanism and mode of action. Pest management science 66: 113–120.
3. Grossmann K, Schmülling T (1995) The effects of the herbicide quinclorac on shoot growth in tomato is alleviated by inhibitors of ethylene biosynthesis and by the presence of an antisense construct to the 1-aminocyclopropane-1-carboxylic acid (ACC) synthase gene in transgenic plants. Plant growth regulation 16: 183–188.
4. Hill B, Moyer J, Inaba D, Doram R (1998) Effect of moisture on quinclorac dissipation in Lethbridge soil. Canadian journal of plant science 78: 697–702.
5. Pan H, Li X, Xu X, Gao S (2009) Phytotoxicity of four herbicides on Ceratophyllum demersum, Vallisneria natans and Elodea nuttallii. Journal of Environmental Sciences 21: 307–312.
6. Zhang S-m, Liu X-x, Xu S-x, Wu K, Liu H-s, et al. (2012) Effects of quinclorac on soil microbes. Jiangsu Journal of Agricultural Sciences 3: 017.
7. Pareja L, Pérez-Parada A, Agüera A, Cesio V, Heinzen H, et al. (2012) Photolytic and photocatalytic degradation of quinclorac in ultrapure and paddy

field water: Identification of transformation products and pathways. Chemosphere 87: 838–844.

8. Pinna MV, Pusino A (2012) Direct and indirect photolysis of two quinolinecarboxylic herbicides in aqueous systems. Chemosphere 86: 655–658.
9. Singh DK (2008) Biodegradation and bioremediation of pesticide in soil: concept, method and recent developments. Indian journal of microbiology 48: 35–40.
10. Lü Z, Min H, Wu S, Ruan A (2003) Phylogenetic and degradation characterization of Burkholderia cepacia WZ1 degrading herbicide quinclorac. Journal of Environmental Science and Health, Part B 38: 771–782.
11. Lü Z, Li Z, Sang L, Min H (2008) Characterization of a strain capable of degrading a herbicide mixture of quinclorac and bensulfuronmethyl. Pedosphere 18: 554–563.
12. LÜ Z, Min H, Li N, Shao T, Ye Y (2006) Variations of bacterial community structure in flooded paddy soil contaminated with herbicide quinclorac. Journal of Environmental Science and Health Part B 41: 821–832.
13. Xu S-x, Zhou J, Huang N, Liu H-s, Feng H-b, et al. (2012) On the way of isolating, identifying and characterization of quinclorac-degrading bacterium HN36. Journal of Safety and Environment 2: 011.
14. Dong; J, Luo; K, Bai L, Zhou X, Zeng A, et al. (2013) Isolation, identification and characterization of an Alcaligenes strain capable of degrading quinclorac Chinese Journal of Pesticide Science 316–322.
15. Fang J, Bai L, Liu M, Zhou X, Ren Z, et al. (2013) Isolation, Identification and Degradation Characteristics of Quinclorac-degrading Strain QC06. Chinese Journal of Biological Control 29: 431–436.
16. Li L, Chen Z, Zhang W, Guo Q, Han Y, et al. (2012) The comparative analysis of residues of quinclorac in soil, deformity and flue-cured tobacco. Agrochemicals 51: 127–129.
17. Wang Y-H, Li X, Zhou X-M, Bai L-Y (2012) Simplified Approach for the Extraction of Quinclorac From Soils. Asian Journal of Chemistry 24.
18. Sheng X, Chen X, He L (2008) Characteristics of an endophytic pyrene-degrading bacterium of Enterobacter sp. 12J1 from Allium macrostemon Bunge. International Biodeterioration & Biodegradation 62: 88–95.
19. Chen X, Sheng X, He L, Jiang C, Sun L, et al. (2008) Isolation and characteristics of a plant growth-promoter in, phenanthrene-degrading endophytic bacteria from plant. Acta Scientiae Circumstantiae 28: 1308–1313.

19

# Environmental versus Anthropogenic Effects on Population Adaptive Divergence in the Freshwater Snail *Lymnaea stagnalis*

**Anthony Bouétard, Jessica Côte, Anne-Laure Besnard, Marc Collinet, Marie-Agnès Coutellec***

INRA, UMR 0985 ESE, Ecology and Ecosystem Health, INRA – Agrocampus Ouest, CS84215, Rennes Cedex, France

## Abstract

Repeated pesticide contaminations of lentic freshwater systems located within agricultural landscapes may affect population evolution in non-target organisms, especially in species with a fully aquatic life cycle and low dispersal ability. The issue of evolutionary impact of pollutants is therefore conceptually important for ecotoxicologists. The impact of historical exposure to pesticides on genetic divergence was investigated in the freshwater gastropod *Lymnaea stagnalis*, using a set of 14 populations from contrasted environments in terms of pesticide and other anthropogenic pressures. The hypothesis of population adaptive divergence was tested on 11 life-history traits, using $Q_{ST}$ -$F_{ST}$ comparisons. Despite strong neutral differentiation (mean $F_{ST}=0.291$), five adult traits or parameters were found to be under divergent selection. Conversely, two early expressed traits showed a pattern consistent with uniform selection or trait canalization, and four adult traits appeared to evolve neutrally. Divergent selection patterns were mostly consistent with a habitat effect, opposing pond to ditch and channel populations. Comparatively, pesticide and other human pressures had little correspondence with evolutionary patterns, despite hatching rate impairment associated with global anthropogenic pressure. Globally, analyses revealed high genetic variation both at neutral markers and fitness-related traits in a species used as model in ecotoxicology, providing empirical support for the need to account for genetic and evolutionary components of population response in ecological risk assessment.

**Editor:** Christopher Joseph Salice, Texas Tech University, United States of America

**Funding:** AB had a PhD fellowship from the French National Institute for Agricultural Research (INRA: CJS Contrat Jeune Scientifique). Experiment and data acquisition were funded by the French National Institute for Agricultural research (INRA: AIP Bioressources and EFPA Department PI). The funders had no role in study design, data collection and analysis, decision to publish, or preparation of the manuscript.

**Competing Interests:** The authors have declared that no competing interests exist.

* Email: marie-agnes.coutellec@rennes.inra.fr

## Introduction

Understanding the causes of genetic divergence within species is a central goal in evolutionary biology. Population evolution is also a question of importance in environmental sciences and conservation biology, and, as a consequence, evolutionary principles are increasingly taken into account in the management of biodiversity and ecosystems [1–3]. Indeed, evolutionary change can be rapid, especially in the current environmental context. Besides evolutionary textbook cases of insecticide resistance [4], environmental stress may trigger fast evolutionary responses [5], and rapid adaptive responses have been documented in ecological situations involving anthropogenic disturbance [6,7]. Furthermore, it is noteworthy that the rate of phenotypic change can itself increase in human-disturbed and toxic environments [8,9]. In spite of this, the potential evolutionary impact of chemicals and toxic substances released by human activities into the environment is still not considered in current procedures of ecological risk assessment. However, strong arguments for mainstreaming such impacts have been formulated for a long time [10–14]. Beyond mutagenic compounds affecting the germline, pollutants have proved to be potential sources of evolutionary impact, as supported empirically by an ever growing number of studies [15–23].

Agriculture is probably one of the main causes of human-induced microevolution by selection, as a direct consequence of genetic resource management, e.g., domestication, selection programs on productive species, or side effects of intensive agricultural practices [24–26]. Furthermore, in the case of pesticides, as substances rarely reach their target species exclusively, non-intentional evolutionary effects are to be expected. Arguably, a substantial loss of biodiversity has resulted from the intensification of arable agriculture over the last five decades [27]. Agriculture, combined with industrial and domestic activities, uses more than one-third of the Earth's accessible renewable freshwater (i.e., approximately 4,430 $km^3$/year in 2006) [28] and often leads to water contamination. About 140 million tons of fertilizers and several million tons of pesticides are applied each year [29]. Natural populations are thus exposed to a diversity of pressures that vary both spatially and temporally, and a succession of wildlife mortality events has marked the evolution of industry and agrichemical practices since the early 20$^{th}$ century [30].

With respect to human-induced stress, some conditions are expected to increase the risk of genetic change in natural populations. For instance, freshwater lentic habitats located within agricultural landscapes are repeatedly contaminated by complex pesticide mixtures, through various modes of transfer from the treated parcels (including aerial drift, run-off, and drainage; see [31]). Although a given pesticide or its metabolites may be only transiently present in water, the succession and combination of different molecules are expected to lead to high environmental variation and recurrent stress. At the population level, if demographic stochasticity is triggered by stressful episodes, random genetic drift and inbreeding will increase, and impede local adaptive processes [32]. Indeed, selection efficiency is positively correlated to effective population size [33], and population response to selection is weaker under high inbreeding, due to linkage disequilibrium and selection interference among loci [34–36]. Also, random drift load will accumulate locally and increase the risk of population extinction [33,37]. Furthermore, population adaptive potential is also predicted to decrease in stressful and changing environments, especially due to genetic erosion [38]. These processes will be exacerbated under limited gene flow, which may result from low dispersal abilities (fully aquatic organisms, *e.g.*, molluscs, crustaceans) or opportunities (weak connectivity among occupied sites, *e.g.*, ponds, temporary ditches) [39]. Also, population tolerance to environmental stress and change will decrease as inbreeding increases, given that inbreeding depression generally worsens under stress [40]. Interestingly, environmental heterogeneity in selection also increases inbreeding depression [41].

Alternatively, faced with stress, populations may adapt in two ways. First, they may adjust through phenotypic plasticity, which should in principle have no evolutionary consequences, although plasticity may itself be under selection [42,43]. Second, genetic adaptation may occur, provided that adapted alleles are present and selection intensity is strong enough to overcome stochastic interference [44].

To sum up, human-induced environmental stress and heterogeneity may alter population evolutionary trajectory in varied and complex ways, the study of which is anything but an easy task. Processes described above will also have consequences on population genetic divergence: both random genetic drift and local adaptation will cause population divergence, whereas environmental stress and heterogeneity may either act in the same way (random drift or varying selection regimes) or on the contrary, trigger similar adaptive responses among populations (uniform selection acting on general stress responsive pathways or traits).

One method commonly used to decipher population genetic divergence in terms of selective and neutral processes, is the $Q_{ST}$-$F_{ST}$ approach [45,46]. The general principle is to partition the total genetic variance at neutral markers and in quantitative traits into within- and between-population components, using Wright's $F_{ST}$ index, and its quantitative analogue $Q_{ST}$, respectively. Under a purely additive determinism, if quantitative traits evolve neutrally, they should lead to similar levels of differentiation as those inferred from neutral markers ($Q_{ST} = F_{ST}$). Basically, $Q_{ST}$ and $F_{ST}$ are compared to test the null hypothesis of neutrality. Differences between the two indices indicate selection on the traits, with two possibilities: homogenizing (uniform stabilizing) selection towards a unique fitness optimum is revealed by $F_{ST}>Q_{ST}$, whereas divergent selection (occurrence of different fitness optima among populations) is reflected by the reverse relationship. This principle has been widely used in the last decade to address population adaptive divergence (see [46] and references therein). It has notably revealed population adaptive divergence as a function of habitat [47], environmental acidification [48], or soil contamination by zinc [49].

In the present study, we applied the $Q_{ST}$-$F_{ST}$ method to test the hypothesis that anthropogenic pressure may act as a selective force and lead to population adaptive divergence, using a set of 14 natural populations of the freshwater snail *Lymnaea stagnalis* (Mollusca, Gastropoda, Panpulmonata, Hygrophila; previously classified in the Sub-Order Basommatophora). Due to its habitat characteristics (lentic systems, often close to agricultural zones) and its limited dispersal ability, this species is a good model to test non-intentional evolutionary effects of pesticides. Moreover, *L. stagnalis* belongs to an ecologically important group, which plays a major role in the transfer of energy across food webs, and can represent up to 20–60% of the total biomass of macro-invertebrates in some freshwater ecosystems [50]. Pesticides and other identified anthropogenic pressures were described qualitatively and considered as contributing to a global and potentially toxic pressure on non-target organisms. In this context, directional selection associated with a particular pesticide was considered irrelevant, and thus not specifically focused. Conversely, we hypothesized that global chronic exposure to various cocktails of pesticides may induce selective response to temporary but recurrent stress. The study aimed at addressing the following questions: (i) is agricultural (pesticide) pressure able to trigger adaptive processes in non-target organisms? (ii) what is the relative strength of such a pressure, compared to other environmental components, including natural and other anthropogenic factors? Indeed, as natural populations in a given level of pesticide pressure are not eco-evolutionary replicates of each other, any other relevant environmental factor should be also considered. Under this rationale, we characterized environments in terms of "pesticide pressure", "other anthropogenic pressures" (roads, urbanized zones), "global pressure" (combination of agricultural and non agricultural anthropogenic pressures), and "habitat" (pond, ditch, channel, as physical features expected to affect population size and isolation). Population divergence patterns were investigated for a set of life-history traits and compared to neutral genetic differentiation.

## Materials and Methods

### Population sampling and characterization

Adult individuals were collected from 14 natural populations during reproduction, in the course of a single campaign (5–12 July 2011) (Fig. 1). Sampling was conducted in public locations that did not require specific authorization, and did not involve endangered or protected species. Locality information and sampling characteristics are summarized in Table 1. Pesticide and other anthropogenic pressures were estimated from land-use patterns observed in the immediate surroundings of sampled sites, and using Google Earth. In the present study, we were not interested in an instant environmental concentration of pesticides at the time of sampling. On the contrary, under the tested hypothesis, we assumed historical contamination (repeated or chronic) as hypothetical selective force driving phenotypic evolution. Our strategy built on the use of landscape geographic information and agricultural land-use data. These have proved to be relevant tools to predict water body contamination by pesticides [51] and to assess pesticide exposure in the field (ditches [52]; ponds [18,53,54]). Moreover, in such aquatic systems, most variation may be explained by close land-use patterns (within a 100 m radius area), as demonstrated for water quality and vegetation complexity [55]. Consistently, we characterized each population's environment in terms of percent coverage by forest or moorland, pasture, crop (including potatoes, corn, orchards, bulb plants), and urban zone (Table 1). By crossing field observations and Google Earth satellite views with informations obtained from farmers on their practices, we estimated coverage proportions using the software *ImageJ* (http://rsbweb.nih.gov/ij/). When available, pesticide concentration data in surface water near the sampled sites were found to be globally consistent with our classification (http://81.93.58.66/bma_nieuw/begin.html; http://www.milieurapport.be/). Finally, populations were also classified according to four criteria: habitat (H: pond, channel, ditch), pesticide pressure (PP: two levels, low vs high), other anthropogenic pressure (OAP: two levels, low vs high), and global environmental pressure, as a combination of pesticide and other anthropogenic pressures (GEP: three levels).

## Molecular analyses

DNA was chelex extracted from haemolymph or foot tissue from 399 wild-caught adults (14 to 33 snails per population). Neutral genetic variation was assessed at 12 microsatellite loci, i.e., A2, A112, B117 [56], 2k11 and 2k27 (Genbank accession: EF208747-EF208748 [57]), and EMLS04, EMLS13, EMLS21, EMLS26, EMLS29, EMLS41, EMLS45 [58], following the protocol described in Besnard et al. [58]. Only individuals with less than three missing genotypes were retained for population genetics analysis.

## Population neutral genetic structure

Mean allele number ($N$), allelic richness ($A_R$), expected heterozygosity $H_E$, and observed heterozygosity $H_O$, were calculated with GENETIX 4.05.2 [59]. The distribution of neutral genetic diversity within and among populations was estimated from Weir and Cockerham's estimators of Wright's $F$ indices [60] using FSTAT 2.9.3.2 [61]. Departures from HWE (heterozygote excess or deficiency) and linkage disequilibria were tested using GENEPOP 4.0.10 [62]. Population differentiation was tested with a permutation test, in which genotypes were permuted among samples (not assuming HWE within samples; see [61]). As *L. stagnalis* is a self-fertile hermaphroditic organism, the selfing rate was estimated per population and statistically compared to zero

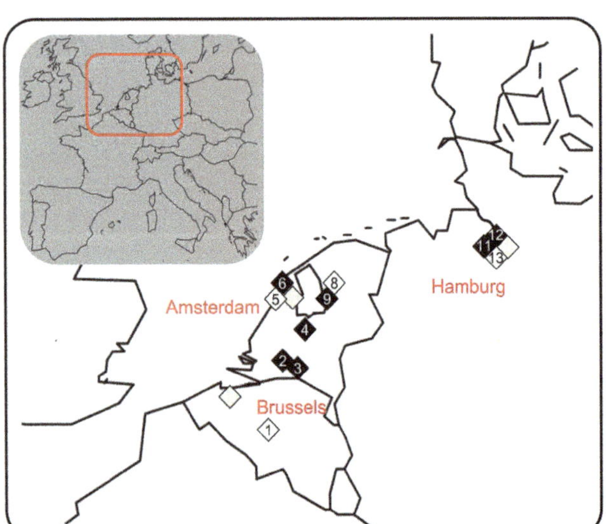


**Figure 1. Location of 14 *L. stagnalis* populations involved in $Q_{ST}$–$F_{ST}$ comparisons.** Populations are coded from 1 to 14. Colours indicate environmental categories (white, grey, and black, for GEP, GEP1, and GEP2, respectively). See Table 1 for details.

using RMES [63]. Effective population size was estimated using the sibship assignment method, as implemented in the software COLONY 2.0.3.0 [64] and assuming inbreeding, male and female polygamic mating systems, and monoecy.

To estimate the number of genetic clusters in our dataset without taking into account any predefined population, we used STRUCTURE 2.2 [65]. Analyses were performed assuming an admixture model and a number of genetic clusters (k) from 1 to 15. Each run started with a burn-in period of 50 000 steps followed by 300 000 Markov Chain Monte Carlo (MCMC) replicates. The most likely number of clusters was determined using the Δk statistic [66] using STRUCTURE HARVESTER [67]. We used DISTRUCT to plot STRUCTURE output data [68].

The effect of environmental factors PP, OAP, GEP, H, and genetic cluster was assessed on population genetics parameters ($A_R$, $H_E$, $F_{IS}$, $F_{ST}$) with a permutation test using FSTAT.

## Common garden experiment

Wild-caught adults were brought to the laboratory and reared under standard conditions at the INRA Experimental Unit U3E (Rennes, France), as previously described [69]. Snails were isolated in plastic vessels filled with 1 L de-chlorinated and charcoal filtered tapwater. They were fed weekly with 1.5 g of organic salad, at each water renewal. Room temperature was maintained at 20±1°C and the photoperiod was 16L/8D. Like other basommatophorans, *L. stagnalis* lays eggs embedded in mucous enveloppes which are deposited and fixed on available substrates (sediment, vegetation, tank walls, etc.). For a given snail, reproduction was followed during 14 days after the first clutch laid in the laboratory. A total of 228 snails over 399 reproduced after 3 months (57%). Families were characterized at various life-history traits (individual growth, female reproduction, hatching success), measured on the laboratory-born progeny ($G_1$) as illustrated on Fig. 2. $G_1$ snails were reared at 20±1°C, under a 14L/10D photoperiod.

## $G_1$ rearing conditions

From hatching to the age of 119 days (roughly corresponding to the end of juvenile growth), $G_1$ snails were reared as groups reflecting clutch origin (n = 427; 1 to 2 clutches per maternal parent). As individuals grew, rearing conditions were adjusted in terms of water volume, individual density, and food supply (see Table S1). Vessels were regularly moved in a randomized way. From the age of 119 days, 7 individuals per clutch were randomly chosen and marked with a honey bee mark (model FC075; diameter: 2.3 mm, weight: 1.8 mg; Ickowicz Apiculture, Bollene, France), and further reared in 30 L tanks (80 individuals per tank), in which they were fed twice a week with organic salad (0.5 g per snail), and from which 2/3 of water was renewed weekly.

## Life history traits: individual growth

From hatching to the age of 56 days, individual size (as inferred from shell height) was measured every two weeks on four randomly chosen juveniles per group. From the age of 63 days, all reared snails were individually measured. Measurements were performed with a stereomicroscope fitted with an ocular micrometer until day$_{63}$, and a digital calliper afterwards. From the age of 119 days, three $G_1$ snails per family (n = 492) were individually followed for growth and reproduction. For these 492 snails, size was measured at the age of 177.71±5.41 days, 191.71±5.41 days, 205.7±5.40 days, and 226.51±5.56 days (owing to variation in clutch age, and because measurements were performed at fixed dates after the age of 119 days), leading to a total of 12 values per individual. Growth was modelled using the Gompertz's model:

**Table 1.** Geographical, environmental, and genetic characteristics of 14 *L. stagnalis* populations involved in $F_{ST}$–$Q_{ST}$ comparisons.

| Localization | Code | Geographic coordinates | Land use | Genetic/Environmental factor | | | | | Population genetics | | | | | | |
|---|---|---|---|---|---|---|---|---|---|---|---|---|---|---|---|
| | | | | GC | H | PP | OAP | GEP | n | *N* | $A_R$ | $H_E$ | $F_{IS}$ | Selfing rate (%) | *Ne* [CI 95%] |
| CASTRICUM<br>(Alkmaar, NL) | CAS<br>(5) | 52°33.17N<br>4°37.22E | 100% moor | West | Pond | 0 | 0 | 0 | 32 | 3.3 | 2.60 | 0.341 | 0.327* | 0.0 | 31<br>[19;58] |
| KUINRE<br>(Nordoostpolder, NL) | KUI<br>(9) | 52°47.67N<br>5°47.69E | 98% forest<br>2% urban | West | Ditch | 0 | 0 | 0 | 25 | 3.8 | 3.33 | 0.428 | 0.207* | 0.0 | 25<br>[14;52] |
| HEDENDORF<br>(Hamburg, DE) | HED<br>(13) | 53°29.47N<br>9°36.55E | 90% forest<br>10% urban | East | Pond | 0 | 0 | 0 | 26 | 2.8 | 2.39 | 0.315 | 0.258* | 0.0 | 32<br>[18;66] |
| OUD-HEVERLEE<br>(Brussels, BE) | OUD<br>(1) | 50°49.39N<br>4°39.29E | 55% forest<br>45% urban | West | Pond | 0 | 0 | 0 | 13 | 3.3 | 3.28 | 0.416 | 0.177 | 40.6* | 26<br>[13;66] |
| SCHOORLDAM<br>(Alkmaar, NL) | SCH<br>(7) | 52°43.13N<br>4°42.2E | 95% pasture<br>5% urban | West | Channel | 0 | 1 | 1 | 26 | 3.3 | 2.89 | 0.324 | 0.225* | 18.5 | 24<br>[13;47] |
| DETSELBERGEN<br>(Gent, BE) | DET<br>(14) | 51°03.29N<br>3°49.35E | 30% forest<br>30% fallow<br>40% urban | West | Pond | 0 | 1 | 1 | 14 | 2.5 | 2.45 | 0.325 | 0.353* | 48.5* | 33<br>[16; 139] |
| BUXTEHUDE<br>(Hamburg, DE) | BUX<br>(10) | 53°29.63N<br>9°42.78E | 80% fallow<br>15% crop<br>5% urban | East | Channel | 0 | 1 | 1 | 27 | 3.2 | 2.84 | 0.355 | 0.094 | 0.0 | 22<br>[12;43] |
| BAARN<br>(Utrecht, NL) | BAA<br>(4) | 52°13.38N<br>5°18.53E | 25% pasture<br>25% crop<br>50% urban | West | Ditch | 1 | 1 | 2 | 22 | 3.9 | 3.54 | 0.448 | 0.308* | 12.0 | 33<br>[18;72] |
| OOSTEIND<br>(Breda, NL) | OOS<br>(2) | 51°38.87<br>4°55.82E | 30% pasture<br>40% crop<br>30% urban | West | Channel | 1 | 1 | 2 | 32 | 4.8 | 4.18 | 0.590 | 0.175* | 0.0 | 198<br>[105; 698] |
| BIEZENMORTEL<br>(Breda, NL) | BIE<br>(3) | 51°38.03N<br>5°9.53E | 20% forest<br>5% pasture<br>75% crop | West | Channel | 1 | 1 | 2 | 30 | 4.4 | 3.67 | 0.452 | 0.168* | 9.1 | 109<br>[64; 259] |
| PUTTEN<br>(Alkmaar, NL) | PUT<br>(6) | 52°45.81N<br>4°39.82E | 30% crop<br>70% urban | West | Ditch | 1 | 1 | 2 | 32 | 2.9 | 2.62 | 0.370 | 0.216* | 18.6 | 37<br>[22;67] |
| EMMELOORD<br>(Nordoostpolder, NL) | EMM<br>(8) | 52°46.27N<br>5°48.26E | 80% crop<br>20% urban | West | Ditch | 1 | 1 | 2 | 24 | 3.7 | 3.32 | 0.456 | 0.252* | 0.0 | 36<br>[20;80] |
| KOENIGSREICH<br>(Hamburg, DE) | KOE<br>(11) | 53°31.6N<br>9°46.29E | 85% crop<br>15% urban | East | Channel | 1 | 1 | 2 | 27 | 5.3 | 4.17 | 0.444 | 0.176* | 9.5 | 48<br>[29;91] |

**Table 1.** Cont.

| Localization | Code | Geographic coordinates | Land use | Genetic/Environmental factor | | | | | Population genetics | | | | | | |
|---|---|---|---|---|---|---|---|---|---|---|---|---|---|---|---|
| | | | | GC | H | PP | OAP | GEP | n | N | $A_R$ | $H_E$ | $F_{IS}$ | Selfing rate (%) | Ne [CI 95%] |
| AGATHENBURG | AGA | 53°34.54N | 100% crop | East | Ditch | 1 | 1 | 2 | 25 | 4.3 | 3.69 | 0.491 | 0.266* | 30.7* | 29 |
| (Hamburg, DE) | (12) | 9°32.82E | | | | | | | | | | | | | [17;53] |

n indicates sample size, N mean number of alleles, $A_R$ allelic richness, $H_E$ expected heterozygosity, $F_{IS}$ inbreeding coefficient, and Ne effective population size. Values significantly different from zero are indicated with an asterisk. GC holds for Genetic cluster and H for Habitat. Environmental pressure is described qualitatively under three alternative factors (PP: Pesticide pressure, OAP: Other Anthropogenic pressure, GEP: Global Environmental Pressure), the levels of which are defined as a function of land use within a radius of 100 m around sample sites (see text for details).

$$L_t = Ae^{-be-kt}$$

where $L_t$ is the shell height at time $t$, $A$ the asymptotic shell length, $b$ a scaling factor related to shell height at $t=0$, and $k$ reflects the growth rate. Using a preliminary subset of 20 individuals, this model performed better than the von Bertalanffy model (lower AIC), and was thus preferred for further growth analysis. Growth parameters were compared among families and populations, as well as size at hatching and size at 119 days.

## Life history traits: reproduction

Once clutches started to be observed in all aquaria, the 492 snails followed for growth were isolated in 200 mL plastic vessels and fed weekly with 1 g of organic salad immediately after water renewal. Several female reproductive traits were measured: time to first oviposition under isolation (with a censoring limit of 30 days), ability to lay eggs (as the proportion of reproductive snails per family), number of clutches and eggs laid during two weeks, clutch size (number of eggs per clutch), and clutch hatching rate.

## Statistical analyses

All traits were analyzed with generalized linear mixed effect models (GLMM, R-package *lme4*) [70], with appropriate error distribution (Gaussian for normal data, Poisson for count data, and binomial for proportions). When necessary, data were log- or BoxCox-transformed, and covariates were included in the model (see Table S2). The model structure was: $Y \sim factor1 + factor2 + factor3 + covariate1 + covariate2 + \ldots + (1|population/family)$.

Fixed effects were tested by model comparison using a log-likelihood ratio test. Fixed factors were: genetic cluster (see STRUCTURE analysis), habitat, and environmental pressure (either pesticide, other anthropogenic, or global pressure). Family was nested within population, and both were treated as random factors. Global tests were followed by post-hoc pairwise comparison tests (Tukey and non parametric equivalent). All statistical analyses were performed using R 2.14.0 (R Core Team 2012).

## $Q_{ST}$ -$F_{ST}$ analysis

$Q_{ST}$ was computed for each trait using the equation [71;72]:

$$Q_{ST} = \frac{(1+f)\ V_b}{(1+f)\ V_b + 2\ V_w}$$

Where $f$ is the inbreeding coefficient, $V_b$ the between population genetic variance and $V_w$ the within population genetic variance. $V_b$ was obtained directly from the population variance component, while for $V_w$, as only the maternal origin could be assessed, family variance was used under the full-sib design ($2V_f$) [73]. Within- and between-population variance components were estimated for each of the 11 studied traits using restricted maximum likelihood (REML) under a simplified mixed model ($Y \sim 1 + (1 | pop/fam)$).

Population quantitative divergence was tested using the method of Whitlock and Guillaume [74], which accounts for variation among neutral loci used to estimate $F_{ST}$, and for sampling error and variation in evolutionary history in $Q_{ST}$ estimation. The method has lower type 1 error rate and better statistical power than previous tests, and seems particularly suited when neutral differentiation is high [46], as in *L. stagnalis* [58,75]. Briefly, a bootstrapping procedure is applied to compare observed values of the difference $Q_{ST}$ - $F_{ST}$ to the $Q_{ST}$ - $F_{ST}$ distribution expected under the neutral hypothesis, as derived from $F_{ST}$ distribution and

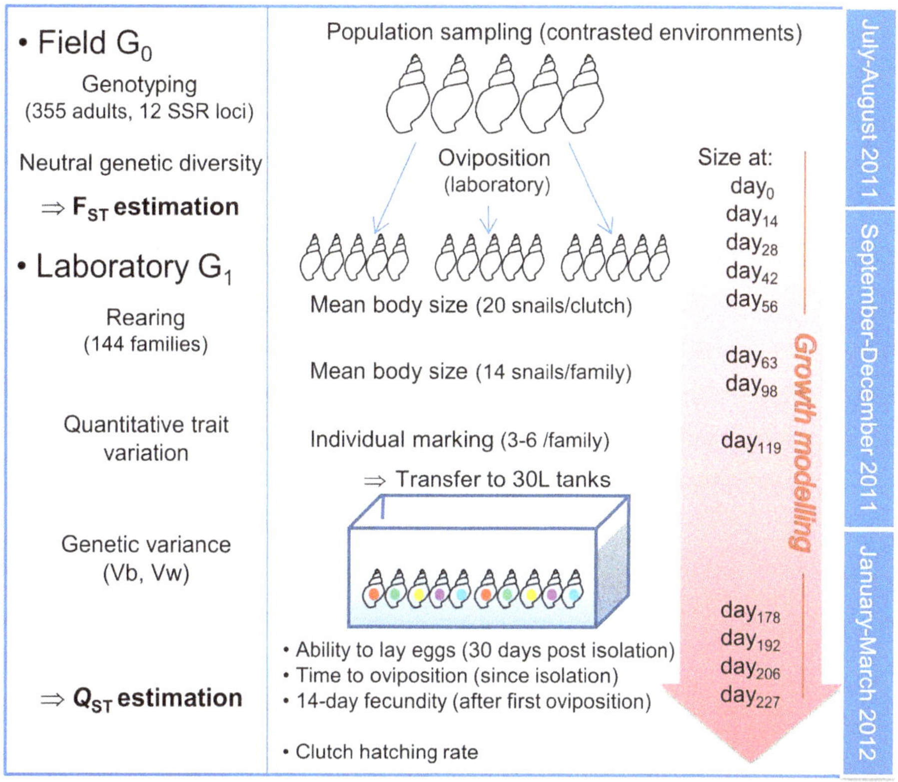


**Figure 2. Schematic overview of the common garden experiment used to investigate population divergence in *L. stagnalis*.**

variance components [60] and $Q_{ST}$ neutral distribution predicted from the Lewontin-Krakauer distribution. Observed values were considered significant when they fell outside the 95% confidence interval of the neutral distribution [74].

## Results

### Population genetics

On 399 individual genotypes, 44 were discarded because of missing or unreadable data at more than three loci. Overall, the observed number of alleles per locus varied from three to 29, mean allelic richness ranged from 1.1 to 9.6 (Table S3), and genetic diversity per locus and sample varied from 0 to 0.86. Significant genotypic linkage disequilibrium was observed in 40 out of 656 comparisons. However, after Bonferroni correction, none of these remained significant. $F_{IS}$-values varied widely across loci and samples, ranging from −0.238 (EMSL41 in BUX) to 1.000 (2k27 and B117 in DET; EMLS13 in HED; EMLS21 in BAA) (Table S3). Genetic parameters estimated per sample over loci reflected discrepancies between populations (Table 1). All populations except BUX and OUD were found significantly inbred (mean $F_{IS}$-value = 0.219, 95% CI [0.147; 0.287]). The selfing rate was significantly different from zero in three populations (OUD, DET, and AGA). Effective population size *Ne* was significantly larger in sites exposed to high pesticide pressure (PP1 > PP0; Kruskal-Wallis test, $P = 0.007$).

Population differentiation was generally high (mean $F_{ST}$-value = 0.291, 95% CI [0.248; 0.334]), as also reflected by pairwise estimates, which were all significant (Table S4), except between the two geographically close populations KUI and EMM. Interestingly, the four pond populations had the longest branches on the NJ-tree, indicating stronger differentiation associated with this habitat (Fig. 3A). From STRUCTURE analysis, individual genotypes clustered into two groups with the highest likelihood, and this was confirmed by the Δk statistic (Fig. 3B). Clusters corresponded to two geographic regions (10 western vs. four eastern populations). Therefore, the effect of "genetic cluster" was tested on life-history traits. No difference in level of genetic diversity, population inbreeding or population differentiation was found between the two clusters (Table 2).

Population genetic parameters were affected by most environmental factors (Table 2). Genetic diversity tended to be greater under higher environmental pressure (GEP2, PP1, OAP1), although the difference was significant under pesticide pressure only ($A_R$, P = 0.007; $H_E$, P = 0.008). Population differentiation was significantly affected by global environmental pressure (GEP, P = 0.016) with a lower value among populations exposed to the highest level of pressure. With regard to specific pressures, lower differentiation was associated with pesticide as well as with other anthropogenic pressures (PP, P = 0.001; OAP, P = 0.031). Habitat affected all tested parameters, when ponds were compared to ditch and channel populations: pond populations were significantly less variable ($A_R$, P = 0.025; $H_E$, P = 0.038), more inbred ($F_{IS}$, P = 0.034) and much more differentiated than were other populations ($F_{ST(pond)} = 0.502$ vs. $F_{ST(ditch\ +\ channel)} = 0.227$, P = 0.002).

### Life history variation

Most $G_1$ traits showed significant heterogeneity at both population and family levels (Tables S2 and S5). No effect of genetic cluster was detected, excepted on early size, which was larger in eastern populations (size at hatching, P = 0.019; growth parameter $b$, P = 0.005; Table 3). Among environmental factors, habitat was the most effective, with six traits or parameters significantly different between systems. Growth parameters indicated that body size at $t_0$ increased significantly from ponds to channels and from channels to ditches, whereas asymptotic size was larger in ponds than in channels and ditches (Fig. S1). With respect to reproduction, snails from ponds had a significantly lower reproductive activity than those from ditches and channels: lower ability to lay eggs, and in reproductive snails, longer time to oviposition. Clutch size decreased significantly from ponds to ditches and from ditches to channels, and clutch hatching rate was lower in pond and ditch populations than in channel ones.

Compared to aquatic system, human pressure appeared much less effective. First, pesticide pressure had no effect, except a marginal one on ability to lay eggs, found greater in exposed populations (P = 0.069). Second, other anthropogenic pressures affected growth parameter b (significantly larger size at $t_0$ under OAP1) and hatching rate (significantly lower under OAP1) (Fig. S2). Interestingly, lower hatching rates were already observed on the clutches laid by $G_0$ snails inhabiting OAP1 sites (720 clutches, GLMM test: $\chi^2 = 11.861$, P<0.001). Third, global pressure affected $b$ (significantly larger snails at $t_0$ under GEP2), and hatching rate, which was significantly impaired in exposed populations (GEP1 and GEP2) relative to reference ones (GEP0).

Several life history traits correlated with population genetic characteristics. First, fecundity was positively correlated with $H_E$ (r = 0.524, P = 0.031; Fig. 4A), and with $A_R$ (r = 0. 489, P = 0.044). Second, a negative correlation was found between population inbreeding and $G_1$ fecundity (r = −0.845, P<0.001; Fig. 4B), ability to reproduce (r = −0.664, P = 0.006), and clutch hatching rate (r = −0.634, P = 0.009) (see Table 1 and Table S5).

### $Q_{ST}$-$F_{ST}$ analysis

Results are summarized on Table 4. Over the set of 11 studied traits or parameters, four traits were found to evolve consistently with neutral expectations (hatching size, growth parameter $k$, number of eggs, clutch size), whereas homogenizing selection was indicated for two traits (growth parameter $b$, clutch hatching rate), and divergent selection for five traits (size at 119 days, asymptotic size $A$, ability to lay eggs, time to oviposition, number of clutches laid).

## Discussion

The hypothesis that human activities may induce population adaptive divergence in *L. stagnalis* was addressed under a common garden experiment involving life history traits and a set of natural populations from contrasted environments. Five traits or parameters over 11 were found under divergent selection, despite a strong neutral genetic structure. The occurrence of two genetic clusters (Eastern and Western populations) had only a weak influence on quantitative genetic divergence, suggesting that population level was the adequate scale to draw selection inferences [76]. Only size at hatching and its corresponding growth parameter ($b$) were affected by this factor (larger hatchlings to the East). Given that early size was found under uniform selection, it was assumed that the genetic divergence associated to geography did not interfere with selection putatively associated with local environmental pressures.

### Selection versus random genetic drift

All traits found to diverge adaptively were adult traits, related to late expressed growth (sub-adult size, asymptotic size $A$) and to reproduction (ability to reproduce, time to oviposition after

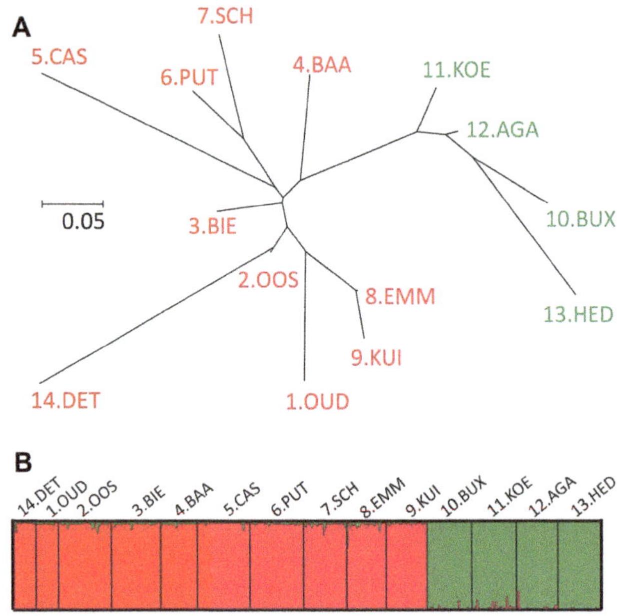


**Figure 3. Population genetic differentiation in *L. stagnalis*.** (A) Bayesian assignment probabilities in STRUCTURE analysis, for k = 2 clusters. Each bar represents an individual. Bar colour indicates the posterior probability that the individual belongs to the cluster of that color. (B) Unrooted Neighbour-Joining tree based on population pairwise $F_{ST}$ values (see Table S4).

isolation, number of clutches). Conversely, uniform selection was the most likely evolutionary hypothesis for early survival and size (hatching success and growth parameter $b$, which is related to size at $t_0$). Globally, inferred patterns, i.e., early (late) traits under homogenizing (divergent) selection, are in lines with those found in another freshwater snail [47,77].

Traditional statistical analyses performed on the $G_1$ revealed that, to the exception of a marginal effect of pesticide pressure on ability to lay eggs (P = 0.069), habitat was the only factor affecting most traits found under divergent selection. Indeed, pond populations were characterized by larger ultimate size (growth parameter $A$), lower ability to lay eggs, and longer time to oviposition after isolation, compared to ditch and channel ones.

Congruent with *L. stagnalis* low dispersion and weak habitat connectivity, pond populations were particularly differentiated (especially CAS and DET, pairwise $F_{ST}$) and presented generally lower genetic diversity and higher inbreeding than channel and ditch populations. Moreover, compared to channel populations, pond and ditch populations were also more inbred (P = 0.009). These aquatic systems are likely to imply stronger population isolation, even among ditches (due to frequent drought). A similar influence of habitat was emphasized in another freshwater snail, *Physa acuta* [78]. Thus, in *L. stagnalis*, although divergent selection was detected for traits also affected by habitat type, genetic drift is likely to highly contribute to divergence among habitats. This result is in line with the detection of significant drift load in experimental populations occupying outdoor close mesocosms [69].

In the common frog *Rana temporaria*, habitat fragmentation correlates with low genetic diversity and high differentiation, and negatively impacts tadpole body size [79], which is a critical trait determining individual fitness [80]. In *L. stagnalis*, although ultimate body size was greater in ponds, the relation between size

**Table 2.** Comparison of *L. stagnalis* population genetic parameters as a function of different grouping factors.

| Population grouping | $A_R$ | $H_E$ | $F_{IS}$ | $F_{ST}$ |
|---|---|---|---|---|
| GEP0 | 2.898 | 0.367 | 0.249 | 0.431 |
| GEP1 | 2.725 | 0.353 | 0.190 | 0.450 |
| GEP2 | 3.599 | 0.468 | 0.214 | 0.197 |
| *P (GEP)* | *0.073* | *0.094* | *0.487* | ***0.016*** |
| PP0 | 2.824 | 0.362 | 0.226 | 0.413 |
| PP1 | 3.599 | 0.468 | 0.214 | 0.197 |
| *P(PP)* | ***0.007*** | ***0.008*** | *0.773* | ***0.001*** |
| OAP0 | 2.898 | 0.367 | 0.249 | 0.431 |
| OAP1 | 3.337 | 0.438 | 0.210 | 0.248 |
| *P(OAP)* | *0.219* | *0.178* | *0.329* | ***0.031*** |
| H Pond | 2.678 | 0.343 | 0.282 | 0.502 |
| H Ditch | 3.300 | 0.436 | 0.246 | 0.221 |
| H Channel | 3.550 | 0.449 | 0.166 | 0.246 |
| *P(H)* | *0.062* | *0.095* | ***0.016*** | ***0.001*** |
| H Pond | 2.678 | 0.343 | 0.282 | 0.502 |
| H Channel+Ditch | 3.425 | 0.443 | 0.203 | 0.227 |
| *P (P vs D+C)* | ***0.025*** | ***0.038*** | ***0.034*** | ***0.002*** |
| H Pond+Ditch | 3.024 | 0.398 | 0.258 | 0.336 |
| H Channel | 3.550 | 0.449 | 0.166 | 0.246 |
| *P (P+D vs C)* | *0.116* | *0.306* | ***0.009*** | *0.271* |
| H Pond+Channel | 3.162 | 0.409 | 0.203 | 0.330 |
| H Ditch | 3.300 | 0.436 | 0.246 | 0.221 |
| *P (P+C vs D)* | *0.686* | *0.641* | *0.285* | *0.178* |
| GC West | 3.187 | 0.422 | 0.229 | 0.244 |
| GC East | 3.272 | 0.410 | 0.196 | 0.162 |
| *P (GC)* | *0.827* | *0.837* | *0.426* | *0.393* |

$A_R$ is allelic richness, $H_E$ expected heterozygosity, $F_{IS}$ inbreeding coefficient, and $F_{ST}$ differentiation index.
GEP holds for Global environmental pressure, PP for Pesticide pressure, OAP for Other anthropogenic pressure, H for Habitat (P = Pond, D = Ditch, C = Channel), GC for Genetic cluster. Permutation-based statistical test (P value, 1000 permutations).

and fitness is not necessarily positive, because of possible energetic trade-offs between growth and reproduction [81]. Indeed, two pond populations (CAS and DET) exhibited both highest mean asymptotic size and lowest fecundity (see Table S5).

The $Q_{ST}$-$F_{ST}$ based test concluded to neutral divergence for size at hatching, growth rate (parameter $k$), and two reproductive traits, clutch size and early fecundity (number of eggs laid during 14 days following first clutch since isolation). On the latter, it seems surprising to find no hint of selection, either uniform or divergent, although this may simply reflect low heritability of fitness traits [82]. In support of this, it is worth noting that fecundity is extremely high in *L. stagnalis* (from adulthood to death, i.e., about one year under standard conditions, a snail produces more than 10 progeny per day), suggesting the possibility of high variation without drastic fitness consequences. Consistently, population growth rate was found to be insensitive to fecundity under experimental conditions (Leslie matrix modelling [83]). Alternatively, apparent neutral evolution might result from a bias in $Q_{ST}$ estimation. Under genetic drift, dominance and epistasis bias $Q_{ST}$ estimates downward [84–86]. As $Q_{ST}$ is traditionally estimated without accounting for these non-additive sources of variation, and given the strong observed global $F_{ST}$ value, divergent selection might have been underestimated for fecundity. On the other side, a hypothetical upward bias in $Q_{ST}$ estimation might result from too high a mutation rate at the markers used to infer $F_{ST}$, as compared to the mutation rate of loci encoding the studied quantitative trait [87,88]. This hypothesis is invalidated by observed $F_{ST}$ values, which were rather high and significant, although they may also suffer from a downward bias. Finally, gene flow may well lead to an apparent pattern of adaptive divergence at a trait under uniform selection, if this trait is correlated to a trait under divergent selection [89]. Again, high population neutral differentiation makes this hypothesis unlikely. Besides methodological causes of bias, it might be also mentioned that (1) true lifespan fecundity was not measured, and (2) as a source of uncontrolled variation, no distinction could be made between outcrossing and potential selfing in the $G_1$.

As reported above, homogenizing selection was indicated for early traits, which is expected to reflect the occurrence of a uniform fitness optimum between the studied populations. $Q_{ST}$ estimation was based on the $G_1$ of wild-caught adult snails, with the exception of survival at hatching (measured on $G_2$). $G_1$ families correspond hence to the hermaphroditic equivalent of isofemale lines, as used for broad-sense heritability estimation [73]. Therefore, these "broad-sense" $Q_{ST}$ estimates are potentially biased by genetic non additive sources of variation (dominance,

**Table 3.** Summary statistics of fixed effects tested on *L. stagnalis* life history traits.

| Trait/parameter | Genetic cluster (W/E) | | Habitat (P/C/D) | | Pesticide pressure (PP0/1) | | Other anthropogenic pressure (OAP0/1) | | Global environmental pressure (GP0/1/2) | |
|---|---|---|---|---|---|---|---|---|---|---|
| | $\chi^2_{df=1}$ | P | $\chi^2_{df=2}$ | P | $\chi^2_{df=1}$ | P | $\chi^2_{df=1}$ | P | $\chi^2_{df=2}$ | P |
| *Growth* | | | | | | | | | | |
| Size at hatching | **6.95** | **0.008** | 3.67 | 0.159 | 0.12 | 0.724 | 0.36 | 0.549 | 0.91 | 0.636 |
| | *W<E* | | | | | | | | | |
| Growth parameter *b* | **9.23** | **0.002** | **17.91** | **<0.001** | 0.44 | 0.506 | **9.60** | **0.002** | **10.20** | **0.006** |
| | *W>E* | | *P>C>D* | | | | *OAP0>OAP1* | | *GP0 = GP1>GP2* | |
| Growth parameter *k* | 1.76 | 0.185 | 0.43 | 0.805 | 0.66 | 0.415 | 0.04 | 0.835 | 1.10 | 0.576 |
| Growth parameter *A* | 0.25 | 0.614 | **8.21** | **0.016** | 0.010 | 0.919 | 3.01 | 0.083 | 3.68 | 0.159 |
| | | | *P>C=D* | | | | *OAP0>OAP1* | | | |
| Size at 119 days | 1.83 | 0.176 | 2.06 | 0.358 | 0.00 | 0.955 | 0.11 | 0.746 | 0.12 | 0.944 |
| *Reproduction* | | | | | | | | | | |
| Ability to lay eggs | 0.11 | 0.745 | **9.45** | **0.009** | 3.30 | 0.069 | 1.75 | 0.186 | 3.96 | 0.138 |
| | | | *P<D=C* | | PP0<PP1 | | | | | |
| Time to oviposition | 2.42 | 0.200 | 4.92 | 0.085 | 0.15 | 0.699 | 1.08 | 0.300 | 1.08 | 0.582 |
| | | | *P>D=C* | | | | | | | |
| Number of clutches | 1.40 | 0.237 | 0.39 | 0.823 | 0.14 | 0.708 | 0.10 | 0.756 | 0.44 | 0.802 |
| Number of eggs | 0.55 | 0.457 | 4.96 | 0.084 | 0.21 | 0.646 | 0.69 | 0.405 | 1.65 | 0.439 |
| | | | *P = D<C* | | | | | | | |
| Clutch size | 2.55 | 0.110 | **11.18** | **0.004** | 0.06 | 0.807 | 1.93 | 0.165 | 3.11 | 0.212 |
| | | | *P<D<C* | | | | | | | |
| Hatching rate | 0.59 | 0.441 | **6.41** | **0.041** | 0.98 | 0.322 | **6.25** | **0.012** | **6.25** | **0.044** |
| | | | *P = D<C* | | | | *OAP0>OAP1* | | *GP0>GP1 = GP2* | |

General model: glmer = Y ~ covar + Genetic cluster + Habitat + Environmental pressure + (1 | pop/fam).
Growth parameters: A is asymptotic size, b is related to size at birth, k to growth rate. *Bold values are statistically significant.*

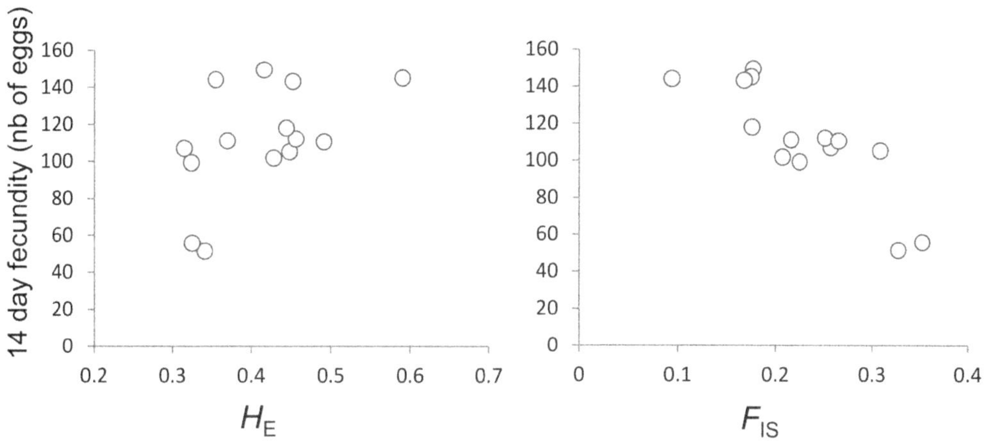


**Figure 4. Correlation plot between 14-days fecundity and (A) population expected heterozygosity, (B) population inbreeding, as based on 14 *L. stagnalis* populations.**

epistasis), and by maternal effects. The latter are likely to operate early in life and tend to recede with age [90]. Hence, in the present study, traits found under uniform selection (early expressed traits) might be more prone to such a bias than traits found under divergent selection (late expressed traits). However, as complex traits seem to vary mostly additively [91,92], we hypothesized, as done in most $Q_{ST}$-$F_{ST}$ comparisons [46], that neglecting these sources of variation would have little consequences.

Alternatively to the uniform selection hypothesis, the pattern exhibited by early traits would result from trait canalization, as a phylogenetically inherited characteristics [86]. Since the pattern was similar to that observed in *Galba truncatula* [77], the criterion of cross-species consistency (supporting the canalization hypothesis) may be met in basommatophorans, but this would need to be confirmed in other species. Further investigation is clearly needed to disentangle both causes of *L. stagnalis* population phenotypic convergence.

## Genetic diversity and fitness

The positive correlation observed between population genetic variability and $G_1$ fecundity, is consistent with previous findings in *L. stagnalis* [93]. Under the assumption that G1 performances reflect the fitness of natural populations, the present result also supports the hypothesis that local drift load can be strong in this species, as previously suggested [69]. Therefore, under high random genetic drift, as indicated here, it might also be asked to what extent $Q_{ST}$ is modified by the expected stochastic evolution of slightly deleterious alleles as compared to pure selection-based predictions (random drift load [37]).

Furthermore, the negative correlation observed between population inbreeding and several traits related to fecundity (ability to reproduce, time to oviposition, early fecundity, and hatching rate) is in line with the occurrence of reduced fitness in inbred populations, and thus of inbreeding depression in these populations. However, as inbreeding was found significant in most populations and because effective population size was usually small, it is suggested that random genetic drift is actually the main cause of these apparent correlations, through reduced genetic diversity and drift load accumulation caused by population isolation and small size [32]. This hypothesis is still strengthened by the fact that population inbreeding correlated negatively with $G_2$ survival at hatching (i.e., irrespective of individual inbreeding level). Inbreeding depression estimated under selfing relative to random outcrossing (ID) is particularly low in *L. stagnalis* [69,93–95], despite a clear preference for outcrossing [96,97]. Thus, although ID was not estimated, it seems unlikely that selfing could be responsible for the observed pattern. In support of this, no correlation was observed between population selfing rate and $G_1$ fecundity.

## Evolutionary impact of anthropogenic pressures

Neutral genetic variation was inflated in populations exposed to anthropogenic pressures, including pesticides. This was reflected in terms of genetic diversity, allelic richness, and effective population size, and through lower genetic differentiation. These results are opposed to those found in experimental populations of *L. stagnalis* exposed to cocktails of pesticides [98]. More generally, they are not consistent with the hypothesis of increased local stochasticity due to anthropogenic pressure (see introduction). High genetic diversity may result either from the maintenance of large population size despite stressful conditions, or from significant gene flow (immigration). Alternatively, undetected population subdivision may also be responsible for the observed patterns, as subdivision is expected to maintain diversity at the global scale. Furthermore, at the same scale, heterozygote deficiencies are also to be expected (Wahlund effect). Therefore, population subdivision into local demes might be responsible for part of the within population fixation observed in the dataset. As a possible impact of anthropogenic pressure, this potential effect should deserve specific attention, e.g., through the scoring of individual location at the time of sampling, and using assignment tests. As already mentioned, the effect of anthropogenic pressures on life history traits was much lower than that of habitat. This was particularly marked for pesticide pressure, which was the primary focus of the study. Three possible causes for this may be discussed: 1) true lack of pesticide effect, 2) effects underestimated due to confounding environmental factors, and 3) experimental conditions too benign to detect adaptive divergence.

Besides the hypothesis of pesticide innocuousness to gastropod populations, one explanation may be that sampling did not reflect the effective magnitude of pesticide pressure. However, as the design included areas of intensive agriculture, and due to the close proximity of water bodies and treated agricultural parcels, this seems unrealistic. Alternatively, evolutionary patterns associated with pesticide exposure may have been masked by other environmental characteristics, an idea supported by the significant

**Table 4.** Summary of $Q_{ST}$-$F_{ST}$ analyses on laboratory G1s of 14 *L. stagnalis* populations.

| Trait/Parameter | N | N*fam* | $V_b$ | $V_f$ | Observed $Q_{ST}$ - $F_{ST}$ | Neutral $Q_{ST}$-$F_{ST}$ 95% CI | P-value left | P-value right | Divergence pattern |
|---|---|---|---|---|---|---|---|---|---|
| *Growth* | | | | | | | | | |
| Hatching size | 427 | 211 | 1.46E-03 | 1.86E-03 | −0.098 | [−0.160; 0.150] | 0.141 | 0.859 | Neutrality |
| Parameter *b* | 492 | 144 | 4.52E-04 | 1.19E-03 | −0.187 | [−0.160; 0.154] | 0.009 | 0.991 | Homogenizing selection |
| Parameter *k* | 492 | 144 | 2.48E-07 | 2.99E-07 | −0.089 | [−0.155; 0.156] | 0.169 | 0.831 | Neutrality |
| Parameter *A* | 492 | 144 | 6.02E-03 | 6.56E-04 | 0.446 | [−0.153; 0.155] | 1.000 | 0.000 | Divergent selection |
| Size at 119 days | 2855 | 183 | 2.26E+01 | 5.07E+00 | 0.285 | [−0.151; 0.155] | 1.000 | 0.000 | Divergent selection |
| *Reproduction* | | | | | | | | | |
| Ability to lay eggs | 492 | 144 | 2.70E+00 | 5.39E-08 | 0.709 | [−0.153; 0.157] | 1.000 | 0.000 | Divergent selection |
| Time to oviposition | 492 | 144 | 2.86E-02 | 8.36E-03 | 0.219 | [−0.141; 0.151] | 0.996 | 0.004 | Divergent selection |
| Number of clutches | 492 | 144 | 1.10E-01 | 1.02E-10 | 0.709 | [−0.160; 0.151] | 1.000 | 0.000 | Divergent selection |
| Number of eggs | 492 | 144 | 8.22E+02 | 5.64E+02 | 0.017 | [−0.162; 0.163] | 0.605 | 0.395 | Neutrality |
| Clutch size | 1537 | 144 | 2.95E+00 | 1.47E+00 | 0.088 | [−0.158; 0.168] | 0.885 | 0.115 | Neutrality |
| Hatching rate | 1537 | 144 | 3.27E-01 | 6.29E-01 | −0.154 | [−0.157; 0.160] | 0.025 | 0.975 | Homogenizing selection |

Analyses were performed using a full-sib design and accounting for population inbreeding.
N = sample size, Nfam = number of families, Vb and Vf indicate between population and family variance, respectively. Gompertz growth parameters: A is asymptotic size, b is related to size at birth, k is related to growth rate. Left and right P-values indicate the probability for $Q_{ST}$-$F_{ST}$ observed value to lie within the neutral distribution (unilateral test based on 1000 bootstraps, $\alpha = 0.025$).

habitat effect observed on most studied traits. Thus, although the number of populations and families was close to the recommended values [72], and despite the ability of mixed effect models to treat unbalanced designs [70], our experiment may lack sufficient statistical power, probably due to the lack of ponds within agricultural zones and of channels in reference sites. Meanwhile, the present study emphasizes the need to take habitat into account in $Q_{ST}$-$F_{ST}$ comparisons [79].

Third, the apparent lack of effect of pesticide pressure on the studied traits might also result from experimental conditions, which were too benign. For example, the trend for enhanced ability to reproduce observed in the $G_1$ from polluted areas, might reflect a specific response of exposed populations to overcome early mortality induced by chemicals (hatching rate: OAP0> OAP1 in $G_1$ and $G_2$), which was unexpected to express under common garden conditions. In the context of evolution under stressful environments, trait expression would be best studied under a gradient of stress conditions, as also recommended for molecular traits and pathways [15]. Stressful conditions may indeed facilitate the detection of divergent selection patterns with respect to pesticide historical exposure, since $Q_{ST}$ has been shown to depend on the experimental environment [48,99].

Finally, the present study also emphasized the occurrence of high genetic variation both at neutral markers and fitness-related traits in a species used as model in ecotoxicology. This finding provides empirical support for the need to account for genetic variation in toxicity testing and, as a perspective, in future procedures of ecological risk assessment [12].

## Supporting Information

**Figure S1** ***L. stagnalis*** **individual growth ($G_1$ snails) according to population habitat.** Gompertz growth curves based on 12 dates of measurement, performed on 492 individuals.

**Figure S2** ***L. stagnalis*** **hatching rate as a function of anthropogenic pressure (OAP, n = 1537 clutches).**

**Table S1** Evolution of *L. stagnalis* G1 rearing conditions as a function of age.

**Table S2** Summary of statistical analyses performed on *L. stagnalis* life history traits.

**Table S3** Population genetic parameters calculated per locus and population on 14 *L. stagnalis* population samples (see materials and methods for explanation of population codes).

**Table S4** Population genetic differentiation estimated among 14 *L. stagnalis* populations, on the basis of 12 SSR loci.

**Table S5** Life-history trait/parameter mean-values (SE) estimated in the laboratory-born $G_1$ from 14 *L. stagnalis* populations used to test patterns of adaptive divergence ($Q_{ST}$-$F_{ST}$ analysis).

## Acknowledgments

The authors thank the U3E staff (INRA, Rennes), and more particularly Maïra Coke for her technical assistance. Our thanks are also due to Udo Hommen for indicating sites in Germany, to Thierry Caquet and Sabrina Le Cam for their help and advices, and to Patrice David and Nico Van Straalen for constructive comments on the manuscript. The authors finally thank Benoît Heurtault, Gaëlle Méheut, Elise Petitpas, and Arnaud Giusti for hospitality and logistic support during sampling.

## Author Contributions

Conceived and designed the experiments: AB MAC. Performed the experiments: AB ALB MC MAC. Analyzed the data: AB JC MAC. Contributed reagents/materials/analysis tools: AB ALB. Contributed to the writing of the manuscript: AB JC MAC.

## References

1. Lankau R, Jørgensen PS, Harris DJ, Sih A (2011) Incorporating evolutionary principles into environmental management and policy. Evol Appl 4: 315–325.
2. Smith TB, Bernatchez L (2008) Evolutionary change in human-altered environments. Mol Ecol 17: 1–8.
3. Hendry AP, Kinnison MT, Heino M, Day T, Smith TB, et al. (2011) Evolutionary principles and their practical application. Evol Appl 4: 159–183.
4. Guillemaud T, Lenormand T, Bourguet D, Chevillon C, Pasteur N, et al. (1998) Evolution of resistance in *Culex pipiens*: allele replacement and changing environment. Evolution 52: 443–453.
5. Hoffmann AA, Hercus MJ (2000) Environmental stress as an evolutionary force. Bioscience 50: 217–226.
6. Reznick DN, Ghalambor CK (2001) The population ecology of contemporary adaptations: what empirical studies reveal about the conditions that promote adaptive evolution. Genetica 112: 183–198.
7. Sänen KR, Laurila A, Merilä J (2003) Geographic variation in acid stress tolerance of the moor frog, *Rana arvalis*. I. Local adaptation. Evolution 57: 352–362.
8. Hendry AP, Farrugia TJ, Kinnison MT (2008) Human influences on rates of phenotypic change in wild animal populations. Mol Ecol 17: 20–29.
9. Pomati F, Nizzetto L (2013) Assessing triclosan-induced ecological and trans-generational effects in natural phytoplankton communities: a trait-based field method. Ecotoxicology 22: 779–794.
10. Depledge MH (1994) Genotypic toxicity - Implications for individuals and populations. Environ Health Perspect 102: 101–104.
11. Leblanc GA (1994) Assessing deleterious ecosystem-level effects of environmental pollutants as a means of avoiding evolutionary consequences. Environ Health Perspect 102: 266–267.
12. Breitholtz M, Rudén C, Ove Hansson S, Bengtsson B-E (2006) Ten challenges for improved ecotoxicological testing in environmental risk assessment. Ecotoxicol Environ Saf 63: 324–335.
13. Bickham JW, Sandhu S, Hebert PDN, Chikhi L, Athwal R (2000) Effects of chemical contaminants on genetic diversity in natural populations: implications for biomonitoring and ecotoxicology. Mutat Res 463: 33–51.
14. Klerks PL (2002) Adaptation, ecological impacts, and risk assessment: insights from research at Foundry Cove, Bayou Trepagnier, and Pass Fourchon. Hum Ecol Risk Assess 8: 971–982.
15. Whitehead A, Roach JL, Zhang S, Galvez F (2011) Genomic mechanisms of evolved physiological plasticity in killifish distributed along an environmental salinity gradient. Proc Natl Acad Sci USA 108: 6193–6198.
16. Varian-Ramos CW, Swaddle JP, Cristol DA (2013) Familial differences in the effects of mercury on reproduction in zebra finches. Environ Pollut 182: 316–323.
17. Belanger-Deschênes S, Couture P, Campbell PGC, Bernatchez L (2013) Evolutionary change driven by metal exposure as revealed by coding SNP genome scan in wild yellow perch (*Perca flavescens*). Ecotoxicology 22: 938–957.
18. Orsini L, Spanier KI, De Meester L (2012) Genomic signature of natural and anthropogenic stress in wild populations of the waterflea *Daphnia magna*: validation in space, time and experimental evolution. Mol Ecol 21: 2160–2175.
19. Oleksiak MF, Karchner SI, Jenny MJ, Welch DBM, Hahn ME (2011) Transcriptomic assessment of resistance to effects of an aryl hydrocarbon receptor (AHR) agonist in embryos of Atlantic killifish (*Fundulus heteroclitus*) from a marine Superfund site. BMC Genomics 12: 263.
20. Bourret V, Couture P, Campbell PGC, Bernatchez L (2008) Evolutionary ecotoxicology of wild yellow perch (*Perca flavescens*) populations chronically exposed to a polymetallic gradient. Aquat Toxicol 86: 76–90.
21. Morgan AJ, Kille P, Sturzenbaum SR (2007) Microevolution and ecotoxicology of metals in invertebrates. Environ Sci Technol 41: 1085–1096.
22. Matson CW, Lambert MM, McDonald TJ, Autenrieth RL, Donnelly KC, et al. (2006) Evolutionary toxicology: population-level effects of chronic contaminant exposure on the marsh frogs (*Rana ridibunda*) of Azerbaijan. Env Health Perspect 114: 547–552.

23. Medina MH, Correa JA, Barata C (2007) Micro-evolution due to pollution: Possible consequences for ecosystem responses to toxic stress. Chemosphere 67: 2105–2114.
24. Yuan JS, Tranel PJ, Stewart CN (2006) Non-target-site herbicide resistance: a family business. Trends Plant Sci 12: 6–13.
25. Via S (1990) Ecological genetics and host adaptation in herbivorous species: the experimental study of evolution in natural and agricultural systems. Annu Rev Entomol 35: 421–446.
26. Jones BA, Grace D, Kock R, Alonso S, Rushton J, et al. (2013) Zoonosis emergence linked to agricultural intensification and environmental change. Proc Natl Acad Sci USA 110: 8399–8404.
27. Memmott J (2009) Food webs: a ladder for picking strawberries or a practical tool for practical problems? Philos Trans R Soc B Biol Sci 364: 1693–1699.
28. Schwarzenbach RP, Escher BI, Fenner K, Hofstetter TB, Johnson CA, et al. (2006) The challenge of micropollutants in aquatic systems. Science 313: 1072–1077.
29. Schwarzenbach RP, Egli T, Hofstetter TB, von Gunten U, Wehrli B (2010) Global water pollution and human health. Annu Rev Environ Resour 35: 109–136.
30. Rattner BA (2009) History of wildlife toxicology. Ecotoxicology 18: 773–783.
31. Brown CD, van Beinum W (2009) Pesticide transport via sub-surface drains in Europe. Environ Pollut 157: 3314–3324.
32. Willi Y, Van Buskirk J, Hoffmann AA (2006) Limits to the adaptive potential of small populations. Annu Rev Ecol Evol Syst 37: 433–458.
33. Lynch M (1995) Mutation accumulation and the extinction of small populations. Am Nat 146: 489–518.
34. Haddrill PR, Halligan DL, Tomaras D, Charlesworth B (2007) Reduced efficacy of selection in regions of the *Drosophila* genome that lack crossing over. Genome Biol 8: 18.
35. B, Morgan MT, Charlesworth D (1993) The effect of deleterious mutations on neutral molecular variation. Genetics 134: 1289–1303.
36. Gordo I, Charlesworth B (2001) Genetic linkage and molecular evolution. Curr Biol 11: 684–686.
37. Whitlock MC, Ingvarsson PK, Hatfield T (2000) Local drift load and the heterosis of interconnected populations. Heredity 84: 452–457.
38. Bijlsma R, Loeschcke V (2012) Genetic erosion impedes adaptive responses to stressful environments. Evol Appl 5: 117–129.
39. Coutellec M-A, Collinet M, Caquet T (2011) Parental exposure to pesticides and progeny reaction norm to a biotic stress gradient in the freshwater snail *Lymnaea stagnalis*. Ecotoxicology 20: 524–534.
40. Fox CW, Reed DH (2011) Inbreeding depression increases with environmental stress: an experimental study and meta-analysis. Evolution 65: 246–258.
41. Long TAF, Rowe L, Agrawal AF (2013) The effects of selective history and environmental heterogeneity on inbreeding depression in experimental populations of *Drosophila melanogaster*. Am Nat 181: 532–544.
42. Pelletier F, Réale D, Watters J, Boakes EH, Garant D (2009) Value of captive populations for quantitative genetics research. Trends Ecol Evol 24: 263–270.
43. Via S, Lande R (1985) Genotype-environment interaction and the evolution of phenotypic plasticity. Evolution 39: 505–522.
44. Hoffmann AA, Willi Y (2008) Detecting genetic responses to environmental change. Nat Rev Genet 9: 421–432.
45. Spitze K (1993) Population Structure in *Daphnia obtusa*: quantitative genetic and allozymic variation. Genetics 135: 367–374.
46. Leinonen T, McCairns RS, O'Hara RB, Merilä J (2013) $Q_{ST}$–$F_{ST}$ comparisons: evolutionary and ecological insights from genomic heterogeneity. Nat Rev Genet 14: 179–190.
47. Chapuis E, Trouvé S, Facon B, Degen L, Goudet J (2007) High quantitative and no molecular differentiation of a freshwater snail (*Galba truncatula*) between temporary and permanent water habitats. Mol Ecol 16: 3484–3496.
48. Hangartner S, Laurila A, Räsänen K (2012) Adaptive divergence in moor frog (*Rana arvalis*) populations along an acidification gradient: inferences from $Q_{ST}$-$F_{ST}$ correlations. Evolution 66: 867–881.
49. Jiménez-Ambriz G, Petit C, Bourrié I, Dubois S, Olivieri I, et al. (2007) Life history variation in the heavy metal tolerant plant *Thlaspi caerulescens* growing in a network of contaminated and noncontaminated sites in southern France: role of gene flow, selection and phenotypic plasticity. New Phytol 173: 199–215.
50. Habdija I, Lajtner J, Belinić I (1995) The contribution of gastropod biomass in macrobenthic communities of a karstic river. Int Rev Gesamten Hydrobiol Hydrogr 80: 103–110.
51. Brown CD, Holmes C, Williams R, Beulke S, van Beinum W, et al. (2007) How does crop type influence risk from pesticides to the aquatic environment? Environ Toxicol Chem 26: 1818–1826.
52. De Zwart D (2005) Ecological effects of pesticide use in The Netherlands: Modeled and observed effects in the field ditch. Integr Environ Assess Manag 1: 123–134.
53. Coors A, Vanoverbeke J, De Bie T, De Meester L (2009) Land use, genetic diversity and toxicant tolerance in natural populations of *Daphnia magna*. Aquat Toxicol 95: 71–79.
54. Cothran RD, Brown JM, Relyea RA (2013) Proximity to agriculture is correlated with pesticide tolerance: evidence for the evolution of amphibian resistance to modern pesticides. Evol Appl 6: 832–841.
55. Declerck S, De Bie T, Ercken D, Hampel H, Schrijvers S, et al. (2006) Ecological characteristics of small farmland ponds: associations with land use practices at multiple spatial scales. Biol Conserv 131: 523–532.
56. Knott E, Puurtinen M, Kaitala V (2003) Primers for nine microsatellites loci in the hermaphroditic snail *Lymnaea stagnalis*. Mol Ecol Notes 3: 333–335.
57. Kopp KC, Wolf K (2007) *Lymnaea stagnalis* microsatellite sequences. Genbank Access EF208747–EF208748.
58. Besnard A-L, Bouétard A, Azam D, Coutellec M-A (2013) Isolation and characterization of three new multiplex sets of microsatellite markers in the hermaphroditic freshwater snail *Lymnaea stagnalis* (Mollusca, Gastropoda, Heterobranchia, Panpulmonata, Hygrophila) using 454-pyrosequencing technology. Mol Ecol Resour 13. doi: http://tomato.bio.trinity.edu/manuscripts/13-1/mer-12-0292.pdf.
59. Belkhir K, Borsa P, Chikhi L, Raufaste N, Bonhomme F (2004) GENETIX 4.05, logiciel sous Windows pour la génétique des populations. Laboratoire Génome, Populations, Interactions, CNRS UMR 5171, Université de Montpellier II, Montpellier (France).
60. Weir BS, Cockerham CC (1984) Estimating F-statistics for the analysis of population structure. Evolution 38: 1358–1370.
61. Goudet J (2001) FSTAT, a program to estimate and test gene diversities and fixation indices (version 2.9.3).
62. Raymond M, Rousset F (1995) GENEPOP (Version 1.2): Population genetics software for exact tests and ecumenicism. J Hered 86: 248–249.
63. David P, Pujol B, Viard F, Castella V, Goudet J (2007) Reliable selfing rate estimates from imperfect population genetic data. Mol Ecol 16: 2474–2487.
64. Jones OR, Wang J (2010) COLONY: a program for parentage and sibship inference from multilocus genotype data. Mol Ecol Resour 10: 551–555.
65. Pritchard JK, Stephens M, Donnelly P (2000) Inference of population structure using multilocus genotype data. Genetics 155: 945–959.
66. Evanno G, Regnaut S, Goudet J (2005) Detecting the number of clusters of individuals using the software STRUCTURE: a simulation study. Mol Ecol 14: 2611–2620.
67. Earl DA, von Holdt BM (2012) STRUCTURE HARVESTER: a website and program for visualizing STRUCTURE output and implementing the Evanno method. Conserv Genet Resour 4: 359–361.
68. Rosenberg NA (2004) DISTRUCT: a program for the graphical display of population structure. Mol Ecol Notes 4: 137–138.
69. Coutellec M-A, Caquet T (2011) Heterosis and inbreeding depression in bottlenecked populations: a test in the hermaphroditic freshwater snail *Lymnaea stagnalis*. J Evol Biol 24: 2248–2257.
70. Pinheiro JC, Bates DM (2002) Mixed Effects Models in S and S-Plus. New York: Springer Verlag New York, LLC.
71. Bonnin I, Prosperi M, Olivieri I (1996) Genetic markers and quantitative genetic variation in *Medicago truncatula* (Leguminosa): a comparative analysis of population structure. Genetics 143: 1795–1805.
72. Goudet J, Büchi L (2006) The effects of dominance, regular inbreeding and sampling design on $Q_{ST}$, an estimator of population differentiation for quantitative traits. Genetics 172: 1337–1347.
73. Lynch M, Walsh B (1998) Genetics and Analysis of Quantitative Traits. Sunderland, MA: Sinauer Associates.
74. Whitlock MC, Guillaume F (2009) Testing for spatially divergent selection: comparing $Q_{ST}$ to $F_{ST}$. Genetics 183: 1055–1063.
75. Kopp K, Wolff K, Jokela J (2012) Natural range expansion and human-assisted introduction leave different genetic signatures in a hermaphroditic freshwater snail. Evol Ecol 26: 483–498.
76. Volis S, Yakubov B, Shulgina I, Ward D, Mendlinger S (2005) Distinguishing adaptive from nonadaptive genetic differentiation: comparison of $Q_{ST}$ and $F_{ST}$ at two spatial scales. Heredity 95: 466–475.
77. Chapuis E, Martin G, Goudet J (2008) Effects of selection and drift on G matrix evolution in a heterogeneous environment: a multivariate $Q_{ST}$-$F_{ST}$ test with the freshwater snail *Galba truncatula*. Genetics 180: 2151–2161.
78. Escobar JS, Nicot A, David P (2008) The different sources of variation in inbreeding depression, heterosis and outbreeding depression in a metapopulation of *Physa acuta*. Genetics 180: 1593–1608.
79. Johansson M, Primmer CR, Merilä J (2007) Does habitat fragmentation reduce fitness and adaptability? A case study of the common frog *Rana temporaria*. Mol Ecol 16: 2693–2700.
80. Bridges CM, Semlitsch RD (2000) Variation in pesticide tolerance of tadpoles among and within species of Ranidae and patterns of amphibian decline. Conserv Biol 14: 1490–1499.
81. Koene JM, Ter Maat A (2004) Energy budgets in the simultaneously hermaphroditic pond snail, *Lymnaea stagnalis*: A trade-off between growth and reproduction during development. Belg J Zool 134: 41–45.
82. Mousseau TA, Roff DA (1987) Natural selection and the heritability of fitness components. Heredity 59: 181–197.
83. Coutellec M-A, Caquet T, Lagadic L (2008) *Lymnaea stagnalis*: the effects of experimental demographic reduction on population dynamics. in: Akçakaya HR, Stark JD, Bridges TS, editors. Demographic Toxicity: Methods in Ecological Risk Assessment. Oxford, UK. pp.152–167.
84. Goudet J, Martin G (2007) Under neutrality, $Q_{ST} < F_{ST}$ when there is dominance in an island model. Genetics 176: 1371–1374.
85. López-Fanjul C, Fernandez A, Toro MA (2003) The effect of neutral nonadditive gene action on the quantitative index of population divergence. Genetics 164: 1627–1633.
86. Lamy J-B, Plomion C, Kremer A, Delzon S (2012) $Q_{ST} < F_{ST}$ as a signature of canalization. Mol Ecol 21: 5646–5655.

87. Le Corre V, Kremer A (2012) The genetic differentiation at quantitative trait loci under local adaptation. Mol Ecol 21: 1548–1566.
88. Edelaar M, Björklund M (2011) If $F_{ST}$ does not measure neutral genetic differentiation, then comparing it with $Q_{ST}$ is misleading. Or is it? Mol Ecol 20: 1805–1812.
89. Guillaume F (2011) Migration-induced phenotypic divergence: the migration-selection balance of correlated traits. Evolution 65: 1723–1738.
90. Wolf JB (2000) Gene interactions from maternal effects. Evolution 54: 1882–1898.
91. Hill WG, Goddard ME, Visscher PM (2008) Data and theory point to mainly additive genetic variance for complex traits. PLoS Genet 4: 1–10.
92. Pujol B, Wilson AJ, Ross RIC, Pannell JR (2008) Are $Q_{ST}$-$F_{ST}$ comparisons for natural populations meaningful? Mol Ecol 17: 4782–4785.
93. Puurtinen M, Knott KE, Suonpää S, Ooik T van, Kaitala V (2004) Genetic variability and drift load in populations of an aquatic snail. Evolution 58: 749–756.
94. Coutellec M-A, Lagadic L (2006) Effects of self-fertilization, environmental stress and exposure to xenobiotics on fitness-related traits of the freshwater snail *Lymnaea stagnalis*. Ecotoxicology 15: 199–213.
95. Puurtinen M, Knott KE, Suonpää S, Nissinen K, Kaitala V (2007) Predominance of outcrossing in *Lymnaea stagnalis* despite low apparent fitness costs of self-fertilization. J Evol Biol 20: 901–912.
96. Koene JM, Montagne-Wajer K, Roelofs D, Ter Maat A (2009) The fate of received sperm in the reproductive tract of a hermaphroditic snail and its implications for fertilisation. Evol Ecol 23: 533–543.
97. Escobar JS, Auld JR, Correa AC, Alonso JM, Bony YK, et al. (2011) Patterns of mating-system evolution in hermaphroditic animals: correlations among selfing rate, inbreeding depression, and the timing of reproduction. Evolution 65: 1233–1253.
98. Coutellec M-A, Besnard A-L, Caquet T (2013) Population genetics of *Lymnaea stagnalis* experimentally exposed to cocktails of pesticides. Ecotoxicology 22: 879–888.
99. Palo JU, O'Hara RB, Laugen AT, Laurila A, Primmer CR, et al. (2003) Latitudinal divergence of common frog (*Rana temporaria*) life history traits by natural selection: evidence from a comparison of molecular and quantitative genetic data. Mol Ecol 12: 1963–1978.

20

# A Country on the Verge of Malaria Elimination – The Kingdom of Saudi Arabia

**Michael Coleman[1]*, Mohammed H. Al-Zahrani[2], Marlize Coleman[1], Janet Hemingway[1], Abdiasiis Omar[1], Michelle C. Stanton[1], Eddie K. Thomsen[1], Adel A. Alsheikh[2], Raafat F. Alhakeem[2], Phillip J. McCall[1], Abdullah A. Al Rabeeah[1,2], Ziad A. Memish[1,2,3]**

**1** Vector Biology Department, Liverpool School of Tropical Medicine, Liverpool, United Kingdom, **2** Public Health Directorate, Ministry of Health, Riyadh, Kingdom of Saudi Arabia, **3** College of Medicine, Alfaisal University, Riyadh, Kingdom of Saudi Arabia

## Abstract

Significant headway has been made in the global fight against malaria in the past decade and as more countries enter the elimination phase, attention is now focused on identifying effective strategies to shrink the malaria map. Saudi Arabia experienced an outbreak of malaria in 1998, but is now on the brink of malaria elimination, with just 82 autochthonous cases reported in 2012. A review of published and grey literature was performed to identify the control strategies that have contributed to this achievement. The number of autochthonous malaria cases in Saudi Arabia decreased by 99.8% between 1998 and 2012. The initial steep decline in malaria cases coincided with a rapid scaling up of vector control measures. Incidence continued to be reported at low levels (between 0.01 and 0.1 per 1,000 of the population) until the adoption of artesunate plus sulfadoxine-pyrimethamine as first line treatment and the establishment of a regional partnership for a malaria-free Arabian Peninsula, both of which occurred in 2007. Since 2007, incidence has decreased by nearly an order of magnitude. Malaria incidence is now very low, but a high proportion of imported cases, continued potential for autochthonous transmission, and an increased proportion of cases attributable to *Plasmodium vivax* all present challenges to Saudi Arabia as they work toward elimination by 2015.

**Editor:** Joseph A. Keating, Tulane University School of Public Health and Tropical Medicine, United States of America

**Funding:** This work was funded by the Kingdom of Saudi Arabia Malaria Ministry of Health who also allowed access to datasets that have not been published and the Innovative Vector Control Consortium for supporting staff from LSTM, as well as unpublished PhD student work of LSTM. The funders had no role in study design, data collection and analysis, decision to publish, or preparation of the manuscript.

**Competing Interests:** The authors have declared that no competing interests exist.

* Email: mcoleman@lstmed.ac.uk

## Introduction

The recent global increase in malaria control efforts has contributed to major reductions in the burden of the disease [1,2]. Since 2007, four countries have eliminated malaria and been certified by the World Health Organization (WHO) as malaria free [3]. Today, 34 countries, including the Kingdom of Saudi Arabia, are actively attempting to eliminate malaria[4].

Malaria control in Saudi Arabia was initiated in 1948 by the Arabian American Oil Company (ARAMCO) in the Eastern province, primarily to protect employees living around the oases [5]. This programme was used by the Saudi Arabian government as the template for a national malaria programme in 1952, [5], which targeted malarious districts across the kingdom and was designed to protect pilgrims en route to the holy sites of Mecca and Medina. Saudi Arabia joined the WHO global malaria eradication effort in 1963 and, by the early 1970s, transmission was arrested in the Eastern and Northern provinces, eliminating malaria in the Palaearctic ecozone [6]. Despite this success, Saudi Arabia switched from malaria eradication to control in 1977 [7], following the worldwide abandonment of the goal of global malaria eradication.

Today in Saudi Arabia, malaria persists in the provinces of Aseer and Jazan [8], both bordering the Republic of Yemen. Following a series of outbreaks (of which the worst was in 1998), malaria control was intensified and the goal of malaria elimination in Saudi Arabia was re-established in 2004, with an elimination target of 2015. After nearly a decade of activity, progress towards this goal is reviewed in this report. Successful strategies and continued challenges are highlighted and discussed so that other countries may benefit from lessons learned in Saudi Arabia.

## Methods

### Data Sources

Data for this review were identified by literature searches of PubMed and general searches using the Google and WHO search engines. Search term strings included 'malaria' AND 'Saudi Arabia'. The bibliographies of selected documents were used to identify additional data and information sources. The National Malaria Control Programme provided access to archived annual reports and statistical data. Malaria data was obtained from the Ministry Of Health annual statistical reports (1980–2012) [9]. Population data was obtained from the Central Statistical Office.

Due to the limited number of data sources available, no formal exclusion criteria were used.

## Analysis

Population figures were extracted from the national censuses in 1992, 2004, and 2010. Population numbers for the interim years were estimated by calculating the average annual population growth between censuses and applying this accordingly. Malaria incidence was calculated as the total number of autochthonous cases divided by the total population.

# Findings

## Vectors and Epidemiology of Malaria

Records of vectors and epidemiology of malaria in Saudi Arabia are limited. A total of 15 Anopheles species have been recorded, five of which are known to be competent malaria vectors: *An. arabiensis, An. sergentii, An. stephensi, An. superpictus* and *An. culicifacies* [8–14]. *An. arabiensis* is currently the only known vector of malaria in Saudi Arabia [10,11], with low reported sporozoite rates (<1%) [11,12]. Abundance of *An. arabiensis* typically peaks following the rains, when multiple breeding sites appear [10], although there is some evidence from Jazan to suggest that irrigation may contribute to malaria risk [12]. With only 40–42% of blood meals in *An. arabiensis* females being of human origin [11,12], vector populations are clearly supported primarily by other host species, most likely domestic cattle, a preferred host of *An. arabiensis* throughout its range. As *An. arabiensis* rests and feeds outdoors as well as indoors, vector control methods targeting the human home, such as indoor residual spraying (IRS) and insecticide treated nets (ITNs, or long-lasting insecticidal nets, LLINs), are not sufficient to control this vector, necessitating the use of additional approaches.

With high aridity and low population density, much of the environment of the Arabian Peninsula is unsuitable for the malaria parasite. However, perennial mountain streams support numerous fertile valleys and oases in the southern provinces of Saudi Arabia, where human communities and vector populations coincide, and from where most malaria cases are reported.

Throughout the 1980s, there were a relatively consistent number of confirmed malaria cases in the country, ranging from approximately 5,000 to 17,000 annually. Cases were largely confirmed by microscopy. In 1990, recognising the importance of imported malaria, Saudi Arabia began to distinguish between autochthonous acquired and imported malaria cases by classifying patients according to their travel history, which revealed that imported malaria constituted the vast majority of annual cases.

Prior to 1998, national incidence ranged from 0.33/1,000 to 0.96/1,000. In 1998, a major outbreak of autochthonous acquired malaria occurred, in which the total number of confirmed cases in the country reached 36,139. This doubled the total annual incidence to 1.87/1,000 compared to0.92/1,000 in the previous year (Figure 1). Certain areas of the country were disproportionately affected by the outbreak, and incidence in the regions of Jazan and Qunfudah increased from 10/1,000 to 20/1,000 during the outbreak and from 12/1,000 to 44/1,000, respectively [9].

The 1998 outbreak triggered an aggressive centrally co-ordinated control campaign, the result of which was a steep decline in autochthonous transmitted malaria until 2004. During the 2007–2008 season, a further decrease of nearly an order of magnitude occurred from 0.02/1,000 to 0.002/1,000 (Figure 2).

As the incidence of autochthonous cases has fallen, the proportion of total imported malaria cases has steadily increased from 11% in 1998 to 98% in 2011. By 2010, only two health regions, Aseer (incidence of 0.001/1,000) and Jazan (incidence of 0.013/1,000), recorded any autochthonous malaria cases, and only two malaria-attributable deaths resulting from autochthonous transmission were recorded nationally in 2011. The latest region in Saudi Arabia to eliminate malaria was Qunfudah, which has reported no autochthonous transmission since 2009 (Figure 3) [13]. Additionally, a shift in the predominant parasite species has occurred in the last seven years, from predominantly *P. falciparum* to a majority of the imported *P.vivax* species (Figure 4) [9].

## Control and elimination activities

The current elimination strategy in Saudi Arabia focuses mainly on: a) targeting high risk areas for sustained preventative measures such as long lasting insecticide treated nets (LLINs) and indoor residual spraying (IRS); b) management of infection through rapid confirmed diagnosis and treatment; c) individual case follow up and reactive surveillance with appropriate treatment and vector control and, d) active case detection at borders with screening and treatment. The details of this strategy are discussed below.

**A. Sustained vector control targeting high-risk areas.** Due to the low incidence of malaria, comprehensive vector control is not considered to be cost effective, and is instead guided by case detection for targeted control measures [14,15]. This maintains coverage of interventions to interrupt transmission in the populations at risk [16,17]. A total of 68 active transmission foci were recorded in Saudi Arabia in 2011 [13]. The vector control programme in these areas used IRS, distribution of LLINs, larviciding and space spraying in an integrated manner.

DDT-based IRS was introduced in the Eastern Province in 1948 by ARAMCO to protect its workforce. DDT use in the ARAMCO and national programmes continued until 1954, when resistance was detected in *An. stephensi*. This prompted a switch in insecticide to dieldrin for three years before dieldrin resistance was detected in 1957 [18]. During this period, malaria cases decreased from 2000 in 1947 to 54 in 1956 [5] and the ARAMCO IRS campaign became the model for a national programme extended to all malarious areas of the country. In 1963, the government and WHO jointly started a pre-eradication campaign, and by the mid 1970's, IRS with DDT or dieldrin was used as the primary strategy to eliminate foci of transmission throughout the eastern and central regions of the country. Despite this, active transmission persisted in southern regions of the country and along the Red Sea coast. Following a decline in IRS efforts in the late 1970's, targeted IRS with DDT was re-introduced in Jazan and other southern regions in the early1980's [19]. IRS was then scaled back in 1987 when the organophosphate, fenitrothion, replaced DDT as the number of autochthonous cases declined.

Following the 1998 epidemic, IRS activities were scaled up again in southern Saudi Arabia. Focused IRS with DDT was carried out in 1999 at the beginning of peak malaria transmission (April to October). However, it became apparent that DDT resistance still existed, necessitating a switch to the pyrethroid lambda-cyhalothrin (10% EC) in 1999 and deltamethrin (25%WB) in 2004. By 2011, IRS with either deltamethrin (Kaothrine 25 WG) or lambda-cyhalothrin (ICON or Demand 10CS) protected 83% of the 3.15 million persons at risk [20] (Table 1).

Treatment of vector breeding sites with insecticide was introduced as a component of integrated vector control in 1971, initially with Paris Green and then with temephos (EC50). Combined with source reduction, weekly larviciding of 10–15 km sections of the affected wadis (seasonal river valleys) with temephos was performed [6]. Following detection of resistance in 2000, temephos was replaced by larvicides using insect growth

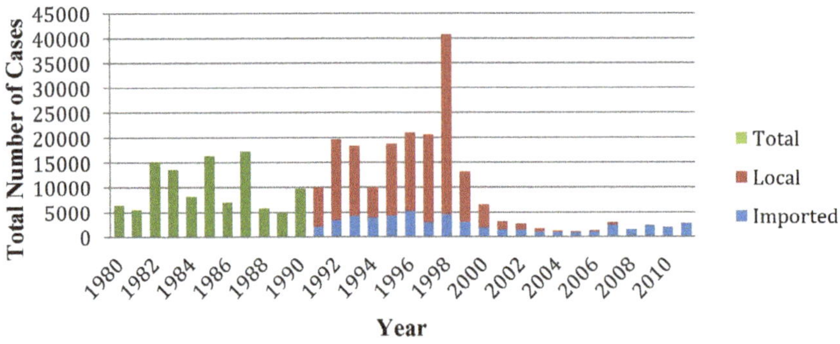


**Figure 1. Number of malaria cases in Saudi Arabia 1980 to 2011.**

regulators, predominantly difubenzuron, methoperene and pyriproxyfen. Larviciding remains one of the primary vector control strategies used in the active transmission foci in Saudi Arabia. Space spraying was used in the early control campaigns to rapidly reduce adult mosquito vector densities around wadis [6]. As this was the only breeding site, this was a successful strategy. Today, space spraying using the pyrethroid insecticides bifenthrin, bioallethrin and cyphenothrin remains a key component of vector control [21]. Currently it is used in response to the detection of a malaria case, when spraying in and around the index home and six nearest neighbours is performed. Following a successful trial of permethrin insecticide treated nets [22], LLINs were added to the national control programme in 1999, with free distribution to all age groups in focal areas. Since 2008, over 680,000 alpha-cypermethrin LLINs (Interceptor, BASF, Germany) have been distributed during the scale up for elimination (Table 1).

Post 1998, the efforts and resources dedicated to malaria control and elimination are shown in the sustained quantities of insecticide used (Figure 5). The current control programme delivers a range of insecticides from different classes, although pyrethroids are predominantly used against the adult insects. Although the importance of insecticide resistance as a major barrier to the success in malaria control and elimination is well known, [23] no routine insecticide susceptibility was undertaken on the malaria vectors in Saudi Arabia.

**B. Rapid confirmed diagnosis and treatment.** Diagnosis and treatment of malaria has been provided free in Saudi Arabia since the inception of the malaria programme [24,25]. All cases are confirmed positive by microscopic examination [26]. However, microscopy facilities are not always available (especially with active case detection). In these situations, a rapid diagnostic test (RDT) is used to confirm cases. A blood smear is always taken and read when conditions permit as a measure of quality assurance.

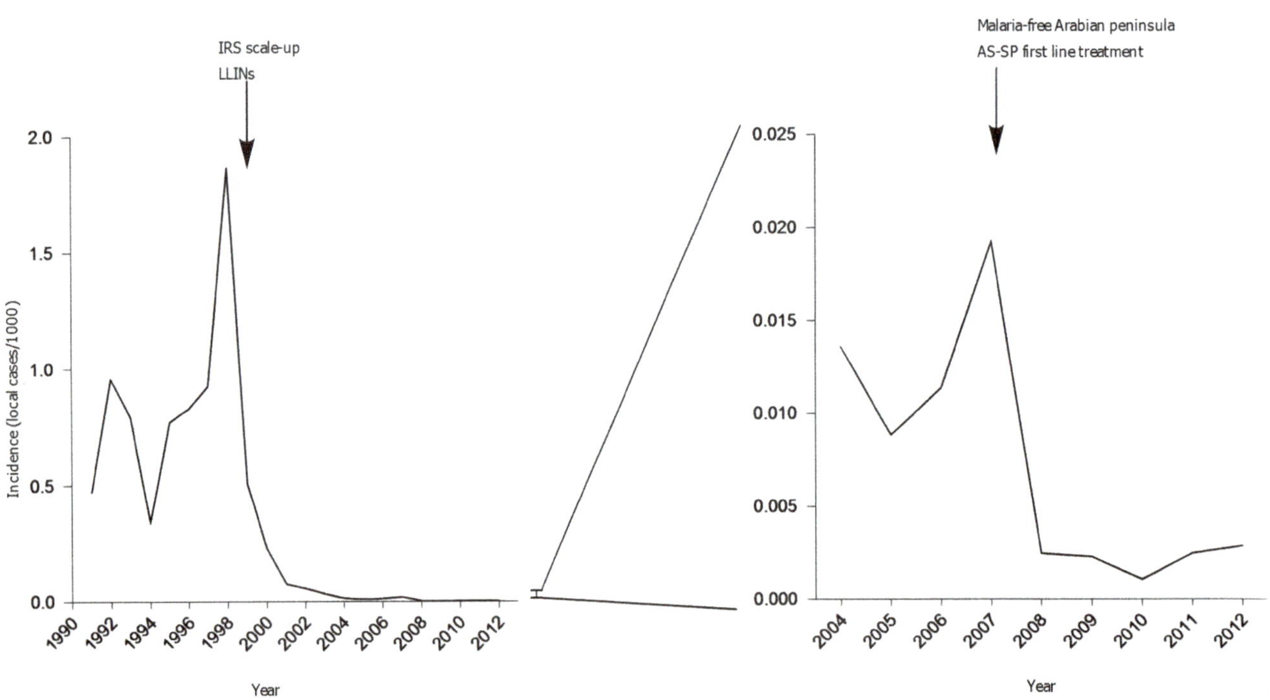


**Figure 2. Annual parasite incidence of autochthonous malaria cases with timing of major control milestones indicated.** IRS: indoor residual spraying, LLIN: long-lasting insecticidal nets, AS-SP: sulfadoxine pyrimethamine. Although the plan for a malaria-free Arabian peninsula was established in 2001, it was not funded until 2007.

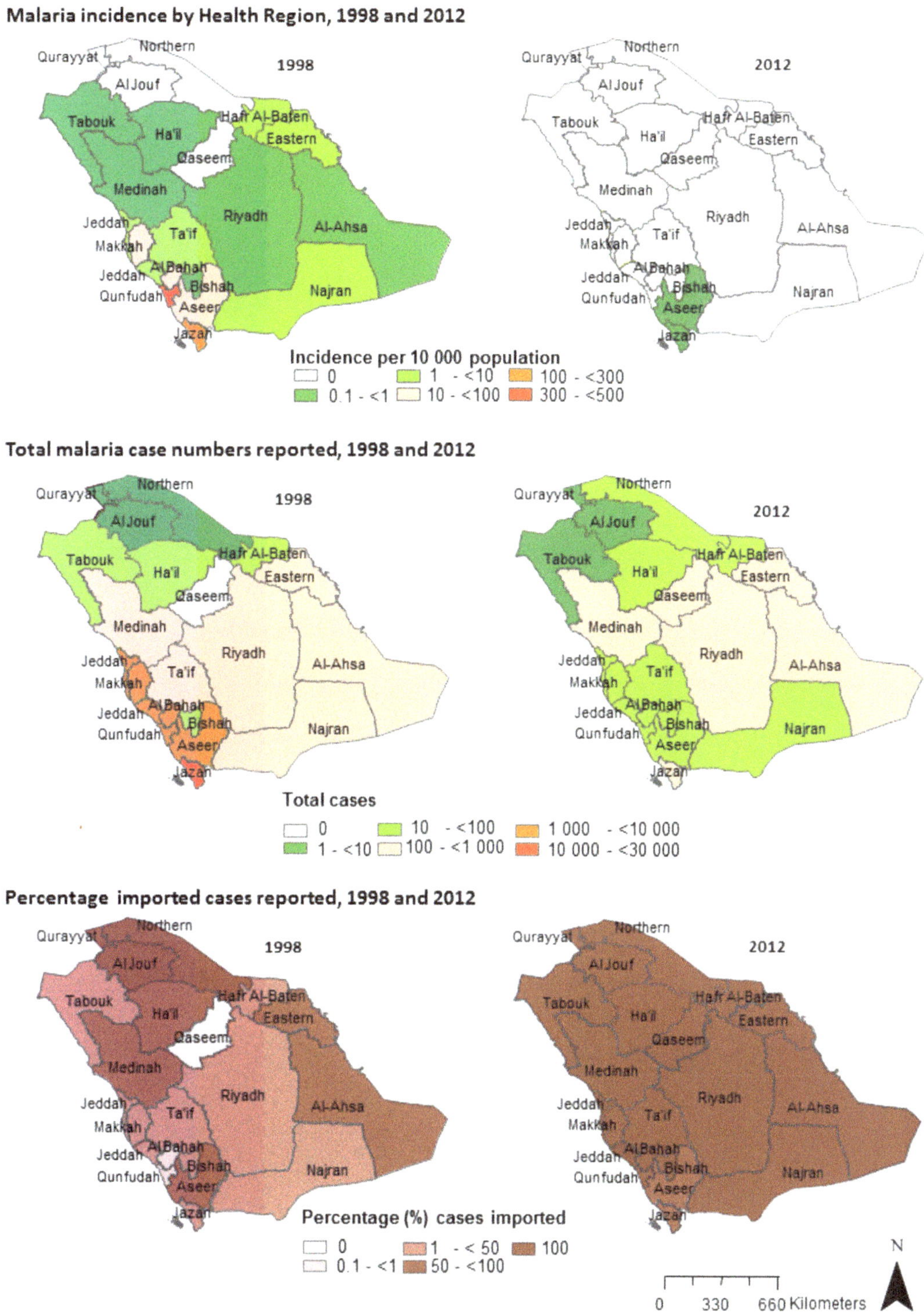


**Figure 3. Maps showing the changes in incidence of autochthonous cases, the total number of cases and percentage of imported cases.**

With the detection of chloroquine resistance in *P. falciparum* [27], artesunate plus sulfadoxine-pyrimethamine was made the first line treatment in 2007, with artemether-lumefanthrine being used for complicated malaria and quinine used for severe complicated malaria. Chloroquine plus primaquine are used for *P. vivax*. As ACTs kill developing gametocytes and reduce malaria transmission [28,29], this treatment strategy is of benefit in areas of low transmission, such as Saudi Arabia, where mature gametocytes may account for residual transmission and outbreaks [30].

**C. Reactive case detection.** Malaria is a notifiable disease in Saudi Arabia and diagnosis is predominantly based on microscopy [25]. Most malaria is identified through passive case detection at

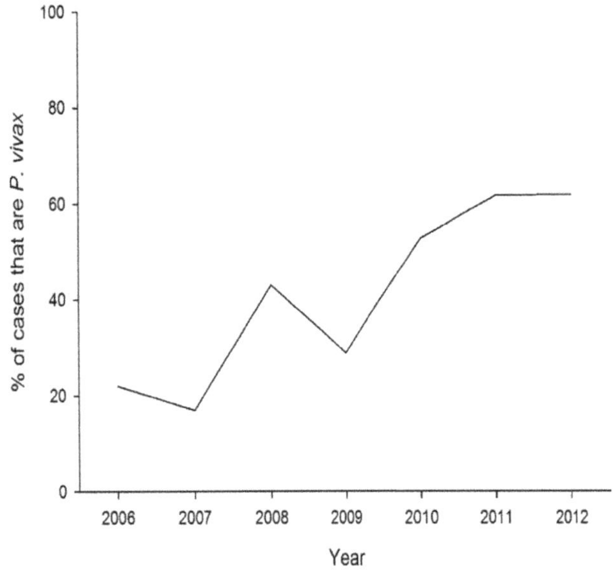


**Figure 4. Proportion of confirmed cases caused by *Plasmodium vivax* from 2006–2012.**

health facilities. Since 1991, all cases are reported within 24 hours to the local malaria centre, using a standard form that triggers a follow-up analysis to determine the probable source of infection. Based on the concept that malaria cases tend to cluster [31–33], reactive case detection focuses on other members of the household and neighbouring houses. Occupants are tested and treated for malaria, and appropriate space spraying and/or larviciding vector control is performed. This rapid action minimises the potential for any onward transmission from an index case. A similar system has been applied successfully in South Africa [34].

**D. Active case detection at the border.** As autochthonous endemicity is reduced or eliminated, controlling imported malaria becomes more important [4,35]. While there has been a marked reduction in the actual number of imported malaria cases in recent years (4,657 and 2,719 cases in 1998 and 2011, respectively), the proportion of imported cases has increased dramatically (11.4% and 97.5% of malaria cases in 1998 and 2011 were imported, respectively) [9]. Therefore, there is a clear risk of human movement reintroducing malaria into areas of elimination and creating outbreaks in Saudi Arabia as seen elsewhere in other countries [36–38].

There are three main sources of imported malaria in Saudi Arabia. First, Saudi Arabia relies on a large expatriate work force, many of which originate from malaria endemic countries in the Middle East, Africa or Asia. Second, Saudi Arabia contains the most important Islamic holy sites, visited by many millions of pilgrims from every country worldwide. The public health importance of imported malaria and other diseases from pilgrims attending the Hajj has long been recognised [39,40], and reactive case detection is used as the primary method of limiting transmission in these groups. Lastly, and perhaps most importantly, Saudi Arabia shares a southern border with a malaria endemic country, the Republic of Yemen [20].

Yemen has the second highest incidence of malaria in the Mediterranean region, with an estimated 40% of paediatric admissions due to malaria in some settings [20,41]. The highest incidences of malaria in Yemen occur in the Tihama coastal plain, a hot and humid geographical region where *An. arabiensis* is common [11,42], which extends north into Saudi Arabia's Jazan province. High levels of human movement occur between both countries with an estimated 3,000 illegal immigrants crossing from Yemen to Saudi Arabia on a daily basis and some 20,000 Saudi Arabians spending weekends in Yemen [43,44].

To limit the risk of imported malaria, plans were established in 2001 for a malaria-free Arabian peninsula by 2020 [38]. The plan was endorsed in 2005 by the Eastern Mediterranean Regional Office of WHO [45] and by the Health Ministers of the Gulf Cooperation Council in 2007, with financial support of $42.7 million [41]. Elimination in the Arabian Peninsula has been largely successful, and only Saudi Arabia and Yemen have yet to achieve malaria free status [46].

Cross border collaboration between Saudi Arabia and Yemen has included the establishment of malaria centres offering free screening and treatment, mostly for Yemenis living in border villages. These are supported by mobile teams that carry out active case detection to target high risk populations living in the border villages [47]. These mobile teams detected ~50% of imported cases in 2010–11 in Jazan. Malaria control in Yemen includes joint Saudi-Yemeni vector control teams, which are responsible for IRS and for space-spraying a 10 km wide buffer zone inside Yemen. Unfortunately, recent political unrest and a deteriorating political situation in Yemen have halted these efforts.

## Discussion

Since the outbreak in 1998, the efforts to control and eliminate malaria in the Kingdom of Saudi Arabia have produced significant gains. The total number of cases has remained consistently low

**Table 1.** Coverage of ITNs and the number of persons protected by IRS.

| Year | Number ITNs distributed | Persons protected by IRS |
|---|---|---|
| 2004 | 460,000 | unknown |
| 2005 | 81,364 | unknown |
| 2006 | 0 | 94,350 |
| 2007 | 0 | unknown |
| 2008 | 250,000 | unknown |
| 2009 | 250,000 | 2,457,965 |
| 2010 | 81,050 | 2,500,000 |
| 2011 | 100,000 | 2,600,000 |

Source: World malaria reports [20,48].

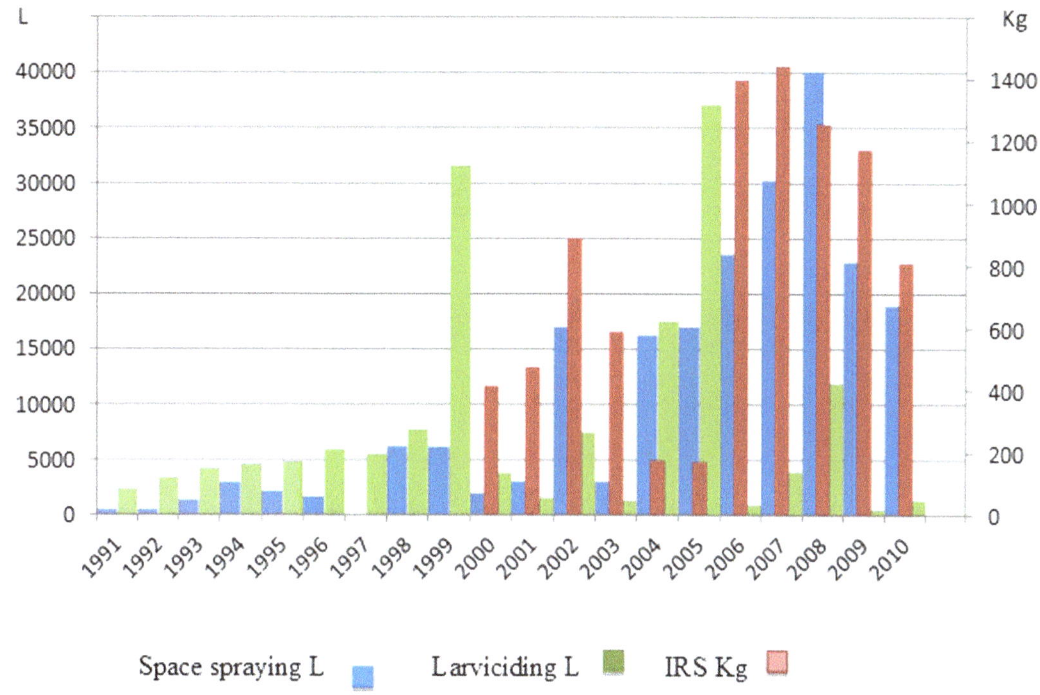


**Figure 5. Amount of insecticide used for vector control from 1991 to 2011.**

since 2001 and the number of autochthonous cases has decreased from 467 in 2007 [48] to just 82 in 2012 [13]. With incidence already well below 0.01 cases per 1,000 persons, elimination of malaria in Saudi Arabia is a realistic goal. However, a number of considerable challenges must be overcome in order to achieve this target.

Migration of humans (and parasites) has been shown to rapidly undermine gains made in malaria control efforts [49,50], and in Saudi Arabia, imported malaria still remains a major problem. The border with Yemen is an area of central concern, and controlling malaria in this region is a major challenge for Saudi Arabia in achieving and maintaining a malaria free status [46]. Autochthonous transmission is limited due to an effective response by the malaria control programme and/or unfavourable conditions for an outbreak to occur. However, the number of autochthonous cases increased to from 29 in 2010 to 69 and 82 in 2011 and 2012, respectively. This, in part, is probably due to the disruption of malaria control activities on the Yemeni side of the border and an increase in both legal and illegal immigrants into Saudi Arabia. Continued importation combined with the presence of vectors means that autochthonous transmission will inevitably occur. Saudi Arabia must find ways to mitigate this risk and ensure that endemic transmission is halted.

Saudi Arabia is likely to experience further difficulties as the proportion of cases attributable to *P. vivax* increases. *P. vivax* is generally less responsive to control than *P. falciparum* for a number of reasons. First, dormant liver stages can lead to relapse. Second, the extrinsic incubation period is shorter. Third, infective gametocytes are produced concurrently with asexual stages. Fourth, parasite densities are often lower than the detection threshold of diagnostic tests. Lastly, adherence to the 7–14 day primaquine regimen to treat liver stages is often poor. All of these factors make *P. vivax* residual transmission more difficult to detect and control.

## The way forward

Regional efforts such as the Lubombo Spatial Development Initiative [35] and the Asia Pacific Malaria Elimination Network [51,52] are excellent examples of where cross border collaborations have succeeded in addressing the challenge of imported malaria. Today, seven countries (Bahrain, Kuwait, Oman, Qatar, Saudi Arabia, the United Arab Emirates and Yemen) remain committed to supporting intensification of malaria control efforts on the Arabian Peninsula [20]. It is clear that reinvigoration of the border collaboration with Yemen will be necessary to achieve a malaria-free Saudi Arabia by 2015. Novel strategies of quickly identifying imported cases should also be considered. Recent work in Swaziland, a country also targeting elimination by 2015, has shown that snowball and time-location sampling can quickly identify networks of individuals at high-risk of bringing malaria into the country [53]. Adoption of this approach in Saudi Arabia would allow for targeted screening and treatment, preventing onward transmission. Additional strategies may include parasite genotyping, which would allow more accurate determination of common sources of infection.

Preventing the growing proportion of P.vivax cases will require implementing new but currently available strategies to quickly identify and treat those infected with the parasite. This may include sensitive, point-of-care diagnostics that are compatible with field conditions, such as a loop-attenuated isothermal amplification assay (LAMP). This assay is as sensitive as PCR detection, and suitable for deployment in remote settings has been shown to be equally as sensitive as PCR, but able to be deployed in remote settings [54].

Preventing autochthonous transmission requires a robust and sophisticated information system that enables timely response to outbreaks. Saudi Arabia aims to strengthen their control programme using emerging information technologies that can inform elimination efforts and maintain the disease free status. It is currently enhancing its surveillance systems through the adoption of a Disease Data Management System, a spatial decision support tool [55] that has a focus on malaria clusters and outbreaks [34,56] with mapping functionality required to support elimination and ongoing surveillance by mapping down to the village and even household level as recommended by WHO [57]. This will enhance the control programme's ability to investigate disease clusters and adjust operational plans accordingly. An independent review recently highlighted the advantages of such a system [58]. The failure to maintain high quality surveillance after near-elimination was a key factor that resulted in the disease returning in India [59] and Madagascar highlands [60].

Malaria elimination is a surprisingly stable state [61]. However, sustained elimination is unlikely to be attributable to continuous vector control. Rather, economic development that reduces vector-human contact and improved health systems are the major drivers in transitioning from malaria-endemic to malaria-free. As such, while improved management of imported malaria, more rapid case detection, and highly targeted interventions should be encouraged in the short-term, investment in health system infrastructure and overall socioeconomic development should be envisioned as the driving force behind malaria elimination in-country and regionally. Given the continued political and financial support from the government, the prospects to attain and sustain elimination in Saudi Arabia are positive.

## Author Contributions

Conceived and designed the experiments: Michael Coleman MHA JH PM ZM. Performed the experiments: Marlize Coleman AO AA AR. Analyzed the data: Michael Coleman MS ET. Contributed reagents/materials/analysis tools: Michael Coleman JH ZM. Wrote the paper: Michael Coleman JH ET ZM.

## References


1. Hay SI, Guerra CA, Gething PW, Patil AP, Tatem AJ, et al. (2009) A world malaria map: Plasmodium falciparum endemicity in 2007. PLoSMed 6: e1000048.
2. Hay SI, Guerra CA, Tatem AJ, Noor AM, Snow RW (2004) The global distribution and population at risk of malaria: past, present, and future. The Lancet Infectious Diseases 4: 327–336.
3. World Health Organization (2013) World Malaria Report 2013. Geneva: WHO.
4. Cotter C, Sturrock HJ, Hsiang MS, Liu J, Phillips AA, et al. (2013) The changing epidemiology of malaria elimination: new strategies for new challenges. Lancet: 10–6736.
5. Daggy RH (1959) Malaria in oases of Eastern Saudi Arabia. AmJTropMedHyg 8: 223–291.
6. Ministry Of Health, Saudi Arabia (1984) Malaria Control Programme in the Kingdom of Saudi Arabia 1983. Malaria Control Services, report.
7. World Health Organization office for the Eastern Mediteranian Region (1986) Joint Government/WHO malaria review misison to Saudi Arabia. WHO Regional Office for the Eastern Mediterranean Region.
8. Malaria Atlas Project (2013) http://www.map ox ac k/explore/countries/SAU/#!mosquito–vectors. Accessed 2013.
9. Kingdom of Saudi Arabia, Ministry of Health (2012) Health Statistcal Year Books 1980 to 2012. Saudi Arabia: Ministry of Health.
10. Abdoon AM, Alshahrani AM (2003) Prevalence and distribution of anopheline mosquitoes in malaria endemic areas of Asir region, Saudi Arabia. East MediterrHealth J 9: 240–247.
11. Al-Sheikh AAH (2004) Studies on the Ecology, Vectorial Role and Population Structure of Anopheles arabiensis in the Tihama Region of Saudi Arabia and Yemen: University of Liverpool, UK.
12. Al-Sheik AA (2011) Larval habitat, ecology, seasonal abundance and vectorial role in malaria transmission of Anopheles arabiensis in Jazan Region of Saudi Arabia. JEgyptSocParasitol 41: 615–634.
13. World Health Organization (2012) World Malaria Report: 2012. WHO Geneva.
14. Smith DL, McKenzie FE, Snow RW, Hay SI (2007) Revisiting the basic reproductive number for malaria and its implications for malaria control. PLoSBiol 5: e42.
15. World Health Organization (2008) Global Malaria Control and Elimination:report of a technical review. WHO Geneva.
16. Muller P, Chouaibou M, Pignatelli P, Etang J, Walker ED, et al. (2008) Pyrethroid tolerance is associated with elevated expression of antioxidants and agricultural practice in Anopheles arabiensis sampled from an area of cotton fields in Northern Cameroon. MolEcol 17: 1145–1155.
17. Pampana E (1969) A testbook for malaria eradication. London: Oxford University Press.
18. Peffly RL (1959) Insecticide resistance in anophelines in eastern Saudi Arabia. BullWorld Health Organ 20: 757–76.: 757–776.
19. World Health Organization (1986) Joint Government/WHO Malaria Review Mission to Saudi Arabia 5–23 July 1986.
20. World Health Organization (2011) World Malaria Report: 2011. Geneva: WHO.
21. Sebai ZA (1988) Malaria in Saudi Arabia. TropDoct 18: 183–188.
22. Jamjoom GA, Mahfouz AA, Badawi IA, Omar MS, al-Zoghaibi OS, et al. (1994) Acceptability and usage of permethrin-impregnated mosquito bed nets in rural southwestern Saudi Arabia. TropGeogrMed 46: 355–357.
23. World Health Organization (2012) Global Plan for Insecticide Resistance Management, in Malaria Vectors. WHO Geneva.
24. Ministry of Health, Saudi Arabia (2008) The National Policy of Malaria Case Management in The Kingdom of Saudi Arabia. Saudi Arabia.
25. Ministry of Health, Saudi Arabia (2012) The National Policy of Malaria Case Management in The Kingdom of Saudi Arabia. Saudi Arabia.
26. Nzila A, Al-Zahrani I (2013) Drugs for the treatment of malaria in the Kingdom of Saudi Arabia. Saudi Med J 34: 569–578.
27. Malik GM, Seidi O, El-Taher A, Mohammed AS (1998) Clinical aspects of malaria in the Asir Region, Saudi Arabia. AnnSaudiMed 18: 15–17.
28. Ding Y, Ortelli F, Rossiter LC, Hemingway J, Ranson H (2003) The Anopheles gambiae glutathione transferase supergene family: annotation, phylogeny and expression profiles. BMCGenomics 4: 35.
29. Craig A, Kyes S, Ranson H, Hemingway J (2003) Malaria parasite and vector genomes: partners in crime. Trends Parasitol 19: 356–362.
30. White NJ (2008) The role of anti-malarial drugs in eliminating malaria. MalarJ 7 Suppl 1:S8. doi: 10.1186/1475-2875-7-S1-S8.: S8-7.
31. Dicko A, Mantel C, Kouriba B, Sagara I, Thera MA, et al. (2005) Season, fever prevalence and pyrogenic threshold for malaria disease definition in an endemic area of Mali. TropMedIntHealth 10: 550–556.
32. Branch O, Casapia WM, Gamboa DV, Hernandez JN, Alava FF, et al. (2005) Clustered local transmission and asymptomatic Plasmodium falciparum and Plasmodium vivax malaria infections in a recently emerged, hypoendemic Peruvian Amazon community. MalarJ 4: 27.: 27.
33. Brooker S, Clarke S, Njagi JK, Polack S, Mugo B, et al. (2004) Spatial clustering of malaria and associated risk factors during an epidemic in a highland area of western Kenya. TropMedIntHealth 9: 757–766.
34. Coleman M, Coleman M, Mabuza AM, Kok G, Coetzee M, et al. (2008) Evaluation of an operational malaria outbreak identification and response system in Mpumalanga Province, South Africa. Malar J 7: 69.: 69.
35. Sharp BL, Kleinschmidt I, Streat E, Maharaj R, Barnes KI, et al. (2007) Seven years of regional malaria control collaboration—Mozambique, South Africa, and Swaziland. AmJTropMedHyg 76: 42–47.
36. Cohen JM, Smith DL, Cotter C, Ward A, Yamey G, et al. (2012) Malaria resurgence: a systematic review and assessment of its causes. MalarJ 11: 122. doi: 10.1186/1475-2875-11-122.: 122–111.
37. Maltezou HC, Tsolia M, Polymerou I, Theodoridou M (2013) Paediatric malaria in Greece in the era of global population mobility. TravelMedInfectDis: 10.
38. Craig MH, Kleinschmidt I, Le SD, Sharp BL (2004) Exploring 30 years of malaria case data in KwaZulu-Natal, South Africa: part II. The impact of non-climatic factors. TropMedIntHealth 9: 1258–1266.
39. Farid MA (1956) The implications of Anopheles sergenti for malaria eradication programmes east of the Mediterranean. BullWorld Health Organ 15: 821–828.
40. Khan AS, Qureshi F, Shah AH, Malik SA (2002) Spectrum of malaria in Hajj pilgrims in the year 2000. JAyubMedCollAbbottabad 14: 19–21.
41. Al-Taiar A, Jaffar S, Assabri A, Al-Habori M, Azazy A, et al. (2006) Severe malaria in children in Yemen: two site observational study. BMJ 333: 827.
42. Alzahrani MH (2006) Impact of Irrigation Systems on Malaria and Rift Valley Fever Transmision in Jizan, Saudi Arabia: University of Liverpool, UK.
43. Meleigy M (2007) Arabian Peninsula states launch plan to eradicate malaria. BMJ 20;334: 117.
44. Meleigy M (2007) The quest to be free of malaria. BullWorld Health Organ 85: 507–508.
45. Atta H, Zamani G (2008) The progress of Roll Back Malaria in the Eastern Mediterranean Region over the past decade. East MediterrHealth J 14 Suppl: S82–9.: S82–S89.
46. Snow RW, Amratia P, Zamani G, Mmundia CW, Noor AM, et al. (2013) The Malaria Transition on the Arabian Peninsula: Progress toward a Malaria-Free Region between 1960–2010. AdvParasitol 82: 205–51. doi: 10.1016/B978-0-12-407706-5.00003-4.: 205–251.
47. Macauley C (2005) Aggressive active case detection: a malaria control strategy based on the Brazilian model. SocSciMed 60: 563–573.
48. World Health Organization (2007) World Malaria Report: 2007. WHO Geneva.

49. Wickramage K, Galappaththy GN (2013) Malaria burden in irregular migrants returning to Sri Lanka from human smuggling operations in West Africa and implications for a country reaching malaria elimination. TransRSocTropMed-Hyg 107: 337–340.
50. Vakali A, Patsoula E, Spanakos G, Danis K, Vassalou E, et al. (2012) Malaria in Greece, 1975 to 2010. EuroSurveill 17: 20322.
51. Alilio MS, Kitua A, Njunwa K, Medina M, Ronn AM, et al. (2004) Malaria control at the district level in Africa: the case of the muheza district in northeastern Tanzania. AmJTropMedHyg 71: 205–213.
52. Hsiang MS, Abeyasinghe R, Whittaker M, Feachem RG (2010) Malaria elimination in Asia-Pacific: an under-told story. Lancet 375: 1586–1587.
53. Koita K, Novotny J, Kunene S, Zulu Z, Ntshalintshali N, et al. (2013) Targeting imported malaria through social networks: a potential strategy for malaria elimination in Swaziland. Malar J 12: 219.
54. Hopkins H, Gonzalez IJ, Polley SD, Angutoko P, Ategeka J, et al. (2013) Highly sensitive detection of malaria parasitemia in a malaria-endemic setting: performance of a new loop-mediated isothermal amplification kit in a remote clinic in Uganda. J Infect Dis 208: 645–652.
55. Eisen L, Coleman M, Lozano-Fuentes S, McEachen N, Orlans M, et al. (2011) Multi-Disease Data Managment System Platform for Vector-Borne Diseases. PLoSNeglTropDis 5: e1016.
56. Coleman M, Coleman M, Mabuza A, Kok G, Coetzee M, et al. (2008) Using the SaTScan method to detect local malaria clusters for guiding malaria control. Malar J Submitted.
57. World Health Organization (2007) Malaria Elimination. A field manual for low and moderate endemic countries. WHO Geneva.
58. Mabuza A, Kok G, Groepe MA, Misiani E, Shandukani M, et al. (2012) Active Case Detection Towards Malaria Elimination: Lessons Learned From Mpumlanga Province, South Africa. Abstract Book 2012. The American Journal of Tropical Medicine and Hygine.
59. Sharma VP (1996) Re-emergence of malaria in India. IndianJMedRes 103: 26–45.: 26–45.
60. Mouchet J, Laventure S, Blanchy S, Fioramonti R, Rakotonjanabelo A, et al. (1997) The reconquest of the Madagascar highlands by malaria. BullSocPatholExot 90: 162–168.
61. Chiyaka C, Tatem AJ, Cohen JM, Gething PW, Johnston G, et al. (2013) Infectious disease. The stability of malaria elimination. Science 339: 909–910.

# 21

# *De Novo* Transcriptome of the Hemimetabolous German Cockroach (*Blattella germanica*)

**Xiaojie Zhou[1,2], Kun Qian[1], Ying Tong[1], Junwei Jerry Zhu[3], Xinghui Qiu[2]*, Xiaopeng Zeng[1]***

**1** Institute of Disinfection and Vector Control, Beijing Center for Disease Control and Prevention, Beijing, China, **2** State Key Laboratory of Integrated Management of Pest Insects and Rodents, Institute of Zoology, Chinese Academy of Sciences, Beijing, China, **3** United States Department of Agriculture, Agricultural Research Service, Lincoln, Nebraska, United States of America

## Abstract

***Background:*** The German cockroach, *Blattella germanica*, is an important insect pest that transmits various pathogens mechanically and causes severe allergic diseases. This insect has long served as a model system for studies of insect biology, physiology and ecology. However, the lack of genome or transcriptome information heavily hinder our further understanding about the German cockroach in every aspect at a molecular level and on a genome-wide scale. To explore the transcriptome and identify unique sequences of interest, we subjected the *B. germanica* transcriptome to massively parallel pyrosequencing and generated the first reference transcriptome for *B. germanica*.

***Methodology/Principal Findings:*** A total of 1,365,609 raw reads with an average length of 529 bp were generated via pyrosequencing the mixed cDNA library from different life stages of German cockroach including maturing oothecae, nymphs, adult females and males. The raw reads were *de novo* assembled to 48,800 contigs and 3,961 singletons with high-quality unique sequences. These sequences were annotated and classified functionally in terms of BLAST, GO and KEGG, and the genes putatively coding detoxification enzyme systems, insecticide targets, key components in systematic RNA interference, immunity and chemoreception pathways were identified. A total of 3,601 SSRs (Simple Sequence Repeats) loci were also predicted.

***Conclusions/Significance:*** The whole transcriptome pyrosequencing data from this study provides a usable genetic resource for future identification of potential functional genes involved in various biological processes.

**Editor:** Yara M. Traub-Csekö, Instituto Oswaldo Cruz, Fiocruz, Brazil

**Funding:** This study was supported by the grants from China Postdoctoral Science Foundation (2013M530727), Beijing Municipal Postdoctoral Scientific Research Foundation (2013ZZ-94) and Public Welfare Industry Scientific Research Foundation of National General Administration of Quality Supervision, Inspection and Quarantine (201310255-02). The funders had no role in study design, data collection and analysis, decision to publish, or preparation of the manuscript.

**Competing Interests:** The authors have declared that no competing interests exist.

* Email: qiuxh@ioz.ac.cn (XQ); xpzeng@126.com (XPZ)

## Introduction

Cockroaches are one of the most ancient and primitive winged insects, which have existed successfully and remained virtually unchanged in body morphology for approximately 350 million years from the Carboniferous to present [1]. There are about 3,500 known species of cockroaches globally, only thirty are considered as household pests. Among them, the German cockroach (*Blattella germanica*) is listed as one of the most important public health related insect pests [2], because it is highly dependent on human for survival and can transmit mechanically a number of pathogenic viruses, fungi, helminths and bacteria (with some exhibiting resistance to antibiotics) [3,4]. For nearly a half century, the worldwide occurrence of allergic respiratory morbidity, especially in childhood, has been considered to be closely related to the German cockroach infestation [5,6]. On the other hand, the German cockroach, a phylogenetically basal species showing a gradual metamorphosis, has also been considered as an important model for scientific studies regarding reproduction, metamorphosis, neurophysiology, behavioral and chemical ecology, allergen biology, nutrition metabolism and insecticide resistance [1].

Despite its medical significance and prominence as a model in entomological study, our understanding about the German cockroach has been largely hindered by the lack of thorough genetic information. Previous efforts in this direction were limited to a few low-throughput EST projects such as subtractive hybridization [7] and cDNA library Sanger sequencing [8]. Until now, only 2,876 short ESTs are publicly available at the NCBI (National Center for Biotechnology Information) website. No complete picture has been achieved even at the scale of a specific gene family. For example, cytochrome P450s (CYPs) have been highly recognized as a super gene family with 36 to 180 genes existed in insect genomes [9], while only 7 CYP genes with a full-length coding region can be accessed for *B. germanica* in GenBank. This scarcity of genetic information in *B. germanica* has also resulted in a paucity of genetic studies and integrated theories for understanding its basic biology.

*B. germanica* was in the list of the 5,000 Insect Genome Project (i5k), a huge project that has been initiated recently. EST or transcriptome sequencing could not only be useful for the assembly and annotation of genomic data, but also as an efficient and feasible alternative approach to obtain genetic information. Next-generation sequencing (NGS) platforms, such as Roche 454 Genome Sequencer FLX System (454), Illumina Genome Analyzer (Solexa) and Applied Biosystems SOLiD system (SOLiD), have been increasingly used and dramatically accelerated biological and biomedical research on a genome-wide scale. The most significant advantage of NGS over traditional Sanger-sequencing is its massively parallel sequencing ability with significant low cost for DNA sequencing [10–12]. The 454 pyrosequencing is very suitable for a non-model species, enabling high efficient *de novo* sequencing, assembly and annotation of expressed genes [13,14], thus has been widely applied in a broad range of arthropod species including *Cimex lectularius* [15,16], *Anopheles funestus* [17], *Aedes aegypti* [18], *Dermacentor variabilis* [19], *Musca domestica* [20], *Sarcophaga crassipalpis* [21], *Cochliomyia hominivorax* [22], *Anastrepha suspensa* [23], *Nilaparvata lugens* [24],*Manduca sexta* [25], *Lutzomyia intermedia* [26], *Amblyomma maculatum* [27], *Corethrella appendiculata* [28] and *Rhodnius prolixus* [29]. For the German cockroach, Illumina sequencing was used to identify microRNAs from the ovaries and whole body of 6$^{th}$ instar nymphs [30]. In the present study, we attempt to explore the transcriptome of *B. germanica* using 454 pyrosequencing. We hope that the generated transcriptome database can serve as a valuable resource for better understandings of the molecular mechanisms underlying important biological processes and the environmental adaptability of the cockroach.

## Materials and Methods

### Ethics Statement

The German cockroach used in the present study is a common indoor insect pest, not an endangered and protected species. No permission was required to sample and collect the German cockroaches from the infested restaurants, where we scheduled routine cockroach density surveillance plan for the public health purpose.

### Cockroaches

Two strains of German cockroach (*B. germanica*) BJ-S and DX-R, were used in this study. Both strains were kept separately in glass jars of the same size (13 liters), coated with petroleum jelly (top 1/5 of the jar) at 26±1°C and 60±10% relative humidity with a photoperiod of 12:12 (L:D) h. Cockroaches were fed with laboratory rodent food (Beijing Huafukang Biotechnology) and water *ad libitum*. The susceptible strain (BJ-S) had been reared in the laboratory without exposure to any insecticide since 1970. A field strain (DX-R) was established from cockroaches collected in a local restaurant (Daxing district, Beijing) in 2011. The comparison in resistance of the two strains to several types of insecticides is shown in Table S1. The inclusion of this field collection in this study was to increase the genetic diversity and to obtain preliminary information about the differentially expressed genes associated with insecticide resistance for future studies. Cockroach samples for RNA extraction and quantification were collected and snap-frozen immediately in liquid $N_2$ and stored at −80°C until use.

### RNA isolation, cDNA library construction and 454 pyrosequencing

In order to obtain a transcriptomic dataset with a coverage as wide as possible, RNA was extracted from pooled samples of 30 maturing oothecae, 30 4$^{th}$ instar nymphs, 30 adult females and 30 adult males (7 days after eclosion) respectively from each strain. Total RNA was extracted using the Trizol reagent (Invitrogen, CA, USA) according to the manufacturer's instruction. The integrity of total RNA was assessed by both 1.4% denaturing formaldehyde agarose gel electrophoresis and the Agilent 2100 Bioanalyzer (Palo Alto, CA, USA) with a minimum integrity value of 8. The quantity of total RNA was determined by NanoDrop 1000 spectrophotometer (Thermo, MA, USA). Equal quantities (10 µL) of total RNA (1 µg/µL) from each of the four life stages were combined to form a RNA pool for each strain. mRNA was isolated by using PolyATtract mRNA isolation systems (Promega, WI, USA) from each total RNA pool. mRNA pools were concentrated by using RNeasy MinElute Cleanup Kit (Qiagen, Valencia, CA) and used as starting material for cDNA library construction.

Two cDNA libraries of BJ-S and DX-R strains were separately constructed from respective pooled mRNA samples for 454 pyrosequencing. Briefly, the mRNA was broken into short fragments in the presence of fragmentation buffer at 94°C for 5 min. These short fragments were used as templates for first-stranded cDNA synthesis using random hexamer primers. Subsequently, second-stranded cDNAs were synthesized using dNTPs, RNaseH and DNA polymerase I. DNA bands (500–800 bp) were excised and purified from agarose gels using the QIAquick Gel Extraction Kit (Qiagen, Valencia, CA). The isolated double-stranded cDNA were blunt-ended using T4 DNA polymerase and T4 polynucleotide kinase (PNK), then ligated to the adapters (Titanium A&B provided in the 454 library kit) with T4 DNA ligase, followed by immobilization on DNA capture beads. DNA capture beads were clonally amplified by

**Table 1.** Summary statistics of the ESTs generated from the German cockroach through pyrosequencing.

| Features | Values |
|---|---|
| Total number of reads | 1,365,609 |
| Average read length | 529 bp |
| Number of reads involved in the assembly | 1,362,260 |
| Total number of unique sequences | 52,761 |
| Total number of contigs | 48,800 |
| Total number of singletons | 3,961 |
| Average unigenes size | 798 bp |
| N50 contig size | 792 bp |

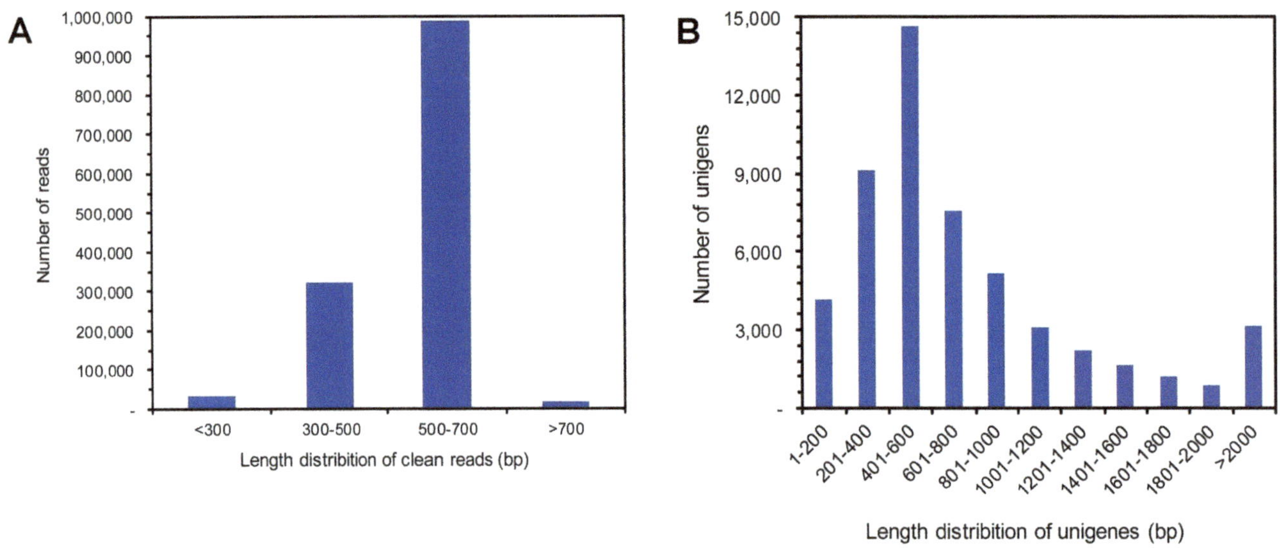


**Figure 1. Length distribution of clean reads (A) produced by 454 pyrosequencing and the assembled unigenes (B).**

emulsion PCR (emPCR) and enriched by removing the waste oil from the beads and selecting for beads with amplified library fragments. The beads were counted using a Beckman Coulter Z1 Particle Counter and loaded into the wells of PicoTiter plate (PTP). A single full PTP was sequenced with a half plate for each library following standard protocols described by Margulies [12] on a 454 GS FLX Titanium instrument (Roche Diagnostics, Indianapolis, IN).

## Sequence preprocessing and *de novo* assembly

The raw 454 reads were produced by the base-calling process which transformed the measured pyroluminescence intensity signals to a sequence of nucleotides. These read can be accessed through NCBI Short Read Archive (SRA) under the accession code SRP042142. Raw reads were preprocessed by in-house developed tools, TagDust [31] and Seqclean [32] to trim the adapter, poly A/T tails and remove low-quality, short read data and contamination sequences. The resultant clean reads from both

**Table 2.** Summary of top 20 abundant assembled unigenes from the German cockroach transcriptome.

| Unigene ID | Length (bp) | Top Blast hit | Identity (%) |
|---|---|---|---|
| Contig6820 | 4019 | 18S ribosomal RNA, *Blattella germanica* | 99 |
| Contig4068 | 6792 | Vitellogenin-2, *B. germanica* | 99 |
| unigene_rep_c6381 | 887 | NADH dehydrogenase subunit 1, *B. germanica* | 100 |
| Contig3091 | 2293 | Transferrin, *Blaberus discoidalis* | 82 |
| Contig3072 | 4114 | Myosin heavy chain, *Musca domestica* | 91 |
| Contig6443 | 1518 | Cytochrome c oxidase subunit 1 (COX1), *B. germanica* | 96 |
| Contig3066 | 1053 | Arginine kinase, *B. germanica* | 99 |
| Contig3115 | 7089 | Apolipophorins, *Locusta migratoria* | 47 |
| Contig2345 | 1323 | Elongation factor 1α (EF-1α), *Cryptocercus punctulatus* | 98 |
| Contig1101 | 1446 | α-Amylase, *B. germanica* | 76 |
| Contig946 | 746 | Tropomyosin-2, *Coptotermes formosanus* | 96 |
| Contig61 | 1120 | β-actin, *Acyrthosiphon pisum* | 100 |
| Contig4872 | 2616 | Elongation factor 2 (EF-2), *Schistocerca gregaria* | 99 |
| Contig594 | 483 | Allergen Bla g 8, *B. germanica* | 100 |
| Contig848 | 996 | Troponin T (TnT), *C. formosanus* | 93 |
| Contig1104 | 1608 | Pyruvate kinase, *Tribolium castaneum* | 80 |
| Contig3092 | 1032 | Aspartic protease Bla g 2, *B. germanica* | 80 |
| Contig611 | 1635 | Cytochrome P450 CYP4G19, *B. germanica* | 100 |
| Contig3123 | 715 | ATP-dependent RNA helicase p62, *T. castaneum* | 85 |
| Contig3084 | 2631 | Paramyosin, *Megachile rotundata* | 84 |

**Table 3.** Genes related to important physiological functions.

| Functional genes | No. of unigenes | Functional genes | No. of unigenes |
|---|---|---|---|
| **Detoxification** | | α-amylase | 46 |
| Cytochrome P450 | 163 | α-glucosidase (maltase) | 5 |
| GSTs | 64 | β-glucosidase | 21 |
| CarE | 12 | Lipase | 41 |
| ATP-binding cassette | 19 | **Immunity and defence** | |
| UDP glycosyltransferases | 5 | β-glucanase | 18 |
| **Insecticide targets** | | Lectin-like protein | 115 |
| Acetylcholinesterase | 3 | **Peritrophic membrane biosynthesis and metabolization** | |
| GABA receptor | 2 | Chitinase | 37 |
| Nicotinic acetylcholine receptor | 5 | Chitin synthase | 5 |
| Sodium channel | 6 | Chitin deacetylase | 3 |
| Glutamate receptor | 12 | Peritrophin (mucin)-like | 19 |
| **Digestion** | | **Olfactory reception** | |
| Serine proteinase all types | 19 | Odorant-binding protein (OBP) | 15 |
| Cysteine proteinase all types | 10 | Chemosensory protein (CSP) | 12 |
| Carboxypeptidase all types | 53 | Odorant receptor (OR) | 2 |
| Aminopeptidase all types | 47 | **Gustatory reception** | |
| Dipeptidyl-peptidase | 2 | Gustatory receptor (GR) | 4 |

strains respectively were combined to assemble into unique sequences by the Newbler [12] assembler programs at default parameters. This Transcriptome Shotgun Assembly project has been deposited at DDBJ/EMBL/GenBank under the accession GBID00000000. The version described in this paper is the first version, GBID01000000.

## Homology searches and EST annotation

The generated unigenes were firstly BLASTx searched against the Swiss-Prot [33] and NCBI non-redundant protein database. The sequences retrieving no BLASTx hit were searched by BLASTn against the NCBI nucleotide collection. Gene ontology (GO), KOG [34] and KEGG [35] analysis were used for functional classification of the annotated ESTs. For gene ontology analysis, both BLAST2GO [36,37] and WEGO [38] were employed. The programs extracted the GO terms associated with homologies identified with BLAST and returned a list of GO annotations represented as hierarchical categories of increasing specificity.

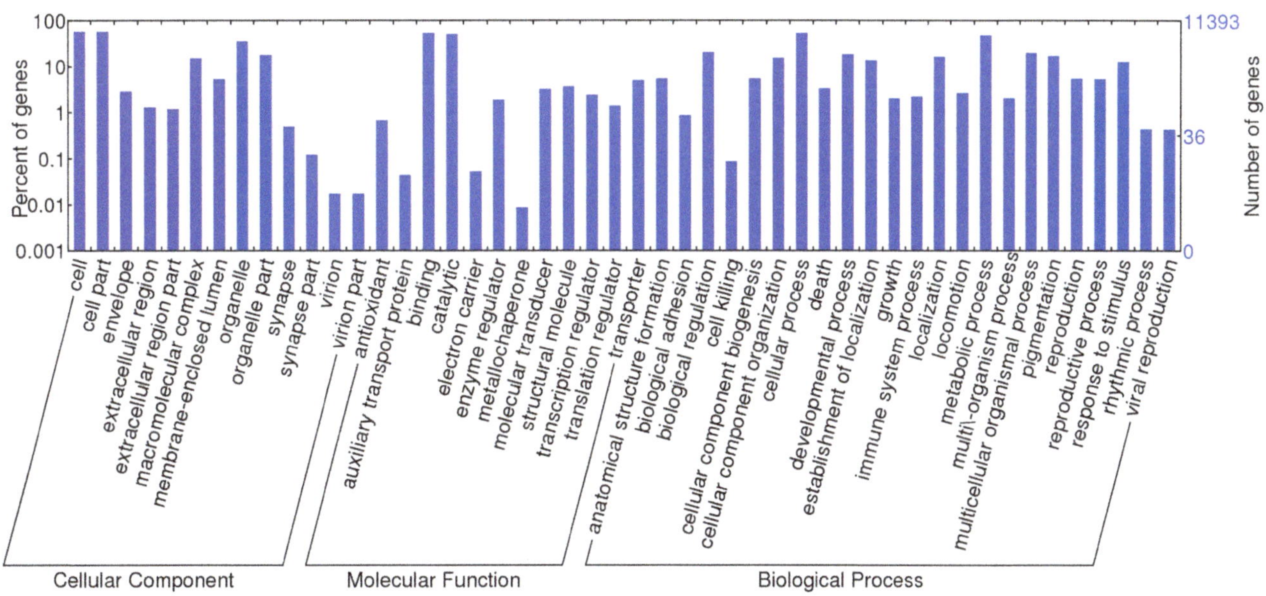


**Figure 2. Gene ontology (GO) terms for the transcriptomic sequences of *Blattella germanica*.**

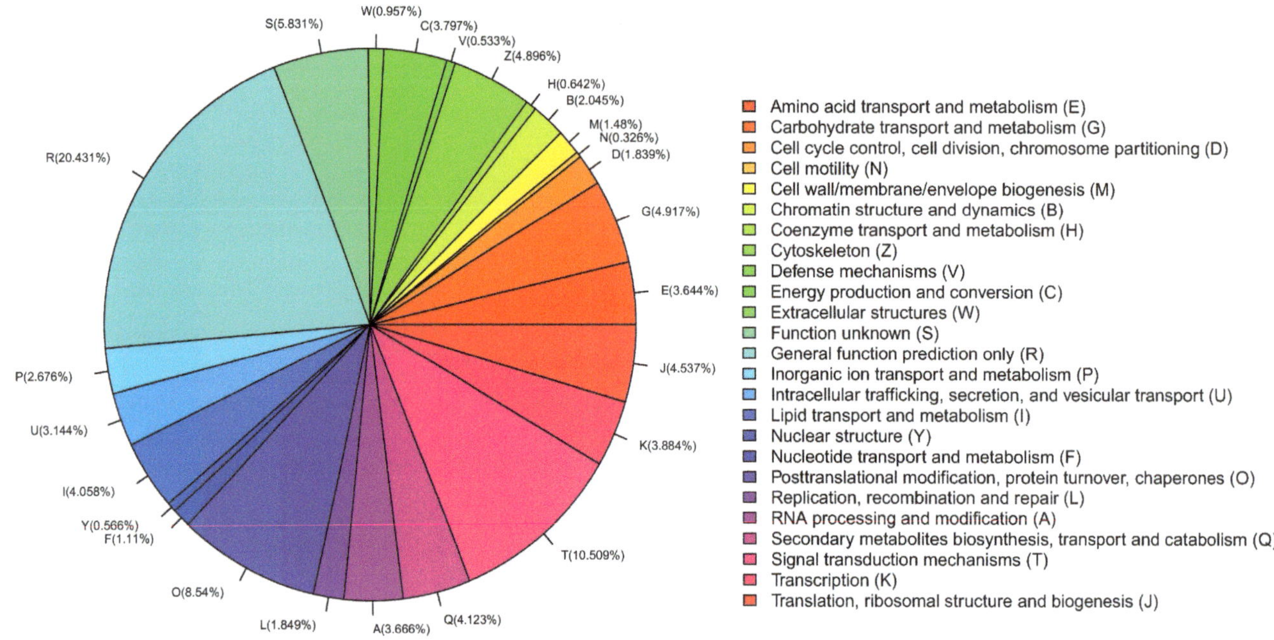


**Figure 3. Eukaryotic orthologous groups (KOG) function classification of 52,761 unigenes of *Blattella germanica*.**

## Identification and analysis of genes of interest

Genes of interest were further manually examined to check for possible frameshifts caused by the 454 sequencing based on the annotated ESTs. All of the manually confirmed protein sequences were used for alignment and phylogenetic analysis. Alignments were implemented using BioEdit, and employed to reconstruct the phylogeny by using the MEGA6 software [39]. The neighbor-joining method was used to create phylogenetic trees with p-distance under the default parameters of MEGA program. Bootstrap analysis of 1000 replications was performed to evaluate the branch strength of each tree.

# Results and Discussion

## Pyrosequencing, assembly and annotation

Using Roche 454 GS-FLX platform, our single full run from two cockroach strains with each strain half-run yielded a total of 1,365,609 raw reads with an average length of 529 bp. After preprocessing the raw data (including adaptor trimming and removal of low-quality reads), 1,362,260 reads remained with a total amount of 42,110,570 bp and an average length of 520 bp. All the clean reads were assembled into 48,800 contigs and 3,961 singletons forming a total of 52,761 high-quality unigenes with an average length of 798 bp (Table 1). Of these unigenes, 12,146 (23.0%) were over 1000 bp in length, and 5504 (36.7%) were between 500–1000 bp. The length distribution of the reads and unigenes is shown in Figure 1.

Through the BLASTx-BLASTn sequential homology search, nearly half (47.7%) of the unigenes could be annotated. When all unique transcripts were aligned against the Swiss-Prot database using BLASTx, a total of 17,779 (33.7%) unique transcripts yielded one or more significant hits. Table 2 lists the characteristics of the top 20 most abundant ESTs (the sequences are in File S1). Besides some housekeeping high-abundant genes, some genes involved in the reproduction (vitellogenin), defence (transferrin), energy metabolism (COX1), detoxification enzyme (CYP4G19) and allergenic proteins also were expressed at high levels. The number of identified genes related to some important physiological functions in the German cockroach is shown in Table 3.

## Functional classifications of annotated ESTs

GO analysis was conducted using BLAST2GO and WEGO programs. Of the 52,761 unigenes, 11,383 could be assigned into 48 functional groups (Figure 2) in three categories. The three categories were further divided into over 100 sub-categories (Figure S1). Among them, cell and cell part in the cellular component, hydrolase and nucleotide binding in the molecular function, cellular processes and metabolic processes in the biological processes represented the major sub-categories. The smallest groups were the metallochaperone in the molecular function category and the virion part in the cellular component. Some unigenes were assigned to multiple categories of GO terms, while others could not be assigned to a given GO term. The biological process terms were associated predominantly with cellular processes such as proteolysis, carbohydrate metabolic processes and oxidation reduction utilization. Similar composition and distribution of unigenes assigned by GO terms have been reported in transcriptomic description from other insects [15,20,40,41].

All unique transcripts were also searched against the KOG database for functional prediction and classification. With some of these unigenes no appropriate KOG annotation, 16,612 KOG annotations were produced and could be classified into 25 molecular families (Figure 3). Among the KOG classifications, the cluster of general function (20.4%) was the largest, followed by translation and signal transduction mechanisms (10.5%), and posttranslational modification, protein turnover, and chaperones (8.5%). The three smallest groups were nuclear structure (0.57%), defense mechanisms (0.53%) and cell motility (0.33%) respectively.

Among all the unigenes, 10,402 sequences with an enzyme classification (EC) number were mapped into 190 KEGG pathways in total (Table S2). Among the pathways identified, the metabolic pathways were found to be the most active, which is similar to the pattern observe *in N. lugens* [24].

## Identification of simple sequence repeats (SSRs) loci

SSRs, also known as microsatellites, are tandem repeated motifs of 1–6 bases and serve as the most important molecular markers in population and conservation genetics, molecular epidemiology and pathology, and gene mapping. By screening all the unigenes, 3601 SSRs loci with di-, tri-, tetra- and penta-nucleotide repeats were identified. Among them, trinucleotide repeats were the most abundant (60.7%), followed by dinucleotide repeats (24.0%), tetranucleotide (13.3%) and pentanucleotide repeats (2.0%). The most frequent motifs were (AAT)n (6.39%) and (TAT)n (5.69%). These SSRs could represent a valuable biomarker resource of *B. germanica*. However, all these putative SSR markers need to be validated to eliminate possible false positives and sequencing errors.

## Transcripts involved in systematic RNA interference pathway

The RNAi process has been fully investigated in *D. melanogaster* and *Caenorhabditis elegans*. Previous studies suggested that systematic RNAi is very efficient in cockroach functional gene study [42]. Diverse proteins such as SID-1 and scavenger receptors (SRs), have been reported to be involved in RNAi process. SID-1 is a protein that transports dsRNA into cells as observed in *C. elegans*; however, no SID-1 orthologue was found in the present transcriptome dataset or other *Dipteran* insects [43]. Over 90% of the dsRNA uptaken into *S2* cells in *Drosophila* is initiated by two SRs, including SR-CI and the scavenger receptor Eater [44]. Gene orthologous to SR-CI was absent herein, and unigene_c62 was identified as Eater, similar to what found in the fruit fly, *Bactrocera dorsalis* [45]. Whether unigene_c62 participates in the uptake of dsRNA in *B. germanica* remains to be investigated.

In general, the dsRNA or short hairpin RNA incorporated into cells is processed into small interfering RNA (siRNA) or microRNA (miRNA) by two distinct Dicer complexes (Dicer-1 and Dicer-2), respectively. Dicer-1 is ATP-independent and prefers to process the stem-loop precursor of miRNA, while Dicer-2 favors long dsRNA as its ideal substrate, and requires ATP hydrolysis for efficient siRNA production [46]. R2D2 can form the Dicer-2/R2D2 complex with Dicer-2, and bind to siRNA to enhance sequence specific messenger RNA degradation mediated by the RNA-initiated silencing complex (RISC). In *Drosophila*, R2D2 acts as a bridge between the initiation and effector steps of the RNAi pathway by facilitating siRNA passage from Dicer to RISC [47]. In the present study, we found four unigenes homologous to two Dicer genes of the German cockroach which were fully sequenced with known function [48], while R2D2 lost in the unigene dataset is probably due to the limitation of sequencing coverage.

The Argonaute 2 (AGO2) has been reported to be another main component of RISC complex involved in cleavage of siRNA-directed mRNA and degradation of the passenger strand in the siRNA duplex [49]. We found two Argonaute genes, AGO1 (unigene_c2208) and AGO2 (unigene_c24957), which show high homology to the counterparts of *T. castaneum* (XM_966202) and *Acyrthosiphon pisum* (XM_001944817). Further molecular cloning and functional analyses of these genes may shed light on their roles in the systemic RNAi pathway in the German cockroach.

## Transcripts encoding detoxification enzymes and insecticide targets

German cockroach has been documented to evolve resistance to insecticide rapidly [50]. The known mechanisms underlying insecticide resistance in the German cockroach include decreased penetration [50], increased detoxification [51], target insensitivity [52] and bait aversion [53]. To screen genes that may evolve insecticide resistance, we mined the current transcriptomic data in order to identify unigenes encoding insecticide targets or detoxification enzymes. As shown in Table 3, a number of sequences homologous to detoxification enzymes including carboxylesterase (CarE), glutathione *S*-transferase (GST), cytochrome P450 and insecticide targets were identified. The mean

**Table 4.** Genes related to innate immunity.

| Tentative annotation | Description | No. of unigenes |
|---|---|---|
| GNBP | Gram negative bacteria binding protein | 4 |
| PGRP | Peptidoglycan recognition protein | 15 |
| SR-b/c | Scavenger receptor which participates in the removal of many foreign substances such as gram (+/−) bacteria | 6 |
| Toll-like receptor | Key receptor in Toll pathway | 16 |
| Imd | Immune deficiency gene encoding death domain-containing protein | 3 |
| LPS-bp | Lipopolysaccharide-binding protein | 21 |
| Iap | Inhibitor of apoptosis | 4 |
| MAPK | Mitogen-activated protein kinase | 2 |
| Prophenoloxidase | Activate the phenoloxidase involved in melanization of pathogens and damaged tissues | 2 |
| Lysozyme | Break the bonds between polysaccharides in peptidoglycan on the bacterial cell walls | 8 |
| Galectin | Function in insect innate immune as a pattern-recognition protein | 3 |
| Cathepsin | Participate in molting, tissue remodeling, embryogenesis and immune evasion | 32 |
| Transferrin | An iron transporter function as antibiotic agent, vitellogenin, and juvenile hormone regulated protein | 20 |
| Defensin | Function as host defense peptide against bacteria, fungi and viruses. | 2 |
| Diptericin | A novel anti-bacterial peptide | 4 |
| Serpin | Regulate the innate immune pathway such as Toll by inhibition of serine proteases | 47 |

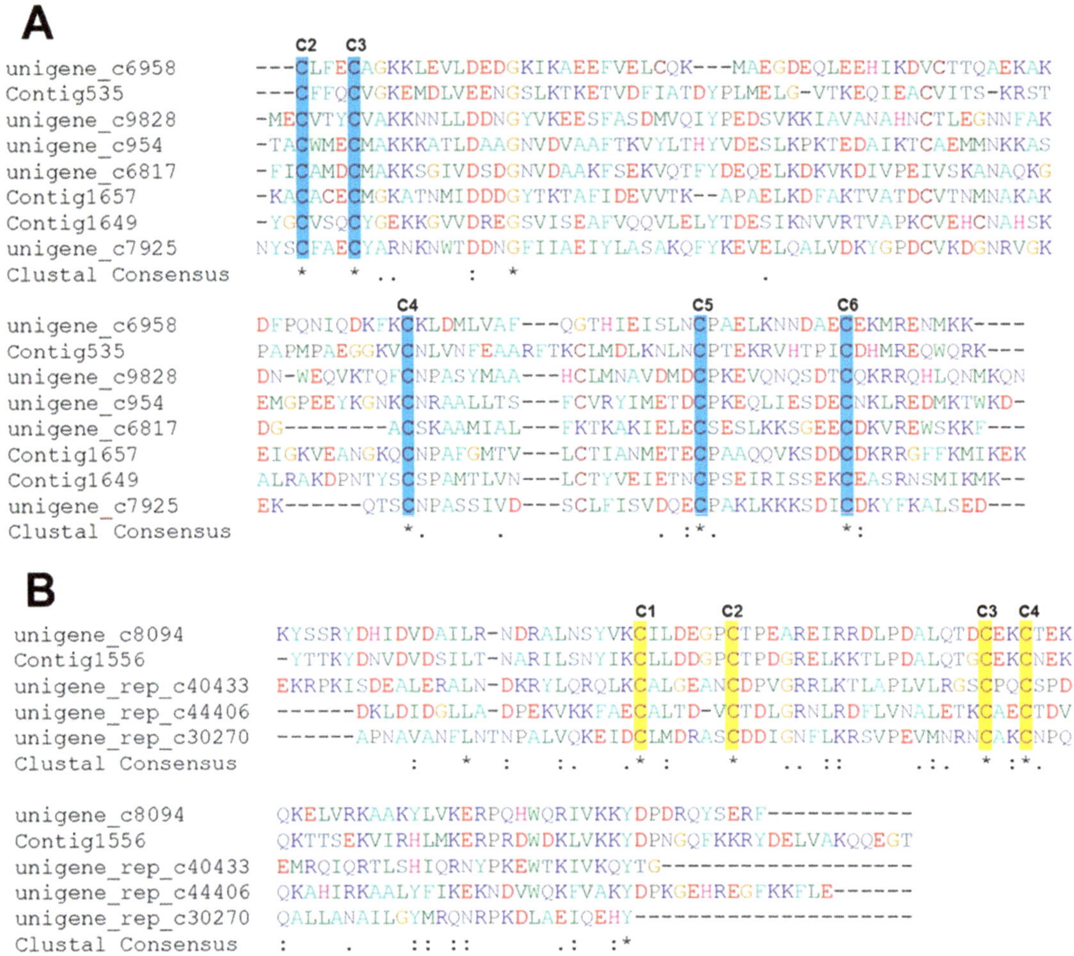


**Figure 4. Multiple sequences alignment of deduced peptide sequences of the putative German cockroach OBPs (A) and CSPs (B) with those from other insect species.** Nearly full-length amino acid sequences are aligned by ClustalX in BioEdit. Residues shaded in blue and yellow colors show the five and four conserved cysteine residues in the alignment of OBPs and CSPs, respectively. Asterisks (*) indicate identical sites in the sequence alignments; colons (:) represent sites with conserved substitutions, and black dots (•) stand for sites with weakly conserved sites.

lengths of these unigenes ranged from 387 bp to 985 bp leading to a relatively reliable annotation.

A total of 163 P450-like transcripts with a size of more than 600 bp were identified and curated manually. This gene number fell within the range of insects that their whole genomes were sequenced [9], covering the 13 P450 ESTs previously deposited in GenBank. Based on the closest BLAST hits in the NCBI nr database, these P450 unigenes were tentatively assigned to appropriate CYP families and clades, containing representatives of all the 4 main insect P450 clans (CYP2, CYP3,CYP4 and mitochondrial). CYP3 ranked as the largest clan, consisting of 45 members belonged to the CYP6 family and 28 genes to CYP9. Members from CYP3 clan appear to share the characteristics of environmental response genes, such as very high diversity and rapid rates of evolution [54]. The CYP4 clan included 39 P450s from the CYP4 family. The remainder belonged to the mitochondrial (15 ones) and CYP2 (22 ones) clan, which might be involved in the ecdysteroid metabolism pathway (CYP301 family) and essential physiological function (CYP303–305 and CYP15 families) respectively. The observation that the members of CYP4, CYP6 and CYP9 families together account for 74.2% of the total P450s indicates that the German cockroach, similar to other omnivorous insects, arms itself with potent capacity of metabolizing various xenobiotics.

A total of 64 GST-related EST sequences were identified in our database. However, only 12 contigs possessing a characteristic motif of CarE were found, which are relatively fewer than those found in other insects. In addition, a number of contigs encoding insecticide target proteins were also identified in the transcriptome, including acetylcholinesterase (AChE) (3 unigenes), nicotinic acetyl choline receptor subunits (nAChRs) (2), gamma-aminobutyric acid (GABA) receptor (2), glutamate receptor (12) and sodium channel (6).

## Transcripts involved in innate immunity

Insects have powerful innate immunity against many pathogens [29]. Cockroaches encounter various types of infectious agents, because they live under diverse unsanitary and unhygienic milieus. It is logical to believe that cockroaches have developed effective innate immunity of protecting themselves against pathogenic microorganisms. Production of diverse antimicrobial peptides (AMPs) plays critical roles in insect immunity [55]. The AMPs are regulated by a balance between two signaling pathways involving the Toll pathway that is activated mainly by fungi and Gram-positive bacteria, and the Immune deficiency (Imd) pathway that is

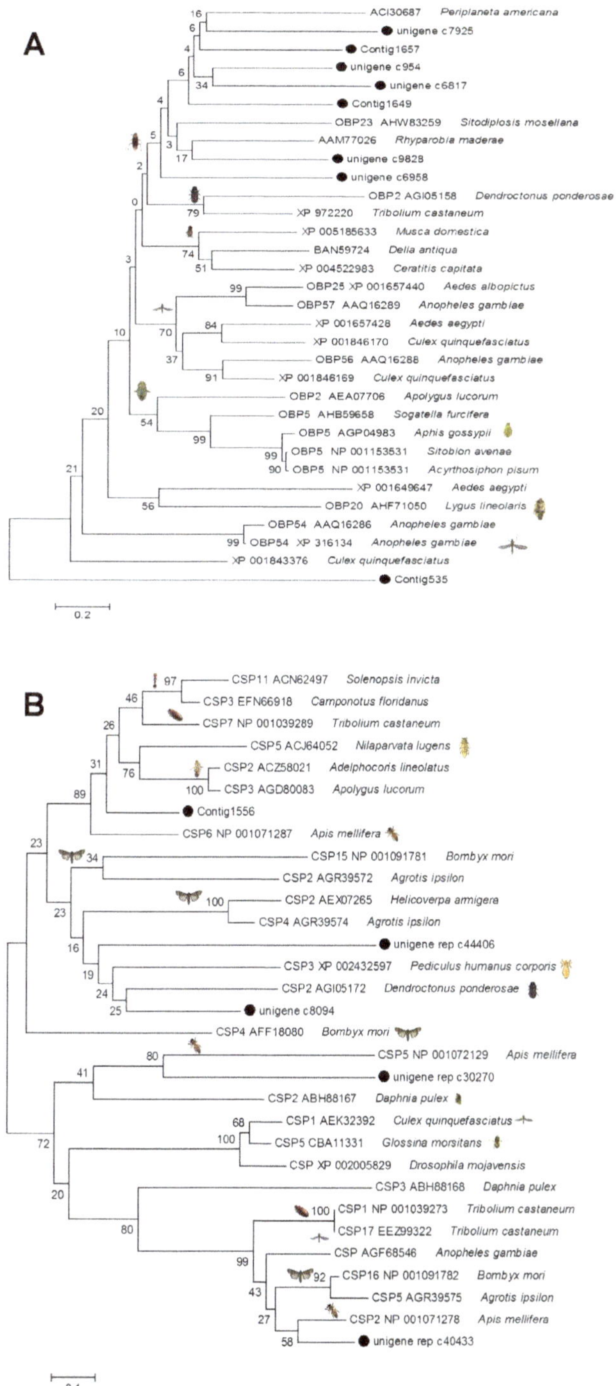


**Figure 5. Phylogeny of the OBPs (A) and CSPs (B) from the German cockroach and their homologs.** The unrooted consensus trees with 1000 bootstrap replicates are generated in MEGA6 [39] using the neighbor-joining method. The tree is drawn to scale, with branch lengths in the same units as those of the evolutionary distances used to infer the phylogenetic tree. All positions containing gaps and missing data are eliminated. GenBank accession numbers and species names of the sequences used here are shown in the phylogenetic trees. German cockroach OBPs and CSPs (marked by •) are in bolds.

activated mainly by Gram-negative bacteria. Toll and Imd cascades also control the majority of the genes regulated by microbial infection in addition to AMP genes, and are involved in nearly all known *Drosophila* innate immune reactions [56]. Previous studies suggest that the brain lysates, fat body or haemolymph of American cockroaches (*Periplaneta americana*) have remarkable antimicrobial activity against different bacteria including *Staphylococcus aureus*, MRSA (Gram-positive methicillin-resistant *S. aureus*), *S. epidermidis* and a neuropathogenic *Escherichia coli* K1 [55].

Dozens of genes identified in our transcriptome dataset may contribute to immune response in *B. germanica* (Table 4), including GNBP, PGRP (Peptidoglycan recognition proteins), SR-b/c, Toll-like receptors, Imd (Immune deficiency gene), Lipopolysaccharide-binding protein (LPS-bp), Iap (Inhibitor of apoptosis), MAPK (mitogen-activated protein kinase), prophenoloxidase, lysozymes, galectin, cathepsins, transferrins, defensins, diptericins and serpins. Most of them have a counterpart in other insects such as *Aedes aegypti* [18], *Musca domestica* [20] and *Nilaparvata lugens* [24]. These genes have been documented to play important roles as immune effect factors against possible pathogenic bacteria [55–57].

## Transcripts involved in olfactory and gustatory perception

We identified gene families that have been implicated in chemosensory reception (Table 3) including odorant binding proteins (OBPs), chemosensory proteins (CSPs), odorant receptors (ORs) and gustatory receptors (GRs). The detailed EST sequences are present in File S1.

OBPs have been proposed to capture and transport hydrophobic chemicals from air to ORs in the lymph of sensory hairs. Fourteen putative OBP transcripts were identified in this study. The number of OBPs identified here was far less than that in *D. melanogaster* (51) [58], *Bombyx mori* (44) [59], *Anopheles gambiae* (69) [60], *Aedes aegypti* (111) [60] and *Culex quinquefasciatus* (109) [60]. We assume that there are still other unidentified OBPs in the German cockroach, although relatively fewer genes encoding OBPs (19 and 5, respectively) were identified in the fire ant *Solenopsis invicta* [61] and the body louse *P. humanus* [62] genomes. An alignment of the deduced German cockroach OBPs (Figure 4A) revealed the most striking six conserved cysteine residues (the first cysteine C1 is absent because of the restriction of assembled unigenes lengths), which are present in the characteristic positions in all known insect OBPs.

CSPs exhibit an expression in the antenna similar to OBPs, and may perform tasks ranging from ontogeny to colony level regulation [63]. CSPs are characterized by only 4 conserved cysteines forming two non-interlocked disulfide bridges, whereas 6 conserved cysteines are paired in three interlocked disulphide bridges in OBPs [64]. Therefore, CSPs may represent a new class of soluble carrier proteins involved in insect chemoreception. The CSP gene family varies in sizes across arthropods. For example, the tick *Ixodes scapularis* has 1 CSP gene, *D. melanogaster* has 4, and larger known repertoires are found in *T. castaneum* (19 genes), *B. mori* (22 genes) and *S. invicta* (21 genes) [65]. No CSPs from the German cockroach have been previously reported. We identified a total of 12 CSP-coding gene fragments in the present study, with 8 CSPs being homologous to CSP7 and CSP18 from *T. castaneum*. The four CSPs with a long coding region identified in the study (Figure 4B) contained the typical four-cysteine signature and a common cysteine sequence motif of $C_1$-$X_{6\text{-}8}$-$C_2$-$X_{16\text{-}21}$-$C_3$-$X_2$-$C_4$ of insect CSPs [66].

In order to gain insight of the relationships between the annotated OBPs/CSPs and their counterparts from other insects, we carried out a phylogenetic analysis using putative amino acid sequences. The resultant tree constructed by the neighbor-joining

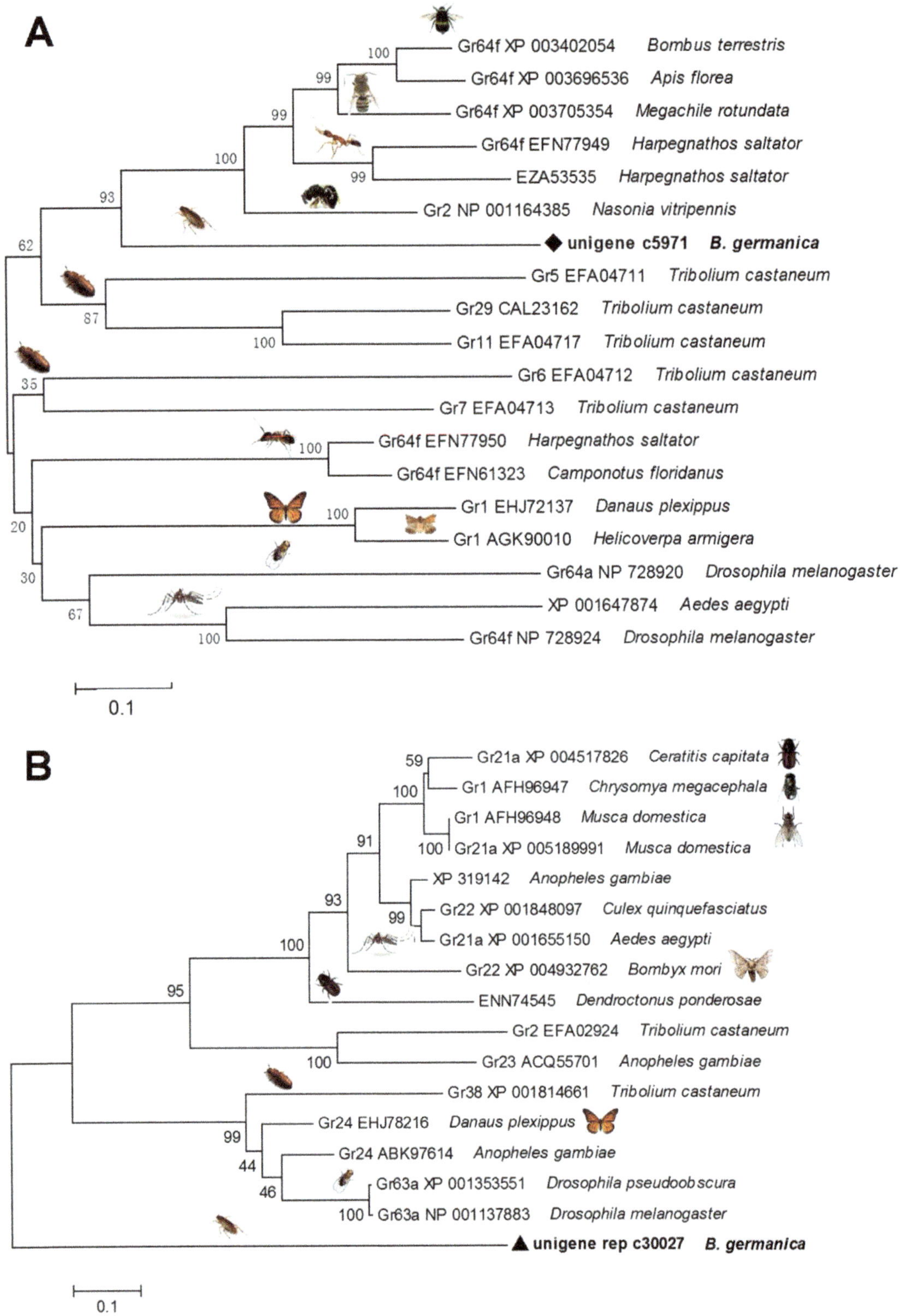


**Figure 6. Phylogeny of two gustatory receptor genes from the German cockroach.** Unigenes c5791 (A) and c rep c30027 (B) are highlighted in bolds and triangles (▲).

method suggests that the 7 identified German cockroach OBPs were clustered into a group comprising related proteins of the other two cockroach species (*P. Americana* and *Rhyparobia Maderae*) and *Sitodiplosis mosellana* (orange wheat blossom midge), whereas the Contig 535 was dispersed at the bottom of the tree (Figure 5A). Consistent with their diverse functions, OBPs have a high sequence divergence (20%–30% of amino acid identity, see OBPs alignment in supporting information) [60,67]. Insect OBPs are likely to be involved in broader physiological functions, not restricted to olfaction [68]. In contrast, CSPs from diverse insect species share high amino acid identities (about 50%–55%, Figure 5B), supporting the view that insect CSPs are highly conserved even across very distant species, and implying important roles they might play in insect physiology [69].

ORs are central to odorant detection in insects. Because of their high sequence variability, the most common method to identify OR-coding genes is to analyze genomic sequence databases [70]. Only 2 OR unigenes were annotated in this transcriptome, far fewer than those from other insects.

It has been reported that *Drosophila* contains a family of 60 gustatory receptor (GR) genes [71]. We identified 4 transcripts encoding homologous gustatory receptor in the cockroach. Two of them (unigene c5971 and unigene rep c30027) with a size of over 500 bp were exploited to analyze their phylogeny. As shown in Figure 6A, unigene c5971 showed homology to several other GRs from different species, such as Gr64f from *D. melanogaster*. Gr64f and other seven proteins (Gr5a, Gr61a and Gr64a–f) have been identified as sugar receptors (SRs) in *D. melanogaster* [72]. *Drosophila* sugar receptors function as multimers, and Gr64f is required broadly as a co-receptor for the detection of sugars [73]. We propose that c5971 gene identified in this study may also be involved in sugar perception of German cockroach. Unigene rep c30027 similarly showed close evolutionary distances with Gr63a and Gr21a (Figure 6B), which are co-expressed in $CO_2$-responsive neurons and play an important role in the fruit fly food-seeking [74], suggesting the involvement of unigene rep c30027 in food seeking in the cockroach. Conservation of these GR sequences between fairly diverged insect species likely reflects indispensable gustatory sensitivities to a particular chemical or set of chemicals, a property that allows us to speculate their potential function.

## Conclusions

This study provides a new genetic data resource helpful for further comprehensive studies on the German cockroach. The information presented here will be useful to improve our understanding about the molecular mechanisms of cockroach immunity, insecticide resistance, chemoreception and gene regulation.

## Supporting Information

**Figure S1 Three GO sub-categories for German cockroach transcriptome.** (A) Biological process, (B) Cellular component and (C) Molecular function.

**Table S1 Bioassay of German cockroach susceptible (BJ-S) and resistant (DX-R) strains to different insecticide ($KT_{50}$, min) using glass test method (WHO, 1963).**

**Table S2 Summary of KEGG pathway types in the transcriptome of German cockroach.**

**File S1 Sequences of CSP, GR, OBP, OR and top 20 abundant unigenes mentioned in the study.**

## Author Contributions

Conceived and designed the experiments: XJZ KQ YT XPZ. Performed the experiments: XJZ KQ. Analyzed the data: XJZ XQ. Contributed reagents/materials/analysis tools: XQ. Wrote the paper: XJZ XQ JJZ XPZ.

## References


1. Appel AG (1995) *Blattella* and related species. In: Rust MK, Owens JM, Reierson DA, editors. Understanding and Controlling the German Cockroach. New York: Oxford Press. pp. 1–19.
2. Ross MH, Cochran DG (1975) The German cockroach, *Blattella germanica*. In: King RC, editor. Handbook of Genetics. New York: Plenum Press. pp. 35–62.
3. Brenner RJ (1995) Economics and medical importance of German cockroaches. In: Rust MK, Owens JM, Reierson DA, editors. Understanding and Controlling the German Cockroach, New York: Oxford Press. pp. 77–92.
4. Ramirez PJ (1989) The cockroach as a vector of pathogenic agents. Bol Oficina Sanit Panam 107: 41–53.
5. Rosenstreich DL, Eggleston P, Kattan M, Baker D, Slavin RG, et al. (1997) The role of cockroach allergy and exposure to cockroach allergen in causing morbidity among inner-city children with asthma. N Engl J Med 336: 1356–1363.
6. Gore JC, Schal C (2007) Cockroach Allergen Biology and Mitigation in the Indoor Environment. Annu Rev Entomol 52: 439–463.
7. Irles P, Bellés X, Piulachs MD (2009) Identifying genes related to choriogenesis in insect panoistic ovaries by Suppression Subtractive Hybridization. BMC Genomics 10: 206.
8. Chung HS, Yu TH, Kim BJ, Kim SM, Kim JY, et al. (2005) Expressed sequence tags analysis of *Blattella germanica*. Korean J Parasitol 43: 149–156.
9. Feyereisen R (2012) Insect CYP Genes and P450 Enzymes. In: Gilbert LI editor. Insect Molecular Biology and Biochemistry. London: Academic Press. pp. 236–345.
10. Shendure J, Ji H (2008) Next-generation DNA sequencing. Nat Biotechnol 26: 1135–1145.
11. Schuster SC (2008) Next-generation sequencing transforms today's biology. Nat Methods 5: 16–18.
12. Margulies M, Egholm M, Altman WE, Attiya S, Bader JS, et al. (2005) Genome sequencing in microfabricated high-density picolitre reactors. Nature 437: 376–380.
13. Vera JC, Wheat CW, Fescemyer HW, Frilander MJ, Crawford DL, et al. (2008) Rapid transcriptome characterization for a nonmodel organism using 454 pyrosequencing. Mol Ecol 17: 1636–1647.
14. Droege M, Hill B (2008) The Genome Sequencer FLX System-longer reads, more applications, straight forward bioinformatics and more complete data sets. J Biotechnol 136: 3–10.
15. Bai X, Mamidala P, Rajarapu SP, Jones SC, Mittapalli O (2011) Transcriptomics of the bed bug (*Cimex lectularius*). PLoS One 6: e16336.
16. Adelman ZN, Kilcullen KA, Koganemaru R, Anderson MA, Anderson TD, et al. (2011) Deep sequencing of pyrethroid-resistant bed bugs reveals multiple mechanisms of resistance within a single population. PLoS One 6: e26228.
17. Gregory R, Darby AC, Irving H, Coulibaly MB, Hughes M, et al. (2011) A *de novo* expression profiling of *Anopheles funestus*, malaria vector in Africa, using 454 pyrosequencing. PLoS One 6: e17418.
18. Price DP, Nagarajan V, Churbanov A, Houde P, Milligan B, et al. (2011) The fat body transcriptomes of the yellow fever mosquito *Aedes aegypti*, pre- and post- blood meal. PLoS One 6: e22573.
19. Jaworski DC, Zou Z, Bowen CJ, Wasala NB, Madden R, et al. (2010) Pyrosequencing and characterization of immune response genes from the American dog tick, *Dermacentor variabilis* (L.). Insect Mol Biol 19: 617–630.
20. Liu F, Tang T, Sun L, Jose Priya TA (2012) Transcriptomic analysis of the housefly (*Musca domestica*) larva using massively parallel pyrosequencing. Mol Biol Rep 39: 1927–1934.
21. Hahn DA, Ragland GJ, Shoemaker DD, Denlinger DL (2009) Gene discovery using massively parallel pyrosequencing to develop ESTs for the flesh fly *Sarcophaga crassipalpis*. BMC Genomics 10: 234.
22. Carvalho RA, Azeredo-Espin AM, Torres TT (2010) Deep sequencing of New World screw-worm transcripts to discover genes involved in insecticide resistance. BMC Genomics 11: 695.
23. Nirmala X, Schetelig MF, Yu F, Handler AM (2013) An EST database of the Caribbean fruit fly, *Anastrepha suspensa* (Diptera: Tephritidae). Gene 517: 212–217.
24. Peng X, Zha W, He R, Lu T, Zhu L, et al. (2011) Pyrosequencing the midgut transcriptome of the brown planthopper, *Nilaparvata lugens*. Insect Mol Biol 20: 745–762.
25. Pauchet Y, Wilkinson P, Vogel H, Nelson DR, Reynolds SE, et al. (2010) Pyrosequencing the *Manduca sexta* larval midgut transcriptome: messages for digestion, detoxification and defence. Insect Mol Biol 19: 61–75.

26. de Moura TR, Oliveira F, Carneiro MW, Miranda JC, Clarêncio J, et al. (2013) Functional transcriptomics of wild-caught *Lutzomyia intermedia* salivary glands: identification of a protective salivary protein against *Leishmania braziliensis* infection. PLoS Negl Trop Dis 7: e2242.
27. Karim S, Singh P, Ribeiro JM (2011) A deep insight into the sialotranscriptome of the gulf coast tick, *Amblyomma maculatum*. PLoS One 6: e28525.
28. Ribeiro JM, Chagas AC, Pham VM, Lounibos LP, Calvo E (2014) An insight into the sialome of the frog biting fly, *Corethrella appendiculata*. Insect Biochem Mol Biol 44: 23–32.
29. Ribeiro JM, Genta FA, Sorgine MH, Logullo R, Mesquita RD, et al. (2014) An insight into the transcriptome of the digestive tract of the bloodsucking bug, *Rhodnius prolixus*. PLoS Negl Trop Dis 8: e2594.
30. Cristino AS, Tanaka ED, Rubio M, Piulachs MD, Belles X (2011) Deep sequencing of organ- and stage-specific microRNAs in the evolutionarily basal insect *Blattella germanica* (*L.*) (Dictyoptera, Blattellidae). PLoS One 6: e19350.
31. Lassmann T, Hayashizaki Y, Daub CO (2009) TagDust: a program to eliminate artifacts from next generation sequencing data. Bioinformatics 25: 2839–2840.
32. Chen YA, Lin CC, Wang CD, Wu HB, Hwang PI (2007) An optimized procedure greatly improves EST vector contamination removal. BMC Genomics 8: 416.
33. Boeckmann B, Bairoch A, Apweiler R, Blatter MC, Estreicher A, et al. (2003) The SWISS-PROT protein knowledgebase and its supplement TrEMBL in 2003. Nucleic Acids Res 31: 365–370.
34. Tatusov RL, Fedorova ND, Jackson JD, Jacobs AR, Kiryutin B, et al. (2003) The COG database: an updated version includes eukaryotes. BMC Bioinformatics 4: 41.
35. Kanehisa M, Araki M, Goto S, Hattori M, Hirakawa M, et al. (2008) KEGG for linking genomes to life and the environment. Nucleic Acids Res 36: D480–4.
36. Conesa A, Götz S (2008) Blast2GO: a comprehensive suite for functional analysis in plant genomics. Int J Plant Genomics 2008: 619832.
37. Gotz S, Garcia-Gomez JM, Terol J, Williams TD, Nagaraj SH, et al. (2008) High-throughput functional annotation and data mining with the Blast2GO suite. Nucleic Acids Res 36: 3420–3435.
38. Ye J, Fang L, Zheng H, Zhang Y, Chen J, et al. (2006) WEGO: a web tool for plotting GO annotations. Nucleic Acids Res 34: W293–7.
39. Tamura K, Stecher G, Peterson D, Filipski A, Kumar S (2013) MEGA6: Molecular Evolutionary Genetics Analysis version 6.0. Mol Biol Evol 30: 2725–2729.
40. Wang XW, Luan JB, Li JM, Bao YY, Zhang CX, et al. (2010) *De novo* characterization of a whitefly transcriptome and analysis of its gene expression during development. BMC Genomics 11: 400.
41. Niu JZ, Dou W, Ding TB, Shen GM, Zhang K, et al. (2012) Transcriptome analysis of the citrus red mite, *Panonychus citri*, and its gene expression by exposure to insecticide/acaricide. Insect Mol Biol 21: 422–436.
42. Garbutt JS, Belles X, Richards EH, Reynolds SE (2013) Persistence of double-stranded RNA in insect hemolymph as a potential determiner of RNA interference success: evidence from *Manduca sexta* and *Blattella germanica*. J Insect Physiol 59: 171–178.
43. Huvenne H, Smagghe G (2010) Mechanisms of dsRNA uptake in insects and potential of RNAi for pest control: a review. J Insect Physiol 56: 227–235.
44. Ulvila J, Parikka M, Kleino A, Sormunen R, Ezekowitz RA, et al. (2006) Double-stranded RNA is internalized by scavenger receptor-mediated endocytosis in *Drosophila* S2 cells. J Biol Chem 281: 14370–14375.
45. Shen GM, Dou W, Huang Y, Jiang XZ, Smagghe G, et al. (2013) *In silico* cloning and annotation of genes involved in the digestion, detoxification and RNA interference mechanism in the midgut of *Bactrocera dorsalis* [Hendel (Diptera: Tephritidae)]. Insect Mol Biol 22: 354–365.
46. Jiang F, Ye X, Liu X, Fincher L, McKearin D, et al. (2005) Dicer-1 and R3D1-L catalyze microRNA maturation in *Drosophila*. Genes Dev 19: 1674–1679.
47. Liu Q, Rand TA, Kalidas S, Du F, Kim HE, et al. (2003) R2D2, a bridge between the initiation and effector steps of the *Drosophila* RNAi pathway. Science 301: 1921–1925.
48. Gomez-Orte E, Belles X (2009) MicroRNA-dependent metamorphosis in hemimetabolan insects. Proc Natl Acad Sci USA 106: 21678–21682.
49. Matranga C, Tomari Y, Shin C, Bartel DP, Zamore PD (2005) Passenger-strand cleavage facilitates assembly of siRNA into Ago2-containing RNAi enzyme complexes. Cell 123: 607–620.
50. Wei Y, Appel AG, Moar WJ, Liu N (2001) Pyrethroid resistance and cross-resistance in the German cockroach, *Blattella germanica* (L). Pest Manag Sci 57: 1055–1059.
51. Valles SM (1998) Toxicological and biochemical studies with field populations of the German cockroach, *Blattella germanica*. Pestic Biochem Physiol 62: 190–200.
52. Liu Z, Valles SM, Dong K (2000) Novel point mutations in the German cockroach para sodium channel gene are associated with knockdown resistance (kdr) to pyrethroid insecticides. Insect Biochem Mol Biol 30: 991–997.
53. Wada-Katsumata A, Silverman J, Schal C (2013) Changes in taste neurons support the emergence of an adaptive behavior in cockroaches. Science 340: 972–975.
54. Oakeshott JG, Johnson RM, Berenbaum MR, Ranson H, Cristino AS, et al. (2010) Metabolic enzymes associated with xenobiotic and chemosensory responses in *Nasonia vitripennis*. Insect Mol Biol 19 Suppl 1: 147–163.
55. Lee S, Siddiqui R, Khan NA (2012) Animals living in polluted environments are potential source of antimicrobials against infectious agents. Pathog Glob Health 106: 218–223.
56. Hoffmann JA, Reichhart JM (2002) *Drosophila* innate immunity: an evolutionary perspective. Nature Immunol 3: 121–126.
57. Kim T, Kim YJ (2005) Overview of innate immunity in *Drosophila*. J Biochem Mol Biol 38: 121–127.
58. Hekmat-Scafe DS, Scafe CR, McKinney AJ, Tanouye MA (2002) Genome-wide analysis of the odorant-binding protein gene family in *Drosophila melanogaster*. Genome Res 12: 1357–1369.
59. Gong DP, Zhang HJ, Zhao P, Xia QY, Xiang ZH (2009) The odorant binding protein gene family from the genome of silkworm, *Bombyx mori*. BMC Genomics 10: 332.
60. Manoharan M, Ng Fuk Chong M, Vaïtinadapoulé A, Frumence E, Sowdhamini R, et al. (2013) Comparative genomics of odorant binding proteins in *Anopheles gambiae*, *Aedes aegypti*, and *Culex quinquefasciatus*. Genome Biol Evol 5: 163–180.
61. Gotzek D, Robertson HM, Wurm Y, Shoemaker D (2011) Odorant binding proteins of the red imported fire ant, *Solenopsis invicta*: an example of the problems facing the analysis of widely divergent proteins. PLoS One 6: e16289.
62. Kirkness EF, Haas BJ, Sun W, Braig HR, Perotti MA, et al. (2010) Genome sequences of the human body louse and its primary endosymbiont provide insights into the permanent parasitic lifestyle. Proc Natl Acad Sci USA 107: 12168–12173.
63. Kulmuni J, Havukainen H (2013) Insights into the evolution of the CSP gene family through the integration of evolutionary analysis and comparative protein modeling. PLoS One 8: e63688.
64. Wanner KW, Willis LG, Theilmann DA, Isman MB, Feng Q, et al. (2004) Analysis of the insect os-d-like gene family. J Chem Ecol 30: 889–911.
65. Vieira FG, Rozas J (2011) Comparative Genomics of the Odorant-Binding and Chemosensory Protein Gene Families across the Arthropoda: Origin and Evolutionary History of the Chemosensory System. Genome Biology and Evolution 3: 476–490.
66. Zhou JJ, Kan Y, Antoniw J, Pickett JA, Field LM (2006) Genome and EST analyses and expression of a gene family with putative functions in insect chemoreception. Chem Senses 31: 453–465.
67. Gu SH, Wang SY, Zhang XY, Ji P, Liu JT, et al. (2012) Functional characterizations of chemosensory proteins of the alfalfa plant bug *Adelphocoris lineolatus* indicate their involvement in host recognition. PLoS One 7: e42871.
68. Hekmat-Scafe DS, Scafe CR, McKinney AJ, Tanouye MA (2002) Genome-wide analysis of the odorant-binding protein gene family in *Drosophila melanogaster*. Genome Res 12: 1357–1369.
69. Pelosi P, Zhou JJ, Ban LP, Calvello M (2006) Soluble proteins in insect chemical communication. Cell Mol Life Sci 63: 1658–1676.
70. Grosse-Wilde E, Kuebler LS, Bucks S, Vogel H, Wicher D, et al. (2011) Antennal transcriptome of *Manduca sexta*. Proc Natl Acad Sci USA 108: 7449–7454.
71. Robertson HM, Warr CG, Carlson JR (2003) Molecular evolution of the insect chemoreceptor gene superfamily in *Drosophila melanogaster*. Proc Natl Acad Sci USA 100: 14537–14542.
72. Kent LB, Robertson HM (2009) Evolution of the sugar receptors in insects. BMC Evol Biol 9: 41.
73. Jiao Y, Moon SJ, Wang X, Ren Q, Montell C (2008) Gr64f is required in combination with other gustatory receptors for sugar detection in *Drosophila*. Curr Biol 18: 1797–7801.
74. Turner SL, Ray A (2009) Modification of $CO_2$ avoidance behaviour in *Drosophila* by inhibitory odorants. Nature 461: 277–281.

# 22

# Market Forces and Technological Substitutes Cause Fluctuations in the Value of Bat Pest-Control Services for Cotton

**Laura López-Hoffman[1,2]*, Ruscena Wiederholt[1,2], Chris Sansone[3], Kenneth J. Bagstad[4], Paul Cryan[5], Jay E. Diffendorfer[4], Joshua Goldstein[6], Kelsie LaSharr[1], John Loomis[7], Gary McCracken[8], Rodrigo A. Medellín[9], Amy Russell[10], Darius Semmens[4]**

**1** School of Natural Resources & the Environment, The University of Arizona, Tucson, Arizona, United States of America, **2** Udall Center for Studies in Public Policy, The University of Arizona, Tucson, Arizona, United States of America, **3** Bayer CropScience, Research Triangle Park, North Carolina, United States of America, **4** United States Geological Survey, Geosciences and Environmental Change Science Center, Denver, Colorado, United States of America, **5** United States Geological Survey, Fort Collins Science Center, Fort Collins, Colorado, United States of America, **6** Department of Human Dimensions of Natural Resources, Colorado State University, Fort Collins, Colorado, United States of America, **7** Department of Agricultural and Resource Economics, Colorado State University, Fort Collins, Colorado, United States of America, **8** Department of Ecology and Evolutionary Biology, University of Tennessee, Knoxville, Tennessee, United States of America, **9** Instituto de Ecología, Universidad Nacional Autónoma de México, Distrito Federal, México, **10** Department of Biology, Grand Valley State University, Allendale, Michigan, United States of America

## Abstract

Critics of the market-based, ecosystem services approach to biodiversity conservation worry that volatile market conditions and technological substitutes will diminish the value of ecosystem services and obviate the "economic benefits" arguments for conservation. To explore the effects of market forces and substitutes on service values, we assessed how the value of the pest-control services provided by Mexican free-tailed bats (*Tadarida brasiliensis mexicana*) to cotton production in the southwestern U.S. has changed over time. We calculated service values each year from 1990 through 2008 by estimating the value of avoided crop damage and the reduced social and private costs of insecticide use in the presence of bats. Over this period, the ecosystem service value declined by 79% ($19.09 million U.S. dollars) due to the introduction and widespread adoption of Bt (*Bacillus thuringiensis*) cotton transgenically modified to express its own pesticide, falling global cotton prices and the reduction in the number of hectares in the U.S. planted with cotton. Our results demonstrate that fluctuations in market conditions can cause temporal variation in ecosystem service values even when ecosystem function – in this case bat population numbers – is held constant. Evidence is accumulating, however, of the evolution of pest resistance to Bt cotton, suggesting that the value of bat pest-control services may increase again. This gives rise to an economic option value argument for conserving Mexican free-tailed bat populations. We anticipate that these results will spur discussion about the role of ecosystem services in biodiversity conservation in general, and bat conservation in particular.

**Editor:** Robert B. Srygley, USDA-Agricultural Research Service, United States of America

**Funding:** This work was funded by a National Science Foundation award (DEB-1118975) to LLH. This work was supported by the U.S. Geological Survey's John Wesley Powell Center for Analysis and Synthesis working group, "Animal Migration and Spatial Subsidies: Establishing a Framework for Conservation Markets." The funders had no role in study design, data collection and analysis, decision to publish, or preparation of the manuscript.

* E-mail: lauralh@email.arizona.edu

## Introduction

The underlying goal of market-based, Payments for Ecosystem Services (PES) approaches to conservation is the creation of monetary incentives for the protection of critical ecological processes such as watershed functioning, pollination and natural pest control [1,2]. Within the conservation community, criticisms about market-based programs range from the ideological – e.g. unease that the approach diminishes nature's intrinsic value [3] – to apprehensions about the nature of the market [3,4]. The latter criticism stems from the worry there will be no reason to protect ecosystems when their services are no longer perceived to be valuable [3,4]. Two issues in particular – volatile market conditions and technological substitutes – are the main source of concern about the compatibility of market-based approaches and biodiversity conservation [4–6].

The first of these concerns is based on the economic principle that as the supply and demand curves for a market good change, the price of that good also changes. It follows that as the price of a market good fluctuates, the value of its ecosystem service inputs also will vary since ecosystem service values are derived from the demands of users of the services, in this case, cotton producers [7]. McCauley [6] illustrated this concern with the anecdote of a Costa Rican coffee plantation that was converted to pineapple production following world-wide declines in coffee prices. The monetary value to coffee production of the pollinators in the surrounding forest fragments previously had been estimated to be $60,000 USD per year [8]. Because pineapples are propagated and not pollinated, as the need for pollination services disappeared, McCauley [6] worried that the rationale for protecting the forest fragments might have disappeared as well.

The second concern about market-based approaches to conservation arises when manufactured capital is substituted for natural capital [9]. The story of the Zapp potato-chip factory in Louisiana (U.S.A.) illustrates this point. The factory once used a nearby wetland to filter its waste. But, as the potato chip business boomed, and as the volume of waste increased, the cost of using the wetland also increased and the company switched to technological forms of wastewater treatment [9].

In our study – using as an example the pest-control services provided by Mexican free-tailed bats (*Tadarida brasiliensis mexicana*) to cotton production in the U.S. – we demonstrate how bat pest-control values have changed in response to both changing market conditions for cotton and the adoption of a technological substitute for the service. To our knowledge, we present one of the first empirical, time-series analyses of the effects of changing market conditions and of the adoption of technological substitutes on the value of an ecosystem service.

As two-thirds of the more than 1,200 extant bat species are insectivorous [10], bats can provide significant complementary pest-control services, particularly by preying on pests early in the growing season before insecticide use has begun and preventing pest outbreaks [11]. Two studies have estimated the monetary value of the pest-control services of Mexican free-tailed bats in reducing crop damage and lowering the costs of insecticide use in cotton [12,13]. Using cotton price and acreage data from the mid-2000s, Cleveland et al. [12] estimated an annual pest-control value in an eight-county region of south-central Texas of $121,000 to $1,725,000 USD. Researchers have subsequently applied the estimates from the above studies to different regions in North America [14,15]. However, no previous study has considered how the value of bat ecosystem services has changed over time in response to shifts in commodity markets or advances in agricultural technology.

Over the last two decades, cotton prices and hectares planted in cotton in the U.S. have declined. The declines are generally attributed to global market forces – including trade barriers falling in the 1990s and increased production of cotton in developing countries [16]. In addition, in 1996, U.S. cotton growers started using Bt (*Bacillus thuringiensis*) cotton, which is transgenically modified to produce proteins that are toxic to susceptible insects [17]. As of 2012, 77% of all cotton grown in the U.S. was Bt-modified (www.ers.usda.gov/Data/BiotechCrops/). Bats have a lower pest-control impact on Bt than on conventional cotton [13]; however, over the past two years, mounting evidence from around the world suggests that insect pests are evolving resistance to Bt-modified crops [18]. While not yet widespread, Bt-resistant pests have been found in the field in India, China, and the U.S., and in laboratory studies [19–23].

Here we investigate the impacts of major global market factors – changes in cotton commodity price, the consequent change in the number of hectares planted with cotton, and the adoption of Bt cotton – on the value of the pest-control services provided by Mexican free-tailed bats in the U.S. We calculate service values each year from 1990 through 2008 by estimating both the value of avoided crop damage and the reduced social and private costs of insecticide use in the presence of bats. We assess pest-control services across all U.S. cotton-producing areas containing major Mexican free-tailed bat roosts. Due to the lack of time-series data on the size of bat populations, we could not consider the impact of fluctuations in population size on the value of bat pest control services; as an alternative we analyzed the sensitivity of pest-control values to changes in bat population size. Currently, there is no market-based approach linking bat pest control services to cotton production, just estimates of value that might allow for the development of such mechanisms. As such, we conclude by addressing implications of our valuation research for the development of incentive-based approaches to the conservation of Mexican free-tailed bats in the context of the anticipated increase in Bt-resistant pests and the future value of bat ecosystem services.

## Results

From 1990 through 2008, the ecosystem service value of cotton pest-control services provided by Mexican free-tailed bats across the southwestern U.S. declined by 79%, from a high of $23.96 million in 1990 to a low of $4.88 million in 2008 (Fig. 1; mean values; values in 2011 USD). The value notably spiked in 1995 (Fig. 1) due to high prices for Pima and Upland cotton and the high number of hectares planted with cotton in that year (only 1991 had more cotton hectares). The mean annual value of pest control by bats over this time period was $12.24 million (s.d. = $6.04 million). The value of pest-control services expressed as a proportion of the total cotton crop value also varied over time, from 28% in 1991 to 6.5% in 2007. The presence of bats in cotton fields precluded, on average, the use of 32,046 kg of insecticide per year and damage to 131,385 kg of cotton per year.

Since Bt cotton produces its own insecticide, bats have less of an impact in controlling pests in Bt cotton than they do in conventional cotton [13]. To illustrate the decreased pest-control value of the bats resulting from the adoption of Bt cotton, we used the mean pest-control values (Fig. 1) to calculate what the pest-control values would have been had Bt not been adopted in 1996 (Fig. 2). To do this, we assumed that conventional cotton (i.e. non-Bt cotton) was planted from 1996 to 2008 (Fig. 2). In 2008, for example, the value of bat pest-control services was $2.66 million dollars (approximately 33%) less than what it might have been had all fields been planted with conventional cotton (Fig. 2). Our calculations also indicate that on a per-hectare basis the number of cotton bolls saved by the use of Bt technology was equivalent on average to the foraging efforts of about 27.5 bats.

To better understand temporal variability in ecosystem service values, we assessed the sensitivity of the annualized pest-control value to three factors: total area planted with cotton, the price of cotton, and the Mexican free-tailed bat population size. Notably, the mean pest-control values were equally sensitive to ecological and economic factors: both a ±10% change in the bat population size and a ±10% change in cotton prices caused a ±9.1% change in the mean pest-control values. However, altering the total area planted with cotton by ±10% only caused a ±0.9% change in the mean pest-control value over time.

## Discussion

The results of this study document that volatile market conditions and technological substitutes can affect the value of an ecosystem service [6]. The value of the pest-control services provided by Mexican free-tailed bats in the U.S. declined by 79% from 1990 through 2008 in response to declining global cotton prices, the consequent reduction in the number of hectares planted with cotton, and the introduction and widespread adoption of Bt cotton, a technological substitute to the natural pest-control services of the bats. Our analysis further indicates that these types of changes in market conditions may have as much impact on the

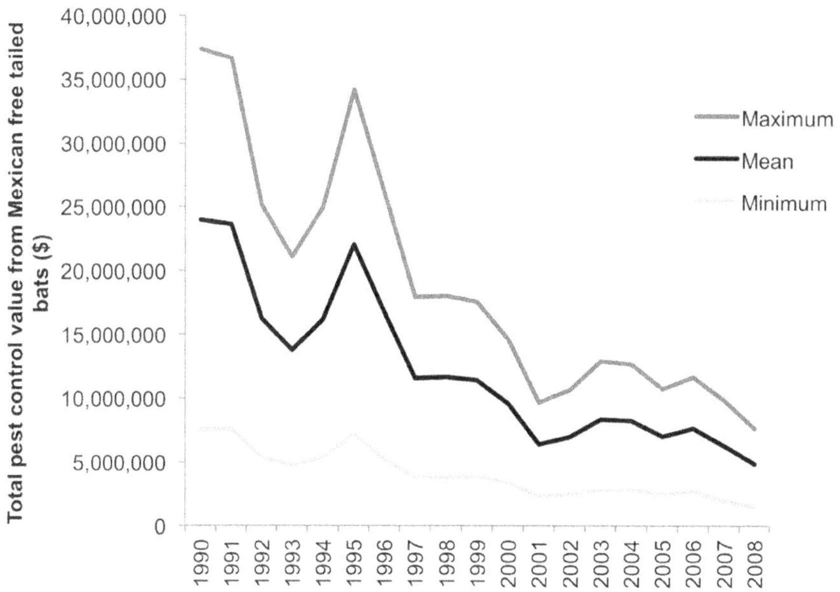


**Figure 1. Cotton pest-control value provided by Mexican free-tailed bats over time.** Maximum and minimum ecosystem service values for pest control represent calculations using the highest and lowest values, respectively, for several model parameters. From 1990 through 2008, the value of cotton pest-control services across the southwestern U.S. declined by 79%, from a high of $23.96 million in 1990 to a low of $4.88 million in 2008 (mean values). Values are indexed to 2011 U.S. dollars.

value of ecosystem services as changes in ecosystem function such as changes in bat population numbers.

Critics of the ecosystem service approach to biodiversity conservation contend that the risk of diminished service values – like the trends we document here – are a fundamental weakness to "economic benefits" arguments for conservation [3,4]. The fact that falling cotton prices and the adoption of Bt cotton caused the value of bat cotton pest-control services to fall appears to confirm this concern. However, as the pest-control service of bats depreciates with investments in technology, so might the depreciation of manufactured capital. A recent meta-analysis of studies from five continents indicates that five of thirteen major insect crop pests have evolved Bt resistance in the last eight years, including cotton bollworms [18]. Resistance to Bt can arise in as

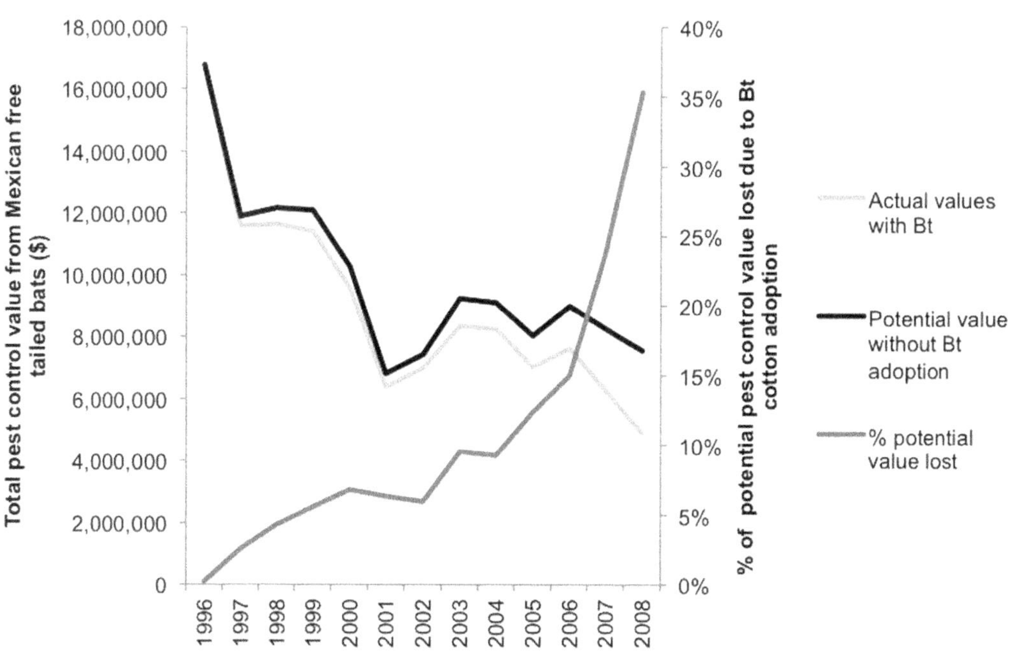


**Figure 2. Decreased pest-control value of the bats resulting from adoption of Bt cotton.** The light gray line shows actual pest control values from 1996 through 2008. The black line shows the potential value of bat pest control services if Bt had not been adopted. The dark gray line shows the percentage of the potential pest control value lost due to the adoption of Bt cotton. In 2008 the value of bat pest control services was $2.66 million dollars (approximately 33%) less than what it might have been.

few as two years depending on local conditions and how carefully growers hew to guidelines for maintaining refuges of non-Bt host plants [18,24]. Further, the efficacy of the second-generation Bt cotton (called pyramids because they produce more than one type of toxin), can be compromised if local pests are resistant to just one of the toxins [18]. This evidence of resistance evolution suggests that Bt may not be a permanent or even long-term solution to pest-related losses in the production of cotton and other crops, and that bats may again play a critical role in pest control. In fact, by preying on the individual insects that survive the Bt toxin – and preventing them from multiplying – bats may provide the additional service of slowing the evolution of resistance to Bt and other insecticides [13]. Indeed, bats and other natural enemies can play an important role in integrated pest management [25]. For example, Hagerty and colleagues [25] showed that Bollgard II cotton, a two-gene Bt product, experienced increased crop damage when natural predators were disrupted. Many agronomic researchers recommend that Bt crops be used in conjunction with other tactics, including natural predators, to avoid pest outbreaks and to delay the evolution of pest resistance [18,25].

### The option value of bats: an argument for conservation?

Option values in environmental economics can take one of two forms. The first is a user's willingness to pay for protecting a resource that is not currently in use so that it might be available for future use [26]. The exact value is influenced by the likelihood of needing the resource and the cost of replacing it should it be lost [27,28]. Alternatively, an option value may be the premium a decision-maker or society is willing to incur to avoid an irreversible loss of a resource by preserving it for future use [29]. In this case, it is possible that the pest-control services of Mexican free-tailed bats will become more valuable again, given that technological replacements for the service run the risk of being temporary. While more work is needed to assess the option value of protecting bats, the notion might provide an intriguing rationale to add to the existing list of reasons for enhancing bat conservation. Conservationists could argue that an investment in protecting bat populations now is an investment in protecting a future stream of potential pest-control services. They might also argue that bats provide pest control free of charge whereas pesticides and Bt cotton are costly to purchase, and their use carries a number of private and social costs. A final point might be that because bats are generalist predators, they provide a broad-spectrum pest control.

This line of reasoning suggests that it might be economically rational for decision-makers to promote bat conservation. However, there is often a disconnection between what is economically profitable and ecologically rational and what is implemented as public policy. Considerable investments have been made in developing Bt and similar pest-control methods and it is likely that such investments will continue [30]. However, we believe that bat conservation is not an alternative, but a rational strategy to use in concert with Bt technology given that neither is completely effective and generalist predators, such as bats, provide broad spectrum pest control and may help slow the evolution of resistance in pest species.

### Have we missed the full value of the ecosystem services provided by Mexican-free-tailed bats?

Critics have long pointed out that it is impossible to identify and measure, let alone value, all the ways functioning ecosystems provide benefits to society [31–33]. This leads us to an important caveat to this study – here we have assessed the impacts of one of 44 insectivorous bat species in the U.S. [34] on a single, albeit major pest (*H. zea*) of only one important crop (cotton). Mexican free-tailed bats are generalist predators and can switch diet preferences very quickly [35,36]. When cotton prices fell and farmers planted other crops instead, the bats likely provided pest-control services to those other crops – indicating that we have not fully accounted for the total value of bats' pest-control services. Finally, a full accounting would consider the other types of services, such as ecotourism, provided by Mexican free-tailed bats [37].

## Conclusion

Valuation of ecosystem services has improved greatly in the last decade [38–41]. There is now a rich and growing literature showing how temporal and spatial variation in ecological functions can cause variation in the economic benefits provided by nature [42–46]. Our contribution is to demonstrate empirically how fluctuations in market conditions also cause temporal variation in ecosystem service values even when ecosystem function is held constant. Just as values from a particular region should not be blindly transferred to other regions [9,31,47,48], our study illustrates that ecosystem service valuations must also consider how changing market conditions and technological substitutes can alter service values over time. At a minimum, results like ours can be used to develop transfer functions for valuing agricultural pest-control services by accounting for the role of changing agricultural prices and practices in the value of these services [49,50].

We also hope that these and similar results will spur discussion about the role of ecosystem services in biodiversity conservation in general, and bat conservation in particular. While the value of bat pest-control services to cotton production in the U.S. did indeed decline in response to global market forces and advances in biotechnology, there is a possibility that as pest resistance to Bt cotton rises, bat pest-control service values will rise again. Although currently there are no market-based approaches linking bat pest control services to cotton production, we wonder if the hope of protecting bats now to preserve the "option value" of future services is a sufficient argument to develop incentive-based approaches to bat conservation? Or, is pinning conservation hopes on the notion of option values risky, since we have already witnessed one cycle of technological capital supplanting natural capital? These are the questions we are left to ponder.

## Materials and Methods

### Pest-control value

We employ the avoided-cost approach used by Cleveland et al. [12] to estimate the value of bat services in reducing crop damage and pesticide use on conventional cotton in an eight-county region of South Texas located west of San Antonio. We expand on the approach of Cleveland et al. [12] by considering the effect of the adoption of transgenic Bt cotton in 1996 on bat service values. Our analysis covers the two decades from 1990 through 2008, allowing us to understand how ecosystem service values vary over time as a function of changes in land-use practices and socio-economic factors. In addition, we estimated bat service values across the southwestern U.S., rather than just a region of Texas. This increased geographic scope includes all cotton-producing areas near major Mexican free-tailed bat roosts in the U.S.

Estimating the avoided costs of crop damage involves the following steps: (a) estimating the number of insects consumed nightly by individual bats; (b) determining the hectares of cotton fields within proximity to bat roosts, which allows us to estimate the number of insects consumed nightly by bats in the area of the fields; and (c) the value of the crops that would have been damaged in the absence of bats. To determine the value of reducing insecticide use, we calculated both the reduced private costs to farmers of applying insecticides, and the reduced cost to society of releasing fewer insecticides into the environment. We modified the method of Cleveland et al. [12] to consider the social costs of only those insecticides that specifically target cotton bollworms (*Helicoverpa zea*) [51].

## Bat population estimates and roost locations

Our study area includes all U.S. counties that produce Pima or Upland cotton and that are located within 50 km (conservatively, the bats' nightly foraging distance [52]) of a major Mexican free-tailed bat roost. We obtained data about roosts (location and bat population censuses) from the U.S. Geological Survey's Bat Population Database [53] and our own literature search (Table S1 in File S1). We only considered large summer roosts (>7,000 individuals) because many smaller roosts lack good geospatial information, and because the combined populations of the largest summer colonies are thought to account for the majority (>99%) of the migratory Mexican free-tailed bat population [54]. These major roosts thus provide a reasonable estimate of the number of bats engaged in pest-control services in the U.S. We used only estimates obtained after 1970 to account for concerns that bat populations may have declined in the 1950s and 1960s due to DDT exposure [55,56]. We assumed that 90% of the adult bats in each colony were female and 10% were male, which is consistent with field data [13]. We did not model changes in the bat population size over time, as the data do not permit time-series analysis [13], but we did analyze the sensitivity of ecosystem service values to a 10% change in bat population numbers (see "Sensitivity analysis of pest-control values").

## Avoided crop damage calculation

**Number of pests consumed.** We first estimated the value of the crops that would have been lost in the absence of bats providing pest-control services. Conventional and molecular analyses show that moths comprise 30–60% of the bats' diet and indicate that each reproductively active female bat consumes 5–10 female adult bollworms (*Helicoverpa zea*) per night during periods of peak bollworm infestation [12,35,36]. Since bollworms also infest other crops in the area or migrate out of the region, we estimated that only 10–20% of the female moths consumed each night (approximately 1.5 individuals per bat) would have dispersed into cotton and laid eggs [12]. Due to high mortality rates during insect development (95–98%), the nightly consumption of 1.5 adult female moths would prevent 5 larvae from developing and damaging cotton crops [12,57]. Bollworm consumption by non-reproductive females and male bats was calculated as 32% lower than reproductive females due to the high metabolic costs of lactation [12,13].

For those bollworm that survive development to the larval stage, a single larva can damage 2–3 bolls of cotton over its lifetime [12]. However, because the value of the cotton bolls declines over the season – bolls produced during the first third of season generate about 50% of the harvest while bolls from the last third generate only 7% [58] – we estimated values separately for each third of the season. Further, because bats prevent damage to fewer bolls in Bt versus conventional cotton, we assumed that bats prevented approximately half (52.6%) the number of larvae from developing in Bt versus conventional cotton [13].

**Cotton locations.** We used data from the U.S. Department of Agriculture's National Agricultural Statistics Service (www.usda.nass.gov) and the National Cotton Council (www.cotton.org) on number of cotton hectares planted per county (Table S2 in File S1). Data on numbers of hectares planted with cotton are at the county level, so we approximated locations of cotton fields using crop potential soil maps for each county. This approach assumes that cotton hectares are uniformly distributed over soils with high cotton potential. We used the U.S. General Soil Maps (STATSGO data) from the USDA Natural Resource Conservation Service (NRCS) for locations of soil types suitable for cotton production. For each year from 1990 through 2008, we assumed that the proportion of the cotton hectares planted per county within foraging distance of the bats was equal to the proportion of suitable cotton-growing soils for each county within their foraging range of 50 km from each roost. We also expected that bats disperse randomly from their roost, such that the percentage of the roost's bat population foraging in each cotton-growing area was equal to the percentage of the area each cotton-growing region composed of a roost's total foraging range. Because bats likely disperse non-randomly from their roosts and concentrate on high quality foraging grounds, our calculation is conservative.

## Cotton prices

We used data on cotton prices from 1990 through 2008 from the National Cotton Council. The prices were adjusted for inflation and reported in 2011 USD (SI Appendix, Table S2).

## Avoided insecticide costs calculation

Private costs savings for insecticides reflect the reduced cost to farmers of purchasing and applying chemicals. Data on costs of cotton insecticide applications from 1990 through 2008 were obtained from the Mississippi State University Department of Entomology and Plant Pathology's databases on cotton losses due to insects (http://www.entomology.msstate.edu/resources/cottoncrop.asp). Social cost savings arise from lowered public health impacts to the farm workers who apply the pesticides, and reduced environmental damage due to loss of beneficial pollinators and groundwater contamination [59]. We ascertained the insecticides in the U.S. that are used predominantly on cotton bollworms [60], and used data from Kovach et al. [61] and from Cornell University's Integrated Pest Management Program [62] to estimate the environmental and toxicological impacts of particular cotton insecticides. We then used a pesticide environmental accounting tool [51] to assign a social-cost value in dollars for each insecticide according to the degree of impact estimates. The pesticide accounting tool calculates detrimental impacts in six categories: human health, ground water contamination, aquatic systems (fish), birds, bees, and other beneficial insects. We used a weighted mean cost of insecticide applications per hectare over time.

## Numbers of insecticide applications avoided

Insecticides are generally applied to cotton fields when bollworm infestations reach a threshold of 20,000–25,000 larvae per hectare. The date at which the threshold is reached, which triggers the first insecticide application, varies by region. Regional estimates of dates of first insecticide application were provided by the following cotton pest experts: C Sansone (Texas, Oklahoma, and Kansas), D Munier (California), J Pierce (New Mexico), and P Ellsworth (Arizona). For fields planted with Bt cotton, the threshold is reached later because the bollworm population growth rate is <10% of that in conventional cotton [63], resulting in a lower number of avoided insecticide applications in Bt cotton. We estimated the number of insecticide applications that were avoided in the bats' presence by calculating the number of times the threshold would have been reached without bat predation from the first date of cotton flowering (and susceptibility to bollworms) to the first date of insecticide application. We used a uniform insecticide application rate of 0.29 kg/ha [60]. Finally, we estimated the value of these avoided applications by summing the private and social costs [12,61]. Data on the cotton season (e.g., mean planting and harvest dates) for different regions were obtained from the USDA's National Agricultural Statistics Service.

### Sources of uncertainty in pest-control estimates

We arrived at high and low estimates of total pest-control services provided by bats using ranges of several parameters for which we did not have accurate estimates. We used the following at their maximum and minimum value: the insecticide application threshold (20,000–25,000 larvae/ha), the number of bolls consumed by a larva over its lifetime (2–3 bolls/larva), and the number of adult female moths dispersing into cotton (0.5–2 individual per bat per night).

### Sensitivity analysis of pest-control values

To better understand factors influencing the ecosystem service values over time, we analyzed the sensitivity of the annualized mean pest-control value over our study period (1990–2008). We altered the following parameters by ±10%: total area planted with cotton, Mexican free-tailed bat population size, and price of cotton. We measured the effect of the parameter alterations on the annualized mean pest control value.

### Impact of Bt on value of pest-control services

Bt cotton was introduced for the Upland variety of cotton in 1996, but is not available for the Pima variety, which accounts for less than 5% of cotton production in the U.S. [64]. Information on the timing of adoption of transgenic Bt Upland cotton was obtained from the Mississippi State University Department of Entomology and Plant Pathology's database on cotton crop losses [64]. To better understand the influence of the adoption of Bt cotton on the bat's pest-control service value, we used the mean pest-control values to calculate potential pest-control values had Bt not been adopted in 1996. To do this, we assumed that conventional cotton (i.e. non-Bt cotton) was planted from 1996 through 2008 and recalculated the pest-control values.

## Supporting Information

**File S1 Contains the following: Table S1 Cotton extent and Mexican free-tailed bat population size per county from 1990 through 2008. Table S2 Upland and Pima cotton price over time.**

## Acknowlegments

We thank D. Munier, J. Pierce, and P. Ellsworth for providing information on insecticide use in cotton. LLH thanks R. Merideth, J.A. Quijada, E. Valdez, C. Kremen, N. Brozovic and an anonymous reviewer for their comments on the paper. Any use of trade product or firm names is for descriptive purposes only and does not imply endorsement by the US Government.

## Author Contributions

Conceived and designed the experiments: RW LLH JL JD CS DS JG. Performed the experiments: RW LLH KL CS KB. Analyzed the data: RW LLH CS JD KB. Contributed reagents/materials/analysis tools: GFM KL RM AR PC. Wrote the paper: LLH RW GFM. Developed the code to analyze the data: RW.

## References

1. Daily G (1997) Nature's services: societal dependence on natural ecosystems. Washington D.C.: Island Press.
2. Pagiola S, Von Ritter K, Bishop J (2004) Assessing the economic value of ecosystem conservation. Washington D.C.: World Bank, Environment Department.
3. Redford KH, Adams WM (2009) Payment for ecosystem services and the challenge of saving nature. Conserv Biol 23: 785–787.
4. Berck P, Ligon E (1996) The swamp and shopping center: an interest rate parable. Ecol Model 92: 275.
5. Rodriguez JP, Beard TD, Bennett EM, Cumming GS, Cork SJ, et al. (2006) Trade-offs across space, time, and ecosystem services. Ecol Soc 11.
6. McCauley DJ (2006) Selling out on nature. Nature 443: 27–28.
7. Brown T, Bergstrom J, Loomis J (2007) Defining, Valuing and Providing Ecosystem Goods and Services. Nat Resour J 47: 331–376.
8. Ricketts TH, Daily GC, Ehrlich PR, Michener CD (2004) Economic value of tropical forest to coffee production. Proc Natl Acad Sci U S A 101: 12579–12582.
9. Plummer ML (2009) Assessing benefit transfer for the valuation of ecosystem services. Front Ecol Environ 7: 38–45.
10. Kunz TH, de Torrez EB, Bauer D, Lobova T, Fleming TH (2011) Ecosystem services provided by bats. Ann N Y Acad Sci1223: 1–38.
11. Wiedenmann RN (2011) Contributions of Robert (Bob) J. O'Niel to the field of predaceous Heteroptera. Biol Control 59.
12. Cleveland CJ, Betke M, Federico P, Frank JD, Hallam TG, et al. (2006) Economic value of the pest control service provided by Brazilian free-tailed bats in south-central Texas. Front Ecol Environ 4: 238–243.
13. Federico P, Hallam TG, McCracken GF, Purucker ST, Grant WE, et al. (2008) Brazilian free-tailed bats as insect pest regulators in transgenic and conventional cotton crops. Ecol Appl 18: 826–837.
14. Boyles JG, Cryan PM, McCracken GF, Kunz TH (2011) Economic importance of bats in agriculture. Science 332: 41–42.
15. Gándara Fierro G, Correa Sandoval A, Hernández Cienfuegos C (2006) Valoracíon económica de los servicios ecológicas que prestan los murciélagos Tadarida brasiliensis como controladores de plagas en el norte de México.
16. Cátedra de Integración Económica y Desarrollo Social. USDA (2013) Cotton: World Markets and Trade.: U.S. Department of Agriculture.
17. Marvier M, McCreedy C, Regetz J, Kareiva P (2007) A meta-analysis of effects of Bt cotton and maize on nontarget invertebrates. Science 316: 1475–1477.
18. Tabashnik BE, Brevault T, Carriere Y (2013) Insect resistance to Bt crops: lessons from the first billion acres. Nat Biotechnol 31: 510–521.
19. Tabashnik BE, Gassmann AJ, Crowder DW, Carriere Y (2008) Insect resistance to Bt crops: evidence versus theory. Nat Biotechnol 26: 199–202.
20. Bagla P (2010) INDIA Hardy Cotton-Munching Pests Are Latest Blow to GM Crops. Science 327: 1439–1439.
21. Tabashnik BE, Wu KM, Wu YD (2012) Early detection of field-evolved resistance to Bt cotton in China: Cotton bollworm and pink bollworm. Journal of Invertebrate Pathology 110: 301–306.
22. Gassmann A (2012) Field-evolved resistance to Bt maize by western corn rootworm: Predictions from the laboratory and effects in the field. J Invertebr Pathol 110: 287–293.
23. Kaur P, Dilawari VK (2011) Inheritance of resistance to Bacillus thuringiensis Cry1Ac toxin in Helicoverpa armigera (Hubner)(Lepidoptera: Noctuidae) from India. Pest manag sci 67: 1294–1302.
24. Kruger M, Van Rensburg JBJ, Van den Berg J (2012) Transgenic Bt maize: farmers' perceptions, refuge compliance and reports of stem borer resistance in South Africa. J Appl Entomol 136: 38–50.
25. Hagerty AM, Kilpatrick AL, Turnipseed SG, Sullivan MJ, Bridges WC (2005) Predaceous arthropods and lepidopteran pests on conventional, Bollgard, and Bollgard II cotton under untreated and disrupted conditions. Environ Entomol 34: 105–114.
26. Weisbrod BA (1964) Collective-Consumption Services of Individual-Consumption Goods. Q J Econ 78: 471–477.
27. Walsh RG, Loomis JB, Gillman RA (1984) Valuing option, existence and bequest demands for nature. Land Econ 60: 14–29.
28. Brookshire DS, Eubanks LS, Randall A (1983) Estimating option prices and existence values for wildlife resources. Land Econ 59: 1–15.
29. Kruitla J, Fisher A (1976) Economics of Natural Environments. Baltimore, MD.: Johns Hopkins University Press.
30. McDougall P (2011) The cost and time involved in the discovery, development and authorisation of a new plant biotechnology derived trait. Consultancy Study for Crop Life International.
31. Spash CL (2008) How much is that ecosystem in the window? The one with the bio-diverse trail. Environ Value 17: 259–284.
32. Peterson MJ, Hall DM, Feldpausch-Parker AM, Peterson TR (2010) Obscuring Ecosystem Function with Application of the Ecosystem Services Concept. Conserv Biol 24: 113–119.

33. Muradian R, Rival L (2012) Between markets and hierarchies: The challenge of governing ecosystem services. Ecosystem Services 1: 93–100.
34. Smithsonian Institute (1980) Bat facts. In: Smithsonian Institute, editors. Science and Technology.
35. McCracken GF, Westbrook JK, Brown VA, Eldridge M, Federico P, et al. (2012) Bats Track and Exploit Changes in Insect Pest Populations. PLoS One 7: e43839.
36. Lee Y-F, McCracken GF (2005) Dietary variation of Brazilian free-tailed bats linked to migratory populations of pest insects. J Mammal 86: 67–76.
37. Bagstad K, Wiederholt R (2013) Tourism values for Mexican free-tailed bat (Tadarida brasiliensis mexicana) viewing. Hum Dimens Wildl 18: 307–311.
38. Barbier EB (2011) Progress and Challenges in Valuing Coastal and Marine Ecosystem Services. Rev Environ Econ Policy 6: 1–19.
39. Turner RK, Paavola J, Cooper P, Farber S, Jessamy V, et al. (2003) Valuing nature: lessons learned and future research directions. Ecol Econ 46: 493–510.
40. Lu Y, Fu B, Feng X, Zeng Y, Liu Y, et al. (2012) A Policy-Driven Large Scale Ecological Restoration: Quantifying Ecosystem Services Changes in the Loess Plateau of China. PLoS One 7: e31782.
41. Kovacs K, Polasky S, Nelson E, Keeler BL, Pennington D, et al. (2013) Evaluating the Return in Ecosystem Services from Investment in Public Land Acquisitions. PLoS One 8: e62202.
42. Kozak J, Lant C, Shaikh S, Wang G (2011) The geography of ecosystem service value: The case of the Des Plaines and Cache River wetlands, Illinois. Appl Geogr 31: 303–311.
43. Barbier EB, Hacker SD, Kennedy C, Koch EW, Stier AC, et al. (2011) The value of estuarine and coastal ecosystem services. Ecol Monogr 81: 169–193.
44. Aburto-Oropeza O, Ezcurra E, Danemann G, Valdez V, Murray J, et al. (2008) Mangroves in the Gulf of California increase fishery yields. Proc Natl Acad Sci U S A 105: 10456–10459.
45. Koch EW, Barbier EB, Silliman BR, Reed DJ, Perillo GME, et al. (2009) Non-linearity in ecosystem services: temporal and spatial variability in coastal protection. Front Ecol Environ 7: 29–37.
46. Bagstad KJ, Johnson GW, Voigt B, Villa F (2013) Spatial dynamics of ecosystem service flows: A comprehensive approach to quantifying actual services. Ecosyst Serv 4: 117–125.
47. Fisher B, Naidoo R (2011) Concerns about extrapolating right off the bat. Science 335: 287.
48. Eigenbrod F, Armsworth PR, Anderson BJ, Heinemeyer A, Gillings S, et al. (2010) Error propagation associated with benefits transfer-based mapping of ecosystem services. Biol Conserv 143: 2487–2493.
49. Bergstrom JC, Taylor LO (2006) Using meta-analysis for benefits transfer: Theory and practice. Ecol Econ 60: 351–360.
50. Loomis JB (1992) The evolution of a more rigorous approach to benefit transfer: benefit function transfer. Water Resour Res 28: 701–705.
51. Leach A, Mumford J (2008) Pesticide Environmental Accounting: A method for assessing the external costs of individual pesticide applications. Environ pollut 151: 139–147.
52. Williams TC, Ireland LC, Williams JM (1973) High altitude flights of the free-tailed bat, Tadarida brasiliensis, observed with radar. J Mammal 54: 807–821.
53. Ellison L, O'Shea T, Bogan M, Everette A, Schneider D (2003) Existing data on colonies of bats in the United States: summary and analysis of the US Geological Survey's bat population database. USGS/BRD/ITR–2003-0003. USGS Information and Technology Report Fort Collins, CO, USA: USGS Fort Collins Science Center. pp. 127–171.
54. McCracken G (2003) Estimates of population sizes in summer colonies in Brazilian free-tailed bats (*Tadarida brasiliensis*). USGS/BRD/ITR–2003-0003. In: O. S. T. B MA, editor editors. Monitoring trends in bat populations of the United States and territories: problems and prospects. Fort Collins, CO, USA: U.S. Geological Survey, Fort Collins, CO. pp. 21–29.
55. Lewis SE (1995) Roost fidelity of bats - a review Journal of Mammalogy 76: 481–496.
56. Betke M, Hirsh DE, Makris NC, McCracken GF, Procopio M, et al. (2008) Thermal Imaging Reveals Significantly Smaller Brazilian Free-Tailed Bat Colonies Than Previously Estimated. J Mammal 89: 18–24.
57. Sansone C, Smith J (2001) Natural mortality of Helicoverpa zea (Lepidoptera: Noctuidae) in short-season cotton. Biol Control 30: 113–122.
58. Sansone CG, Isakeit T, Lemon R, Warrick B (2002) Texas cotton production: Emphasizing integrated pest management. In: T. C. E. Publication, editor editors. College Station, TX.
59. Pimentel D, Acquay H, Biltonen P, Rice S, Silva M, et al. (1991) Environmental and economic effects of reducing pesticide use. BioScience 42: 750–760.
60. Gianessi L, Reigner N (2006) Pesticide use in U.S. crop production. Washington D.C.: CropLife Foundation.
61. Kovach J, Petzoldt C, Degni J, Tette J (1992) A method to measure the environmental impact of pesticides. New York Food Life 129: 1–8.
62. Cornell University's Integrated Pest Management Program (2012) New York State Integrated Pest Management Program.
63. Jackson RE, Bradley Jr JR, Van Duyn JW (2003) Field performance of transgenic cottons expressing one or two *Bacillus thuringiensis* endotoxins against bollworm, *Helicoverpa zea* (Boddie). J Cotton Sci 7: 57–64.
64. Mississippi State University Department of Entomology and Plant Pathology United States of America, http://www.entomology.msstate.edu/resources/cottoncrop.asp, Accessed 11 January 2014.

# 23

# Natural Enemies Delay Insect Resistance to Bt Crops

**Xiaoxia Liu[1,2], Mao Chen[2¤a], Hilda L. Collins[2], David W. Onstad[3¤b], Richard T. Roush[4], Qingwen Zhang[1], Elizabeth D. Earle[5], Anthony M. Shelton[2*]**

**1** Department of Entomology, China Agricultural University, Beijing, China, **2** Department of Entomology, Cornell University, New York State Agricultural Experiment Station, Geneva, New York, United States of America, **3** Department of Entomology, University of Illinois, Urbana, Illinois, United States of America, **4** Melbourne School of Land and Environment, University of Melbourne, Victoria, Australia, **5** Department of Plant Breeding and Genetics, Cornell University, Ithaca, New York, United States of America

## Abstract

We investigated whether development of resistance to a *Bt* crop in the presence of a natural enemy would be slower than without the natural enemy and whether biological control, in conjunction with a *Bt* crop, could effectively suppress the pest population. Additionally, we investigated whether insecticide-sprayed refuges of non-*Bt* crops would delay or accelerate resistance to the *Bt* crop. We used a system of *Bt* broccoli expressing Cry1Ac, a population of the pest *Plutella xylostella* with a low frequency of individuals resistant to Cry1Ac and the insecticide spinosad, and a natural enemy, *Coleomegilla maculata*, to conduct experiments over multiple generations. The results demonstrated that after 6 generations *P. xylostella* populations were very low in the treatment containing *C. maculata* and unsprayed non-*Bt* refuge plants. Furthermore, resistance to *Bt* plants evolved significantly slower in this treatment. In contrast, *Bt* plants with no refuge were completely defoliated in treatments without *C. maculata* after 4–5 generations. In the treatment containing sprayed non-*Bt* refuge plants and *C. maculata*, the *P. xylostella* population was low, although the speed of resistance selection to Cry1Ac was significantly increased. These data demonstrate that natural enemies can delay resistance to *Bt* plants and have significant implications for integrated pest management (IPM) with *Bt* crops.

**Editor:** Nicolas Desneux, French National Institute for Agricultural Research (INRA), FRANCE

**Funding:** This project was supported by the Biotechnology Risk Assessment Program Competitive Grant no. 2008-33120-19536 from the USDA National Institute of Food and Agriculture and the Special Research Projects for Developing Transgenic Plants (2011zx08011-002) in China. The funders had no role in study design, data collection and analysis, decision to publish, or preparation of the manuscript.

**Competing Interests:** The authors have declared that no competing interests exist.

* E-mail: ams5@cornell.edu

¤a Current address: Monsanto Company, Chesterfield, Missouri, United States of America
¤b Current address: Dupont Agricultural Biotechnology, Wilmington, Delaware, United States of America

## Introduction

The commercialization of plants expressing insecticidal crystal (Cry) proteins from *Bacillus thuringiensis* (*Bt*) for insect management has revolutionized agriculture and become a major tool for integrated pest management (IPM) programs [1–2]. In 2011, *Bt* crops were grown on nearly 70 million ha in 27 countries in 2012 [3]. *Bt* crops have provided economic benefits to growers and reduced the use of other insecticides [1,4–6], suppressed pest populations on a regional basis [7–9], conserved natural enemies [10] and promoted biological control services in agricultural landscapes [6]. However, the development of insect resistance is a major threat to the sustainable use of *Bt* crops [11–12].

Since *Bt* crops were first commercialized in 1996, there is evidence that three lepidopteran pests have evolved resistance to *Bt* crops in the open field [13–15] and one case of a coleopteran pest [16]. Resistance to *Bt* plants is a serious concern, but the relatively few number of cases is in stark contrast to many cases of resistance to conventional insecticides, which has occurred much more rapidly [17]. Commonly proposed reasons for the few confirmed cases of resistance to *Bt* plants are the high dose of *Bt* proteins expressed in plants and the use of refuges of non-*Bt* plants that can serve as a pool of *Bt* susceptible alleles in the population [18–19].

Another possible reason for the relatively few cases of resistance to *Bt* plants could be their safety to natural enemies that help suppress pest populations. Numerous studies have investigated the effects of *Bt* crops and Cry proteins on natural enemies (predators and parasitoids) in the laboratory and field [20–21]. A meta-analysis has confirmed the safety of *Bt* proteins, especially when compared to traditional insecticides [10]. When negative effects on natural enemies have been observed with *Bt* proteins, they appear to be due to the poor quality of the host and not the Cry protein [22], but see Desneux et al., 2010 [23]. The safety of several *Bt* proteins has been verified in tritrophic studies conducted with *Bt*-resistant or non-susceptible herbivores that avoided the problems of prey-quality in some previous studies [22]. Allowing *Bt*-resistant hosts to ingest *Bt* proteins and then feeding the hosts to natural enemies (both predators and parasitoids) has revealed no effects on the natural enemies [24–26]. However, some reports continue to suggest natural enemies may be harmed by *Bt* proteins [27], but these have been challenged [28].

The conservation of natural enemies by the use of *Bt* plants could also influence the development of resistance to *Bt* crops. This question was first studied by Gould et al. [29] in their conceptual and mathematical models on tritrophic interactions of a plant, an herbivore and a natural enemy. Their simplest conclusion was that natural enemies that increase differential fitness between susceptible and resistant phenotypes on host plants will accelerate resistance; those that decrease the differential will delay resistance. Johnson et al. [30–31] carried out controlled studies of a parasitoid and a pathogenic fungus that attack *Heliothis virescens* on *Bt* tobacco and concluded that the parasitoid would likely delay the development of resistance to transgenic tobacco, while the pathogen would likely promote the development of resistance. Mallampalli et al. [32] discovered that different prey species of a generalist predator had different effects on the development of resistance by *Leptinotarsa decemlineata* to *Bt* potato: the presence of one prey species delayed resistance while the other accelerated resistance. Heimpel et al. [33] reported that the form of egg mortality could influence the rate of resistance, but the importance of egg mortality depended on other ecological processes in the pest population. Other simulation models reported that natural enemies could slow insect resistance to *Bt* crops or *Bt* pesticides [34–36]. As this summary indicates, the literature contains suggestions that natural enemies could delay or accelerate resistance, depending on whether there is a differential impact on susceptible or resistant phenotypes.

In the present study we used a unique system, composed of broccoli plants transformed to express Cry1Ac protein, a population of *Plutella xylostella*, a global pest of crucifers [37] with a low frequency of resistant individuals to Cry1Ac and the insecticide spinosad, and the predaceous ladybird beetle, *Coleomegilla maculata*, to conduct a multigenerational study in the greenhouse. Our objectives were to determine: (1) if a natural enemy can delay the development of insect resistance to a *Bt* crop; and (2) if biological control in conjunction with *Bt* crops can effectively suppress the pest population. In addition, to simulate field-realistic conditions for both the predator and prey, we sprayed refuges with insecticide in some treatments, but not others, and observed those effects on the development of insect resistance.

## Results

### Population Density of *P. xylostella*

The predator, presence of a refuge, and use of a spray on the refuge each influenced the population dynamics of *P. xylostella* per *Bt* plant over the 6 generations of the experiment. A repeated-measures ANOVA, with generation and treatment as factors, yielded a significant effect for generations ($F_{3.164} = 77.101$, $P<0.001$), for treatments ($F_4 = 31.788$, $P<0.001$), and the interaction term generation*treatments ($F_{12.656} = 16.250$, $P<0.001$). During the 1st generation, few *P. xylostella* were found on *Bt* plants (Fig. 1A), and there were no significant differences between treatments using one-way ANOVA ($F_4 = 0.258$, $P = 0.900$). By the 2nd generation, there was an average of 7 *P. xylostella* larvae and pupae per *Bt* plant in the treatment with only *Bt* plants, but the number of *P. xylostella* was still about 1 per plant in the other treatments ($F_4 = 3.767$, $P = 0.026$). *Bt* plants were completely defoliated and control failure was evident in the *Bt* plant-only treatment at the 3rd generation when the number of *P. xylostella* had increased to 51 per *Bt* plant ($F_4 = 8.667$, $P = 0.001$).

At the 4th generation, the number of *P. xylostella* increased to 12 per *Bt* plant in treatment Bt+R (75% *Bt* plants +25% non-*Bt* refuge plants) and 29 per *Bt* plant in Bt+SR (75% *Bt* plants +25% spinosad-sprayed non-*Bt* refuge plants), significantly higher than those in treatments Bt+R+Cm (75% *Bt* plants +25% non-*Bt* refuge plants + predator) and Bt+SR+Cm (75% *Bt* plants +25% spinosad-sprayed non-*Bt* refuge plants + predator) ($F_4 = 21.294$, $P<0.001$). There was still only about 1 *P. xylostella* per *Bt* plant in treatment Bt+R+Cm at the 5th and 6th generations, significantly lower than other treatments (5th generation: $F_4 = 59.203$, $P<0.001$; 6th generation: $F_4 = 61.164$, $P<0.001$). Most importantly, over the 6 generations of the test, only treatment Bt+R+Cm maintained <2 *P. xylostella* per *Bt* plant at each generation, suggesting the important role that the predator played in maintaining a low pest population.

When spinosad was used (treatment Bt+SR+Cm), the pest population was also maintained at a low level, except for a flare-up in the 5th generation when the *P. xylostella* population was >40 per *Bt* plant. Without the use of the predator (Bt+R), the pest population gradually rose and peaked at 61 per plant by the 6th generation, compared to being maintained at about 1 for all generations when the predator was used (Bt+R+Cm). When the predator was replaced by the insecticide spinosad (Bt+SR), the pest population increased more rapidly and peaked at 107 per plant in the 6th generation, again showing the strong and lasting benefit of the predator in maintaining a low pest population.

Examining the pest population on the refuge plants in the cages provided another indication of the overall performance of the treatments. Using repeated measures analysis, there were significant differences for generations ($F_{3.740} = 29.341$, $P<0.001$), for treatments ($F_4 = 60.288$, $P<0.001$), and the interaction term generation*treatments ($F_{14.960} = 10.071$, $P<0.001$). On refuge plants, the density of *P. xylostella* varied between 5–120 *P. xylostella* per plant between generations and treatments (Fig 1B). Only the populations in Bt+SR+Cm, which combined *Bt* plants, spinosad and the predator, were consistently the lowest and did not exceed 20 per plant in any generation. Use of the predator alone (Bt+R+Cm) maintained a relatively low and stable pest population over all 6 generations. In treatments of 100% non-*Bt* refuge plants, *P. xylostella* densities exceeded 100 per plant at each generation despite keeping only 3 defoliated plants with their larvae and pupae in each cage to reduce the overall population. Therefore, we did not include the treatment of only non-*Bt* plants in Figure 1B.

### Population Density of *C. maculata*

In treatments Bt+R+Cm and Bt+SR+Cm, predator populations generally increased as the pest, *P. xylostella*, populations increased. A repeated-measures ANOVA yielded a significant effect for generations ($F_{1.838} = 11.873$, $P = 0.002$), for treatments ($F_1 = 9.410$, $P = 0.022$), and the interaction term generation*treatments ($F_{1.838} = 11.225$, $P = 0.003$). Only a few *C. maculata* adults were found in the 1st generation because only 3 pairs were released at the start of experiment (Fig. 2A). Predator populations on *Bt* plants in treatment Bt+R+Cm remained about 1 per plant because of the low pest population, especially on *Bt* plants through the 6th generation (Fig. 1A). In treatment Bt+SR+Cm, predator populations remained about 1 per plant until the 5th generation when they increased to >4.5 in the 5th and 6th generations (Fig. 2A). Mean values for the 5th and 6th generations by the independent-test between the two treatments, respectively, differed significantly (G5: $t_{(6)} = 4.562$, $P = 0.004$; G6: $t_{(6)} = 6.268$, $P = 0.001$). This likely maintained the pest population on the *Bt* plants at a low population (Fig. 1A), although resistance increased (Table 1).

Predator populations were generally higher on the refuge plants (Fig. 2B) than on the *Bt* plants (Fig. 2A), likely reflecting the higher pest density on these plants. For the three treatments Bt+R+Cm, Bt+SR+Cm, and non-Bt+Cm, there were no significant differences in predator density in most generations. The repeated measures showed significant effects for generations ($F_{3.460} = 12.234$, $P<0.001$) and the interaction term generation*treatments ($F_{6.920} = 3.241$, $P = 0.012$), but not for the treatments ($F_2 = 2.181$, $P = 0.175$).

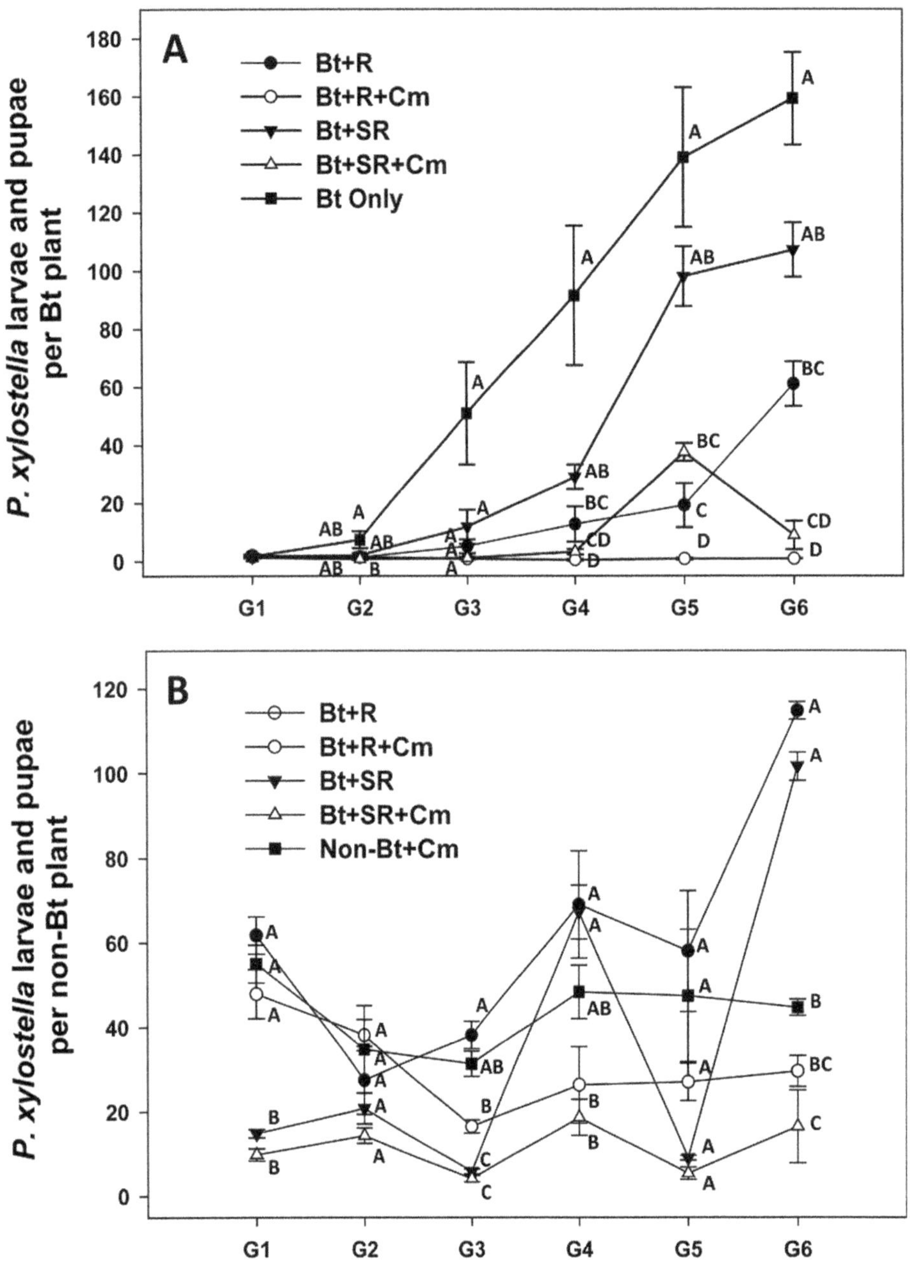


**Figure 1.** ***Plutella xylostella*** **populations on** ***Bt*** **plants (A) and on non-*Bt* refuge plants (B) in greenhouse cages (Means ± SEM).** Bt+R: 75% *Bt* and 25% non-*Bt* refuge plants; Bt+R+Cm: 75% *Bt* and 25% non-*Bt* refuge plants with *C. maculata*; Bt+SR: 75% *Bt* and 25% spinosad-sprayed non-*Bt* refuge plants; Bt+SR+Cm: 75% *Bt* and 25% spinosad-sprayed non-*Bt* refuge plants with *C. maculata*; Bt Only: 100% *Bt* plants only, Non-Bt+Cm: 100% non-*Bt* refuge plants with *C. maculata*. Different letters within the same generations denote significant differences ($P<0.05$, one-way ANOVA, Games-Howell test for G3 and G6 in A, G1, G3 and G5 in B, other comparsions by Tukey HSD test).

## Resistance to Cry1Ac

Survival of *P. xylostella* larvae indicates resistance and, on *Bt* broccoli, it was significantly affected when analyzed using repeated-measures for generations ($F_{2.714} = 35.792$, $P<0.001$), for treatments ($F_6 = 81.199$, $P<0.001$), and the interaction term generation*treatments ($F_{16.282} = 6.719$, $P<0.001$). Resistance levels reached 74.9% after only 3 generations in the treatment with *Bt* plants only, significantly greater than the survival in other treatments ($F_6 = 50.049$, $P<0.001$) and >2x the next highest treatment of Bt+SR (75% *Bt* plants + 25% spinosad-sprayed non-*Bt* refuge plants) (Table 1). The high rate of resistance in the treatment with only *Bt* plants was sustained through the 6$^{th}$ generation. In treatment Bt+SR, resistance also developed rapidly, with 56.0% *P. xylostella* surviving at the 4$^{th}$ generation, significantly greater than the treatments Bt+R, Bt+R+Cm, Bt+SR+Cm and the control cages (Non-Bt Only and Non-Bt+Cm) ($F_6 = 33.319$, $P<0.001$). The survival rates were 4–7% in treatment Bt+R+Cm (with predators) at all generations, but increased to 32.5% and 39.5% in treatment Bt+R (without predators) at the 5$^{th}$ and 6$^{th}$ generations, respectively. There were significant differences among treatments (5$^{th}$ generation: $F_6 = 26.302$, $P<0.001$; 6$^{th}$ generation: $F_6 = 38.850$, $P<0.001$) with <10% survival in

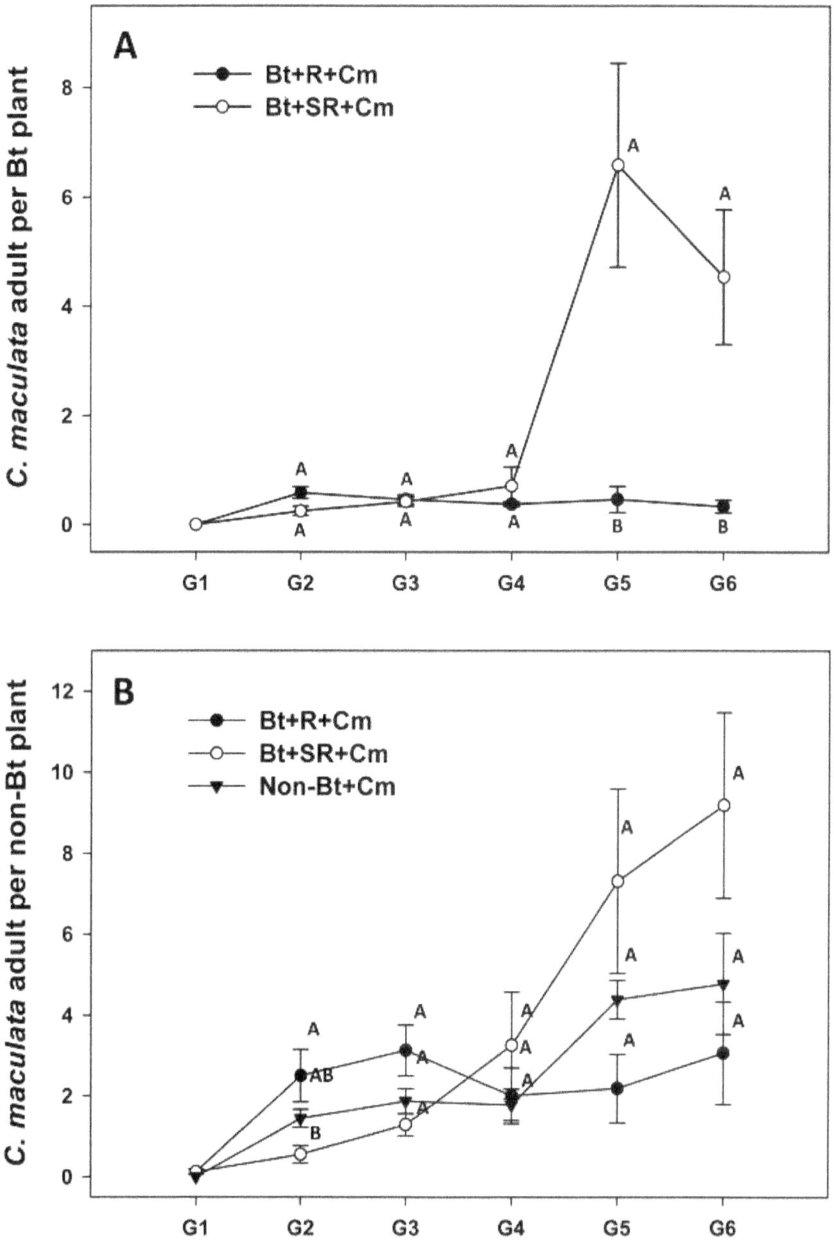


**Figure 2. *Coleomegilla maculata* population on *Bt* plants (A) and on non-*Bt* refuge plants (B) in greenhouse cages (Means ± SEM).** Bt+R+Cm: 75% *Bt* and 25% non-*Bt* refuge plants with *C. maculata*; Bt+SR+Cm: 75% *Bt* and 25% spinosad-sprayed non-*Bt* refuge plants with *C. maculata*; Non-Bt+Cm: 100% non-*Bt* refuge plants with *C. maculata*. Different letters within the same generations denote significant differences ($P<0.05$, one-way ANOVA, Tukey HSD test).

Bt+R+Cm, Non-Bt Only and Non-Bt+Cm, and >70% survival in Bt+SR, Bt+SR+Cm and Bt Only, while Bt+R had nearly 40% survival in the 6$^{th}$ generation. These significant differences at the 6$^{th}$ generation among Bt+R, Bt+R+Cm and Bt+SR+Cm highlight the ability of the predator to slow the rate of resistance to *Bt* plants. Treatment Bt+R+Cm, which included the predator, had the lowest survival (6.0%) while the treatment without the predator (Bt+R) had 39.5% survival. Including the predator but using spinosad (Bt+SR+Cm), which has some toxicity to Cm [38], increased the development of resistance to *Bt* plants as indicated by the 72.8% survival (Table 1).

## Resistance to Spinosad

In the treatments in which spinosad was used (Bt+SR and Bt+SR+Cm), only low levels of survival of *P. xylostella* were detected. Survival of individuals removed from cages and fed spinosad-treated leaves was only about 1% at the 4$^{th}$ and 6$^{th}$ generations.

# Discussion

A large body of literature has evaluated the potential ecological risks of *Bt* crops to non-target organisms including natural enemies of insect pests [1,2,22,39]. However, as is the case with conventional insecticides, the interaction of natural enemies and

**Table 1.** Survival on Cry1Ac leaf (%) (Means ± SEM) of *Plutella xylostella* larvae from adults taken from cages.

| Treatments | Generation | | | |
|---|---|---|---|---|
| | G3 | G4 | G5 | G6 |
| Bt+R | 3.3±1.86 A | 19.0±6.61 B | 32.5±14.55 BC | 39.5±14.06 B |
| Bt+R+Cm | 4.7±3.18 A | 7.0±3.06 AB | 5.0±3.51 AB | 6.0±2.86 A |
| Bt+SR | 29.3±4.01 B | 56.0±4.74 C | 87.5±1.44 D | 83.8±3.07 D |
| Bt+SR+Cm | 3.3±1.93 A | 20.7±2.19 B | 36.8±6.85 C | 72.8±3.28 D |
| Bt Only | 74.9±5.46 C | 72.0±6.76 C | 90.5±3.97 D | 93.5±1.19 D |
| Non-Bt Only | 1.0±0.58 A | 1.3±0.33 A | 1.0±0.58 A | 3.7±1.33 A |
| Non-Bt+Cm | 0.7±0.33 A | 1.0±0.58 A | 4.7±1.86 AB | 2.3±1.86 A |

Bt+R: 75% *Bt* and 25% non-*Bt* refuge plants; Bt+R+Cm: 75% *Bt* and 25% non-*Bt* refuge plants with *C. maculata*; Bt+SR: 75% *Bt* and 25% spinosad-sprayed non-*Bt* refuge plants; Bt+SR+Cm: 75% *Bt* and 25% spinosad-sprayed non-*Bt* refuge plants with *C. maculata*; Bt Only: only *Bt* plants; Non-Bt Only: only non-*Bt* plants; Non-Bt+Cm: only non-*Bt* plants with *C. maculata*.
Different letters within the same column denote significant differences ($P<0.05$, one-way ANOVA, Tukey's test).

resistance development in their hosts has received far less attention.

Our present results indicate that the predator, *C. maculata*, combined with non-*Bt* and unsprayed refuge plants, delayed resistance in the *P. xylostella* population to *Bt* broccoli plants, when compared to treatments without the predator. Some theoretical work predicted a similar conclusion, that natural enemies could slow the development of insect resistance to *Bt* crops or *Bt* sprayed on crops [30,34–36]. The present work, conducted over multiple generations, adds empirical evidence to the theory that natural enemies can slow the development of resistance to *Bt* crops.

Arpaia et al. [35] reported that the distribution of *C. maculata* in potato fields is driven by the prey's density but, in spite of this aggregation, a significant number of them occur in areas where food density is low. In the present study, we observed some *C. maculata* adults and larvae on *Bt* plants, although the prey density was very low on these plants. Still the likelihood of a *Bt*-resistant *P. xylostella* individual reaching adulthood on transgenic plants will be lowered if there are natural enemies on *Bt* plants, and this, in turn, will slow the rate at which the resistance alleles increase in frequency. In our tests, *C. maculata* were found on *Bt* plants (Fig. 2B) where they likely encountered *Bt*-resistant *P. xylostella* and removed them, thus eliminating resistance alleles from the population. This is the likely cause of the slower rate of resistance in the treatment with the refuge and predator. Although spinosad is considered relatively non-toxic to natural enemies, it does have a toxic effect on *C. maculata*, although much less than "harder insecticides" [38].

Widespread planting of *Bt* crops places strong selection pressure on pest populations, which could result in resistance to *Bt* and control failure [11,17–19]. Currently, an accepted cornerstone of resistance management programs for *Bt* crops in the United States, Australia, and elsewhere, is a refuge consisting of plants free of *Bt* insecticidal proteins, thus allowing *Bt*-susceptible alleles to persist in the population. When resistance is recessive, as has been the case so far for many key pests, then matings between resistant individuals emerging from *Bt* plants and those from refuge plants will greatly reduce the possibility of creating homozygous resistant offspring that can damage *Bt* plants [17]. In the United States, the Environmental Protection Agency (EPA) mandates non-*Bt* refuges of different sizes for different *Bt* crops.

Empirical data have demonstrated the usefulness of refuges for *Bt* crops. Tang et al. [40] reported that pest population growth was influenced by refuge size, with the highest populations occurring in treatments that had either no refuge plants or all refuge plants. That study also confirmed other models [41] that the development of resistance was inversely proportional to the size of the refuge. Most importantly, Tang et al. [40] demonstrated that a balance could be struck by refuge sizes that would delay resistance while at the same time limit the growth of the pest population and damage to the crop, as has been observed in the field [7,9].

In the present study, results indicate that refuges slow resistance, but a sprayed refuge could accelerate it. These findings are in line with our previous field study [42]. In the treatment with only *Bt* plants without refuge plants, control failure and high insect densities were observed in the 3rd generation (two replications) and the 4th generation (the other two replications). Plants in the treatment of spinosad-sprayed non-*Bt* broccoli plants (Bt+SR) were completely defoliated between the 4th and 5th generations.

In the treatment (Bt+SR+Cm) in which the predator and spinosad were used, 36.8% of the pest population survived on *Bt* plants by the 5th generation and by the 6th generation that had increased to 72.8% (Table 1). Spinosad killed most susceptible *P. xylostella* on refuge plants, and not enough susceptible individuals mated with the resistant individuals on *Bt* plants. Therefore, resistance development was quick although there were *C. maculata* in the treatment (Bt+SR+Cm). However, the pest populations were very low in this treatment with 16.5 *P. xylostella* larvae per refuge plant (Fig. 1A) and 8.9 per *Bt* plant (Fig. 1B) at the 6th generation. In the 6th generation, we found >100 *C. maculata* in each cage. This indicates that the high predator population controlled *P. xylostella* density to a low level, despite the fact that the pest had developed resistance to the *Bt* toxin.

While farmers are concerned with reducing the likelihood of resistance, they are more immediately concerned with lowering the pest population to avoid crop injury. It is well worth noting that in our experiments we not only saw the lowest rate of resistance development in the prey when the predator was not decimated by the use of an insecticide, but also the lowest and least fluctuating pest population on *Bt* plants (Fig. 1A) and low and stable pest populations on non-*Bt* plants in the refuge (Fig. 1B). Thus, our data suggest that farmers can have sustainable management of pests if they combine *Bt* plants with biological control.

In conclusion, this study provides empirical evidence to confirm the theory that natural enemies can delay resistance development to *Bt* plants, but also demonstrates that it can do so while maintaining a low pest density and low crop damage. Non-*Bt* refuges are necessary to delay resistance to *Bt* plants, but spraying refuges could accelerate resistance if sprays reduce the function of important biological control agents of the pest. We suggest that host-plant resistance with *Bt* plants and biological control can be fully compatible within an overall integrated pest management (IPM) program. Our results have significant implications for IPM and insect resistance management (IRM) for *Bt* crops.

## Methods

### Insects

Three strains of *Plutella xylostella* were used to create a hybrid population for the cage tests: the susceptible Geneva 88 (SS), the Cry1Ac-resistant (Cry1Ac-RR), and the spinosad-resistant (Pearl-RR) [24,40,43] strains. The hybrid population was created by releasing 100 F1 RS1 (G88 female × Cry1Ac-RR male), 100 RS2 (G88 female × Pearl-RR male) and 300 G88 moths into a cage.

**Table 2.** Experimental treatments.

| Group | Treatments | Replications |
|---|---|---|
| Bt+R | 75% Bt plants and 25% non-Bt refuge plants | 4 |
| Bt+R+Cm | 75% Bt plants and 25% non-Bt refuge plants and *C. maculata* | 4 |
| Bt+SR | 75% Bt plants and 25% non-Bt refuge plants treated with spinosad | 4 |
| Bt+SR+Cm | 75% Bt plants and 25% non-Bt refuge plants treated with spinosad and *C. maculata* | 4 |
| Bt Only | 100% Bt plants only | 4 |
| Non-Bt Only | 100% Non-Bt refuge plants | 3 |
| Non-Bt+Cm | 100% Non-Bt refuge plants and *C. maculata* | 3 |

R: refuge.
SR: refuge with spinosad.
Cm: *C. maculata*.

The total number of moths in the cage was 500 with a 1:1 ratio for female and male moths from each strain. Eggs were collected from the cage and put on artificial diet to rear F1 larvae. About 1,000 moths from F1-F4 were used to produce a synthetic population. F5 pupae were used in the selection experiments. The expected allele frequency of the synthetic population (square root of survival rate) was 0.1 for Cry1Ac and spinosad resistance. The mean survival of F5 larvae was 0.033 on Cry1Ac plants and 0.025 on spinosad-sprayed plants. Therefore, the actual initial allele frequency at the start of the experiment was estimated to be 0.057 (square root of 0.033) for Cry1Ac resistance and 0.050 (square root of 0.025) for spinosad resistance [43]. While this is higher than would be expected in the field initially when *Bt* crops are released (perhaps $10^{-3}$ or lower [41]), we expected that these initial frequencies would allow us to see differences among the treatments in a reasonable time frame.

Larvae and adult *C. maculata* were obtained from DuPont Pioneer (Johnston, IA) and maintained in a climatic chamber at 27±1°C, 65±5% RH and 16:8 L: D at Cornell University's Department of Entomology at Geneva, NY. Both larvae and adults were reared on decapsulated eggs of brine shrimp, *Artemia franciscana*, (Brine Shrimp Direct, Ogden UT) and 1.5% agar solution provided separately as a water source.

## Transgenic broccoli plants and insecticide

We used *Brassica oleracea* producing high levels of Cry1Ac for our *Bt* broccoli plants [44]. *Bt* broccoli plants with 8 true leaves were used, and analysis by ELISA indicated that the Cry1Ac protein level was 12.33±1.62 µg/g fresh leaf tissue (n = 7). To ensure the activity of the *Bt* broccoli, the plants were screened with *P. xylostella* neonates of F1 heterozygotes (G88 × Cry1Ac-RR) when plants were 4 to 5 wk old [43]. The Cry1Ac plants that killed 100% of neonates of F1 heterozygotes (SS x Cry1Ac-RR), indicating high levels of expression in the *Bt* plants, were used in the greenhouse experiments. Non-*Bt* broccoli (cv. Packman) was used in the refuge.

A commercial formulation of spinosad (SpinTor 2 SC, 240 g [AI]/Liter) was used. Refuge plants in the treatments B1 and B2 were sprayed in the cages using a small hand-held sprayer. The Cry1Ac plants were covered during the sprays to avoid drift. The concentration we used was 90 ppm, which was the lowest field dose listed on the insecticide label. The insecticide-treated refuges were sprayed at the 1st, 3rd and 5th generations in order to keep the *P. xylostella* population in check in the cages.

## Experimental designs

The selection experiment was conducted in greenhouses at Cornell University's New York State Agricultural Experiment Station and followed the general procedures used in previous studies [40,43]. Each cage was 1.8 m long×0.9 m wide×1.7 m high and constructed of nylon netting. Adults, but not larvae, could easily move between the different broccoli types, which were separated by a nylon-netting barrier (0.9 m high). This arrangement simulated adjoining fields with frequent inter-field movement by adults but negligible movement of larvae.

Seven treatments were included in the experiment (Table 2). There were 16 plants total in each cage (12 *Bt* plants plus 4 non-*Bt* refuge plants). 250 F4 pupae of the synthetic *P. xylostella* population were released into each cage. Three pairs of 1-week old *C. maculata* were released into cages of the treatments Bt+R+Cm, Bt+SR+Cm and non-Bt+Cm when the *P. xylostella* larvae were 2nd instars. During the pupal period, all old plants were cut at the base of the stem, and plants and pots (which might have pupae on them) were kept in the cages for a week in order to allow adults to emerge. New plants were introduced into the cage to provide foliage for egg laying or existing larvae that had defoliated plants. There were four replications (cages) of *Bt*-plant treatments (Bt+R, Bt+R+Cm, Bt+SR, Bt+SR+Cm and Bt Only) and three of the controls (Non-Bt Only and non-Bt+Cm).

## Data collection

Older larvae (primarily 3rd or 4th instars) and pupae of *P. xylostella*, and pupae and adults of *C. maculata* on broccoli plants were counted every generation when larval and pupal densities peaked. To test for resistance, 30–40 larvae from non-*Bt* refuge plants were collected from each cage at the 3rd, 4th and 5th generations. The larvae were reared on diet in the laboratory, and then adults were allowed to mate and eggs were harvested and tested as described below. At the 6th generation, we collected at least 60 pupae on refuge plants from most of the cages. The survival of 2nd instars derived from the pupae collected in each cage was tested on Cry1Ac broccoli leaf disks in 30-ml plastic cups [43]. For each cage, a total of 100 larvae in 10 replications were tested on *Bt* broccoli, and non-*Bt* broccoli was used as a control. We also tested survival of 2nd instars derived from B1 and B2 treatments on spinosad-dipped broccoli leaf disks in 30-ml plastic cups at the diagnostic dose of 10 ppm [45]. A total of 100 larvae were tested (10 replications, 10 larvae/rep) for each cage. For both Cry1Ac plants and spinosad treatments, survival was determined after 3 days at 27±1°C, 50±10% RH and 16:8 L:D photoperiod.

## Statistical analysis

All statistical analyses were conducted using SPSS 17.0 Windows [46]. Descriptive statistics are given as mean values and standard errors of the mean. Because the data fit the assumptions for parametric analysis, the repeated measures ANOVA was used with the factors of generations and treatments for analysis of the population of *P. xylostella* on *Bt* and non-*Bt* plants and the survival on Cry1Ac leaves of *P. xylostella* larvae from adults taken from cages. The survival rates were transformed by square root before analysis. For each generation, the differences of resistance development and the population of *P. xylostella* and *C. maculata* were analyzed by one-way ANOVA, and means were compared by Tukey HSD, if the data fit homoscedasticity, and Games-Howell test if not. Differences of the mean values in the population of *C. maculata* per *Bt* plants between the treatments of refuge + *C. maculata* and sprayed-refuge + *C. maculata* were examined by independent t-tests. In all tests $P$ values $<0.05$ were considered significant.

## Acknowledgments

We thank Yen Mei Cheung for technical assistance.

## Author Contributions

Conceived and designed the experiments: XXL MC DO RR AMS. Performed the experiments: XXL MC AMS. Analyzed the data: XXL MC AMS. Contributed reagents/materials/analysis tools: HLC EDE. Wrote the paper: XXL AMS. Discussed and helped analyze the results: QWZ.

## References


1. Shelton AM, Zhao J-Z, Roush RT (2002) Economic, ecological, food safety, and social consequences of the deployment of Bt transgenic Plants. Annu Rev Entomol 47: 845–881.
2. Romeis J, Shelton AM, Kennedy GG (2008) Integration of insect-resistant genetically modified crops within IPM programs. Dordrecht, Springer. 441 pp.
3. James C (2012) Global status of commercialized transgenic crops: Bt cotton, ISAAA Briefs No. 44. Ithaca, NY: International Service for the Acquisition of Agri-biotech Applications.
4. Qaim M, Pray CE, Zilberman D (2008) Economic and social considerations in the adoption of Bt crops. In: Romeis J, Shelton AM, Kennedy GG, editors. Integration of insect-resistant genetically modified crops within IPM programs. Dordrecht: Springer. pp 329–356.
5. Kathage J, Qaim M (2012) Economic impacts and impact dynamics of Bt cotton in India. Proc Natl Acad Sci USA 109: 11652–11656.
6. Lu Y, Wu K, Jiang Y, Guo Y, Desneux N (2012) Widespread adoption of Bt cotton and insecticide decrease promotes biocontrol services. Nature 487: 362–365.
7. Carrière Y, Ellers-Kirk C, Sisterson M, Antilla L, Whitlow M, et al. (2003) Long-term regional suppression of pink bollworm by *Bacillus thuringiensis* cotton. Proc Natl Acad Sci USA 100: 1519–1523.
8. Wu KM, Lu YH, Feng HQ, Jiang YY, Zhao J-Z (2008) Suppression of cotton bollworm in multiple crops in China in areas with Bt toxin-containing cotton. Science 321: 1676–1678.
9. Hutchison WD, Burkness EC, Mitchell PD, Moon RD, Leslie TW, et al. (2010) Areawide suppression of European corn borer with Bt maize reaps savings to non-Bt maize growers. Science 330: 222–225.
10. Naranjo SE (2009) Impacts of Bt crops on non-target organisms and insecticide use patterns. CAB Reviews: Perspectives in agriculture, veterinary science, nutrition and natural resources 4: No.011. DOI: 10.1079/PAVSNNR20094011.
11. Gould F (1998) Sustainability of transgenic insecticidal cultivars: integrating pest genetics and ecology. Annu Rev Entomol 43: 701–726.
12. Onstad DW (2008) The future of insect resistance management. In: Onstad DW, editor. Insect resistance management: biology, economics and prediction. London: Academic Press. pp. 289–300.
13. Van Rensburg JBJ (2007) First report of field resistance by stem borer, *Busseola fusca* (Fuller) to Bt-transgenic maize. S African J Plant Soil 24: 147–151.
14. Storer NP, Babcock JM, Schlenz M, Meade T, Thompson GD, et al. (2010) Discovery and characterization of field resistance to Bt maize: *Spodoptera frugiperda* (Lepidoptera: Noctuidae) in Puerto Rico. J Econ Entomol 103: 1031–1038.
15. Dhurua S, Gujar GT (2011) Field-evolved resistance to Bt toxin Cry1Ac in the pink bollworm, *Pectinophora gossypiella* (Saunders) (Lepidoptera: Gelechiidae) from India. Pest Manag Sci 67: 898–903.
16. Gassmann AJ, Petzold-Maxwell JL, Keweshan RS, Dunbar MW (2011) Field-evolved resistance to Bt maize by western corn rootworm. PLoS ONE 6(7): e22629.
17. Bates SL, Zhao JZ, Roush RT, Shelton AM (2005) Insect resistance management in GM crops: past, present and future. Nat Biotechnol 23: 57–62.
18. Tabashnik BE, Van Rensburg JBJ, Carrière Y (2009) Field-evolved insect resistance to Bt crops: definition, theory, and data. J Econ Entomol 102: 2011–2025.
19. Huang F, Andow D, Bushman LL (2012) Success of the high-dose/refuge resistance management strategy after 15 years of Bt crop use in North America. Entomol Exp Appl 140: 1–16.
20. Wolfenbarger LL, Naranjo SE, Lundgren JG, Bitzer RJ, Watrud LS (2008) Bt crops effects on functional guilds of non-target arthropods: a meta-analysis. PLoS ONE 3(5): e2284. DOI:10.1371/journal.pone.0002118.
21. Desneus N, Bernal JS (2010) Genetically modified crops deserve greater ecotoxicological scrutiny. Ecotoxicology 19: 1642–1644.
22. Romeis J, McLean MA, Shelton AM (2013) When bad science makes good headlines: the case of Bt maize and regulatory bans. Nat Biotechnol 37: 386–7.
23. Desneus N, Ramirez-Romero R, Bokonon-Ganta AH, Bernal JS (2010) Attraction of the parasitoid *Cotesia marginiventris* to host (*Spodoptera frugiperda*) frass is affected by transgenic maize. Ecotoxicology, 19: 1183–1192.
24. Chen M, Zhao JZ, Collins HL, Earle ED, Cao J, et al. (2008) A critical assessment of the effects of Bt transgenic plants on parasitoids. PLoS ONE 3(5): e2284. DOI: 10.1371/journal.pone.0002284.
25. Liu XX, Chen M, Onstad D, Roush R, Shelton AM (2011) Effect of Bt broccoli and resistant genotype of *Plutella xylostella* (Lepidoptera: Plutellidae) on development and host acceptance of the parasitoid *Diadegma insulare* (Hymenoptera: Ichneumonidae). Transgenic Res 20: 887–897.
26. Tian JC, Collins HL, Romeis J, Naranjo SE, Hellmich RL, et al. (2012) Using field-evolved resistance to Cry1F maize in a lepidopteran pest to demonstrate no adverse effects of Cry1F on one of its major predators. Transgenic Res 21: 1303–1310.
27. Lovei GL, Andow DA, Arpaia S (2009) Transgenic insecticidal crops and natural enemies: a detailed review of laboratory studies. Environ Entomol 38: 293–306.
28. Shelton AM, Naranjo S, Romeis J, Hellmich RH, Wolt J, et al. (2009) Setting the record straight: a rebuttal to an erroneous analysis on transgenic insecticidal crops and natural enemies. Transgenic Res 18:317–322.
29. Gould F, Kennedy GG, Johnson MT (1991) Effects of natural enemies on the rate of herbivore adaptation to resistant host plants. Entomol Exp Appl 58: 1–14.
30. Johnson MT, Gould F, Kennedy GG (1997) Effects of natural enemies on relative fitness of *Heliothis virescens* genotypes adapted and not adapted to resistant host plants. Entomol Exp Appl 82: 219–230.
31. Johnson MT, Gould F, Kennedy GG (1997) Effect of an entomopathogen on adaptation of *Heliothis virescens* populations to transgenic host plants. Entomol Exp Appl 83: 121–135.
32. Mallampalli N, Gould F, Barbosa P (2005) Predation of Colorado potato beetle eggs by a polyphagous ladybeetle in the presence of alternate prey: potential impact on resistance evolution. Entomol Exp Appl 114: 47–54.
33. Heimpel GE, Neuhauser C, Andow DA (2005) Natural enemies and the evolution of resistance to transgenic insecticidal crops by pest insects: the role of egg mortality. Environ Entomol 34: 512–526.
34. Chilcutt CF, Tabashnik BE (1999) Simulation of integration of *Bacillus thuringiensis* and the parasitoid *Cotesia plutellae* (Hymenoptera: Braconidae) for control of susceptible and resistant diamondback moth (Lepidoptera: Plutellidae). Environ Entomol 28: 505–512.
35. Arpaia S, Gould F, Kennedy G (1997) Potential impact of *Coleomegilla maculata* predation on adaptation of *Leptinotarsa decemlineata* to Bt-transgenic potatoes. Entomol Exp Appl 82: 91–100.
36. Gassmann AJ, Stock SP, Carrière Y, Tabashnik BE (2006) Effect of entomopathogenic nematodes on the fitness cost of resistance to *Bt* toxin *Cry1Ac* in pink bollworm (Lepidoptera: Gelechiidae). J Econ Entomol 99: 920–926.
37. Talekar NS, Shelton AM (1993) Biology, ecology and management of the diamondback moth. Annu Rev Entomol 38: 275–301.
38. Liu XX, Chen M, Collins HL, Onstad D, Roush R, et al. (2012) Effect of insecticides and *Plutella xylostella* (Lepidoptera: Plutellidae) genotype on a predator and parasitoid and implications for the evolution of insecticide resistance. J Econ Entomol 105: 354–362.
39. Lundgren JG, Gassmann AJ, Bernal J, Duan JJ, Ruberson J (2009) Ecological compatibility of GM crops and biological control. Crop Prot 28: 1017–1030.
40. Tang JD, Collins HL, Metz TD, Earle ED, Zhao JZ, et al. (2001) Greenhouse tests on resistance management of Bt transgenic plants using refuge strategies. J Econ Entomol 94: 240–247.
41. Roush RT (1998) Two toxin strategies for management of insecticidal transgenic crops: can pyramiding succeed where pesticide mixtures have not? Phil Trans Royal Soc Lond B 353: 1777–86.
42. Shelton AM, Tang JD, Roush RT, Metz TD, Earle ED (2000) Field tests on managing resistance to Bt-engineered plants. Nat Biotechnol 18: 339–342.
43. Zhao JZ, Cao J, Collins HL, Bates SL, Roush RT, et al. (2005) Concurrent use of transgenic plants expressing a single and two *Bacillus thuringiensis* genes speeds

insect adaptation to pyramided plants. Proc Natl Acad Sci USA 102: 8426–8430.
44. Metz TD, Roush RT, Tang JD, Shelton AM, Earle ED (1995) Transgenic broccoli expressing a *Bacillus thuringiensis* insecticidal crystal protein: implications for pest resistance management strategies. Mol Breeding 1: 309–317.
45. Zhao J-Z, Collins HL, Shelton AM (2010) Testing insecticide resistance management strategies: mosaic versus rotations. Pest Manag Sci 66: 1101–1105.
46. SPSS (1998) SPSS User's Guide. Chicago, SPSS Inc.

# PERMISSIONS

All chapters in this book were first published in PLOS ONE, by The Public Library of Science; hereby published with permission under the Creative Commons Attribution License or equivalent. Every chapter published in this book has been scrutinized by our experts. Their significance has been extensively debated. The topics covered herein carry significant findings which will fuel the growth of the discipline. They may even be implemented as practical applications or may be referred to as a beginning point for another development.

The contributors of this book come from diverse backgrounds, making this book a truly international effort. This book will bring forth new frontiers with its revolutionizing research information and detailed analysis of the nascent developments around the world.

We would like to thank all the contributing authors for lending their expertise to make the book truly unique. They have played a crucial role in the development of this book. Without their invaluable contributions this book wouldn't have been possible. They have made vital efforts to compile up to date information on the varied aspects of this subject to make this book a valuable addition to the collection of many professionals and students.

This book was conceptualized with the vision of imparting up-to-date information and advanced data in this field. To ensure the same, a matchless editorial board was set up. Every individual on the board went through rigorous rounds of assessment to prove their worth. After which they invested a large part of their time researching and compiling the most relevant data for our readers.

The editorial board has been involved in producing this book since its inception. They have spent rigorous hours researching and exploring the diverse topics which have resulted in the successful publishing of this book. They have passed on their knowledge of decades through this book. To expedite this challenging task, the publisher supported the team at every step. A small team of assistant editors was also appointed to further simplify the editing procedure and attain best results for the readers.

Apart from the editorial board, the designing team has also invested a significant amount of their time in understanding the subject and creating the most relevant covers. They scrutinized every image to scout for the most suitable representation of the subject and create an appropriate cover for the book.

The publishing team has been an ardent support to the editorial, designing and production team. Their endless efforts to recruit the best for this project, has resulted in the accomplishment of this book. They are a veteran in the field of academics and their pool of knowledge is as vast as their experience in printing. Their expertise and guidance has proved useful at every step. Their uncompromising quality standards have made this book an exceptional effort. Their encouragement from time to time has been an inspiration for everyone.

The publisher and the editorial board hope that this book will prove to be a valuable piece of knowledge for researchers, students, practitioners and scholars across the globe.

# LIST OF CONTRIBUTORS

**Duk-Hee Lee**
Department of Preventive Medicine, School of Medicine, Kyungpook National University, Daegu, Korea BK21 Plus KNU Biomedical Convergence Program, Department of Biomedical Science, Kyungpook National University, Daegu, Korea

**Lars Lind**
Department of Medical Sciences, Cardiovascular Epidemiology, Uppsala University Hospital, Uppsala, Sweden

**David R. Jacobs Jr.**
Division of Epidemiology and Community Health, School of Public Health, University of Minnesota, Minneapolis, Minnesota, United States of America
Department of Nutrition, University of Oslo, Oslo, Norway

**Samira Salihovic and Bert van Bavel**
MTM Research Center, School of Science and Technology, Örebro University, Örebro, Sweden

**P. Monica Lind**
Department of Medical Sciences, Occupational and Environmental Medicine, Uppsala University, Uppsala, Sweden

**Guangsheng Li, Yongmo Wang and Guoan Zhang**
Hubei Insect Resources Utilization and Sustainable Pest Management Key Laboratory, College of Plant Science and Technology, Huazhong Agricultural University, Wuhan, Hubei, P.R. CHINA

**Biao Liu**
Nanjing Institute of Environmental Science, Ministry of Environmental Protection of the People's Republic of China, Nanjing, P.R. CHINA

**Mateus R. Campos, Agna Rita S. Rodrigues, Wellington M. Silva, Tadeu Barbosa M. Silva, Vitória Regina F. Silva and Herbert Alvaro A. Siqueira**
Departamento de Agronomia (Entomologia), Universidade Federal Rural de Pernambuco, Recife, PE, Brazil

**Raul Narciso C. Guedes**
Departamento de Entomologia, Universidade Federal de Viçosa, Viçosa, MG, Brazil

**Christoph Sandrock**
Agroscope, Swiss Bee Research Centre, Berne, Switzerland

**Matteo Tanadini**
Seminar for Statistics, ETH Zurich, Zurich, Switzerland

**Lorenzo G. Tanadini**
Division of Biostatistics, Institute for Social and Preventive Medicine, University of Zurich, Zurich, Switzerland

**Aline Fauser-Misslin and Peter Neumann**
Agroscope, Swiss Bee Research Centre, Berne, Switzerland
Institute of Bee Health, Vetsuisse Faculty, University of Berne, Berne, Switzerland

**Simon G. Potts**
School of Agriculture, Policy and Development, University of Reading, Reading, United Kingdom

**Jian-wei Xu, Hui Liu and Ren-hua Nie**
Yunnan Institute of Parasitic Diseases, Puer, China

**Yuan-mei Liao**
Health School of Ganzhou, Jiangxi Province, Ganzhou, China

**Joshua Havumaki**
Foundation for Innovative New Diagnostics, Cointrin/Geneva, Switzerland

**Nina Cedergreen**
Department of Plant and Environmental Sciences, University of Copenhagen, Frederiksberg, Denmark

**Shi-Hua Sun**
Department of Epidemiology and Health Statistics, Shandong University School of Public Health, Jinan, China

**Cun-Xian Jia**
Department of Epidemiology and Health Statistics, Shandong University School of Public Health, Jinan, China
Shandong University Center for Suicide Prevention Research, Jinan, China

**Dorte H. Højland, Karl-Martin Vagn Jensen and Michael Kristensen**
Department of Agroecology, Aarhus University, Slagelse, Denmark

**Ying Li**
Key Laboratory of Optoelectronic Information and Sensing Technologies of Guangdong Higher Education Institutes, Jinan University, Guangzhou, Guangdong, China
Pre-university Department, Jinan University, Guangzhou, Guangdong, China

**Xiao Ma and Minglu Zhao**
Department of Optoelectronic Engineering, Jinan University, Guangzhou, Guangdong, China

**Pan Qi**
Department of Electronics Engineering, Guangdong Communication Polytechnic, Guangzhou, Guangdong, China

**Jingang Zhong**
Key Laboratory of Optoelectronic Information and Sensing Technologies of Guangdong Higher Education Institutes, Jinan University, Guangzhou, Guangdong, China
Department of Optoelectronic Engineering, Jinan University, Guangzhou, Guangdong, China

**Michele La Merrill**
Department of Environmental Toxicology, University of California Davis, Davis, California, United States of America
Department of Preventive Medicine, Mount Sinai School of Medicine, New York, New York, United States of America
Metabolism Institute, Mount Sinai School of Medicine, New York, New York, United States of America

**Emma Karey**
Department of Environmental Toxicology, University of California Davis, Davis, California, United States of America

**Erin Moshier**
Department of Preventive Medicine, Mount Sinai School of Medicine, New York, New York, United States of America

**Claudia Lindtner**
Metabolism Institute, Mount Sinai School of Medicine, New York, New York, United States of America
Division of Endocrinology, Diabetes and Bone Disease, Mount Sinai School of Medicine, New York, New York, United States of America

**Michael R. La Frano**
West Coast Metabolomic Center, University of California Davis, Davis, California, United States of America
Department of Nutrition, University of California Davis, Davis, California, United States of America

**Christoph Buettner**
Metabolism Institute, Mount Sinai School of Medicine, New York, New York, United States of America
Division of Endocrinology, Diabetes and Bone Disease, Mount Sinai School of Medicine, New York, New York, United States of America

**John W. Newman**
West Coast Metabolomic Center, University of California Davis, Davis, California, United States of America
Department of Nutrition, University of California Davis, Davis, California, United States of America
Obesity and Metabolism Research Unit, USDA-ARS-Western Human Nutrition Research Center, Davis, California, United States of America

**Ashutosh K. Srivastava, Satyajeet Rai, M. K. Srivastava, M. K. R. Mudiam and L. P. Srivastava**
Pesticide Toxicology Laboratory, CSIR-Indian Institute of Toxicology Research, (Council of Scientific and Industrial Research, Govt. of India), Mahatma Gandhi Marg, Lucknow, India

**M. Lohani**
Department of Biotechnology, Integral University, Lucknow, India

**Ranjan Ramasamy, Pavilupillai J. Jude, Thabothiny Veluppillai, Thampoe Eswaramohan and Sinnathamby N. Surendran**
Department of Zoology, Faculty of Science, University of Jaffna, Jaffna, Sri Lanka

**Jie Liu, Ying Xiao, Meiying Hu and Guohua Zhong**
Laboratory of Insect Toxicology, and Key Laboratory of Pesticide and Chemical Biology, Ministry of Education, South China Agricultural University, Guangzhou, P.R. China

**Yue He**
Guangdong Zhuhai Supervision Testing Institute of Quality and Metrology, Zhuhai, P.R. China

**Shaohua Chen**
Guangdong Province Key Laboratory of Microbial Signals and Disease Control, South China Agricultural University, Guangzhou, P.R. China

**Kristin Lees**
Faculty of Life Sciences, University of Manchester, Manchester, United Kingdom

**Maria Musgaard, Siros Suwanmanee and Philip Biggin**
Department of Biochemistry, University of Oxford, Oxford, United Kingdom

**David Sattelle and Steven David Buckingham**
Wolfson Institute for Biomedical Research, Department of Medicine, University College London, London, United Kingdom

**Sylvie Cornelie, Marie Rossignol, François Mouchet, Fabrice Chandre and Françoise Mathieu-Daudé**
Maladies Infectieuses et Vecteurs, Ecologie, Génétique, Evolution et Contrôle (MIVEGEC), UMR IRD 224-CNRS 5290-UM1-UM2, Institut de Recherche pour le Développement (IRD), Montpellier, France

**Martial Seveno, Edith Demettre and Philippe Marin**
Institut de Génomique Fonctionnelle (IGF), UMR CNRS 5203-INSERM 661-UM1-UM2, Montpellier, France

**Vincent Corbel, Franck Remoué and Innocent Djègbè**
Maladies Infectieuses et Vecteurs, Ecologie, Génétique, Evolution et Contrôle (MIVEGEC), UMR IRD 224-CNRS 5290-UM1-UM2, Institut de Recherche pour le Développement (IRD), Centre de Recherche Entomologique de Cotonou (CREC), Cotonou, Bénin

**Noa Simon-Delso**
Beekeeping Research and Information Centre, Louvain la Neuve, Belgium
Environmental Sciences, Copernicus Institute, Utrecht University, Utrecht, The Netherlands

**Gilles San Martin and Louis Hautier**
Plant Protection and Ecotoxicology Unit, Life Sciences Department, Walloon Agricultural Research Centre, Gembloux, Belgium

**Etienne Bruneau, Laure-Anne Minsart and Coralie Mouret**
Beekeeping Research and Information Centre, Louvain la Neuve, Belgium

**Claudia Husseneder and Dawn M. Simms.**
Department of Entomology, Louisiana State University Agricultural Center, Baton Rouge, Louisiana, United States of America

**Min Liu and Kun Luo**
Institute of Pesticide Science, Hunan Agricultural University, Changsha, China
College of Plant Protection, Hunan Agricultural University, Changsha, China

**Yunsheng Wang and Feng Luo**
College of Plant Protection, Hunan Agricultural University, Changsha, China
School of Computing, Clemson University, Clemson, South Carolina, United States of America

**Aiping Zeng and Xiaomao Zhou**
Institute of Pesticide Science, Hunan Agricultural University, Changsha, China

**Lianyang Bai**
Hunan academy of agricultural sciences, Changsha, China

**Anthony Bouétard, Jessica Côte, Anne-Laure Besnard, Marc Collinet and Marie-Agnès Coutellec**
INRA, UMR 0985 ESE, Ecology and Ecosystem Health, INRA - Agrocampus Ouest, CS84215, Rennes Cedex, France

**Michael Coleman, Phillip J. McCall, Marlize Coleman, Janet Hemingway, Abdiasiis Omar, Michelle C. Stanton and Eddie K. Thomsen**
Vector Biology Department, Liverpool School of Tropical Medicine, Liverpool, United Kingdom

**Mohammed H. Al-Zahrani, Adel A. Alsheikh and Raafat F. Alhakeem**
Public Health Directorate, Ministry of Health, Riyadh, Kingdom of Saudi Arabia

**Abdullah A. Al Rabeeah**
Vector Biology Department, Liverpool School of Tropical Medicine, Liverpool, United Kingdom
Public Health Directorate, Ministry of Health, Riyadh, Kingdom of Saudi Arabia

**Ziad A. Memish**
Vector Biology Department, Liverpool School of Tropical Medicine, Liverpool, United Kingdom
Public Health Directorate, Ministry of Health, Riyadh, Kingdom of SaudiArabia
College of Medicine, Alfaisal University, Riyadh, Kingdom of Saudi Arabia

**Xiaojie Zhou**
Institute of Disinfection and Vector Control, Beijing Center for Disease Control and Prevention, Beijing, China
State Key Laboratory of Integrated Management of Pest Insects and Rodents, Institute of Zoology, Chinese Academy of Sciences, Beijing, China

**Kun Qian, Ying Tong and Xiaopeng Zeng**
Institute of Disinfection and Vector Control, Beijing Center for Disease Control and Prevention, Beijing, China

**Junwei Jerry Zhu**
United States Department of Agriculture, Agricultural Research Service, Lincoln, Nebraska, United States of America

**Xinghui Qiu**
State Key Laboratory of Integrated Management of Pest Insects and Rodents, Institute of Zoology, Chinese Academy of Sciences, Beijing, China

**Pei Wang**
School of Mathematics and Information Sciences, Henan University, Kaifeng, China
School of Electrical and Computer Engineering, RMIT University, Melbourne, Victoria, Australia

**Jinhu Lü**
Institute of Systems Science, Academy of Mathematics and Systems Science, Chinese Academy of Sciences, Beijing, China

**Xinghuo Yu**
School of Electrical and Computer Engineering, RMIT University, Melbourne, Victoria, Australia

**Kelsie LaSharr**
School of Natural Resources & the Environment, The University of Arizona, Tucson, Arizona, United States of America

**Laura López-Hoffman and Ruscena Wiederholt**
School of Natural Resources & the Environment, The University of Arizona, Tucson, Arizona, United States of America,
Udall Center for Studies in Public Policy, The University of Arizona, Tucson, Arizona, United States of America

**Chris Sansone**
Bayer CropScience, Research Triangle Park, North Carolina, United States of America

**Kenneth J. Bagstad, Jay E. Diffendorfer and Darius Semmens**
United States Geological Survey, Geosciences and Environmental Change Science Center, Denver, Colorado, United States of America

**Paul Cryan**
United States Geological Survey, Fort Collins Science Center, Fort Collins, Colorado, United States of America

**Joshua Goldstein**
Department of Human Dimensions of Natural Resources, Colorado State University, Fort Collins, Colorado, United States of America

**John Loomis**
Department of Agricultural and Resource Economics, Colorado State University, Fort Collins, Colorado, United States of America

**Gary McCracken**
Department of Ecology and Evolutionary Biology, University of Tennessee, Knoxville, Tennessee, United States of America

**Rodrigo A. Medellín**
Instituto de Ecología, Universidad Nacional Auto´noma de México, Distrito Federal, México

**Amy Russell**
Department of Biology, Grand Valley State University, Allendale, Michigan, United States of America

**Qingwen Zhang**
Department of Entomology, China Agricultural University, Beijing, China

**Xiaoxia Liu**
Department of Entomology, China Agricultural University, Beijing, China,
Department of Entomology, Cornell University, New York State Agricultural Experiment Station, Geneva, New York, United States of America

**Mao Chen, Hilda L. Collins and Anthony M. Shelton**
Department of Entomology, Cornell University, New York State Agricultural Experiment Station, Geneva, New York, United States of America

**David W. Onstad**
Department of Entomology, University of Illinois, Urbana, Illinois, United States of America

**Richard T. Roush**
Melbourne School of Land and Environment, University of Melbourne, Victoria, Australia

**Elizabeth D. Earle**
Department of Plant Breeding and Genetics, Cornell University, Ithaca, New York, United States of America

# Index